Solid Waste Management
Volume 1
Chemical Approaches

Editors

Garima Chauhan
Chemical and Materials Engineering
University of Alberta
Edmonton, AB, Canada

and

Surajbhan Sevda
Department of Biotechnology
National Institute of Technology Warangal
Telangana, India

CRC Press is an imprint of the
Taylor & Francis Group, an **informa** business
A SCIENCE PUBLISHERS BOOK

Cover image provided by the first editor, Garima Chauhan.

First edition published 2024
by CRC Press
2385 NW Executive Center Drive, Suite 320, Boca Raton FL 33431

and by CRC Press
4 Park Square, Milton Park, Abingdon, Oxon, OX14 4RN

Library of Congress Cataloging-in-Publication Data (applied for)

ISBN: 978-1-032-03900-8 (hbk)
ISBN: 978-1-032-03901-5 (pbk)
ISBN: 978-1-003-18960-2 (ebk)

DOI: 10.1201/9781003189602

Typeset in Times New Roman
by Radiant Productions

Preface

What is the first thing that comes to mind when someone refers to '*waste*'? Most common search engine results of the word '*waste*' will invoke images of discarded things in a garbage bin, empty plastic bottles, and food/organic waste in a landfill. In essence, the term 'waste' identifies goods or material that are perceived to be mostly valueless. However, objects that are perceived to be waste based on consumers' object valuation can be redefined to create value. This requires a multitude of efforts using different strategies in waste prevention and management. This volume pulls together many of the leading researchers working on the topic, particularly, waste electrical and electronic equipment (WEEE), plastic waste and agro-residue waste, to provide new research insights into the field of waste valorization.

Broadly speaking, research in the area of solid waste management is motivated by intellectual curiosity toward defining '*waste*'. Currently, various potential sources of waste generation, composition of the waste, hazards associated with their disposal, and development of solid waste regulations are being widely explored. However, one of the primary sources of solid waste, the unhindered consumerism, has not been discussed sufficiently. The objectives of this volume are centered on the waste that is generated primarily in our households, such as mobile phones, computers, laptops, plastic bags, food waste, and garden waste, owing to lack of consumer awareness. Emerging requirement of alternative energy resources and increasing amount of domestic wastes define the scope of this volume. Research practices in the field of 'waste to energy' are compiled to make readers aware about the recent trends in solid waste management.

Chapter 1 discusses the term '*waste*' from the perspective of consumers. Behavioral aspects, in addition to the economic incentives, play a crucial role in the household waste-disposal choices. The authors of this chapter suggest that understanding of '*wastefulness*' in society is dominated by consumers' valuation of objects on a spectrum of valuable to waste. Chapter 2 of this volume covers the solid wastes generated recently during global outbreak of coronavirus disease 2019 owing to the excessive usage of facemasks, hand sanitizers, and personal protective equipment kits. Authors discuss various chemical approaches for the disposal of these types of waste. In addition, new developments in the waste management policies and regulations have been highlighted.

Exponential increase in the generation of electronic waste has been a matter of concern for last two decades. Furthermore, *work-from-home* arrangements owing to quarantine policies during the COVID-19 pandemic led to an increased purchase of electronic equipment, which could cause an evident increase in WEEE in the coming years. Chapters 3–6 of this volume cover aspects of WEEE management, including disassembly and pretreatment of WEEE, recovery of metals and energy resources from WEEE, and eco-design strategies for efficient recycling. In addition, WEEE contain a significant fraction of plastics. Chapter 7 provides a detailed review on various physicochemical pathways to recover plastics from WEEE and convert the recovered plastics into valuable energy resources. Chapters 8–10 include discussion on several chemical approaches used for plastic valorization. The research work presented in Chapters 11–15 explores conversion of agro-residue waste into fuels and energy resources using various approaches, including Fischer-Tropsch synthesis, bio-based adsorbents, and thermochemical conversion of sludge.

We believe that this volume will provide a comprehensive overview of various chemical approaches for treatment of solid wastes. Covering a variety of interdisciplinary topics on waste treatment and resource recovery makes the book one for all that serves as an excellent reading material for engineers, science scholars, entrepreneurs, and organizations who are working in the field of waste management.

The editors gratefully acknowledge the hard work and patience of all the authors who have contributed to this book. In addition, we would like to extend our gratitude to all experts across the globe, who contributed in reviewing the scientific accuracy and relevance of the materials.

Garima Chauhan
Surajbhan Sevda

Contents

1

WASTE

IMPACT, TRENDS, INTERVENTIONS, AND FUTURE GOALS

Saurabh Rawal[1,*] and *Bijit Ghosh*[2]

1. INTRODUCTION

The age of consumerism is characterized by an unprecedented rise in the acquisition of goods and materials. Never in the history of humankind have people ever owned as much as they do today. One of the most alarming outcomes of this unhindered consumerism is waste. The concept of waste is not a new one and has been around for a long time. Its etymology can be traced to the latin word *vastus* which means void or the idea that a thing seems to be empty. In the case of consumer products, emptiness refers to the perceived absence of value in an object and the object itself is referred to as waste. Most common references to the word 'waste' will invoke images of things inside the garbage bin, empty food packets and plastic bottles lying in a dump, broken furniture and many more. Depending on the context, such an object has also been called by many other names like rubbish, garbage, trash and junk. But in essence, the term waste has been broadly used to identify goods or material that are perceived to be mostly valueless. Several scholars have corroborated with this object-valuation appraisal of waste. Thompson's (1979) rubbish theory elaborated this idea by proposing that objects move on a spectrum of valuation with valued and unvalued elements (or rubbish) representing two ends. Strasser (2002) argues that the historical understanding of waste in America has been a result of sorting and identifying products as useful or useless. This approach to wastefulness has also found resonance across different societies and cultures (Siniawer 2018). Overall, consumers' object valuation, on a spectrum of valuable to waste, has dominated our understanding of wastefulness in society.

Although valuation is an inherent part of identifying an object as waste or not, reducing waste demands a more behavioral appraisal of the problem. Knowing what behaviors and choices result in a wasteful society can provide a framework that is relatable for consumers and useful for researchers and policy makers. Moreover, behavioral indicator of waste, or object valuation, is not directly expressed

[1] Assistant Professor, Wilfrid Laurier University, Waterloo, ON, N2L 3C5.
[2] Alberta School of Business in Marketing, University of Alberta, Edmonton, AB, T6G 2R6.
* Corresponding author: srawal@wlu.ca

by consumers and is primarily revealed through their act or decision to discontinue ownership. From that standpoint, waste is a result of the *discarding of goods and materials*. We refer to this dispositional conceptualization as the traditional view of waste. Other words may be used to describe this behavior (e.g., trashing, throwing away in the garbage, etc.), but in essence the traditional view identifies any act of disposition that results in a loss of a product's utility from active consumption. In this view, wastefulness of a society or group of people (e.g., a country) can be studied by examining the quantity of discarded goods and materials, e.g., tonnage of solid municipal waste (Kaza et al. 2018). The greater the quantity of discarded goods and materials, the greater is the wastefulness of that society and vice-versa.

Today, there is an abundance of literature as well as policies in support of the traditional view but most of what we know is implied using examples and consumer lay-beliefs rather than explicitly defined. For instance, in his treatise on the wasteful American lifestyle, Packard (1960) discussed how after the end of the Second World War consumer purchases increased drastically, primarily as a result of the *throwaway culture*, which encouraged consumers to regularly throw away old possessions to buy new ones. Author referred to this practice as wasteful and corporations and governments, as promoters of the culture, being the *waste makers*. Several consumer researchers also support the traditional view. On similar lines, Harrell and McConocha (1992) found that consumers consider throwing away goods as a wasteful and irresponsible practice. Others have also discussed throwing away a valuable product as equivalent to throwing away (i.e., wasting) money (Coulter and Ligas 2003, Arkes 1996). More recently, Haws et al. (2012) used an alternative way to understand wastefulness. The authors studied consumer motivation to retain products, instead of discarding them, and found that a consumer's product retention tendency is positively associated with avoiding wastefulness and therefore, implying the opposite, that discarding possessions is wasteful.

Outside of consumer related research, several scholars and policy makers also employ the traditional perspective in studying waste and designing interventions to reduce it. For instance, Cheyne and Purdue (1995) define waste as an output that the owner does not want. Others have defined waste as an output with zero and possibly negative economic value or an object that has served its purpose (Lox 1994). Policy makers have also developed directives generally defining waste as objects or substances that are discarded, intended to be discarded, or obligated to be discarded (e.g., by law) (European Council 1975–2008, OECD2008, UNEP 2015). Pongracz and Pohjola (2004) define waste as an output that is not useful to its owner or an output that does not have an owner. Put simply, wastefulness has been identified and thus defined in terms of the unwanted material output. Behaviorally, scholars continue to rely on the act of discarding as a necessary condition for defining wasteful consumer behavior.

Clearly, the traditional view's reliance on discarding, as a necessary condition for appraising a behavior as wasteful, is pervasive across academia as well as policy making. Although it is difficult to pinpoint its causal origin, we suggest that there are several reasons that have led to the sustained use of the discarding as a criterion for wastefulness. From a practical standpoint, discarded goods are often less likely to be accessible or retrievable for use. At the household level, once trash is collected it quickly ends up at a place (e.g., landfill, incinerator, etc.), from where it is either permanently lost or rendered useless. In other words, discarding something is likely to result in a permanent loss of utility and thus making the act a good candidate for evaluating and measuring wastefulness. Moreover, even when some discarded goods are accessible to others (e.g., a piece of usable furniture kept by the community garbage bin), people are less likely to be interested in them because the owner's discarding of goods symbolizes lack of product usefulness or social denigration of the product (Brough and Isaac 2020, Weber et al. 2013). Additionally, consumer aversion and disgust toward trash is further likely to decrease consumer interest in retrieving and using discarded goods. Another reason for the emphasis on disposition is that it provides a convenient way to quantify wastefulness (e.g., tonnage of municipal waste) and thus compare and evaluate the progress of waste reduction programs. Last but not the least, focusing solely on disposition as means of defining wasteful consumer behavior supports individual rights, which are central to the modern democratic society. These rights are reflected in every aspect of life, including our consumption-driven economies. Dispositional wastefulness does not deprive consumers of their larger freedom to buy whatever they can afford, choose to use, not use or even destroy their possessions (McCaffery 2000). It is when consumers decide to get rid of their possessions that they are encouraged to be less wasteful by disposing of their possessions in a manner that leaves the least ecological footprint.

There are numerous environmental challenges which human beings face today like climate crisis and global warming, endangerment of species and pollution of water and air. Many of them are the consequences of the commercial developments of agriculture, industries, and modernisation of cities. However, solid waste is different as it is something which is mostly generated by the consumer and not the firm. Although the per capita volume is insignificant in terms of the big picture, the waste which consumers produce adds up to huge consequences. It is unsightly, produces foul smell, pollutes the freshwater supplies, and adds up to the airborne toxins, thereby causing diseases and becoming the breeding ground for pests. Following are the most common sources of waste.

1.1 Food or organic materials

Food-related waste accounts for roughly 44% of all the waste produced in the world. Also, the amount of food that is wasted every year is a third of the total amount of food that is produced. Clearly, food waste is massive source of waste in society. Food waste can be understood as the food which was intended for human consumption but is discarded without being eaten. Unfortunately, as the organic matter in the food decomposes, it produces a significant amount of greenhouse gases which in turn increases global warming and accelerates climate change. Moreover, when there is wastage of food, the human, technological and financial resources that were used for the growth, manufacturing, and distribution are also wasted. When equated in terms of global carbon footprint, food-related waste creates approximately 4.4 billion tonnes of carbon dioxide annually. Organic waste can be defined as the waste originating from households and institutions which is biodegradable and compos table. Few examples of the same are food leftovers, lawn trimmings, soiled-papers and cardboards, human hair and biosolids.

1.2 Electronic waste

Electronic waste, popularly referred to as e-waste, is another common source of waste. E-waste can be defined as any electrical or electronic product that is discarded, often as a result of the product becoming damaged, non-functional, or outdated. Such products and related materials are dumped along with garbage or fed into a recycling stream. Unfortunately, the recycling rate for e-waste is abysmally low (approximately 5%) and thus most of the e-waste tends to get dumped into landfills or even oceans. Although new-age electronic items are safe to use, most of them contain a combination of toxic materials like mercury, lead, and cadmium. When the materials get buried at a landfill, these chemical substances can leach into the ground and contaminate water supplies, such as groundwater. Such leaching at a high rate could result in the endangerment to the health of the people and animals alike. Not only the people who work with e-waste but also the people who live near a dumping ground experience negative effects on their health.

1.3 Clothing waste

Textile waste can be understood at two levels: (1) pre-consumer (i.e., materials discarded at the production stage), and (2) post-consumer (materials discarded after use) waste. As pre-consumer textile waste tends to be reused and repurposed by manufacturers to reduce costs, it is post-consumer textile waste that tends to be a source of concern. The latter generally consists of garments which have been discarded and household items which are made up of textiles. Often these items may be worn out, damaged, or perceived to be out of fashion to be used and thus discarded. During this practice, consumers ignore the detrimental effects of the clothes on our environment. Textile production consumes not only precious resources, such as water and cash crops, but also chemicals and inks which can be harmful for the local ecology.

1.4 Packaging waste

One of the most widespread causes of environmental degradation is packaging related waste. Majority of packaging wastes constitutes of materials like glass, aluminium, steel, paper, wood and worst of all, plastic. Unlike most other forms of material consumption, the purpose of packaging tends to be limited to the carrying of goods and materials (e.g., a toothpaste carrying paste for consumption). Once this purpose is fulfilled, consumers are unlikely to find ready alternative uses for the packaging. Consequently, packaging related waste is a pervasive outcome of most forms of material consumption, and it is not surprising to find such waste even in the remotest of places (e.g., forests) in the form of litter. Materials employed in developing packaging materials are frequently laced with harmful chemicals and inks. So, incinerating them releases harmful compounds, such as vinyl chloride, chlorofluorocarbon, and hexane, into the atmosphere. Moreover, incineration practices can result in forest fires, which disrupts ecological diversity on a larger scale. A lot of plastic waste is dumped into the river finding their way into the oceans. Reports suggest 344 types of marine animals like whales, fish, turtles, and seabirds entangle themselves in plastic waste and at least 233 ingest them (Ritchie and Roser 2018). This further degrades the environment as these animals eventually die from physical harm or starvation, disrupting entire ecosystems. Therefore, dumping of packaging materials, such as plastics, has frighteningly harmful effect on the environment, especially because it is not easily biodegradable. Plastic materials can take over 200 years to degrade. This means that all the plastic that has been produced in human history, is still present in environment.

1.5 Chemical waste

Similar to clothing wastes, chemical waste can also be industrial as well as household. Household chemical waste can be referred to as the discarded or unused portions of the household products which contain toxic chemicals. These are normally in products which bear labels of "Warning", "Toxic", "Flammable", etc. Some examples of chemical wastes are batteries, insecticides, cooking oils, drain cleaners, detergents, and fire extinguishers. These wastes cannot be dumped into the regular garbage as it may result in leaching and poisoning of the groundwater supplies. Landfills are also not designed for treating chemical wastes. The synthetic liner can be destroyed by the chemicals thereby making it ineffective. Incinerating them can create more problems as toxic chemicals are released into air, thereby increasing air pollution. These chemicals, if air-borne, may react with the water droplets, transforming into acid rain. People also pour the liquid waste down the drain which ultimately pollutes the groundwater.

Furthermore, these chemical wastes also affect the balance of the ecosystem. One of the large-scale effects in the marine ecosystem is eutrophication. It is a result of an increase in nutrients, engendering an algal bloom. This results in the decrease of oxygen in the water and thereby affecting the food chain. Additionally, it causes murkiness in the water, stopping the sunlight from reaching the plants at the seabed and also block the gills of the fish. Chemical wastes negatively affect animals higher in the food chain as well. Drinking water exposed to these wastes results in genetic defects, diseases, and potentially, the loss of life.

Summing up, it can be said that, although the sources of these wastes look very small at the outset but if not treated in a proper way, these can cause life threatening consequences and permanent damage to the planet. The waste which is generated and dumped in an improper way gives rise to many consequences which have a negative effect. Some of the consequences are mentioned in the following sections.

2. POSSIBLE CONSEQUENCES

2.1 Climate and environmental damage

Most of the organic waste when decomposed produces methane and carbon dioxide. Both are potent greenhouse gases and can exacerbate global warming and climate change. Similarly, plastic, which is made from organic compounds like ethylene and propylene, when incinerated,also causes a significant increase in the carbon dioxide volume of the atmosphere. Inorganic waste does not directly contribute to global warming by emitting greenhouse gases via decomposition. However, when incinerated, they can release greenhouse gases. Apart from greenhouse emissions being directly into the atmosphere, waste can cause environmental degradation as a result of inefficient consumption of virgin natural resources like oil, metal ores, wood, and water among many others, the processing of which ultimately increases the carbon footprint. Marine ecosystems are also significantly affected by discarded non-biodegradable materials. Recent investigations have revealed parts of the deep ocean, including the Mariana Trench,[1] have been exposed to plastic waste. Clearly, the extent of pollution in the ocean is rampant. One of the key issues associated with plastic waste is endangering of marine life. Discarded plastic either gets entangled around sea creatures (e.g., turtles, fishes, sea arthropods, etc.), often choking these creatures to death or gets eaten as food leading to death, of marine life, from starvation.

2.2 Resource depletion

As waste increases, the amount of non-renewable resources available for consumption decreases. This is a natural outcome of living in a planetary ecosystem with finite and limited resources. Such repercussions of inefficient consumption may not be felt or experienced by current generations. However, the future resource availability for key resources, such as fresh water, has already begun to look bleak. Moreover, creation of waste not only exhausts resources that make up the product but also results in the gradual depletion of the all the resources that go into making the product. Reports on material consumption found that for every can of waste which a consumer dumps, 87 cans worth of manufacturing materials, used to prepare the products and packaging, were also wasted (Materials Flow | U.S. Geological Survey 2018). In the same manner, discarding food also wastes the agricultural (pesticide, insecticide, water) and industrial (manpower, land, fossil fuels) resources. Unfortunately, considerations about the exhaustion of resources tends to receive limited attention due to conflation of concepts like waste and pollution. For instance, waste from the discarding plastic straws is often appraised in terms of the hazardous nature of plastic and its effect on environmental and human health rather than the loss of plastic materials, which could have been used for other purposes. This is another reason why waste needs to be perceived and understood as a potential source of resources over extracting virgin materials for the manufacturing of new products. For example, sustainable management of food is an approach which aims at re-channelizing food scraps for animal feed.

2.3 Public health

Both organic and chemical waste pose an imminent threat to the health of living beings. Organic waste in and around our surroundings is a common sight. The decomposition of organic waste is carried by pests and microbial pathogens, which turn the waste site into a breeding ground for deadly

[1] Details can be accessed on https://www.nationalgeographic.org/article/plastic-bag-found-bottom-worlds-deepest-ocean-trench/.

infections and diseases. This is one reason why landfill sites are situated away from the cities or other living shelters. However, rapid urbanization has started to bring these sites closer to people, in turn putting them at the risk of acquiring and spreading contagious diseases. Further, handling of the waste or even coming into close contact, can result in serious infections. Hence, workers in the waste management industry and rag pickers are the most vulnerable to this threat. Children who normally play around in fields are also in danger of coming into close contact with hazardous waste. Not only can children acquire various diseases, but their physical well-being can be affected as well. For example, children exposed to waste disposal sites exhibit a reduced rate of growth in their height and weight (Ocampo et al. 2008).

Exposure to chemical waste can also produce detrimental effects on human health. When people are exposed directly it can cause chemical poisoning, which can spread both through the air as well water sources. The significant amounts of cadmium, lead, and mercury in the household chemical products often leach into the groundwater below and pollute the dwindling drinking water supplies. When they decompose, they also release harmful gases into the air and regular exposure to these gases can cause breathing difficulties and other chronic conditions for humans and animals alike. Few of the most dreaded diseases known to mankind like typhoid, cholera, HIV, and hepatitis have the connection to improper disposal of waste.

2.4 Ecological and wildlife disruption

Waste is not only harmful for the environment and humans, but it also negatively affects different species of animals and plants. The strong sense of smell attracts animals to litter in search of food. As they cannot easily differentiate between waste and food, they can be exposed to hazardous materials, like heavy metals or plastic, which are known to cause poisoning and even death. Animals can even get their heads stuck in jars and bottles, causing them to roam about without food for days, or become easy prey to other animals. On other occasions, animals also suffocate to death when these containers block airways. Many marine species like the turtle and fish get entangled into plastic bags and ropes that limit their mobility to find food, eventually dying of starvation. Finally, while organic garbage can be a source of fertilizer for plants, other synthetic waste materials can have disastrous effects on plant growth. When poisonous chemicals are absorbed by plants and subsequently consumed by herbivores, it can lead to their death. Overall, the detrimental effects of consumer waste on the fauna and flora of a region can disrupt natural ecological balance. For example, easy availability of certain preys has the potential to endanger those animals in the region. Drastic shifts in food chains, predation cycles, and poisoning local resources has serious consequences for local and neighboring ecology.

3. WASTE PREVENTION AND MANAGEMENT

Complex problems tend to require complex solutions. The problem of waste is no different. Counteracting the harm caused by waste on humans, climate and animals, requires a multitude of efforts using different strategies in waste prevention and management. Most of these approaches can be principally summarised using the concept of circular economy. Basically, the traditional notion of perceiving objects that are discarded (or intended to be discarded) as valueless (and thus waste) ignores that value does not reside in objects per se, rather it is an outcome co-created by operand resources (i.e., objects) as well as operant resources (i.e., skills and knowledge of people) (Vargo and Lusch 2004). This implies that objects that are perceived to be waste can be put to use to create value, provided people are able and willing to invest their skills and knowledge. Circular economy is a model of the consumption process that is driven by three principles:

- Eliminate waste and pollution
- Circulate the products and materials (at their highest value)
- Regenerate nature

3.1 Reduce

One of the strategies of waste reduction is reducing consumption. Past research has found a positive correlation between income and amount of waste generated. Perhaps that explains why the most affluent countries, despite having strong policies and interventions in place to reduce waste, tend to be the most wasteful. High income levels increase acquisition of goods. Given everything that is acquired needs to be disposed of at some point in time, one keyway to reduce waste is by reducing unnecessary and mindless acquisition of new products. This entails lessening the quantity, weight, size, or volume of the products we acquire. Product here does not simply refer to the object acquired for consumption (e.g., food, furniture, electronics) but also materials used for packaging the object. Therefore, the intention and objectives of packaging must be re-examined. Another way to reduce consumption entails the use of concentrates. The use of concentrates especially in the logistics eliminates the quantity of the product and accompanying packaging and therefore is more sustainable and economical. In the context of services, another great example of reducing waste is using carpools instead of personal vehicles. This reduces carbon emissions per capita and also reduce the resources behind the manufacturing of new automobiles.

3.2 Repair

Waste can also be reduced by repairing what we own instead of discarding items that stop working. Products can break down in a multitude of circumstances, including increasing age, malfunctioning components, and accidents among many others. Throwing away such products is a natural choice for consumers. However, a lesser discussed option is to repair damaged products, when possible. Repairing existing products to fulfill consumer needs and wants is a two-pronged strategy because not only does it reduce discarding of items into oceans and landfills but also reduces the extent to which new resources are used from the environment. The key challenge with repairing of products is the understanding of or knowledge about a product that can help people determine what and how it needs to be repaired. Thus, promoting repairing of products cannot be done solely by consumers themselves. At this point, firms selling those products need to act as enablers of a repair promoting economy. They can provide knowledge in terms of manuals for simple repair jobs for their products or can provide consumers access to repair centers for a nominal cost. Other independent organizations can also aid in the promotion of repairing. For instance, there are repair cafes in places like Amsterdam and Melbourne that work on a non-commercial basis, only accepting small donations. Not only do these cafes contribute in addressing the problem of waste but since these are voluntary, people with relatively low incomes can also be a part of them. Thus, repairing eliminates waste and also reduces the acquisition of additional resources from the environment.

3.3 Repurpose

Another strategy to reduce waste is to repurpose the product in which a product can be put to use for a purpose other than what it was originally intended for by the manufacturer or consumer. Repurposing can be done in two primary ways. First, consumers can consider repurposing a product by focusing on its aesthetic value. Using candy wrappers to make an art project (e.g., decorating a vase) is a classic example of repurposing materials to create something of aesthetic value. Here the colorful wrappers, which were originally intended to hold the candy, is now being used to enhance the look of a vase. Second, consumers can consider repurposing a product for its functional utility. Case in point is repurposing old or torn clothes for mopping. Apart from being used for wearing, clothes, once they become old or damaged, can be used to absorb spills because cotton can easily absorb liquids. Another common example of functional repurposing is repurposing empty yogurt containers to hold dry food produce, such as lentils or rice. Regardless of the method of repurposing, it focuses on seeing a product or materials through the creative lens to understand how else it can be used. Non-profit organizations can consider quick repurposing of goods donated to them and sell them for a premium. Research has shown highlighting a product's history in the form of a story (e.g., how a fishing net became a produce holder) increases consumer willingness to buy repurposed products.

3.4 Redistribute

Redistribution is another key option for reducing waste. Consumers often discard goods with market value because those goods are perceived as unwanted or not useful by its owner. However, others in the second-hand goods market may be willing to acquire a used product because it is usually cheaper than buying a new one. So instead of discarding or indefinitely storing goods, which can be used by others, consumers should consider disposing of those goods to potential users in the marketplace. This can mean selling to potential buyers or donating to potential users or charities. Moreover, consumers can even consider providing temporary access of the product to others for use. Practices such as renting, lending, and sharing fall under the last method of redistribution. Thus, the unused utility in a product is put to use by someone in society via permanent transfer or ownership or by providing temporary access to the product for use. This is the essence of redistribution of goods and materials to reduce waste in society. Redistribution has several other benefits as well. For instance, donations of goods to non-profit organizations can help deprived populations with essential commodities, such as clothes and food, and thus promotes societal well-being. Additionally, selling products in the second hand marketplace can help consumers recuperate at least part of the money that was spent on the unnecessary purchase of the product, in turn, promoting individual level financial well-being. Many online portals, such as Amazon, eBay, Kijiji, Facebook, and Craigslist, provide listing space for people to sell products that are usable but are no longer being used by their owners.

3.5 Recycle

While reducing the discarding of goods and materials is important, households all over the world increasingly continue to discard stuff they do not want. To stop the discard from being dumped into landfills, oceans, and incinerators, policymakers aggressively promote waste management practices, such as recycling. Among the different methods to reduce and manage waste, recycling is the most common one. In this method, recyclable materials are collected separately from the regular garbage and trash, and then broken down in different ways like grinding and melting among others. One of the key challenges in recycling is that it must conserve the essential elements and not expend a lot of energy. In that regard, the recycling of aluminium cans is one of the best examples of this method. Basically, aluminium cans when recycled derives almost the entire metal used in its production and a relatively small amount of energy is used in the process. Governments all around the globe heavily promote recycling of paper, metal, and plastic. As part of their Sustainable Development Goals program, United Nations evaluates the success of its waste management policies and interventions by examining the extent to which different countries recycle part of the municipal waste. An important part of accomplishing high levels of recycling is promoting the use of recyclable materials in the manufacturing process. All kinds of plastics are not recyclable. So, companies are redesigning their production in a way such that their products and materials can be easily recycled. Consumers should also keep in mind that all types of wastes are not recyclable and, therefore, segregating waste types according to their recyclability is crucial. For example, laundry detergent bottles and soda bottles are both made up of plastic but cannot be recycled together. Recycling therefore is a focused task which can produce wonderful results in the management of discarded output.

3.6 Responsible discarding

Due to lack of education, awareness, and time, consumer waste is commonly disposed of in irresponsible ways. Encouraging consumers to dispose of household waste in more responsible ways in critical in tackling the problem of waste. One of the most important practices in responsible discarding is the segregation of waste at source (i.e., in the house). When waste is segregated into streams which make it easily manageable for waste collectors, it is possible to attenuate the repercussions of waste in society. Consumers need to segregate waste into biodegradable (e.g., food scraps, garden and lawn trimmings) and non-biodegradable (e.g., metals, plastic, furniture, etc.). Then consider using biodegradable waste for composting or donating to community gardens. The non-biodegradable waste can be carefully segregated into recyclable and non-recyclable waste and disposed of in their respective bins. Finally, any hazardous waste (e.g., strong chemicals) should be disposed of as per local legal regulations. At the least, this approach ensures that the negative impact of waste is minimized. Composting organic waste reduces the greenhouse emissions from slow degradation of such waste in landfills. Further, adequately planned composting will be less likely to attract pests and rodents and thus contain the spread of germs and disease.

Following local rules in disposing hazardous substances like acids and oils will lead to an efficient and effective waste management system and likely keep dangerous chemicals away from commonly accessed resources (e.g., ground water). E-waste is another type which contains chemicals like cadmium, mercury and lead which can give rise to air and water pollution. Separating electronic components, to the extent possible, and disposing them of into hazardous or recyclable streams can contain the damage to the environment, wildlife, and human health. By responsible segregation and disposal, the wastes can find their respective destinations and hence not pollute the environment. Therefore, it can be said that the circular economic framework is well and truly something that the governments and businesses around us are promoting and practising. It is now on the consumers to adapt it into their daily lives so that the hazardous effects on the environment can be reduced. In the next section, we explore what the future looks like and what can be done in order to address the problem of waste.

4. REDEFINING WASTE

As noted earlier, we continue to employ a consumer-centric perspective of waste, wherein a wastefulness is appraised from the viewpoint of the consumer. While the approach may seem convenient, there are reasons to review how we define wastefulness. At the forefront, it is the discrepancy between identifying waste only when items are discarded and the prioritizing of waste prevention over waste management practices. If waste is solely identified based on what we discard, then how can we prevent waste from occurring. Take the case of a piece of usable and unused furniture or a kitchen appliance which was used initially but has not been used at all in the last 2 years. Although consumers may continue to own these without using, especially under the pretext of waste avoidance (from discarding them), there is a high likelihood that eventually these products will be discarded. Products can undergo natural degradation over time or can even lose psychological value in a fast-paced world. Either way, the product is likely to be discarded and waste could not be prevented because it was appropriately identified.

Finally, focusing only on discarded waste also puts undue emphasis on pollution rather than waste. Pollution is a worrisome downstream consequence of waste and often takes the center stage in the discussion of waste. For example, discarded plastic materials are primarily seen as a matter of concern because plastic can destroy or disturb wildlife that comes in contact with plastic waste. Consequently, policy makers and firms aim to replace plastic materials with more biodegradable materials with similar functionality. Although this can reduce plastic pollution, using biodegradable materials for packaging simply replaces one form of waste with another. In short, our current understanding of waste may not be adequate in helping us resolve the problem of waste and reviewing what it means to be wasteful can be a good starting point.

4.1 Awareness and education related challenges

Although consumers express an aversion to appear wasteful, as the foregoing section detailed, there is still a long way to go in educating consumers about the consequences of wasteful society and waste reduction practices. Public awareness programs should be more widespread, where people are educated about the consequences of improper waste management. For example, piling organic waste and water for long near living communities is harmful as it is the breeding ground for pests, viruses, fungi, and bacteria, all of which may cause serious diseases. Consumers should be made aware of long-lasting effects, like decomposition of plastic increases the carbon footprint, which in turn exacerbates global warming. Local and regional government bodies should regularly advertise the proper ways of disposal and segregation of waste. Last but not the least, schools should teach children about waste management practices to create a sustainable future.

4.2 Motivational challenges

Public education and awareness programs can only increase the knowledge of the consumers. However, consumer practices can be only changed when the consumer brings about a change in their lifestyle. Most of the time consumers are resistant to change because of the motivational factors and public policies. For example, recycling has been considered as a low priority task by consumers (Langley 2012) and any task which is low priority is not performed when it is inconvenient. Therefore, public recyclers should be close to the recycling facilities, and they should be uncomplicated to operate. Any sort of incentive always builds motivation. Incentives can be in the form of financial or moral. Consumers can be provided with small monetary remunerations when they recycle a certain volume or weight of waste in a month. In terms of moral incentives, households can be provided certain badges if they manage to recycle a certain amount of waste. This will not only motivate the consumers from a social standpoint, but the gamification of the process will also encourage them to recycle more waste. Consumers have also started to utilise their idle resources and opt for the sharing economy. Airbnb is one of the prime examples of the same. In the domains of fashion, social media influencers opting for sustainable products will motivate the consumers to reciprocate the same. Redesigning old clothes in an innovative way is always a good way to reduce waste. Thus, motivation should come from both intrinsic and extrinsic sources for consumers to change their lifestyle.

4.3 Behavioral nudges

Behavioral nudges can be defined as innovative alternatives to the government policies, incentives and laws which influence the consumer's lifestyle (Ebeling and Lotz 2015). Therefore, it can be said that nudges are a cost-effective method of obtaining the same results. However, the implementation of the same is not as simple as it sounds. This is primarily because behavioral nudges are a very customisable option and cannot be same for all consumer mind set and lifestyle like a public policy is.

Nudges can be of three types: cognitive, affective, and behavioral, and all three can be employed for improving waste management. Let's take an example of food wastage. Cognitive nudges reduce food wastage by implying to consumers about a healthy lifestyle and therefore indirectly not encouraging them to overload their plates. Affective nudges are basically used in the form of verbal cues and displays to prompt consumers to consider buying misshapen products, which would be discarded if no buyers are found. Behavioral nudge is a process by which the habits of the consumers can be changed most effectively. For example, reducing the plate size in a buffet can help cut down on food wastage by a significant margin. Previous research has shown that if consumers are encouraged to take a second helping, they do not overload their plates in the first one (by thinking about how much they want when they are hungry) and they take less the second time (as they know exactly how much they need when they are a bit full).

REFERENCES

Arkes, H.R. 1996. The Psychology of Waste. Journal of Behavioral Decision Making 9(3): 213–224.

Brough, A. and Isaac, M. 2020. Symbolic Disposal, in NA—Advances in Consumer Research Bagchi, R., Block, L., Lee, L. and Duluth, M.N. (eds.). Association for Consumer Research, Volume 47: 158–163.

Cheyne, I. and Purdue, M. 1995. Fitting definition to purpose: The search for a satisfactory definition of waste. Journal of Environmental Law 7(2): 149–168.

Coulter, R. A. and Ligas, M. 2003. To Retain or to Relinquish: Exploring The Disposition Practices of Packrats and Purgers. in NA—Advances in Consumer Research. Keller, P.A. and Rook, D.W. and Valdosta, G.A. (eds.). Association for Consumer Research, Volume 30: 38–43.

Ebeling, F. and Lotz, S. 2015. Domestic uptake of green energy promoted by opt-out tariffs. Nature Climate Change 5(9): 868–871.

Harrell, G.D. and McConocha, D.M. 1992. Personal factors related to consumer product disposal tendencies. Journal of Consumer Affairs 26(2): 397–417.

Haws, K.L., Naylor, R.W., Coulter, R.A. and Bearden, W.O. 2012. Keeping it All Without Being Buried Alive: Understanding Product Retention Tendency. Journal of Consumer Psychology 22(2): 224–236.

Kaza, S., Yao, L., Bhada-Tata, P. and Van Woerden, F. 2018. What a Waste 2.0: A Global Snapshot of Solid Waste Management to 2050. Urban Development; Washington, DC: World Bank. License: Creative Commons Attribution CC BY 3.0 IGO.

Langley, J. 2012. Is Green a Grey Area? Sustainability and inclusivity: Recycling and the ageing population. The Design Journal 15(1): 33–56.

Lox, F. 1994. Waste Management-Life Cycle Analysis of Packaging. Brussel: Vrije Universiteit Brussel.

Materials Flow|U.S. Geological Survey, 2018. https://www.usgs.gov/centers/national-minerals-information-center/materials-flow

McCaffery, E.J. 2000. Must We Have The Right To Waste? http://dx.doi.org/10.2139/ssrn.246407

Ocampo, C.E., Pradilla, A. and Méndez, F. 2008. Impact of a Waste Disposal Site on Children Physical Growth. Colombia Médica, 39(3): 253–259.

Packard, V. 1960. The Waste Makers. D. McKay Co., New York.

Pongrácz, E. and Pohjola, V.J. 2004. Re-Defining Waste, The Concept of Ownership and The Role of Waste Management. Resources, Conservation and Recycling 40(2): 141–153.

Ritchie, H. and Roser, M. 2018. Plastic Pollution. Our World in Data. https://ourworldindata.org/plastic-pollution

Siniawer, E.M. 2018. Waste: Consuming Postwar. Cornell University Press, Japan.

Strasser, S. 2002. Waste and want: A Social History of Trash. Holt Paperbacks, ISBN : 978-080-506-5121.

Thompson, M. 1979. Rubbish Theory: The Creation and Destruction of Value. Oxford University Press.

Vargo, S.L. and Lusch, R.F. 2004) Evolving To a New Dominant Logic for Marketing. Journal of Marketing 68(1): 1–17.

Weber, V., Argo, J. and Berger, J. 2013. Getting Rid of Possessions to Get Back at People: Rejection and Consumer Disposal Choices. in NA—Advances in Consumer Research. Simona Botti and Aparna Labroo and Duluth, M.N. (eds.). Association for Consumer Research, Volume 41: 656–657.

2

INSIGHTS INTO COVID-19 WASTE MANAGEMENT

SOURCES, COMPOSITION, DISPOSAL AND CHALLENGES

Nidhi Kushwaha,[1] *Debarun Banerjee,*[1] *Ejaz Ahmad,*[2,*] *Shireen Quereshi*[3] and *K.K. Pant*[3]

1. INTRODUCTION

Waste management sector worldwide is facing acute challenges due to huge quantity of waste generated from anthropogenic activities, urbanization and exponential societal growth (Haque et al. 2021b). In addition, recent COVID-19 pandemic has brought a devastating effect to existing challenges of waste management. Indeed, pandemic has affected 168 million people globally between December 2019 to August 2021 based on data available as on 30th August 2021 (https://covid19.who.int/). However, it is worth mentioning that COVID-19 has caused both positive and negative impact on the environment and humankind (Fig. 1). For example, restricted movement, less mobility and controlled industrial operation have caused a significant decrease in CO_2 and NO_2 concentrations in European and Asian countries (Zambrano-Monserrate et al. 2020). In contrast, solid waste generation has increased drastically because of excessive usage of face masks, hand sanitizers and PPE kits (Benson et al. 2021). Thus, solid waste generation during pandemic has shown a huge adverse effect especially in developing countries. In general, primary waste management practices in such countries include random dumping, littering and scavenging due to poor awareness among waste collection workers and inadequate services (Matete and Trois 2008).

[1] Department of Fuel, Minerals and Metallurgical Engineering, Indian Institute of Technology (ISM), Dhanbad-826004, India.
[2] Department of Chemical Engineering, Indian Institute of Technology (ISM), Dhanbad-826004, India.
[3] Department of Chemical Engineering, Indian Institute of Technology Delhi-110016, India.
* Corresponding author: ejaz@iitism.ac.in

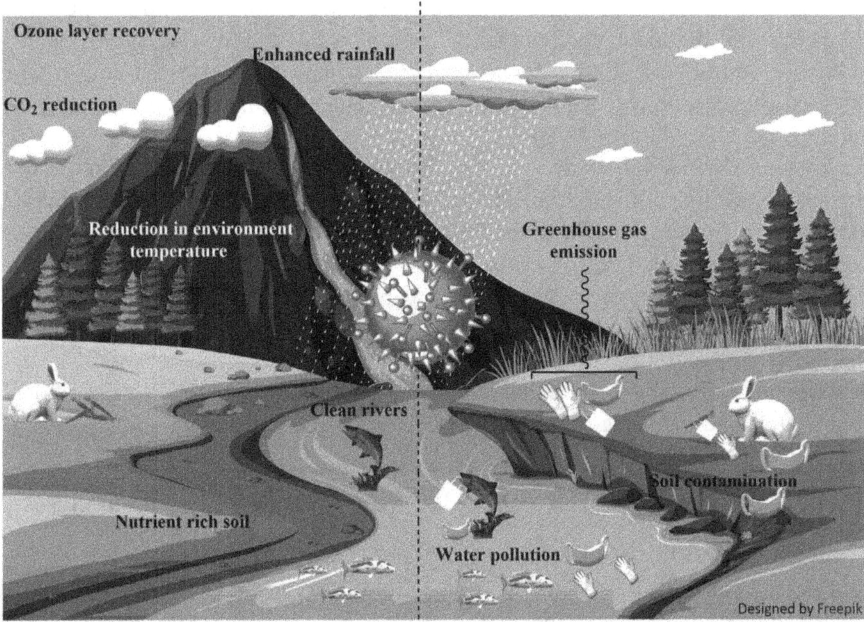

Figure 1. Positive and negative effects of COVID-19 pandemic (Background pic courtesy: Freepik).

Noteworthy that COVID-19 spreads through human interaction, bodily fluids, cough, respiratory droplets and infected surfaces (Das et al. 2020).Thus, use of face mask is mandatory for general public whereas health workers necessarily use personal protective equipment (PPE) (Klemeš et al. 2020). These PPE kits which include face masks, face shields, gloves, apron, hair caps and respirators have very short span of life; thus, they need to be changed frequently (Hantoko et al. 2021). Indeed, China reported increase in use of face mask to 116 million per day which is almost 12 times of usual quantity averaging two mask per day per person (https://www.scmp.com/economy/global-economy/article/3074821/coronavirus-chinas-mask-making-juggernaut-cranks-gear). As a matter of fact, use of one mask per day for global population could end up in global consumption of 129 billion face masks and 65 billion gloves per day (https://www.who.int/news/item/03-03-2020-shortage-of-personal-protective-equipment-endangering-health-workers-worldwide). Besides, waste produced from treatment of COVID-19 infected person in home quarantine and medical facilities also comes under the category of COVID-19 waste. Moreover, implementation of social distancing has also resulted in the use of PVC (polyvinyl chloride) sheets in public places such as shops and pharmacies to cut direct contact with local public. Similarly, excessive use of plastics for food packaging, single use plastic bags, sanitizer bottles, tissue papers and disinfectant bottles have created an upsurge in solid waste generation (Vanapalli et al. 2021). In a combined statistics, amount of waste generated during pandemic is about three to six times greater than municipal solid waste (MSW) generation in normal situation (Dharmaraj et al. 2021).

It is evident that demand and production of plastic has increased dramatically during COVID-19 pandemic. For example, Singapore alone generated 1400 tonnes of plastic waste from packaging material in one month (Adyel 2020). Use of plastic is widespread from industries to commercial and medical areas to households. On the contrary, mismanagement of waste generated pose great environmental concerns. Therefore, with continued unmanaged plastic waste it has been estimated that around 12 billion metric tonnes of plastic waste will end littered in natural environment or be poorly land filled which will be a major challenge to combat climate change and meet carbon emission targets (Geyer et al. 2017, Patrício Silva et al. 2020). Also,waste plastic can lead to flood due to accumulation and blocking drainage system in urban areas. Post flood consequences are in the form of breeding of bacteria and viruses which cause serious health issues (Krystosik et al. 2020). Besides, plastics have resulted in soil degradation and poor crop development in agricultural land (Changrong et al. 2014). Moreover, atmospheric driving forces such as wind and river water currents or waste water treatments plants have made way for plastic to enter aquatic systems which is hindering the natural ecosystem of aquatic life (Ajith et al. 2020).

Due to prolonged pandemic situation in previous years, plastic waste generation was overlooked since human safety was prioritized over environmental health (Prata et al. 2019). Increase in waste generation in past two years due to COVID-19 requires major attention for disposal and management. In addition, corona virus has been observed to sustain on surfaces for few hours to several days (Hantoko et al. 2021). The virus can persist on metals such as copper for 4 hours, stainless steel for 2–3 days, wood and paper for 4–5 days and plastics and glass for 5 days (Kampf et al. 2020, Penteado and Castro 2021). In this regard, waste from medical facility, homes and commercial places needs revised management and reimplementation of policies to avoid any secondary transmission of virus through waste and handle the huge amount of plastic waste generated.

2. SOURCES OF COVID-19 WASTE AND THEIR COMPOSITION

It is noteworthy that COVID-19 waste consists of several discarded items such as face mask, hand gloves, PPE kits, sanitizer bottles and food packaging materials, etc., generated from household, medical facilities and commercial facilities as shown in Fig. 2. Depending on the type of COVID-19 waste, its chemical composition and building block structure varies drastically. Subsequent sections discuss about individual composition of different constituents of the COVID-19 waste and their disposal techniques which is briefly described by Fig. 3.

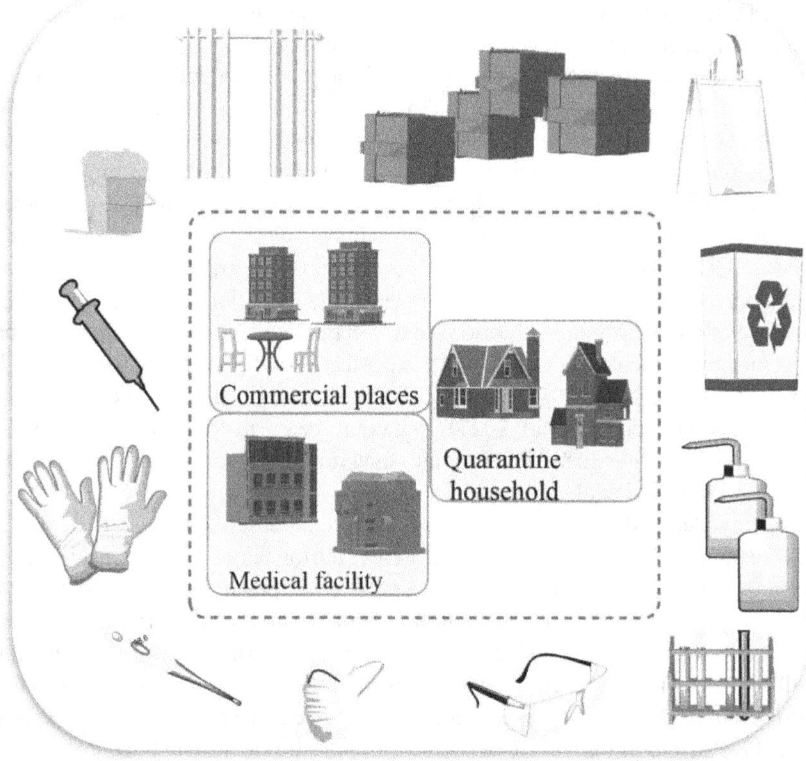

Figure 2. Sources of COVID-19 solid waste.

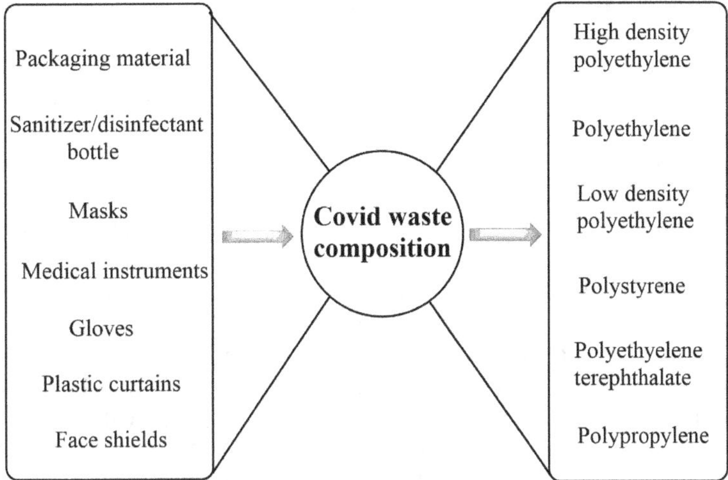

Figure 3. Composition of COVID-19 solid waste.

2.1 Medical gloves: Sources and composition

In general, medical gloves are used for clinical purposes mostly by healthcare workers to avoid transfer of infection from patient to doctor and vice versa. Owing to fact that use of gloves cuts direct contact between skin, surfaces and people, it has become a safety wear among humankind during COVID-19 pandemic to avoid virus infection. Although supplementary protection provided by gloves to common population except medical professionals is debatable, use of gloves in public has increased during COVID-19 pandemic. According to World Health Organization (WHO), about 67 million gloves are required each month because of COVID-19 (Purnomo et al. 2021a, World Health Organization (WHO) 2020a). According to a survey in Poland, use of gloves among common population during cleaning or working increased from 0.58 pairs/person/week to 5.4 pairs/person/week within two month of appearance of SARS-Cov-2 (Jędruchniewicz et al. 2021). In this regard, mainly three types of gloves are being used during pandemic: latex, nitrile and folic gloves. Nevertheless, these three constitute 94.6% of total gloves where as remaining are less used vinyl gloves. Thus, major raw material used for production of gloves are natural rubber latex (NRL), nitrile butadiene rubber (NBR), polyvinyl chloride (PVC), polyethylene (PE), styrene butadiene rubber (SBR), neoprene and polyvinyl alcohol (PVA) (Yew et al. 2019). Latex gloves are produced from natural rubber obtained from *Hevea brasiliensis* tree by tapping tree bark. The rubber structure is altered by additional processes such as chlorination and polymer coating to enhance the properties of gloves (Yip and Cacioli 2002). Latex gloves are disposable and free of pollutants since these are obtained from a natural source. Nevertheless, these have also been reported to exhibit allergic reaction in large populations (Preece et al. 2021). Therefore, nitrile gloves' demand has increased which are manufactured from polymerization of butadiene and acetonitrile. The nitrile gloves are widely used because these are non-allergic, resist chemical reactions and possess high durability (Yew et al. 2019). In addition, foil gloves are being used in the food services, deli counters, and petrol pumps. Foil gloves are not advised for medical purpose since these are made from polyethylene which is toxic in nature and have poor physical and mechanical strength (Jędruchniewicz et al. 2021). Moreover, vinyl gloves are also used rarely which are made from polyvinylchloride. These gloves undergo chlorination at high concentrations and heat treatment, and thus instil chlorinated dioxins and HCl in environment (Glas et al. 2014). Overall, there is huge increase in gloves waste due to single use.

2.2 Face masks: Sources and composition

The transmission of COVID-19 is fastest particularly from the infection carrying aerosols which are released when an infected person sneezes. Elachola et al., discovered that face masks have potential to prevent spread of virus in a community (Elachola et al. 2020). It is also recommended by WHO and different nations to wear face mask in public to reduce transmission of virus. As a result, demand for face mask has increased exponentially. It is reported that 89 million masks each month were required globally in 2020 in response to COVID-19 (Parashar and Hait 2021). Another market research estimated that global demand of masks will increase from 14.6 billion units in 2019 to 33.4 billion units in 2023 with an annual increase rate of 23% (Purnomo et al. 2021b). In general, face masks are polymeric material produced from polyethylene, polypropylene, polycarbonate, polyurethane, polyacrylonitrile, polyester and polystyrene. In general, face masks consist of three layers - an inner layer of soft fibres, middle layer with melt blown filter which is main layer for filtering made from conventional fabrication of micro and nano fibers and lastly outer layer which is water resistant and appears coloured made from nonwoven fibres (Fadare and Okoffo 2020). It is worth mentioning that during COVID-19 all types of masks available were accepted for use and they were categorized into homemade cloth masks, surgical/medical masks and respirators. Originally, surgical masks were used by healthcare workers but it has been used very commonly by common people lately to prevent infections. Surgical masks are flat and pleated in appearance which are designed to give balanced filtration with a capacity of filtering 3 μm droplets and breath ability,while respirators are similar to surgical masks with identical properties and safety standards with an additional advantage of filtering particles as small as 0.75 μm (Purnomo et al. 2021b). Celina et al. has reported that face masks can be sterilized using disinfectants, alcohol-based liquids and by heat washing resulting in increasing their reusability which can essentially reduce polymeric waste (Celina et al. 2020). However, the demand and utilization of mask is still very high (Celina et al. 2020). Thus, further increase in waste arising from discarded masks is expected.

2.3 Protective clothing sources and composition

Protective clothing, predominantly gowns, are used by healthcare workers to avoid transmission of infectious fluid or solid material from patients and vice-versa (Gordon et al. 2020). Protective clothing as classified by Food and Drug Administration (FDA) includes nonsurgical gowns, surgical gowns and surgical isolation gowns (https://www.thomasnet.com/articles/other/how-to-make-protective-gowns-for-coronavirus-covid-19/). Nonsurgical gowns are class I devices used in medical tasks where risk factor is very minimal. It covers the body area as much is required for the task to protect from any bodily fluids,while surgical gowns are class II devices which are used to prohibit transfer of microorganisms. It covers the body from top of shoulder to knees and arms from cuffs to elbow. Surgical isolation gowns are used where risk of contamination is extremely high. It covers almost entire body to keep it isolated from any kind of infected fluid and solid material (https://www.fda.gov/medical-devices/personal-protective-equipment-infection-control/medical-gowns). Notably, isolation gowns are made from nonwoven materials such as plastic films using chemical thermal and mechanical fibre bonding technology in order to prevent fluid penetration and provide strength. They are typically made from synthetic fibres such as polyethylene, polypropylene and polyester (Kilinc 2015). Like face masks and gloves, an exponential increase in waste arising from discarded protective clothing is also observed.

2.4 Goggles and face shields: Sources and composition

Protection of eyes, nose and mouth from particulate matter and bodily fluids through spray or splash to healthcare personnel can be provided by goggles and face shields. Goggles are made from PVC (for the frame), polycarbonate (PC) (for lens) and rubber (for headband) (Purnomo et al. 2021b). Face shield is a single use head mounted device that provides full face coverage from above forehead to neck protecting eyes, nose and mouth (Chaturvedi et al. 2020). During COVID-19 pandemic, demand for face shields has increased as it is now being used by common people especially people having direct contact with others such as receptionist, sales person, waiter/waitress and in flights. In general, face shields are produced using additive manufacturing (3D printing technology) using polylactic acid as raw material. On the contrary, additive manufacturing is proved to be a slow production process requiring skilled labour and produce shields of high costs of $7.3/shield (Amin et al. 2020). Therefore, for mass production of face shields, novel manufacturing process using simple tools was employed. The new method achieved production of 30 face shields per person per hour without any experienced workforce which costs as low as $0.55/shield (Shokrani et al. 2020). The raw material required are PVC sheets for the visor, polyurethane foam for facial support, elastic rubber for headband and simple office tools such as scissors and staplers (Shokrani et al. 2020). Overall, goggles and face shields also contribute significantly in increase of total solid waste generated during COVID-19 pandemic.

2.5 COVID-19 test kits

The analysis of number of COVID-19 positive cases is due to availability of testing kits. Essentially, millions of tests worldwide during pandemic has been taken and the number will keep increasing until complete eradication of virus from community. The greater number of tests taken directly relates to the number of single-use testing kits. COVID-19 tests are classified into three categories; Nucleic acid amplification test (NAAT), Antigen test and Antibody test (https://www.fda.gov/consumers/consumer-updates/coronavirus-disease-2019-testing-basics). Statistical data of test taken from the onset of pandemic to the current situation in different countries has been tabulated in Table 1 from which it can be deduced that in some countries such as United States, United Kingdom, Russia and Italy, COVID-19 tests have been taken by an individual more than once (https://ourworldindata.org/grapher/full-list-total-tests-for-covid-19). Although, tests taken per person in India is relatively less, yet the total number of tests taken recorded for India is second highest (408.14 million) followed by United Kingdom by the end of June, 2021.

Table 1. Statistical data of number of tests taken during COVID-19 pand emic.

S. No.	Country	Test taken in March, 2020	Test taken in June, 2021 (million)	Test taken per person
1.	United States	347	466.20	1.40
2.	India	6500	408.14	0.30
3.	United Kingdom	155174	205.85	3.03
4.	Russia	46414	148.98	1.02
5.	Italy	4324	71.21	1.17
6.	Germany	129291	63.74	0.76
7.	Spain	930230	45.65	0.97
8.	Canada	166	36.60	0.96
9.	Brazil	62985	31.72	0.15
10.	Australia	143056	20.30	0.80
11.	Japan	565	15.52	0.12
12.	South Africa	106	12.91	0.21

2.6 Sanitizer and disinfectant bottles: Sources and composition

Another preventative measure during COVID-19 pandemic is maintaining hand hygiene. It is mandatory for healthcare workers who are dealing with infectious patients. In hospitals, cleaning hands using soap or alcohol-based sanitizers should be done according to 'My 5 moments for hand hygiene' (Sax et al. 2007). Particularly, it rules out necessary moments for healthcare personnel for hand sanitization which are before touching patient, before cleaning procedure, after exposure to infection, after patient check-up and after contacting patient surroundings. Besides, use of sanitizer has become common among general public. It is observed that people use sanitizer before and after any kind of public contact which is a good practise to break COVID-19 transmission chain. Moreover, surface disinfectants are now used domestically to keep surfaces and packages decontaminated. WHO recommends effective sanitizer to contain 60–80% alcohol following European Norm 1500 and ASTM E-1174 standards (WHO 2020). In general, sanitizer and disinfectant bottles are made from PET and PE and the caps are made from PP. A study estimated that in Bangladesh, 49 million pieces of empty sanitizer bottles are discarded which accounts for 30.2 tonnes of daily waste due to sanitizer bottles (Haque et al. 2021a). Similarly, a drastic increase in waste arising due to discarded sanitizer and disinfectant bottles is witnessed.

2.7 Plastic sheets and packaging material: Sources and composition

Although social distancing has been implemented during COVID-19 pandemic to contain spread of virus, some places such as banks, restaurants, shops, markets, offices and schools needed more measures to be taken. Therefore, plastic curtains are widely used for safety and isolation of one person from another. In schools and offices, desks are surrounded by plastic curtains to avoid direct contact of fellowmates. Similarly, shops and banks have put up plastic curtains as a barricade between customers and staff. Furthermore, during lockdown people have relied on online services for food and groceries. Accordingly, hygienic behavioural changes in people and takeaway services have increased use of packaging material which has increased waste plastic generation. Statistically, World Economic Forum (WEF 2020) has observed 14% and 40% respective increase in sales of plastic packaging material in US and Spain. Overall consumption trend of plastic during COVID-19 estimates 5.5% annual growth rate of plastic packaging products which will eventually end up as waste (Parashar and Hait 2021). It is evident that all preventative measures taken have generated huge amount of waste which is dominantly composed of plastic polymer. However, another precautionary measure includes pause on recycling activities for COVID-19 infectious waste which has created major challenges to waste management sector.

3. CLASSIFICATION OF COVID-19 WASTE AND DISPOSAL PRACTICES

Typically, COVID-19 waste can be classified into two categories- (1) infectious waste and (2) non-infectious waste. Infectious waste according to the U.S. Environment Protection Agency (EPA) include (i) equipment, instruments, fomites and utensils used in the treatment room of patient who has been diagnosed of having communicable disease, (ii) laboratory waste such as pathological sample, tissue, syringe and disposable fomites and (iii) surgical operating room waste. In general, waste is termed infectious when it contains pathogens that can cause disease to spread (Lee and Huffman 1996). On the contrary, non-infectious waste exclusively do not contain any disease-causing pathogen and do not have the potential to cause any danger to the community. It includes packaging material, paper, files, plastic and office material. The prime sources of infectious wastes are basically healthcare facilities and self-isolated quarantined houses whereas non-infectious waste arises from commercial waste, packaging waste and regular household waste.

4. DISPOSAL PRACTICES

It is evident that COVID-19 has drastically changed composition of municipal solid waste and institutional solid waste. On account of major challenges ahead of waste management sector, it is necessary to modify waste disposal policies to avoid spread of virus through existing disposal techniques. In general, COVID-19 waste disposal practices can be divided into 4 stages that include (1) waste collection and pre-treatment (2) waste segregation and storage (3) waste collection and transportation and (4) final disposal (Peng et al. 2020).

4.1 *Waste collection and pre-treatment*

In general, most of waste during pandemic is generated from regular households, quarantine households, regular hospitals and COVID-19 medical facilities. Therefore, waste generated from home quarantine and COVID-19 facilities should be collected following newly issued protocols for waste disposal. These protocols are implemented to avoid spread of virus from improperly managed waste (Mol and Caldas 2020). It is stated that all COVID-19 waste should be considered as non-recyclable to avoid any chance of contamination in the process and transmission of infection any further (ADB 2020). Therefore, waste treatment facilities are suffering from increased incoming load during pandemic. A general rule implemented by waste authorities in several nations suggests that waste from medical facility should be labelled as 'COVID-19 infectious waste' and packed in double layer yellow or red bag (http://www.uppcb.com/pdf/Guidelines_190320.pdf). The bags should be pre-treated before placing them into storage bins. Chlorine disinfectant (0.5% or 5000 ppm) should be sprayed to disinfect the surface of bags. Chlorine in lower concentration or low pH has a very low shelf life. Therefore, concentrations of chlorine must be between 0.1 to 0.5% to achieve effective disinfection (World Health Organization 2020a). Besides disinfection, other pre-treatment method for waste from laboratories such as specimen and pathogen containing fomites must be packed in double layer bags and sealed properly (Chattopadhyay 2017). Thereafter, bags should be autoclaved at 121–134°C for 110 minutes or high temperature above 70°C for 5 minutes. Similarly, steam sterilization can be done using microwave technology at 100°C in which heat is produced due to friction along with grinding in moist conditions after which temperature reaches 150°C to achieve sterilization (Safe management of wastes from health-care activities a summary 2017). Heat pre-treatment is effective and short time procedure to inactivate coronavirus, thereby thwarting any potential risk of virus transmission. Similarly, for quarantine household facilities, COVID-19 infected waste should be separated beforehand , packed in a sealed double layered bag and placed in red colour or specially marked bins for COVID-19 waste (Nghiem et al. 2020a). The pre-treatment techniques used to disinfect waste produced in hospital and sterilize medical instruments are given in Table 2 and discussed in subsequent sections. These methods are efficient in eliminating fungi, virus, bacteria and spores.

4.1.1 High temperature steam disinfection

In general, high temperature steam disinfection is done in an autoclave where latent heat released by water vapour is used to disintegrate protein and coagulate infectious pathogens to achieve disinfection. The process operates at 134°C at 220 kPa for at least 45 min. The process is accompanied by crushing in two ways, i.e., (i) either steam treatment followed by crushing or (ii) simultaneous steam treatment and crushing. According to Ministry of Ecology and Environment (MEE), China, the logarithmic killing value of thermophilic lipobacillus spores should be greater than 5 to ensure complete disinfection (Wang et al. 2020). At the end, product is disinfected with moisture content up to less than 20% which can be disposed safely. No additional chemicals are required for disinfection and all surfaces remain well exposed to heat in the autoclave. However, autoclave treatment allows

Table 2. Disinfection techniques for COVID-19 infected waste.

Process	**Operating conditions**	**Advantages**	**Limitations**
High temperature steam disinfection (autoclave)	Pressure: 220 kPa Temperature: 134°C Time: 45 min	Low investment Low operating cost No environmental pollution	Low waste volume reduction. Produce toxic volatile compounds for specific waste.
High temperature dry disinfection	Pressure: 300Pa Temperature: 180–200°C Time: 20 min Stirring speed: 30 rotation/min	Low investment Low operating cost Does not corrode instruments Powders and oils can be sterilized	Vacuum maintenance requires external energy. Cloth, rubber and plastic cannot be sterilized.
Microwave disinfection	Temperature: 177–540°C Time: 45 min Microwave frequency: 915 or 2450 MHz	Effective on wide range pathogens No residue Deodorizing effect and cleanliness	High installation cost Low waste handling capacity. Not suitable for metal waste.
Chemical disinfection	Time: 120 min pH: 11–12.5	Low capital investment Low operating cost Low in pollution	No volume reduction.

only 70% of total feed capacity in each batch due to which several runs are required to disinfect total waste. Also, it does not provide higher volume reduction compared to other thermochemical disposal methods (Thind et al. 2021). Moreover,it retains moisture which can corrode instruments while disinfecting. Furthermore, it cannot disinfect all types of COVID-19 waste possibly due to production of toxic volatile compounds during operation (Teng et al. 2015).

4.1.2 High temperature dry heat disinfection

Dry heat treatment for waste disinfection has been proposed as an alternative to high steam disinfection to prevent retention of water and corrosion of instruments. In this process, waste is subjected to a negative pressure and high temperature after high intensity grinding. In general, minimum of 20 minutes residence time is required for effective transfer of heat to the waste for destruction of pathogens (Haque et al. 2021b). The sterilization operation occurs at 180–200°C and 300 Pa, nearly vacuum pressure. Dry heat treatment process is suitable for infectious and injury waste and does not cause any corrosion to instruments. However, this technique is not widely used due to extensive energy requirement to maintain vacuum and challenges associated with managing high temperature. Also, cloth, rubber and plastic cannot be sterilized using dry heat treatment method which makes it a less preferred choice for disposal of COVID-19 waste (Ilyas et al. 2020).

4.1.3 Microwave disinfection

Microwave disinfection is also a high temperature treatment which kills pathogens. According to MEE, China, microwave disinfection can achieve logarithmic values of killing hydrophilic viruses, parasites and mycobacteria of propagules, which should be no less than 6 and logarithmic value of killing bacillus spores which should be more than 4. The process operates at a temperature of 177–540°C for depolymerization due to microwave frequency. Microwave frequency of range 915–2450 MHz increases internal energy of material leading to vibration and rubbing of molecules (Wang et al. 2020). An inert atmosphere to prevent combustion of materials with oxygen is created under which disinfection takes place due to high temperature. The process can be an on site operation that avoids transportation of infectious waste, there is no residue production, has low energy requirement and is environmentally benign. However, installation costs are high which makes it less popular in developing countries and waste handling capacity of microwave is low. Furthermore, it is not suitable for treating metal wastes (Ilyas et al. 2020).

4.1.4 Chemical disinfection

Besides face masks and empty sanitizer bottles, fraction of infectious waste generated from hospitals has increased drastically. Thus, some chemicals which possess potential to kill pathogens and bacterial spores to achieve decontamination of materials are used. In this regards, sodium hypochlorite, calcium hypochlorite and chlorine dioxide of typically 2000 mg/L concentration are generally used for disinfection. These chemicals are usually non-corrosive, colourless and inodorous in nature, safe, water soluble and shows physical as well as chemical inertness as shown in Table 3 (Wang et al. 2020). Conventionally, chemical disinfection is done together with mechanical crushing. Firstly, disinfecting chemicals are added to crushed waste materials which remains effective for a sufficient time. The process decomposes organic waste, kills microorganisms and provides sterilization under low effective concentration of disinfectant. However, it is noteworthy that chemical treatment only decontaminates the material and does not provide any reduction in waste volume (Thind et al. 2021). Therefore, it is only a preventative measure to capture the spread of virus. For example, Barcelo et al. mentioned that vaporous H_2O_2 has been effective in sterilization of polymer-based material (Barcelo 2020). Therefore, PPE kits used by medical healthcare workers can be reused after sterilization via chemical disinfection method that will eventually help in minimizing the amount of waste generated each day (Barcelo 2020). Similarly, disinfection using alcohol-based solution is a common practice among common population and quarantine household for reusing masks and gloves which altogether reduce the polymer waste generation. However, due to huge amount of waste generation, chemical disinfection method remains insufficient; therefore, other thermochemical methods which are faster are used for waste disposal.

4.2 Waste segregation and storage

In general practice, waste from regular household is stored as recyclable waste and residual wastes. On the contrary, waste management advisory during pandemic states that quarantine household need not segregate the waste into recyclables and residuals. Since recycle of COVID-19 waste has been prohibited, waste in quarantine household should be packed as mixed waste in double layered bags. According to Association for Cities and Region for Sustainable Resource Management (ACR+), waste from home quarantine should be placed in special bins established for COVID-19 waste (https://www.acrplus.org/en/municipal-waste-management-covid-19). The waste from COVID-19 bins is to be collected in an interval of 72 hours (after the probable life span of virus in environment) to ensure safety of the waste collecting personnel (ADB 2020). In medical facilities, it is advised to immediately segregate COVID-19 waste from regular waste before packing into bags. The COVID-19 infectious waste bags must be placed in a separate area made for storage of infected waste away from interaction of public. Herein, the waste is stored for 48–72 hours before collection by the waste collection personnel (Kapoor et al. 2020). It is to mention here that for continuous working of chain of tasks from storage, collection, transportation to disposal, coordination between health authorities and waste sector personnel is very crucial.

4.3 Waste collection and transportation

To ensure safe collection of waste from community area and medical facility, it is important to implement the operation of special vehicles exclusively for COVID-19 waste collection. Thus, waste collection vehicles must be sterilized with 70% concentrated alcohol solution, each vehicle should

Table 3. List of chemical disinfectants and their uses (Centers for Disease Control and Prevention 2013).

Disinfectant	Material for disinfection	Actions	Quantity %	Limitations
Ethyl Alcohol	Oral or rectal thermometers Pagers Scissors Stethoscope	Bactericide, Tuberculocide, Fungicide, Virucide.	60–90	Bleach rubber and plastic tiles Damage tonometer tips Swell and harden rubber Flammable
Sodium hypochlorite	Tonometer Disinfection of Floors Household bleach	Bactericide, Tuberculocide, Fungicide, Virucide, Sporicide.	0.25	Corrosiveness in metals at > 500 ppm Inactivation by organic matter Release toxic chlorine gas
Formaldehyde	Hemodialyzers Renal transplant unit Anatomy laboratories	Bactericide, Tuberculocide, Fungicide, Virucide, Sporicide	37	Irritating fumes Pungent odour Prolonged exposure can cause respiratory and skin irritation
Glutaraldehyde	Endoscopes Dialyzers Spirometry tubing Respiratory equipment	Bactericidal, Virucidal, Fungicidal	≥ 2	Toxic Expensive Exposure causes dermatitis and pulmonary symptoms
Hydrogen peroxide	Surfaces Soft contact lenses Fabrics Ventilators Endoscope	Bactericidal, Virucidal, Sporicidal, Fungicidal	3–15	Damage tonometer Discoloration of metal surface
Iodophors	Hydrotherapy tanks Thermometers Endoscopes	Tuberculocidal Fungicidal Virucidal Bactericidal	0.01	Cannot be used on hard surfaces
Ortho-phthalaldehyde		Bactericidal, Sporicidal	0.21–0.5	Skin staining
Peracetic acid	Endoscope Arthroscope Surgical Dental equipments	Bactericidal, Virucidal, Sporicidal, Fungicidal	0.05–3.5	Tarnishes metal Highly unstable compound
ortho-phenylphenol and ortho-benzyl-para-chlorophenol	Surfaces Non critical medical devices	Bactericidal, Fungicidal, Virucidal, and Tuberculocidal	2–5	Cannot be used in nurseries
Quaternary ammonium compounds	Cystoscopes or cardiac catheters surfaces (floor, wall and furniture)	Fungicidal, Bactericidal, and Virucidal	95	-

have specific route and the driver and waste picker must be trained according to the guidelines. The working personnel must use PPE kits during waste collection. Moreover, the timings of waste collection vehicles should be adjusted to avoid any interaction with local community in the rush hour. The waste collecting chamber of vehicle must be maintained at a temperature of 4°C for medical waste (Waste Management during the COVID-19 Pandemic 2020). Overall, vehicle should be closed and sealed so that waste can be isolated from environment and reach final disposal site without any spread of infection. Furthermore, the vehicle must be disinfected every time after loading and unloading of waste. Besides COVID-19 waste, regular waste from households and commercial places have increased drastically. For example, takeaway services have increased the amount of packaging and organic waste (surplus food, vegetables and fruits peels, etc.), during lockdown. Thus, it is required for the waste collection department to increase frequency of waste collecting trucks to ensure complete collection of waste from local community, public places and commercial places (offices, restaurants and hotels).

4.4 Disposal of COVID-19 waste

COVID-19 waste disposal is an important issue to be focused since it cannot be disposed as regular MSW. In developing nations, informal waste disposal in unsanitary land fills and open dumpsite is a common practice which must be avoided while handling COVID-19 waste. Additionally, activities of scavengers and waste pickers should be prohibited in these areas. These practices accelerate spread of infection in the community. To contain the spread of virus from infectious waste, it is important that it should be sterilized or disinfected before processing it. As a standard practice, COVID-19 waste is not recycled or recovered to avoid transmission of virus through it. Thus, mixed waste also contains hazardous components which cannot be directly disposed. It is observed that high temperature process is effective in killing the pathogens. Also, COVID-19 waste is predominantly composed of plastic and polymer waste (PP, PE, PVC, PET, PS, polyurethane, etc.), which can be easily decomposed at higher temperatures. Therefore, common practices for disposal currently employed are thermal treatment such as incineration, pyrolysis and gasification. The waste collected from medical facilities and quarantine households are taken into thermal treatment facility. In here, high temperature process disposes waste and the ash produced is disposed in land fills. To be noted, COVID-19 waste can be treated in the existing waste facility to kill virus. However, existing facilities are not designed to receive infectious waste which makes workers very vulnerable to infection. Therefore, existing facilities need to be modified for receiving and storing infectious waste to limit its exposure to working personnel. The possible methods for disposal of COVID-19 waste have been discussed briefly in subsequent sections and illustrated in Fig. 4.

4.4.1 Incineration

Incineration of waste is a widely employed technology which is in use since early 1970s (Arvanitoyannis 2012). Ilyas et al. reported incineration to be an effective technique for sterilization and killing pathogens (Ilyas et al. 2020). Typically, it is a combustion process at 850–1100°C which is used to handle high volume waste more than 10 tonnes/day. Accordingly, incineration reduces waste volume by 85–90% of original waste and converts waste into ash. Thus, this process eliminates transportation difficulties and saves land fill space (Windfeld and Brooks 2015). However, large amount of additional fuel is required to maintain such high temperature. Furthermore, COVID-19 waste is primarily composed of polymeric compounds such as PE, PP, PET and PS, their combustion results in emission of H_2, CO_2, CO, CH_4 and lighter hydrocarbons even at very low temperatures of 500°C (Verma et al. 2016). Also, complete combustion of PVC leads to formation of chlorinated dioxins and hydrochloric acid. This results in corrosion in boiler tubes and formation of harmful compounds such as polycyclic aromatic hydrocarbons (PAHs), polychlorinated dibenzo-p-dioxins

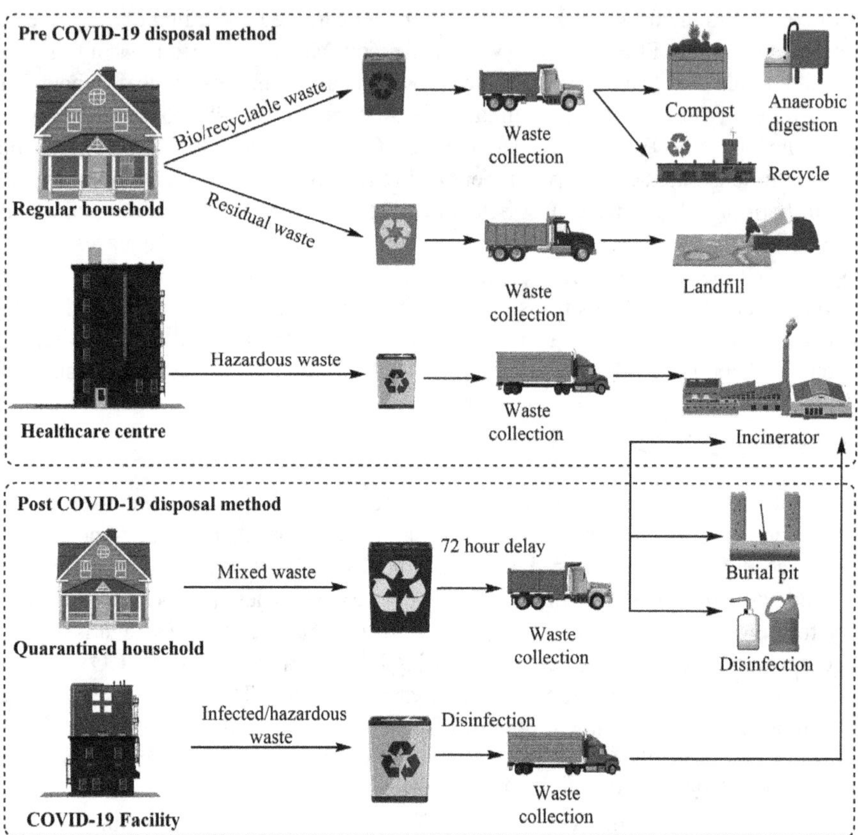

Figure 4. Solid waste disposal methods: Pre and post COVID-19 scenario (Hantoko et al. 2021).

(PCDD), polychlorinated dibenzofurans (PCDF), chlorophenols, and chlorobenzenes (Font et al. 2010). Thind et al. demonstrated pollutant emission from incineration of biomedical waste in Indian facility. Accordingly, the emission of pollutants were in the order NOx > CO >SOx> PM (particulate matter) > HCl > Cd > Pb > Hg > PCBs (polychlorinated biphenyls) > Ni > Cr > Be > As (Thind et al. 2021). Pacyna et al. reported 13% mercury emission due to incineration of medical and municipality waste in North America, while Weir et al. reported 9% atmospheric mercury production in Canada (Pacyna et al. 2006). However, design modification such as installation of bag filters, carbon filters, water or alkali scrubber, catalytic converter, gas quencher and proper flue gas treatment can limit the pollutant emissions, thereby, increasing the operating cost (Purnomo et al. 2021b).

4.4.2 Gasification

The prevailing drawbacks of waste incineration motivated the scientific community to develop new methods for waste disposal. In this regard, gasification is another high temperature method for waste disposal. Essentially, gasification takes place at 800–1500°C in presence of gasifying agents (Thomson et al. 2020, Yang et al. 2021). Gasification process is a sequence of complex reactions of drying, pyrolysis, combustion and reduction. It is the combustion step of waste material that provides energy for subsequent steps to take place. Notably, waste produced during COVID-19 pandemic from regular household and restaurants have high moisture content; thus, pre-treatment of waste to reduce moisture is necessary. On the contrary, gasification process is inclusive of drying step from the heat provided by combustion step and does not need any prior pre-treatment. Drying occurs at low temperature of 100°C which removes inherent water but no thermal decomposition and emission of volatile matter takes place.

Drying: Organic/ Wet waste + Heat → Dry waste + H_2O (1)

Following drying, pyrolysis takes place at 200–500°C in the absence of oxidising agent. The products obtained from pyrolysis are char, liquid tar and gases, the composition of which depends upon the operating conditions. Under 300°C, hydrocarbons are condensed to form tar which is an exothermic process. However, above 300°C, external heat is needed to be supplied for further conversion of feedstock into liquid and gaseous product.

Pyrolysis: Dry feedstock + heat → Char + Volatiles + Tar (2)

In the combustion zone, post pyrolysis volatile gases are oxidized in presence of gasifying agents. These chemical reactions are exothermic in nature which produce heat up to 1100–1500°C and produce gas which is composed of CO, H_2, CO_2 and H_2O. Gasification agent can be air, oxygen, steam and CO_2 provided that except air, other agents do not contribute in NO_x production and enhance production of CO and H_2 rich syngas. Primary products of gasification are CO, H_2 and CO_2, while depending upon feedstock and raw gas contamination, other gases such as CH_4, HCl and tar can also be produced (Sikarwar et al. 2016). Herein, solid carbonaceous fuel and hydrogen combines with oxygen to produce CO_2, water vapours and heat.

Oxidation: $C + O_2 \rightarrow CO_2 + 406$ KJ/gmole (3)

$H + O_2 \rightarrow H_2O + 242$ MJ/g mole (4)

In the reduction zone, a number of endothermic complex reactions takes place to produce combustible gaseous products CO_2, CO, H_2, and CH_4. The temperature in reduction process is 1000°C, high enough to decompose tar. However, particular operating temperature and pressure must be set in reduction step to enhance the production of H_2 (Sansaniwal et al. 2017)

Reduction: $C + CO_2 \rightarrow 2\,CO - 172.6$ kJ/g. mole (Boudourd reaction) (5)

$C + H_2O \rightarrow CO + H_2 - 131.4$ KJ/g. mole (Water gas reaction) (6)

$CO + H_2O \rightarrow CO_2 + H_2 + 42.3$ KJ/g. mole (Water gas shift reaction) (7)

$C + 2H_2 \rightarrow CH_4 + 75$ KJ/g. mole (Hydrogasification reaction) (8)

Interestingly, flexibility of feedstock for gasification serves as an advantage for waste disposal process. However, gasification of plastic waste accompanies higher tar formation which hinders the easy operation of internal combustion engine and power production. Moreover, the process is not suitable for PVC, PET and rubber based waste disposal (Salaudeen et al. 2018).

4.4.3 Pyrolysis

The issue of tar production in gasification process can be resolved by alternative thermochemical treatment pyrolysis. Additionally, it does not produce any harmful or poisonous gases such as in incineration. It is performed at both low and high temperature with or without catalyst depending upon the choice of product (Ansari et al. 2021). Based on higher heating temperatures, there are plasma pyrolysis (up to 9730°C) and high heating pyrolysis (up to 540 to 8300°C) (Cai and Du 2021). Using high temperature pyrolysis at about 982–1093°C, complete destruction of potential infection vectors is achieved. However, the conventional pyrolysis takes place at temperatures from 200–500°C in absence of air (Dai et al. 2014, Ahmad et al. 2019).

4.4.3.1 Low temperature pyrolysis

Low temperature pyrolysis is also called torrefaction operated in a temperature range of 180–300°C. As a result of low temperature, solid char is the major product obtained from low temperature pyrolysis. However, COVID-19 infectious waste consists of both dry waste such as PPEs and packaging material as well as wet waste such as organic waste from quarantine household. Thus, both high or low temperature pyrolysis processes can be utilized depending upon feedstock composition. The solid char obtained as major product is accompanied by liquid and gaseous products. Accordingly, an increase in pyrolyzer temperature leads to decrease in solid product content and increase in liquid product. For example, Poerschmann et al. reported the decrease in char content from 93 wt% to 75 wt% when temperature is increased from 180 to 250°C. Moreover, it was observed that polyaromatic hydrocarbon (PAH) separated from solid and dissolved in liquid product with no sign of polychlorinated dibenzo-p-dioxins (PCDD) and dibenzofurans (PCDF) compounds.

4.4.3.2 Medium temperature pyrolysis

Medium temperature pyrolysis occurs at temperature higher than 300°C reaction temperature. Thus, formation of liquid oil is dominant and measured nearly 75 wt% or higher. Ahmad et al. demonstrated pyrolysis as an effective technique to dispose polymer-based waste (Ahmad et al. 2015). The study showed decomposition of HDPE at 300°C yields 80% liquid oil. Kumar et al. stated that an increase in temperature from 400 to 550°C slightly reduces HDPE liquid product yield (Kumar and Singh 2011). This was due to reduction in residence time with the presence of less-cracked high molecular wax or highly cracked non condensable volatiles. On the contrary, LDPE showed higher liquid yield of 95% at higher temperature 500°C than 75% yield at 430°C (Matsuzawa et al. 2004). Herein, the liquid product is comprised of linear paraffins and alpha-olefins of molecular weight ranging from C_1 to C_{24}. Thus, the liquid oil had properties similar to diesel and gasoline fuel and suitable for blending in regular transportation fuel. Similarly, pyrolysis of PP takes place by chain rupture which resulted in increased liquid yield from 70–82% on increasing temperature from 300–500°C. Fakhrhoseini and Dastanian suggested that low temperature with slow heating rates prevented secondary cracking of components leading to formation of gaseous products (Fakhrhoseini and Dastanian 2013). It was also observed that PET converts to compounds of low energy value and produces more char. Furthermore, pyrolysis profile of PS was identical to PP with slight increase in liquid product from 97–98.7% on increasing temperature from 425–600°C. This was due to the fact that liquid product was composed of styrene monomer, dimer and trimer along with monoaromatics such as benzene, toluene, ethylbenzene and alpha-methylstyrene (Liu et al. 2000). Of these, styrene trimer are high boiling fraction above 350°C due to which liquid yield up to 600°C increased and decreased thereafter due to formation of gas and coke from low boiling fraction (Miskolczi et al. 2011). On the contrary, decomposition of PVC via pyrolysis yields high content of aromatic, it is accompanied by hydrochloric acid which makes its commercial application a challenging task (Miranda et al. 1999).

4.4.3.3 High temperature pyrolysis

High temperature pyrolysis is also called plasma pyrolysis where heat is produced using electric current leading to formation of plasma torch at 9730°C. Plasma pyrolysis serves an advantage by eliminating toxic dioxins, furans and pyrene. The gases produced from plasma pyrolysis are usually inert. Moreover, less residual ash is formed than incineration, total waste volume reduced by 95 wt% and provides complete destruction of COVID-19 waste safely (Punčochář et al. 2012).

4.4.4 Chemical recycling

Besides recovering energy from disposed waste, production of value-added chemicals is also of significant interest. Chemical recycling is a process where polymeric waste can be degraded to its monomer or materials which can be reused as fuels and chemicals (Pifer and Sen 1998). Essentially, chemically recycling is an alternative method for polymers such as PET, PE, and PP since they have compositional similarity to hydrocarbons of petroleum fraction.

4.4.4.1 Chemical recycling of PET

In general, chemical recycling of PET is carried out via its depolymerization into monomers terephthalic acid (TPA), dimethyl terephthalate (DMT), bis (hydroxylethylene) terephthalate (BHET), and ethylene glycol (EG) (Ragaert et al. 2017). The major processes involved in the chemical recycling are methanolysis, glycolysis, hydrolysis, aminolysis and hydrogenation depending upon the type of chemicals used for recycling purpose. Primarily, PET is degraded via transesterification with glycol at 180–250°C into BHET which has application in plasticizer, corrosion inhibitors and chain extender. However, overall process is kinetically slow; thus, transesterification catalysts are employed in higher concentration for the reaction (George and Kurian 2014). Alternatively, use of sub or super critical fluid as solvent accelerates monomer formation from waste plastic with higher selectivity. In this regard, depolymerisation of PET using water under neutral, acidic or basic condition at high temperature has confirmed formation of TPA and EG. However, reaction was slow and low purity TPA is formed because of weak nucleophilic water. To overcome the drawback of water, Thiounnet al. observed that reaction of PET in methanol solvent at 180–280°C and elevated pressure of 20–40 atm may lead to the formation of DMT and EG without catalyst (Thiounn and Smith 2020). It is worth mentioning that critical temperature of methanol is lower than that of water and depolymerized products remain stable at lower temperature. In addition to this, methanol also served both as a reactant and a solvent (Goto 2009).

4.4.4.2 Chemical recycling of PE and PP

Chemical recycling of PE involves catalytic pyrolysis. Notably, liquid product obtained from pyrolysis are complex mixtures of molecules similar to those found in petroleum. Indeed, primarily liquid products are produced due to the presence of polymeric aromatics and olefins (Chattopadhyay et al. 2016). In general, catalytic pyrolysis is thermal degradation process in the presence of catalyst in which catalyst lowers the activation energy required for a reaction, thereby increasing the reaction rate. Thus, high quantity of liquid yields can be achieved at low temperature by consuming less energy than thermal pyrolysis operations at large scale (Zunic and Peter 2018). Santos et al. observed that catalytic pyrolysis in the presence of zeolite catalyst with large pore size can yield higher molecular weight liquid product (Santos et al. 2018). In contrast, less acidic sites lead to formation of gases and low molecular weight liquid products. Also, chemical routes for PE depolymerisation involves functionalization of saturated hydrocarbon backbone followed by depolymerisation via small molecule dehydrogenation–metathesis sequence to long-chain substrates using metal catalysts such as Rh_2O_7–Al_2O_3 (Jia et al. 2016). In addition to PE, PP can also be depolymerized using pyrolysis through C-C bond cleavage (Jin et al. 2012). Besides, depolymerisation of PP can also be achieved by plasma ionization which converts it into 94% propylene gas (Tang et al. 2003).

4.4.4.3 Chemical Recycling of PVC

Similar to PE and PP, PVC can also be decomposed via catalytic pyrolysis but only after dechlorination pre-treatment. Notably, PVC does not display significant density difference from other polymer which leads to unseparated PVC pyrolysis which releases corrosive HCl fumes and harmful dioxins (Bhaskar et al. 2003). Therefore, it is necessary to dechlorinate PVC by heating to 300°C for 60 minutes. The chlorine content reduces by 75% though higher processing cost increases because of pre-treatment. Thus, several methods such as mechanochemical process using planetary ball mill grinder with CaO have been reported to reduce pre-treatment cost (Rahimi and Garciá 2017). Herein, reaction of CaO and HCl removes chlorine, thus producing easily washable $CaCl_2$ salt as by product. In addition, solvent based recovery of PVC can also be used as a cost effective method (Grause et al. 2017). In this regard, solvents which can dissolve PVC are tetrahydrofuran (THF), methyl ethyl ketone (MEK), cyclopentanone, cyclohexanone and N,N-dimethylformamide (DMF).

4.4.5 Waste burial pits

Although significant advances have been made in the field of thermochemical and chemical recycling techniques, many of developing countries still follow waste disposal in "on site burial pits". It is an alternative solution for managing COVID-19 waste to ensure no spread of infection. In general, burial pits are deep hole of maximum cross section of 2 m by 3 m (Behera 2021). The bottom of pit is lined with clay or geosynthetic liner to prevent any contamination of groundwater from hazardous waste. It requires also covering daily dumped waste with soil or soil lime. Finally, when the pit is filled it is covered with cement or mortar mixture cover. An additional layer of 50 cm soil is added together with earth mound all over it to prevent rainwater seeping. Also, it is fenced to prevent any exposure to humans or animals (Haque et al. 2021c). The on site burial serves advantage of refraining exposure of toxic material to environment during long transport routes. Indeed, it is a cost effective technique practised in various countries that lack infrastructure and technology (Sharma et al. 2020).

5. CHALLENGES AND REGULATIONS

Waste generated during COVID-19 pandemic throughout the world has challenged the infrastructure of developed as well as developing countries. Several countries have enforced regulations in response to emergency of waste disposal. It is to be noted that waste handling is a critical problem, not only due to its amount, but the fact that majority of waste is infectious and has potential to spread the virus. In regards to this fact, the challenges faced by several countries are given as follows:

(1) Lack of complete implementation of 3S methodology, that is, sorting, separation and storage, among households and medical centres for easy operation of recycling and disposal facilities. Thus, desired results from 3S methodology are not obtained which incapacitates the waste facility (United Nation Environment Programme 2020).

(2) Exponential increase in amount of waste has necessitated the capacity enhancement of existing facilities for waste management (United Nation Environment Programme 2020).

(3) In developing countries, the most common practice of waste disposal is land filling due to lack of infrastructure. Thus, disposal and land filling of COVID-19 infectious waste as per new guidelines may become a challenging task (Tripathi et al. 2020).

(4) Establishment of temporary recyclable waste storage facility to encourage people to segregate waste at home can be a quick solution but may face challenges such as public acceptance, legal aspects, economic issues and emission control (Yousefi et al. 2021).

In addition, major challenge is to define policies and implement them based on analysis and survey of existing gap between disposal capacity and waste generation. Essentially, measures are taken keeping balance of several parameters such as services provided to citizens, maintain safety of workers, endure illegal practise and continuing sorting process. Moreover, there are different hurdles faced by developing countries especially which has a huge population (Anwer and Faizan 2020). These challenges include

(1) Inculcating awareness among people to segregate domestic regular waste and infectious waste such as gloves, masks, disinfectants and thermometers and accordingly label "COVID-19 infectious waste".

(2) Providing proper training to the working staff for collection and handling of regular and infectious waste differently.

(3) Ensuring regular supply of disinfectant and PPEs to waste workers. However, there still persists freehold waste collectors in many colonies who cannot be reached by the municipality for proper guidance.

(4) Wastewater discharged from healthcare facilities contains virus; therefore, operators and staff at wastewater treatment plants remain at high risk of infection.

(5) City outskirts and villages have no access to municipality services and lack monitoring and verification services.

5.1 Regulations for handling COVID-19 solid waste

On account of challenges ahead of several countries, steps had been taken by every country to minimize the issues caused due to improper waste management. In these regards, governments and centralized authorities have issued policies and regulations that should be followed by each country on account of COVID waste management. These policies are given below as follows:

(1) United Nations Environment Protection (UNEP)

- Map out sources that are generating COVID-19 infectious waste such as quarantine facilities, quarantine households and COVID testing laboratories.
- Waste workers to be made aware of risk of exposure to infections. Special training be provided to waste collectors and truck drivers regarding disinfecting trucks after each waste picking cycle. Furthermore, all the working staff should be aware of keeping the work environment healthy by maintaining social distancing and working from home in case of any COVID-19 symptoms.
- Both formal and informal sectors should be approached for assigning work individually and make proper coordination to ensure safe waste collection and disposal. Moreover, authorities should strengthen informal waste collection sector and provide them with proper safety equipment.
- Ensure adequate supply of bins, trollies and bags in medical facility for waste collection. Additionally, it is required to find gaps in capacity of waste generation and waste disposal and accordingly enhance disposal methods.
- As an emergency response, encourage burial pits for waste disposal which will increase the capacity of waste disposal from medical facility (Waste Management during the COVID-19 Pandemic 2020).

(2) The Association of Cities and Regions for Sustainable Resource Management (ACR+)

ACR+ is an international network that contributes to waste management sector of Europe and other international cities to establish sustainable development. The guidelines issued by the authority are as follows (Jean-Benoît Bel 2021):

- Provision of separate collection services in COVID-19 effected household and medical quarantine facilities. The waste collection to be delayed 72 hours for safety of workers (Nghiem et al. 2020b).
- Allocation of different staff and staff levels to specific areas for waste collection and tracking.
- Classification of waste into high priority (medical or residual waste from contaminated household), medium priority (dry recyclable waste) and low priority (bulky or garden waste).
- Implement online booking for waste collection to ensure separate facility for waste collection from contaminated household (Bel 2020).

(3) Public Health of England (PHE)

Public Health of England has issued instructions in favour of households and medical facilities to maintain safe disposal of waste as follows:

- For households, it is advised that residual and recyclable waste must be segregated and stored in double layer black bag for 72 hours before dumping in the municipality bins.
- Residual, food and fly tipping waste collection are marked as high priority waste and collection frequency must be increased. Residual bags in environment can transmit virus if left for several weeks. Similarly, food waste will start decomposing aerobically due to climatic conditions and release bioaerosol. Moreover, local authorities should keep roads, highways and sidewalks clear of litter and refuse as they will cause public health concerns.
- Households should not mix recyclables with residual waste. This will cause an increase in disposal costs. Services might be infrequent but it should restore to normal once pandemic is controlled.
- Establishment of household waste recycling centres to provide public with a space to deposit household recycling waste at all times. It is initiated with an understanding that waste cannot be stored at houses for longer time (https://www.gov.uk/government/publications/coronavirus-covid-19-advice-to-local-authorities-on-prioritising-waste-collections/guidance-on-prioritising-waste-collection-services-during-coronavirus-covid-19-pandemic).

(4) Central Pollution Control Board (CPCB), India

According to CPCB, waste management regulations implemented in India are as follows:

- Use of colour coded bins for waste segregation in medical facility. The colour of bins administered are yellow (for laboratory chemicals, expired medicines, dressing material, contaminated fomites), red (for recyclable contaminated waste such as bottle, gloves, goggles, etc.), white (for sharp objects such as scalpel, needle, blades, etc.), and blue (for glassware) (Ibrahim Dincer Pouria Ahmadi 2019).
- COVID infected waste must be labelled COVID-19 infected waste along with surface of bags disinfected using 1% sodium hypochlorite.
- Infectious waste produced from home quarantine facilities must be closed tightly in yellow bags and hand ed over to authorized waste collector or deposited at designate centre for waste collection operated by urban local bodies (http://www.uppcb.com/pdf/Guidelines_190320.pdf).

(5) Ministry of Ecology and Environment (MEE),China

- Ecological and environmental department should work in coordination with health departments to prevent spread of virus due to collection, transportation, storage and disposal of waste.
- Make prior preparations for managing COVID waste in regions which are not yet affected to avoid mismanagement when the situation builds up.
- Medical waste should be collected and disposed on priority basis.
- On site emergency response facilities near medical institutions should be set up. The choice of place for such facility should be away from source of drinking water and residential area.

(6) World Health Organization (WHO)

- Segregation of waste to be done at source in different colour bins and COVID infected waste be pre-treated on site and then disposed.
- Mandatory use of PPE for healthcare workers and other personnel who by any means are in contact with the infected source such as waste collecting workers.

- Additional waste treatment methods such as autoclave or high temperature incinerator to be installed to support operation of existing facilities.
- In regions with limited or no waste facility, the COVID infected waste from households should be buried in pits to ensure safe disposal and littering of used tissue and masks should be strictly avoided.

(7) European Union (EU)

- The infected waste from quarantine household should be packed and closed in separate bags. However, they can be collected together with regular waste and no special treatment or disposal method is necessary for COVID infected waste.
- Emergency pre-treatment and disposal of COVID waste in on site facility should be done only by ensuring no harm to environment and safety of workplace.
- Regular changing and cleaning of PPE should be ensured to contain the spread of virus to working personnel.
- Spreading awareness and right information on waste collection and disposal and safety of every individual should be circulated regularly.

(8) International Committee of Red Cross (ICRC)

ICRC prepared guidelines for mass fatality response plan and technical recommendation for workers associated with it. The directives include aspects for handling dead bodies infected from COVID-19 which have potential to spread virus. The guidelines focused on the following in detail (Finegan et al. 2020):

- Providing PPE and disinfectants to workers handling dead bodies to lower the risk for them.
- Necessary measures for documentation of large amount of death cases. Also, the traceability from recovery of dead bodies to its storage and disposal to avoid misplacement of corresponding information.
- Autopsies be conducted following the protocol of wearing double surgical gloves, waterproof apron, impermeable gown FFP3 mask or NIOSH-certified disposable N-95 respirators.
- Decontamination of remains or possession of the deceased must be done prior to handing it over to relatives.
- Emergency response team must be prepared to address fatalities apart from those in medical facility.
- Establishment of multi-agency approach including religious authorities, medical administrators, municipal services and crematorium for integrated response.

6. SUMMARY

The chapter aims to bring up the most challenging issue, i.e., solid waste management during COVID-19 pandemic and possible disposal methods. Indeed, an exponential growth in usage of facemasks, hand gloves, PPE kits and sanitizer bottles have been witnessed in last two years. As a result, burden on waste processing facilities increased drastically, thereby necessitating the implementation of strict measures and new regulations for solid waste collection and disposal. The formulation of new strategies for infected waste disposal includes 72 hours delayed collection of waste, separation of COVID infected waste at the source of generation, and making separate bins for infectious and non-infectious waste. Also, medical facilities are required to practice several disinfection methods for infectious waste such as microwave, autoclave treatment and chemical disinfection. Indeed, several nations have brought forward many policies for smooth operation of waste management sector during pandemic, though they lack the coordination with private waste collectors and scavengers who have the potential to spread the virus. The authors recommend to use artificial intelligence for solid waste management which can create a database to monitor possible amount of waste generation, coordination among waste collection vehicles and their timing of visiting localities, coordination with private waste sectors and amount of waste disposed to understand the practices being undertaken by the public community and waste sectors.

Acknowledgement

EA would like to thank Indian Institute of Technology (ISM) Dhanbad for institutional grant for solid waste management work. EA and KKP would also like to thank Delhi Research Implementation and Innovation (DRIIV), Office of PSA to the Government of India for financial support of solid waste management project.

List of abbreviations

PPE	:	Personal protective equipment
MSW	:	Municipal solid waste
PVC	:	Polyvinyl chloride
PE	:	Polyethylene
LDPE	:	Low density polyethylene
HDPE	:	High density polyethylene
PP	:	Polypropylene
PS	:	Polystyrene
PC	:	Polycarbonate
PET	:	Polyethylene terephthalate
PVA	:	Polyvinyl alcohol
NRL	:	Nitrile rubber latex
SBR	:	Styrene butadiene rubber
NBR	:	Nitrile butadiene rubber
PCDD	:	Polychlorinated dibenzo-p-dioxin
PCDF	:	Polychlorinated dibenzofurans
PAH	:	Polyaromatic hydrocarbon
PM	:	Particulate matter
EG	:	Ethylene glycol
BHET	:	Bis (hydroxylethylene) terephthalate
DMT	:	Dimethyl terephthalate
TPA	:	Terephthalic acid
THF	:	Tetrahydrofuran
DMF	:	Dimethylformamide
MEK	:	Methyl ethyl ketone
EPA	:	Environment protection agency
FDA	:	Food and drug administration
UNEP	:	United Nations environment protection
ACR+	:	Association of cities and regions for sustainable resource management
ADB	:	Asian development bank
CPCB	:	Central pollution control board
WHO	:	World health organization
ICRC	:	International Committee of Red Cross
PHE	:	Public health of England
MEE	:	Ministry of ecology and environment
EU	:	European union

WEF	:	World economic forum
NAAT	:	Nucleic acid amplification test
ASTM	:	American society for testing and materials
ML	:	Machine learning

References

ADB. 2020. Managing Infectious Medical Waste during the COVID-19 Pandemic. 2.

Adyel, T.M. 2020. Accumulation of plastic waste during COVID-19. Science (80-). 369(6509): 1314–1315.

Ahmad, E., Vani, A. and Pant, K.K. 2019. Overview of fossil fuel and biomass-based integrated energy systems: co-firing, co-combustion, co-pyrolysis, co-liquefaction and co-gasification. *In*: Fuel Process Energy Util. [place unknown]; pp. 15–30.

Ahmad, I., Ismail Khan, M., Khan, H., Ishaq, M., Tariq, R., Gul, K. and Ahmad, W. 2015. Pyrolysis study of polypropylene and polyethylene into premium oil products. Int. J. Green Energy. 12(7): 663–671.

Ajith, N., Arumugam, S., Parthasarathy, S., Manupoori, S. and Janakiraman, S. 2020. Global distribution of microplastics and its impact on marine environment—a review. Environ Sci Pollut Res. 27 (21):25970–25986.

Amin, D., Nguyen, N., Roser, S.M. and Abramowicz, S. 2020. 3D Printing of Face Shields During COVID-19 Pandemic: A Technical Note. J. Oral Maxillofac. Surg. [Internet]. 78(8): 1275–1278. https://doi.org/10.1016/j.joms.2020.04.040

Ansari, K.B., Hassan, S.Z., Bhoi R. and Ahmad E. 2021. Co-pyrolysis of Biomass and Plastic Wastes: A Review on Reactants Synergy, Catalyst Impact, Process Parameter, Hydrocarbon Fuel Potential, COVID-19 Waste Management. J Environ Chem Eng [Internet].:106436. https://doi.org/10.1016/j.jece.2021.106436

Arvanitoyannis, I.S. 2012. Waste Management for Polymers in Food Packaging Industries. First Edit. [place unknown]: Elsevier Inc. http://dx.doi.org/10.1016/B978-1-4557-3112-1.00014-4

Association of cities and regions for sustainable resource management, (ACRPlus) 2020. Municipal Waste Management and COVID-19. [Internet]. https://www.acrplus.org/en/municipal-waste-management-covid-19

Barcelo, D. 2020. An environmental and health perspective for COVID-19 outbreak: Meteorology and air quality influence, sewage epidemiology indicator, hospitals disinfection, drug therapies and recommendations. J Environ Chem Eng [Internet]. 8(4): 104006. https://doi.org/10.1016/j.jece.2020.104006

Behera, B.C. 2021. Challenges in handling COVID-19 waste and its management mechanism: A Review. Environ Nanotechnology, Monit Manag [Internet]. 15 (February):100432. https://doi.org/10.1016/j.enmm.2021.100432

Bel, J. 2020. Implementation Systems Thinking [place unknown].

Benson, N.U., Fred-Ahmadu, O.H., Bassey, D.E. and Atayero, A.A. 2021. COVID-19 pandemic and emerging plastic-based personal protective equipment waste pollution and management in Africa. J. Environ Chem. Eng. [Internet]. 9(3): 105222. https://doi.org/10.1016/j.jece.2021.105222

Bhaskar, T., Uddin, M.A., Kaneko, J., Kusaba, T., Matsui, T., Muto, A., Sakata, Y. and Murata, K. 2003. Liquefaction of mixed plastics containing PVC and dechlorination by calcium-based sorbent. Energy and Fuels 17(1): 75–80.

Cai, X. and Du, C. 2021. Thermal Plasma Treatment of Medical Waste. [place unknown]: Springer US. https://doi.org/10.1007/s11090-020-10119-6

Celina, M.C., Martinez, E., Omana, M.A., Sanchez, A., Wiemann, D., Tezak, M. and Dargaville, T.R. 2020. Extended use of face masks during the COVID-19 pandemic - Thermal conditioning and spray-on surface disinfection. Polym Degrad Stab [Internet]. 179: 109251. https://doi.org/10.1016/j.polymdegradstab.2020.109251

Centers for Disease Control and Prevention. 2013. Guideline for Disinfection and Sterilization in Healthcare Facilities, 2008; Miscellaneous Inactivating Agents. CDC website [Internet]. (May): 9–13. http://www.cdc.gov/hicpac/Disinfection_Sterilization/10_0MiscAgents.html

Changrong, Y., Wenqing, H., Turner, N., Enke, L., Qin, L. and Shuang, L. 2014. Plastic-film mulch in Chinese agriculture: Importance and problems. World Agric [Internet]. 4(May): 32–36. https://www.researchgate.net/publication/296353247

Chattopadhyay, J., Pathak, T.S., Srivastava, R. and Singh, A.C. 2016. Catalytic co-pyrolysis of paper biomass and plastic mixtures (HDPE (high density polyethylene), PP (polypropylene) and PET (polyethylene terephthalate)) and product analysis. Energy [Internet]. 103: 513–521. http://dx.doi.org/10.1016/j.energy.2016.03.015

Chattopadhyay, S. 2017. Sample Collection Information Document for Pathogens. [place unknown].

Chaturvedi, S., Gupta, A., Krishnan, S.V. and Bhat, A.K. 2020. Design, usage and review of a cost effective and innovative face shield in a tertiary care teaching hospital during COVID-19 pandemic. J. Orthop. [Internet]. 21(July): 331–336. https://doi.org/10.1016/j.jor.2020.07.003

Coronavirus: China's mask-making juggernaut cranks into gear, sparking fears of over-reliance on world's workshop|South China Morning Post. [accessed 2021 May 29]. https://www.scmp.com/economy/global-economy/article/3074821/coronavirus-chinas-mask-making-juggernaut-cranks-gear

Coronavirus Disease 2019 Testing Basics. US food drug Adm [Internet]. [accessed 2021 Jun 30]. https://www.fda.gov/consumers/consumer-updates/coronavirus-disease-2019-testing-basics

Dai, Q., Jiang, X., Jin, Y., Wang, F., Chi, Y., Yan, J. and Xu, A. 2014. Investigation into advantage of pyrolysis over combustion of sewage sludge in PCDD/FS control. Fresenius Environ. Bull. 23: 550–557.

Das, O., Neisiany, R.E., Capezza, A.J., Hedenqvist, M.S., Försth, M., Xu, Q., Jiang, L., Ji, D. and Ramakrishna, S. 2020. The need for fully bio-based facemasks to counter coronavirus outbreaks: A perspective. Sci Total Environ [Internet]. 736: 139611. https://doi.org/10.1016/j.scitotenv.2020.139611

Dharmaraj, S., Ashokkumar, V., Pandiyan, R., Halimatul Munawaroh, H.S., Chew, K.W., Chen, W.H. and Ngamcharussrivichai, C. 2021. Pyrolysis: An effective technique for degradation of COVID-19 medical wastes. Chemosphere [Internet]. 275: 130092. https://doi.org/10.1016/j.chemosphere.2021.130092

Elachola, H., Ebrahim, S.H. and Gozzer, E. 2020. COVID-19: Facemask use prevalence in international airports in Asia, Europe and the Americas, March 2020. Travel Med Infect Dis [Internet]. 35(March): 101637. https://doi.org/10.1016/j.tmaid.2020.101637

Fadare, O.O. and Okoffo, E.D. 2020. Covid-19 face masks: A potential source of microplastic fibers in the environment. Sci Total Environ [Internet]. 737: 140279. https://doi.org/10.1016/j.scitotenv.2020.140279

Fakhrhoseini, S.M. and Dastanian, M. 2013. Predicting pyrolysis products of PE, PP, and PET using NRTL activity coefficient model. J. Chem. 2013: 7–9.

Finegan, O., Fonseca, S., Guyomarc'h, P., Morcillo Mendez, M.D., Rodriguez Gonzalez, J., Tidball-Binz, M. and Winter, K.A. 2020. International Committee of the Red Cross (ICRC): General guidance for the management of the dead related to COVID-19. Forensic Sci Int Synerg [Internet]. 2: 129–137. https://doi.org/10.1016/j.fsisyn.2020.03.007

Font, R., Gálvez, A., Moltó, J., Fullana, A. and Aracil, I. 2010. Formation of polychlorinated compounds in the combustion of PVC with iron nanoparticles. Chemosphere [Internet]. 78(2): 152–159. http://dx.doi.org/10.1016/j.chemosphere.2009.09.064

George, N. and Kurian, T. 2014. Recent developments in the chemical recycling of postconsumer poly(ethylene terephthalate) Waste. Ind. Eng. Chem. Res. 53(37): 14185–14198.

Geyer, R., Jambeck, J.R. and Law, K.L. 2017. Production, use, and fate of all plastics ever made. Sci. Adv. 3(7): 25–29.

Glas, D., Hulsbosch, J., Dubois, P., Binnemans, K. and Devos, D.E. 2014. End-of-life treatment of poly(vinyl chloride) and chlorinated polyethylene by dehydrochlorination in ionic liquids. ChemSusChem. 7(2): 610–617.

Gordon, C., Lecturer, S., Thompson, A., Nurse, P.C., Clinical, D. and Groups, C. 2020. during the COVID-19 pandemic. 29(13).

Goto, M. 2009. Chemical recycling of plastics using sub- and supercritical fluids. J Supercrit Fluids. 47(3): 500–507.

Grause, G., Hirahashi, S., Toyoda, H., Kameda, T. and Yoshioka, T. 2017. Solubility parameters for determining optimal solvents for separating PVC from PVC-coated PET fibers. J Mater Cycles Waste Manag. 19(2): 612–622.

Guidance on prioritising waste collection services during coronavirus (COVID-19) pandemic—GOV.UK. [accessed 2021 Jul 13]. https://www.gov.uk/government/publications/coronavirus-covid-19-advice-to-local-authorities-on-prioritising-waste-collections/guidance-on-prioritising-waste-collection-services-during-coronavirus-covid-19-pandemic

Guidelines for Handling, Treatment and Disposal of Waste Generated during Treatment/Diagnosis/ Quarantine of COVID-19 Patients. 2016.

Hantoko, D., Li, X., Pariatamby, A., Yoshikawa, K., Horttanainen, M. and Yan, M. 2021. Challenges and practices on waste management and disposal during COVID-19 pandemic. J. Environ. Manage [Internet]. 286 (February): 112140. https://doi.org/10.1016/j.jenvman.2021.112140

Haque, M.S., Sharif, S., Masnoon, A. and Rashid, E. 2021. SARS-CoV-2 pandemic-induced PPE and single-use plastic waste generation scenario. Waste Manag Res.

Haque, M.S., Uddin, S., Sayem, S.M. and Mohib, K.M. 2021a. Coronavirus disease 2019 (COVID-19) induced waste scenario: A short overview. J. Environ. Chem. Eng. [Internet]. 9(1): 104660. https://doi.org/10.1016/j.jece.2020.104660

Haque, M.S., Uddin, S., Sayem, S.M. and Mohib, K.M. 2021b. Coronavirus disease 2019 (COVID-19) induced waste scenario: A short overview. J Environ Chem Eng. 9(1).

How to Make Protective Gowns for Coronavirus/COVID-19. [accessed 2021 Jun 29]. https://www.thomasnet.com/articles/other/how-to-make-protective-gowns-for-coronavirus-covid-19/

Ibrahim Dincer Pouria Ahmadi, M.A.R. 2019. Bio-medical_Waste_Management_Rules_2016. J Chem. Inf. Model. 53(9): 1689–1699.

Ilyas, S., Srivastava, R.R. and Kim, H. 2020. Disinfection technology and strategies for COVID-19 hospital and bio-medical waste management. Sci. Total Environ. [Internet]. 749: 141652. https://doi.org/10.1016/j.scitotenv.2020.141652

Jean-Benoît Bel, P.M. 2021. The impact of the covid-19 pandemic on municipal waste management systems.

Jędruchniewicz, K., Ok, Y.S. and Oleszczuk, P. 2021. COVID-19 discarded disposable gloves as a source and a vector of pollutants in the environment. J. Hazard Mater. 417(April): 125938.

Jia, X., Qin, C., Friedberger, T., Guan, Z. and Huang, Z. 2016. Efficient and selective degradation of polyethylenes into liquid fuels and waxes under mild conditions. Sci. Adv. 2(6).

Jin, H., Gonzalez-Gutierrez, J., Oblak, P., Zupančič, B. and Emri, I. 2012. The effect of extensive mechanical recycling on the properties of low density polyethylene. Polym. Degrad. Stab. 97(11): 2262–2272.

Kampf, G., Todt, D., Pfaender, S. and Steinmann, E. 2020. Persistence of coronaviruses on inanimate surfaces and their inactivation with biocidal agents. J. Hosp. Infect. [Internet]. 104(3): 246–251. https://doi.org/10.1016/j.jhin.2020.01.022

Kapoor, K., Calisesi, F., Kapp, N., Takeuchi, N. and Jinno, S. 2020. Strategy Guidance : Solid Waste Management Response to COVID-19 [Internet].: 1–8. https://unhabitat.org/sites/default/files/2020/05/un-habitat_strategy_guidance_swm_reponse_to_covid19.pdf

Kilinc, F.S. 2015. A Review of Isolation Gowns in Healthcare: Fabric and Gown Properties. J. Eng. Fiber Fabr. 10(3): 155892501501000.

Klemeš, J.J., Fan, Y. Van, Tan, R.R. and Jiang, P. 2020. Minimising the present and future plastic waste, energy and environmental footprints related to COVID-19. Renew Sustain Energy Rev. 127(April).

Krystosik, A., Njoroge, G., Odhiambo, L., Forsyth, J.E., Mutuku, F. and LaBeaud, A.D. 2020. Solid Wastes Provide Breeding Sites, Burrows, and Food for Biological Disease Vectors, and Urban Zoonotic Reservoirs: A Call to Action for Solutions-Based Research. Front. Public Heal. 7(January): 1–17.

Kumar, S. and Singh, R.K. 2011. Recovery of hydrocarbon liquid from waste high density polyethylene by thermal pyrolysis. Brazilian J. Chem. Eng. 28(4): 659–667.

Lee, C.C. and Huffman, G.L. 1996. Medical waste management/incineration. J. Hazard Mater. 48(1–3): 1–30.

Liu, Y., Qian, J. and Wang, J. 2000. Pyrolysis of polystyrene waste in a fluidized-bed reactor to obtain styrene monomer and gasoline fraction. Fuel. Process. Technol. 63(1): 45–55.

Matete, N. and Trois, C. 2008. Towards Zero Waste in emerging countries - A South African experience. Waste Manag. 28(8): 1480–1492.

Matsuzawa, Y., Ayabe, M., Nishino, J., Kubota, N. and Motegi, M. 2004. Evaluation of char fuel ratio in municipal pyrolysis waste. Fuel. 83(11–12): 1675–1687.

Medical Gowns|FDA. [accessed 2021 Jun 29]. https://www.fda.gov/medical-devices/personal-protective-equipment-infection-control/medical-gowns

Mirand a, R., Yang, J., Roy, C. and Vasile, C. 1999. Vacuum pyrolysis of PVC I. Kinetic study. Polym. Degrad. Stab. 64(1): 127–144.

Miskolczi, N., Borsodi, N., Buyong, F., Angyal, A. and Williams, P.T. 2011. Production of pyrolytic oils by catalytic pyrolysis of Malaysian refuse-derived fuels in continuously stirred batch reactor. Fuel Process Technol [Internet]. 92(5): 925–932. http://dx.doi.org/10.1016/j.fuproc.2010.12.012

Mohsin Anwer. 2020. Solid Waste Management in India Under COVID19 Pandemic: Challenges and Solutions. Int. J. Eng. Res. V9(06): 4–7.

Mol, M.P.G. and Caldas, S. 2020. Can the human coronavirus epidemic also spread through solid waste? Waste Manag. Res. 38(5): 485–486.

Nghiem LD, Morgan B, Donner E, Short MD. 2020a. The COVID-19 pandemic: Considerations for the waste and wastewater services sector. Case Stud Chem Environ Eng. 1:100006.

Nghiem LD, Morgan B, Donner E, Short MD. 2020b. The COVID-19 pandemic: Considerations for the waste and wastewater services sector. Case Stud Chem Environ Eng [Internet]. 1 (April):100006. https://doi.org/10.1016/j.cscee.2020.100006

Pacyna EG, Pacyna JM, Steenhuisen F, Wilson S. 2006. Global anthropogenic mercury emission inventory for 2000. Atmos Environ. 40 (22):4048–4063.

Parashar, N. and Hait, S. 2021. Plastics in the time of COVID-19 pandemic: Protector or polluter? Sci. Total Environ. [Internet]. 759: 144274. https://doi.org/10.1016/j.scitotenv.2020.144274

Patrício Silva, A.L., Prata, J.C., Walker, T.R., Campos, D., Duarte, A.C., Soares, A.M.V.M., Barcelò, D. and Rocha-Santos, T. 2020. Rethinking and optimising plastic waste management under COVID-19 pandemic: Policy solutions based on redesign and reduction of single-use plastics and personal protective equipment. Sci. Total Environ. [Internet]. 742: 140565. https://doi.org/10.1016/j.scitotenv.2020.140565

Peng, J., Wu, X., Wang, R., Li, C., Zhang, Q. and Wei, D. 2020. Medical waste management practice during the 2019–2020 novel coronavirus pandemic: Experience in a general hospital. Am J Infect. Control. 48(8): 918–921.

Penteado, C.S.G. and Castro, M.A.S. de. 2021. Covid-19 effects on municipal solid waste management: What can effectively be done in the Brazilian scenario? Resour Conserv. Recycl [Internet]. 164 (June 2020): 105152. https://doi.org/10.1016/j.resconrec.2020.105152

Pifer, A. and Sen, A. 1998. Chemical recycling of plastics to useful organic compounds by oxidative degradation. Angew. Chemie. - Int. Ed. 37(23): 3306–3308.

Prata, J.C., Patr, A.L., Mouneyrac, C., Walker, T.R., Duarte, A.C. and Rocha-santos, T. 2019. Solutions and Integrated Strategies for the Control and Mitigation of Plastic and Microplastic Pollution. Int. J. Environ. Res. Public Health. 16(2411): 1–19.

Preece, D., Lewis, R. and Carré, M.J. 2021. A critical review of the assessment of medical gloves. Tribol - Mater Surfaces Interfaces [Internet]. 15(1): 10–19. https://doi.org/10.1080/17515831.2020.1730619

Punčochář, M., Ruj, B. and Chatterjee, P.K. 2012. Development of process for disposal of plastic waste using plasma pyrolysis technology and option for energy recovery. Procedia Eng. 42(August): 420–430.

Purnomo, C.W., Kurniawan, W. and Aziz, M. 2021a. Technological review on thermochemical conversion of COVID-19-related medical wastes. Resour. Conserv. Recycl. [Internet]. 167(September 2020): 105429. https://doi.org/10.1016/j.resconrec.2021.105429

Purnomo, C.W., Kurniawan, W. and Aziz, M. 2021b. Technological review on thermochemical conversion of COVID-19-related medical wastes. Resour Conserv Recycl [Internet]. 167(January): 105429. https://doi.org/10.1016/j.resconrec.2021.105429

Ragaert, K., Delva, L. and Van Geem, K. 2017. Mechanical and chemical recycling of solid plastic waste. Waste Manag. 69: 24–58.

Rahimi, A.R. and Garciá, J.M. 2017. Chemical recycling of waste plastics for new materials production. Nat. Rev. Chem. 1: 1–11.

Safe management of wastes from health - care activities a summary 2017. https://apps.who.int/iris/bitstream/handle/10665/259491/WHO-FWC-WSH-17.05-eng.pdf;jsessionid=BE197A8BAB73EC864CA3573E15D4F0E6?sequence=1

Salaudeen, S.A., Arku, P. and Dutta, A. 2018. Gasification of plastic solid waste and competitive technologies. [place unknown]: Elsevier Inc. http://dx.doi.org/10.1016/B978-0-12-813140-4.00010-8

Sansaniwal, S.K., Pal, K., Rosen, M.A. and Tyagi, S.K. 2017. Recent advances in the development of biomass gasification technology: A comprehensive review. Renew Sustain Energy Rev. 72(December 2016): 363–384.

Santos, B.P.S., Almeida, D., Marques, M. de F.V. and Henriques, C.A. 2018. Petrochemical feedstock from pyrolysis of waste polyethylene and polypropylene using different catalysts. Fuel. 215(November 2017): 515–521.

Sax, H., Allegranzi, B., Uçkay, I., Larson, E., Boyce, J. and Pittet, D. 2007. My five moments for hand hygiene: a user-centred design approach to understand , train, monitor and report hand hygiene. J. Hosp. Infect. 67(1): 9–21.

Sharma, H.B., Vanapalli, K.R., Cheela, V.S., Ranjan, V.P., Jaglan, A.K., Dubey, B., Goel, S. and Bhattacharya, J. 2020. Challenges, opportunities, and innovations for effective solid waste management during and post COVID-19 pandemic. Resour Conserv Recycl [Internet]. 162(July): 105052. https://doi.org/10.1016/j.resconrec.2020.105052

Shokrani, A., Loukaides, E.G., Elias, E. and Lunt, A.J.G. 2020. Exploration of alternative supply chains and distributed manufacturing in response to COVID-19; a case study of medical face shields. Mater Des[Internet]. 192: 108749. https://doi.org/10.1016/j.matdes.2020.108749

Shortage of personal protective equipment endangering health workers worldwide. [accessed 2021 May 29]. https://www.who.int/news/item/03-03-2020-shortage-of-personal-protective-equipment-endangering-health-workers-worldwide

Sikarwar, V.S., Zhao, M., Clough, P., Yao, J., Zhong, X., Memon, M.Z., Shah, N., Anthony, E.J. and Fennell, P.S. 2016. An overview of advances in biomass gasification. Energy Environ. Sci. 9(10): 2939–2977.

Tang, L., Huang, H., Zhao, Z., Wu, C.Z. and Chen, Y. 2003. KINETICS, CATALYSIS , And REACTION ENGINEERING Pyrolysis of Polypropylene in a Nitrogen Plasma Reactor. Distribution.: 1145–1150.

Teng, H., Bao, Z., Jin, D. and Li, Y. 2015. The Key Problem and Solution of Medical Waste High-temperature Steam Treatment. Proc 2015 Asia-Pacific Energy Equip. Eng. Res. Conf. 9(August 2006): 349–353.

Thind, P.S., Sareen, A., Singh, D.D., Singh, S. and John, S. 2021. Compromising situation of India's bio-medical waste incineration units during pandemic outbreak of COVID-19: Associated environmental-health impacts and mitigation measures. Environ. Pollut. [Internet]. 276: 116621. https://doi.org/10.1016/j.envpol.2021.116621

Thiounn, T. and Smith, R.C. 2020. Advances and approaches for chemical recycling of plastic waste. J. Polym. Sci. 58(10): 1347–1364.

Thomson, R., Kwong, P., Ahmad, E. and Nigam, K.D.P. 2020. Clean syngas from small commercial biomass gasifiers; a review of gasifier development, recent advances and performance evaluation. Int J Hydrogen Energy [Internet]. 45(41): 21087–21111. https://doi.org/10.1016/j.ijhydene.2020.05.160

Total COVID-19 tests. [accessed 2021 Jun 30]. https://ourworldindata.org/grapher/full-list-total-tests-for-covid-19?tab=table

Total population in millions, 2021. United Nations Popul Fund [Internet]. [accessed 2021 Aug 30]. https://www.unfpa.org/data/world-population-dashboard

Tripathi, A., Tyagi, V.K., Vivekanand, V., Bose, P. and Suthar, S. 2020. Challenges, opportunities and progress in solid waste management during COVID-19 pandemic. Case Stud Chem Environ Eng [Internet]. 2 (October):100060. https://doi.org/10.1016/j.cscee.2020.100060

United Nation Environment Programme. 2020. Covid-19 waste management factsheet. United Nations Environ Program.: 2–3.

Vanapalli, K.R., Sharma, H.B., Ranjan, V.P., Samal, B., Bhattacharya, J., Dubey, B.K. and Goel, S. 2021. Challenges and strategies for effective plastic waste management during and post COVID-19 pandemic. Sci Total Environ [Internet]. 750:141514. https://doi.org/10.1016/j.scitotenv.2020.141514

Verma, R., Vinoda, K.S., Papireddy, M. and Gowda, A.N.S. 2016. Toxic Pollutants from Plastic Waste- A Review. Procedia Environ. Sci. [Internet]. 35: 701–708. http://dx.doi.org/10.1016/j.proenv.2016.07.069

Wang Jiao, Shen, J., Ye, D., Yan, X., Zhang, Y., Yang, W., Li, X., Wang Junqi, Zhang, L. and Pan, L. 2020. Disinfection technology of hospital wastes and wastewater: Suggestions for disinfection strategy during coronavirus Disease 2019 (COVID-19) pandemic in China. Environ Pollut [Internet]. 262:114665. https://doi.org/10.1016/j.envpol.2020.114665

Waste Management during the COVID-19 Pandemic. 2020.

WHO. 2020. Water, sanitation, hygiene, and waste management for SARS-CoV-2, the virus that causes COVID-19. Interim Guid [Internet]. (29 July): 1–11. https://www.who.int/publications/i/item/water-sanitation-hygiene-and -waste-management-for-the-covid-19-virus-interim-guidance

WHO Coronavirus (COVID-19) Dashboard|WHO Coronavirus (COVID-19) Dashboard With Vaccination Data. [accessed 2021 May 28]. https://covid19.who.int/

Windfeld, E.S. and Brooks, M.S.L. 2015. Medical waste management—A review. J Environ Manage [Internet]. 163: 98–108. http://dx.doi.org/10.1016/j.jenvman.2015.08.013

World Health Organization (WHO). 2020. Cleaning and Disinfection of Environmental Surfaces in the context of COVID-19: Interim guidance. Who. (May): 7.

World Health Organization (WHO). 2020a. Shortage of Personal Protective Equipment Endangering Health Workers Worldwide [Internet]. https://www.who.int/news/item/03-03-%0A2020-shortage-of-personal-protective-equipment-endangering-health-workers-w%0Aorldwide

Yang, Y., Liew, R.K., Tamothran, A.M., Foong, S.Y., Yek, P.N.Y., Chia, P.W., Van Tran, T., Peng, W. and Lam, S.S. 2021. Gasification of refuse-derived fuel from municipal solid waste for energy production: a review. Environ. Chem. Lett. [Internet]. 19(3): 2127–2140. https://doi.org/10.1007/s10311-020-01177-5

Yew, G.Y., Tham, T.C., Law, C.L., Chu, D.T., Ogino, C. and Show, P.L. 2019. Emerging crosslinking techniques for glove manufacturers with improved nitrile glove properties and reduced allergic risks. Mater Today Commun [Internet]. 19: 39–50. https://doi.org/10.1016/j.mtcomm.2018.12.014

Yip, E. and Cacioli, P. 2002. The manufacture of gloves from natural rubber latex. J Allergy Clin Immunol. 110(2): S3–S14.

Yousefi, M., Oskoei, V., Jonidi Jafari, A., Farzadkia, M., Hasham Firooz, M., Abdollahinejad, B. and Torkashvand, J. 2021. Municipal solid waste management during COVID-19 pandemic: effects and repercussions. Environ. Sci. Pollut. Res.: 32200–32209.

Zambrano-Monserrate, M.A., Ruano, M.A. and Sanchez-Alcalde, L. 2020. Indirect effects of COVID-19 on the environment. Sci Total Environ [Internet]. 728: 138813. https://doi.org/10.1016/j.scitotenv.2020.138813

Zunic, B. and Peter, S. 2018. World's largest Science. Technology and Medicine (2016): 267–322.

3

TOWARDS EFFICIENT METHODS IN E-WASTE DISASSEMBLING
THE CIRCULAR ECONOMY APPROACH

Piotr Nowakowski

1. INTRODUCTION

Waste electrical and electronic equipment (WEEE), also known in the literature as e-waste, is an important source of secondary raw materials. This equipment includes materials that are easy to recycle like metals, plastics, glass, or other metals (Goodship and Stevels 2012). The hazardous substances present in the equipment constitute a serious threat to health and the environment. Meeting the requirements of effective and sustainable waste management, a concept of a circular approach towards resources conservation, has appeared in the economy (Korhonen et al. 2018). It greatly facilitates waste management and the recovery of raw materials or energy from the WEEE stream. The coronavirus pandemic announced by World Health Organization (Time 2020) also increased the number of equipments sold in the market due to the need to use information and telecommunications tools and devices (Statista 2020). Working and learning from home by the individuals increased purchases of the necessary equipment, which in the future will contribute to an increase in the volume of equipments withdrawn from use and the need to develop them.

The circular economy (CE) approach (Fig. 1) allows for greater care for the natural environment and resource management (Ning 2001). One of the commonly disposed of categories of waste from households is e-waste. Taking into consideration relatively easy collection, disassembling, and recycling of WEEE, it can fulfill the requirements of CE. However, it is necessary to use appropriate technologies when disassembling that equipment. The level of recycling and recovery of individual raw materials should be maximized with high process efficiency and ensuring proper separation of valuable or hazardous materials. This chapter presents the basic methods of WEEE dismantling, along with the necessary machinery and material separation processes, which are essential to obtain high purity material fractions after disassembling e-waste. New challenges and attempts to increase disassembly effectiveness for new products are also presented in the discussion section.

Silesian University of Technology, Faculty of Transport and Aviation Technologies, 40-019 Katowice ul. Krasińskiego 8, Poland.
Email: Piotr.Nowakowski@polsl.pl

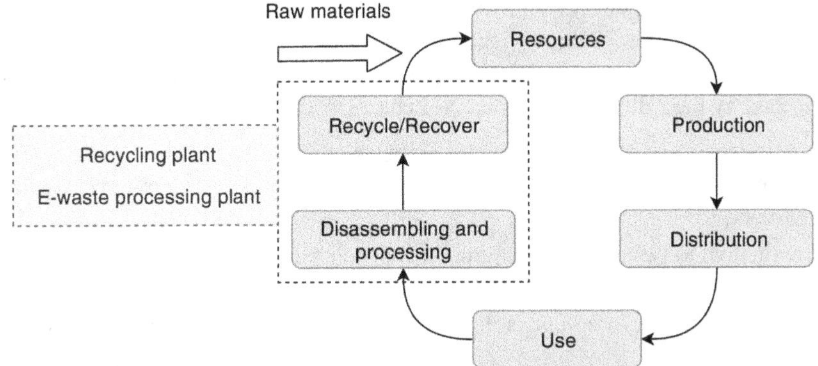

Figure 1. Circular economy approach for electrical and electronic equipment.

2. FRAMEWORK OF DISASSEMBLING WASTE ELECTRICAL AND ELECTRONIC EQUIPMENT

WEEE is a valuable source of raw materials, as well as reusable parts. The end of life marks the transition to the dismantling and recycling phases. Disposed of electric and electronic devices have a high recycling potential that can be fully utilized after proper disassembly (Cucchiella et al. 2015). As a result, it will be possible to use raw materials in the recycling process or to include parts or components for reuse. The general flow chart for making decisions on disassembling WEEE is shown in Fig. 2.

Some of the equipment is fully reusable, such a situation applies only to a small group of end-of-life electrical and electronic equipment, most often Information Technologies (IT) equipment. The

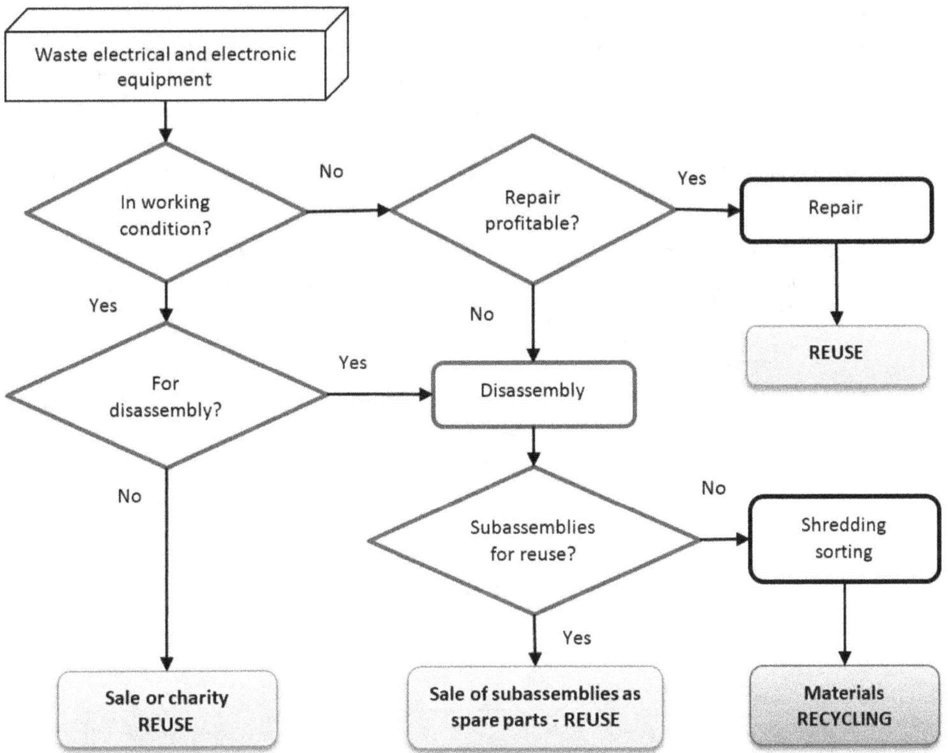

Figure 2. General decision scheme in ending the use of technical appliances.

disassembly decision should be taken into consideration if there are components or parts suitable for repair or reuse (Parajuly et al. 2019, Xavier and Adenso-Díaz 2015). The remaining used equipment goes to the disassembly process line. Disassembly is, to some extent, the opposite of assembly, but it is distinguished by the following features (Nof et al. 1997):

- the complexity of the operation to be performed or the high degree of difficulty of disassembly—the process becomes unprofitable then and destructive disassembly must be carried out,
- the main goal to be achieved is to maximize profits and minimize the impact on the natural environment,
- dismantled parts are often unusable, with damage caused during disassembly.

Full disassembly includes operations necessary to disconnect both detachable and non-detachable connections. Partial disassembly involves disassembling the assemblies to replace the fast-wearing parts (it can take the form of an operational replacement of fast-wearing parts, e.g., a printer cartridge), without disconnecting the main assemblies from the base part of the machine. General considerations and tasks related to the dismantling system are shown in Fig. 3. The first step is disassembly planning that includes the selection of qualified personnel, technological possibilities of a plant, and capability to process specified categories of waste equipment. The effectiveness will depend on the capability of the technological line (Scharke 2003). Assuming the right choice of technology, the most important is to maintain the material flow of waste electrical equipment for disassembly. Large fluctuations in the supply of e-waste equipment place a heavy financial burden on a disassembling plant operator. The problem here is the over sizing of the installation which, while reducing the material flow, causes losses in the form of unused processing capacity concerning the funds invested in plant equipment. Properly selected parameters allow for the implementation of tasks related to the disassembly process and obtaining economic and environmental benefits.

Dismantling plays a key role in the recycling process. Knowing how to dismantle a given product is essential for economically viable recycling and following the principles of environmental protection. To achieve high efficiency, the dismantling process must be carefully planned (Dyckhoff et al. 2013). It is also necessary to exchange information about the connections used in the construction and material composition, taking into account hazardous substances. In the disassembly of e-waste, two basic methods of removing parts or materials from an end-of-life product are distinguished:

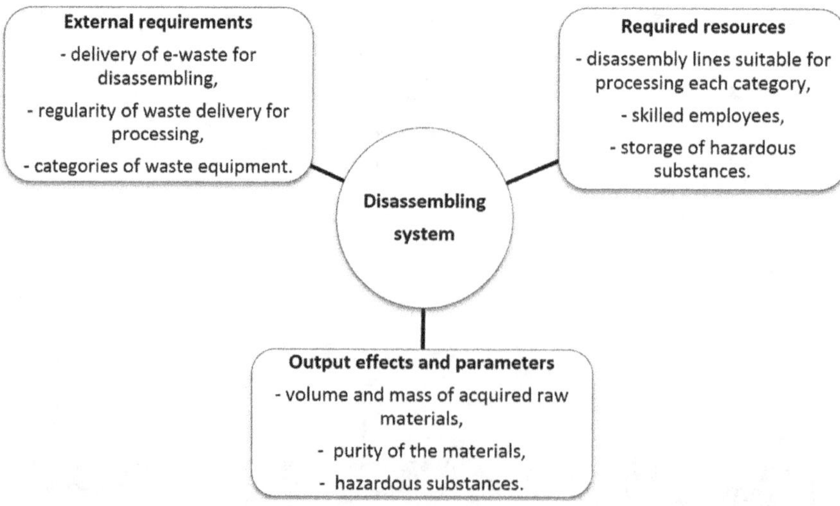

Figure 3. Disassembly tasks and environment in recycling process.

destructive disassembly (mechanical processing) and non-destructive disassembly (mostly manual or robotized). Most frequently, mechanical processing is shredding or grinding to obtain fragmented materials for further processing (Kuo 2006).

The non-destructive method aims to recover parts or components for reuse or refurbishment. Often it is carried out manually. In this method, although whole parts or subassemblies are obtained, they can later go to mechanical processing to reduce the volume of the material for transport purposes. Parts that can be reused should be functionally tested, which increases the cost of the entire process and is therefore rarely used. Broader use of parts from disassembly can be found, for example, in the IT (Ongondo and Williams 2011, Williams et al. 2008).

3. METHODS OF E-WASTE DISASSEMBLY

For electrical and electronic equipment in the first phase of disassembly in the European Union, the procedure set out in the WEEE Directive is applied (European Commission 2012). The first step in dismantling is the removal of hazardous substances and materials. In electronic equipment, these are substances listed in the RoHS Directive (including electrolytic capacitors, mercury-containing components, batteries, accumulators, printed circuit boards) (European Union 2011). In further stages, the components are shredded and fragmented to facilitate their separation, reduce the volume for transport purposes, and finally separate materials for the shipment from recycling plants (Kang and Schoenung 2005).

3.1 Manual disassembly

Disposal of waste equipment should be carried out in accordance with the recommendations contained in the annex to the WEEE Directive and should take into account the removal of the following components or assemblies (Morselli et al. 2009):

- accumulators and batteries,
- discharge lamps, toners, cartridges,
- printed circuit boards with an area of more than 10 cm^2,
- liquid crystal display (LCD) screens, cathode ray tubes (CRT),
- elements containing polybrominated biphenyls (PBB), polychlorinated biphenyl (PCB) (electrolytic capacitors),
- items containing mercury and radioactive substances.

At the manual disassembly stage, it is essential to open the housings of electrical or electronic products and separate them from the rest. It is also important to separate other large parts, as well as to separate valuable elements, which include components made of non-ferrous metals (copper, aluminum), ferrous metals, as well as components containing precious metals—e.g., motherboards, processors, etc. Manual disassembly is used for large home appliances. This applies to both very technologically advanced plants and simple plants focused only on this type of disassembly. The most commonly used manual dismantling tools include a hammer, cutter, drill, screwdriver, tongs, or angle grinder. In order to achieve higher efficiency, pneumatic tools are also used with the possibility of a quick exchange of working tips. The advantages of manual disassembly are the low capital expenditure on tools and staff (Chatterjee and Kumar 2009). The employees do not have to be people who have special qualifications, because the activities they perform with the use of basic tools are simple procedures. However, the disadvantages include low efficiency, and no possibility of obtaining smaller fractions with detailed selection.

3.2 Mechanical processing in disassembly of e-waste

Mechanical processing is used in specialized disassembly plants. The operators at the disassembly line sort the devices and then place them onto the conveyor. There are various configurations of disassembly lines but commonly the entire process is fully automated. Transportation of the materials is performed mostly by conveyors and each material after final separation ends up in a metal container or big bag. The process starts with shredding, grinding, or crushing and goes through a configuration of separators (Williams 2006). The general scheme is shown in the diagram in Fig. 4 (Hester and Harrison 2009).

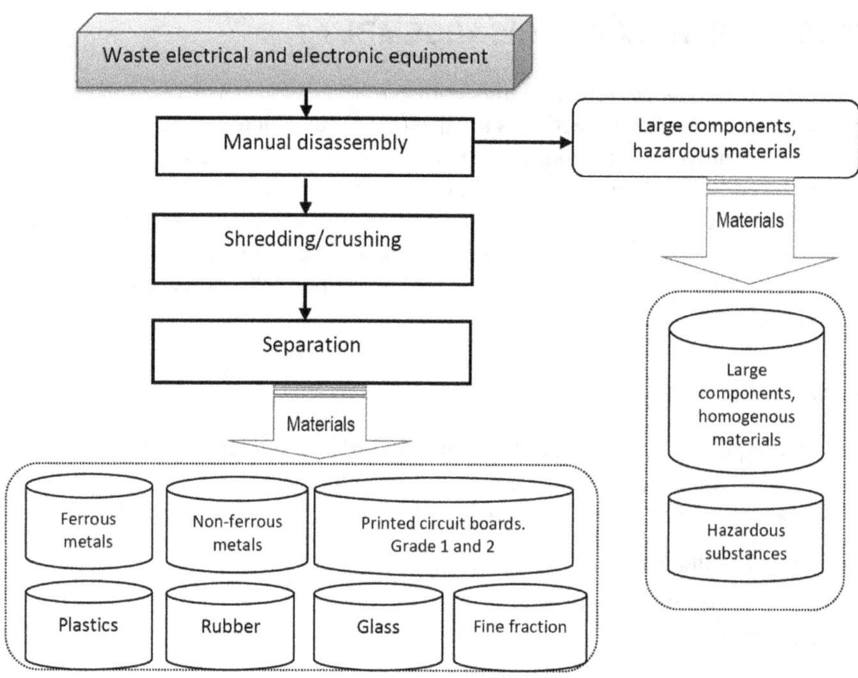

Figure 4. Generic scheme of raw materials recovery from WEEE.

4. PROCESSING MACHINES USED IN E-WASTE DISASSEMBLY PLANTS

The main task for waste processing machines is the fragmentation of large WEEE components or entire devices, in the case of small-sized equipment (small home appliances, toys, small IT equipment, etc.). Fragmentation has two main goals: to reduce the volume of the material stream by obtaining particles grains with a specific fraction size and to prepare the shredded or fragmented material for separation on machines installed further on the processing line. In the shredding of WEEE elements and components, two-stage (and more) shredding is used. This is to obtain the most appropriate size of the fraction to be separated (Goodship and Stevels 2012). The size of the grains after grinding is also important for transport purposes. The volume that the material will fill in the container intended for transport to the recycling plant depends on this. The basic machines used in the fragmenting WEEE are mills (including hammer mills), shredders, and crushers (Pichtel 2014).

4.1 Hammer mills

Hammer mills are used in larger processing plants, especially in the processing of general-purpose scrap, including end-of-life vehicles, and the components from large household appliances (washing machines, dishwashers, kitchens), dispensing machines, etc. The rotational speed of the rotor shaft ranges from about 400 rpm to over 3000 rpm (Kaya 2016). For milling large plastic housings, including fragile elements—refrigerator cabinets, plastic parts, etc.—hammer mills with a vertical shaft are used. Hammers, knives, rods, or chains are mounted on the shaft. It depends on the type of material to be shredded (Bilitewski et al. 1997). The material is loaded from above, employing the conveyor, through the charging hopper, into the working chamber of the machine. Falling freely, it encounters rotating elements mounted on the shaft, which break it or crush it.

At the bottom of the machine, there can be a special system of knives or cutters, which allows the material to be pulverized, until very fine grains are obtained. In this arrangement, the mills are called pulverizing mills. For two-stage processing, there is a stage I—comminution in a hammer mill and stage II—pulverization in a powder mill (pulverizer) (Kaya 2016). Such a process can be carried out using one or two machines; after that, the shredded material is loaded on-to conveyor. The rotational speed of the shaft, on which hammers, rods, or chains are attached, ranges from 500–700 rpm.

4.2 Shredders

Shredders (or shear shredders) are quiet in operation, consume low energy, and produce relatively low noise. The material is fed through the hopper into the working chamber. In this chamber, cutting knives are mounted on a rotating shaft. The number of shafts can be one, two shafts, or four (Hester and Harrison 2009). The batch material is torn, cut, or crushed and falls on the screen—it is a sieve with a mesh size corresponding to the assumed grain size of the fraction, located below the chamber with the rollers. From there, it can go to a container or onto a conveyor belt. It is dependent on machine performance and processing line configuration. The thickness of the knives determines the size of the material shreds obtained. The shafts of two- and four-shaft shredders work in opposite directions so that the used equipment is pulled into the area of the knives' operation. The rotational speed is determined individually depending on the shredded material and is usually from 10–50 rpm. In the event of jamming of the shredded e-waste equipment or component, it is possible to reverse the direction of rotation. Single shaft shredders, as a rule, have a rotation speed of 30–50 rpm, while double shaft shredders, from 8–20 rpm (Kaya 2016). Two types of drives are used in their construction—electric and hydraulic. The first one is easier to build and maintain, the latter provides better overload protection.

4.3 Crushers

Single-shaft crushers are equipped with a working chamber in which the shaft rotates. There are knives on it, whose task is to fragment the material (Prasad et al. 2019). Unlike shredders, the shaft of the working part operates at a higher rotational speed in the ranges from 70–130 rpm, and for high-speed ones: 300–600 rpm. In the operation of crushers, the material through the hopper goes to the working chamber, where the shaft rotates. Cutters are mounted on it, but unlike shredders, the width of the knife most often corresponds to the width of the working chamber. These machines are suitable for grinding homogeneous materials resulting from the dismantling of WEEE. The best results are obtained with brittle materials such as plastics or glass.

5. MATERIALS AND FRACTION SEPARATION AFTER PRELIMINARY PROCESSING OF E-WASTE

The next step in the processing that follows the fragmentation of the parts and components of WEEE is focused on the separation of individual raw materials. This makes it possible to obtain material for sale of recycling plants. In the separation of fractions of various or homogeneous materials, there are three processes (Jess and Wasserscheid 2019):

- classification—for a chemically homogeneous material, separation only by grain size,
- separation—separation from a multi phase system of two or more products that differ from the starting material quantitatively or qualitatively in terms of at least one feature (e.g., ferromagnetic material from the fragmented WEEE stream),
- sorting—separation of a multi-component mix of materials into fractions that differ in certain physical properties (e.g., particle size, density, magnetic and electrical properties, etc.).

The basic machines for automated material separation include magnetic separators, eddy current separators, gravity separators, electrostatic separators, cyclone separators, X-rays or optical analyzers and separators (Letcher and Vallero 2011). In most cases, the fractions after separation are of a high degree of purity. Groups of mixed materials, e.g., plastics, can end up in incineration plants. Fractions containing non-ferrous and precious metals should be processed in plants that have the appropriate refining technologies for maximum revenue and recovery rate.

5.1 Magnetic separation

Magnetic separators with generated magnetic fields provide a sufficiently high force acting on ferromagnetic particles. Magnetic elements in such separators are made of ferrite and neodymium magnets (Cui and Zhang 2008). The shredded material fraction, while moving on the conveyor, enters the area of the magnetic field, and then ferromagnetic parts are attracted by the magnet. In the place of the disappearance of the magnetic field, the material fall by gravity into the container placed under the device (Fig. 5). Separation takes place mainly by positioning the separator transversely to the axis of the belt conveyor.

This way, after separation, the material goes directly to the container placed next to the belt conveyor. Another way to configure the mounting of the separator is to place the magnetic element on or above the face drum.

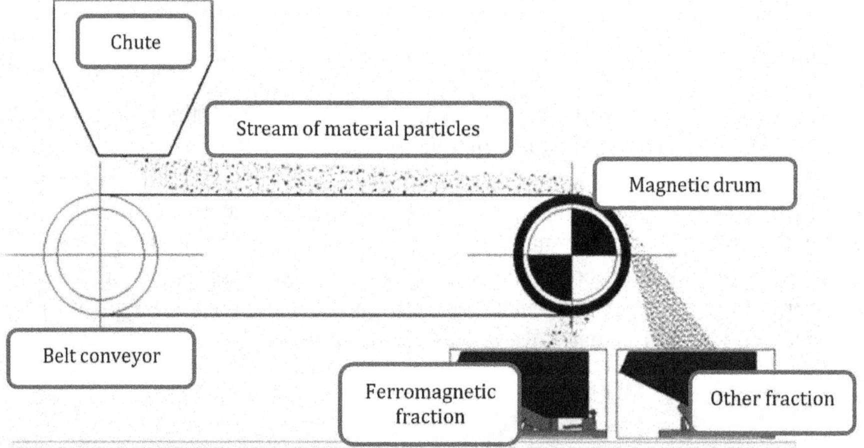

Figure 5. Scheme and principle of operation of magnetic separator.

5.2 Eddy current separation

In this type of separator, in the front drum of the conveyor, there is a rapidly rotating system of permanent magnets generating constant magnetic fields (Fig. 6).

In non-ferrous metal particles, they cause strong eddy currents. They create their magnetic fields counteracting the external field, so that non-ferrous metals are rejected from the remaining material stream (Jujun et al. 2014). The rejected metals go to the container located further from the front drum of the conveyor, and the remaining material (mainly plastics) goes to the tank or the belt located directly behind the drum. In this type of separation, it is necessary to homogenize the grain size of the material. This can be done after separation from the sieves.

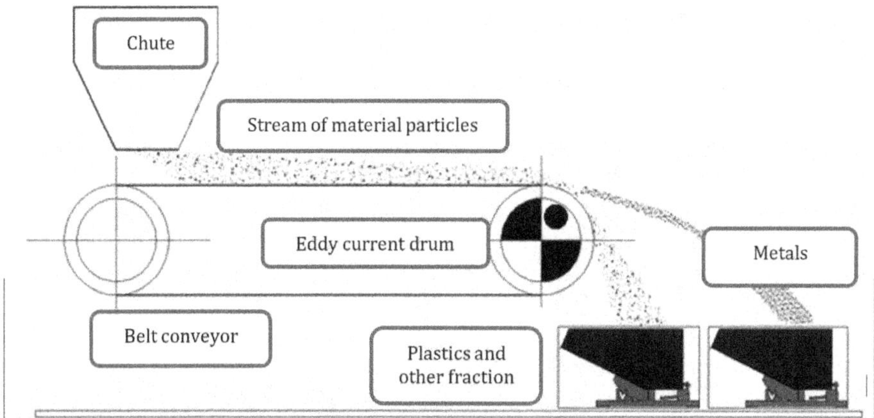

Figure 6. Scheme and principle of operation of eddy current separator.

5.3 Gravity separators

Gravity separators are used for sorting materials with medium and fine grain. The resilient support of the sifter plate allows for forcing a free oscillating motion with a circular trajectory, caused by the inertial drive (Fig. 7). The table of the separator where the material is dumped is equipped with screens. Forced vibrations and inclination of the screen plate cause tossing, screening, and particle displacement (Das et al. 2009).

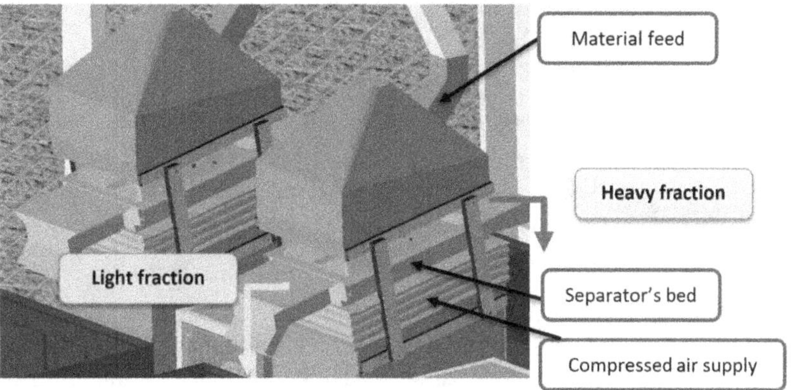

Figure 7. Principle of operation of gravity separator.

Homogeneous materials of similar density accumulate in various parts of the screened separator's bed. In other design solutions of vibration separators, the air is blown from below on the material on the perforated plate of the separator. The use of an appropriate degree of blowing causes aeration and liquefaction of the material, which results in a more precise degree of separation.

5.4 Electrostatic separation

In electrostatic separation, the properties of various plastic particles are used, resulting from the differences in the accumulation of electric charge by them. It occurs (Fig. 8) due to the interaction of electric field on charged particles that are in the field (Xue et al. 2013).

This field is generated by the electrodes supplied with high voltage, and the charge accumulates on the surface of the material particles. The charge is obtained by electrifying the particles, which can be derived from their interaction. Then they can also be imparted by the metal surface of the conveyor that evenly distributes the material onto the separator drum. Other parameters to acquire the high effectiveness of this process are conductivity and electric permittivity.

Depending on the conditions and method of electrification, the accumulated charge may have a different value and polarity. For this reason, part of the material will be attracted by the field created and will go to the first hopper, and part will go to the intermediate. The residue sticks to the drum which is grounded. These are the particles that have been neutralized on the drum and fall into the third hopper. It is additionally cleaned with a special roller.

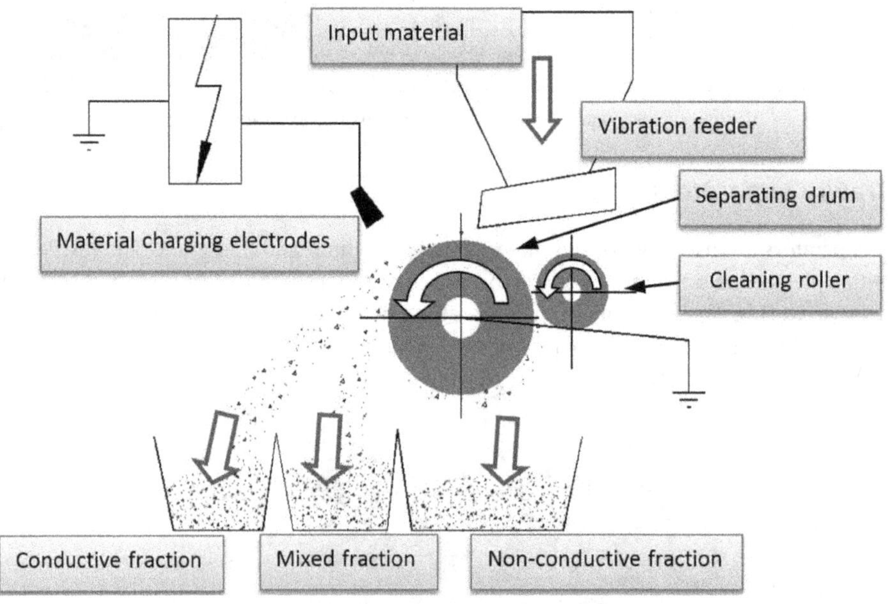

Figure 8. Scheme and principle of operation of electrostatic separator.

5.5 Cyclone separation

The cyclone separator consists of a housing, a drive, and collectors that suck in the fine fraction and discharge it, e.g., to a filter. It is characterized by low power consumption and low noise level. Dusty air, light fractions including foil, foam, etc., are sucked directly into the separator, where it is evenly distributed. In the separator chamber, it is subjected to the action of a vertical air stream generated by a large fan. Light fractions are sucked up, and heavy fractions fall for further processing, onto a conveyor or into an appropriate container (Kaya 2016).

5.6 X-ray and optical separators

For X-ray or optical separation, fragmented materials are evenly distributed on the conveyor belt. Above it, there is an analyzer equipped with an X-ray transmitter and receiver (Fig. 9). The energy emitted by the X-ray source passes through a controlled piece of the material moving on the belt. The receiver measures the energy level depending on the chemical composition of the part. Identification of the type of material (including metals) is carried out after analyzing the signal coming from the sensor. The information read by the receiver is processed by the control computer, which decides to turn on the pneumatic sorter located just behind the front drum. After opening the pneumatic valve, the compressed air is blown out and a specific particle of material is thrown into the further hopper.

The remaining material, which does not meet the assumed separation criterion, falls freely into the container located at the end of the conveyor (Habich 2007). Other types of separators (optical, shape recognition, etc.), operate on a similar principle, differing in the type of sensor used to recognize the shape or the type of material on the conveyor. Table 1 summarizes main mechanical processes and sorting techniques described in this section.

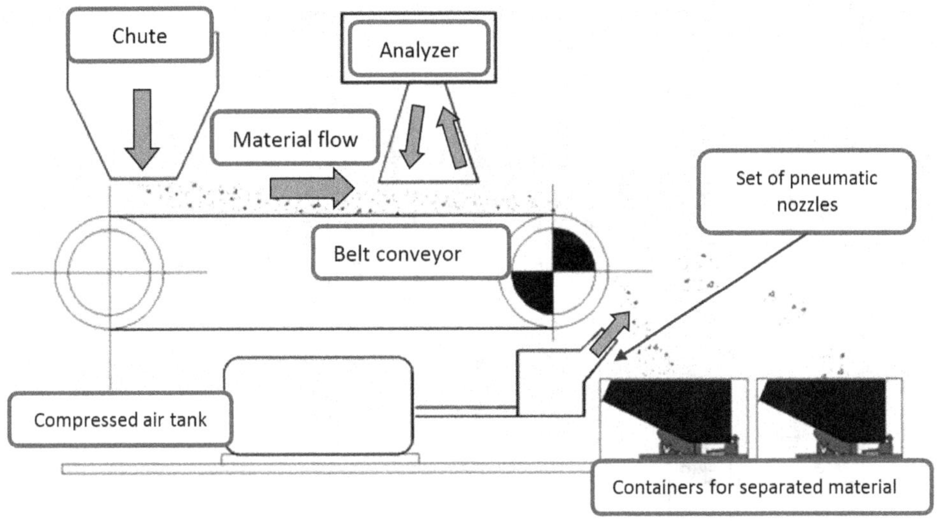

Figure 9. The principle of operation of X-ray and optical separators.

6. CHARACTERISTICS OF DISASSEMBLING FOR SELECTED CATEGORIES OF E-WASTE EQUIPMENT

6.1 Disassembly of cooling appliances

The method of processing large household appliances begins with manual disassembly of easily removable components, followed by disassembly of the components of the interior of the device. For refrigeration equipment, however, the procedure is more complex (Nowakowski 2020). This is due to the use of refrigerants. They are present in the installation of refrigerators, coolers, freezers, and air-conditioning devices, and are also used in polyurethane insulation foam production. Dismantling

Table 1. Main parameters of mechanical treatment of e-waste.

Type of machine	**Principle and purpose of the process**	**Purpose of operation**	**Size of equipment, parts or grains**	**Grain size after process***	**Capacity***	**Power consumption***
Shredder	Shredding and tearing whole waste products	Reduction of size of material's parts or breaking one item into smaller parts for easier separation	Fine to medium parts size. Size depends on screen size under shredder's shafts	15–40 mm 20–30 mm 15 mm	1 shaft shredder 800 kg/h 2 shaft shredder—1500 kg/h 4 shaft shredder 400 kg/h	22 kW 30 kW 15 kW
Grinder/crusher	Crushing glass, plastics, or cast elements	Reduction of grain size or glass/ plastics size for further process	Fine to medium grain size depending on screen size	10 mm	1200 kg/h	220 kW
Pulverizer	Pulverizing fragile and brittle parts	Preparation of homogenous powder material for further separation	Fine and very fine (powder like) fraction	Depending on mesh size	350 kg/h	50 kW
Hammer-mill	Breaking brittle parts and materials, rounding sharp metal parts	Crushing large brittle parts (mainly plastic or cast metal)	Fine or medium grains depending on screen size under shafts of hammer-mill	10–20 mm	800 kg/h	30 kW
Cable granulator	Cutting cables into small pieces for further process	Small cut parts are easy for separation on gravity separators (plastic/metal)	Small cylindrical grains of cable	0.2–20 mm	800 kg/h	52 kW
Magnetic separator	Separation of ferrous metals from other non-magnetic fractions		Small, medium or even large parts (sheet metal of housings or cabinets)		Belt width 900 mm	6 kW
Eddy current separator	Separation of non-ferrous metals from plastics		Medium or small parts		Belt width 700 mm	7 kW

Note- *—the parameters depend on the manufacturer and configuration of a machine must be carried out on lines specially prepared for this purpose. They must allow the complete and safe removal of the refrigerants and oil from devices. This should be ensured by appropriate equipment for suction of fluids, as well as the use of vacuum chambers in which the refrigerants from the cooling system and polyurethane foam are captured (Keri 2019).

Figure 10 shows a general scheme of disassembly of cooling equipment on a specialized line. The technology used in the plant may differ depending on the type of refrigerant applied. However, the disassembly steps are similar. After manual separation of parts, removal of hazardous materials and refrigerants are applied.

The waste refrigerators and freezers are placed on a tilting bench and after locking it on the bench, the cooling system is punctured at the lowest point. Removing refrigerants takes place utilizing a centrally controlled system to prevent the escape of refrigerants.

Waste oil from compressors and the refrigerants are usually utilized by combustion. The identification of refrigerant is possible after reading the information from the product plate of the disassembled device. Next, the compressor, evaporator, and condenser are disassembled and collected in appropriate containers. Depending on the capabilities of the processing plant, the aggregates are stored and prepared for further processing in another plant. After dismantling large components, the remaining cabinets go to the shredder. It takes place in two stages—in the first phase,

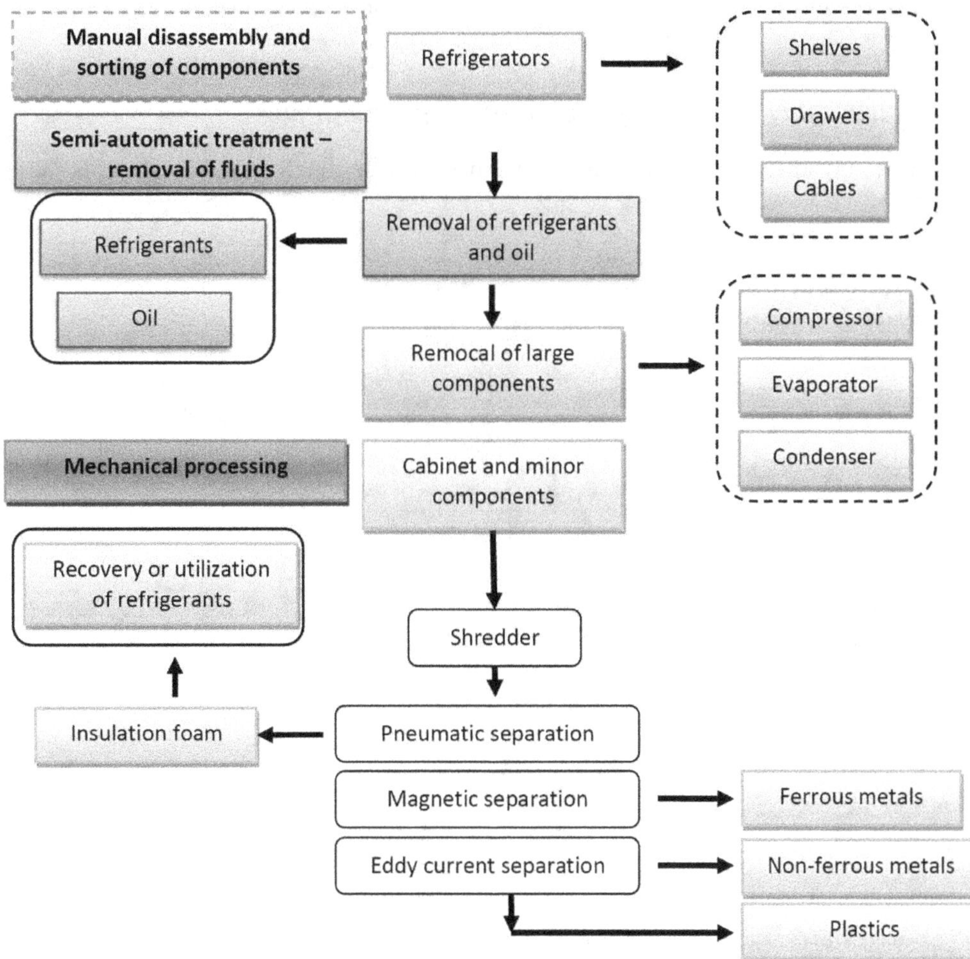

Figure 10. Generic scheme of disassembly and recycling of cooling equipment.

shredding to obtain the size of the fragments of 60–100 mm, and then to the grain size of the fraction 0–20 mm. Shredding is performed in shredders with horizontal axes, bar mills, or chain mills with vertical axis. Polyurethane foam and plastics, ferrous and non-ferrous metals are routed separately for further processing. At this stage, the polyurethane foam is degassed. During this process, refrigerants contained in the foam are released and therefore the grinding is carried out in a closed system housing with negative pressure and filled with an inert gas—nitrogen. Foam separation can take place in a pneumatic (cyclone) separator, where the shreds of foam are blown out and directed to the machine, where the foam is pressed and thus the remnants of cooling agents are removed, and polyurethane in the form of pellets is intended for further processing or use as fuel (Yang et al. 2012). When pressing the foam, a significant reduction in the volume of the raw material, up to 90%, is obtained.

The magnetic separator reclaims ferromagnetic fraction from the material flow. The belt conveyor transports the remaining material to the eddy current separator, to separate non-ferrous metals from plastics. If a gravity separator is present, it is possible to separate the aluminum fraction and the copper. Then the materials are stored in containers for shipment to recycling plants.

The method of disassembling other large equipment, e.g., washing machines, dishwashers, dryers, etc., is similar to that of cooling equipment. In this type of equipment, there are no hazardous substances like fluorocarbons and other gases, and therefore there is no gas or oil removal step. This shortens and facilitates the dismantling process. It should be noted that in the electronic system of the control module, there are precious metals in trace amounts: silver, gold, palladium. Hazardous substances include printed circuit boards with capacitors and lead-containing components.

6.2 Disassembly of information technology equipment

The disassembly of IT equipment begins with a manual stage and the entire flowchart is shown in Fig. 11.

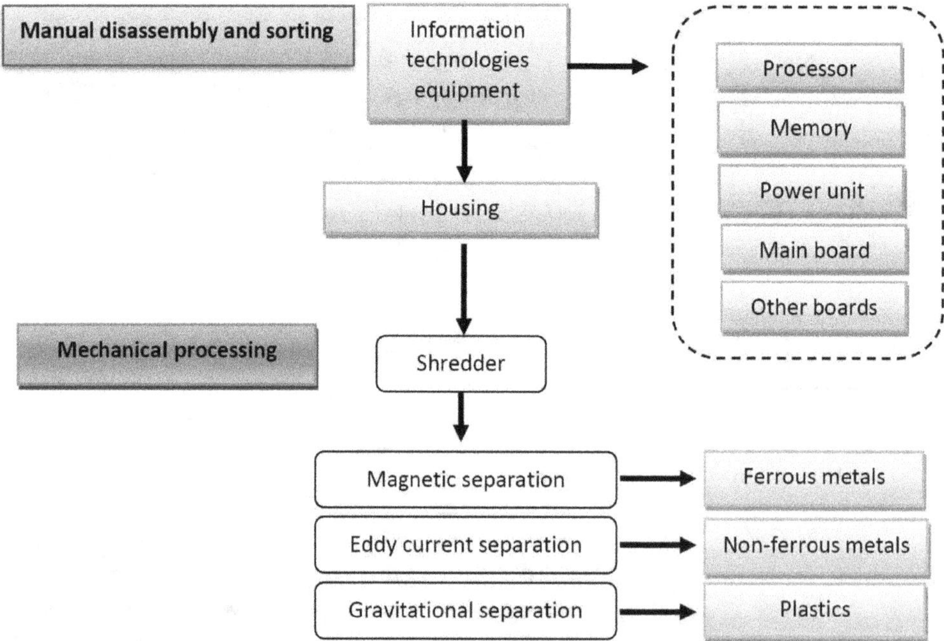

Figure 11. Disassembly scheme of information technology equipment (personal computers, notebooks).

Before this happens, this type of equipment can be verified and checked for functionality. If it is functional, it can be returned to the market for reuse. Also, many components obtained from disassembly can be reused. Their verification is quite simple. Most often they appear on the aftermarket as spare parts. Computer equipment belongs to the category of e-waste, the most abundant in valuable raw materials including precious metals: gold, silver, palladium, as well as rare earth metals, the concentration of which in selected parts used in the construction is significant (Cui and Zhang 2008). For this reason, the disassembly of this group of devices is carried out in detail. The ease of separating the components of computer sets allows them to be grouped not only according to the content of precious metals but also rare earth metals.

6.3 Disassembly of small home appliances

Small home appliances are very diverse in terms of materials used in the construction and their disassembly in a plant with high processing capacity usually takes place according to the scheme shown in Fig. 12.

Despite the division of small-size equipment into several categories according to the WEEE Directive (toys, telecommunications equipment, electric tools, and small household appliances), selection into subgroups is difficult due to the different nature of collections from individual customers and companies. In some groups of equipment, plastics dominate (e.g., dryers, electric kettles); in other metals (e.g., power tools), and telecommunications or photographic devices (e.g., cameras, cameras, mobile phones), there are also precious and rare earth metals. When equipment arrives in a disassembly facility in a large container in a mixed state, the manual disassembly or selection phase is often omitted due to the difficulty of manual separation and the high cost of the labor. Table 2 includes a summary of main technologies, special processes and output fractions for the main categories of e-waste.

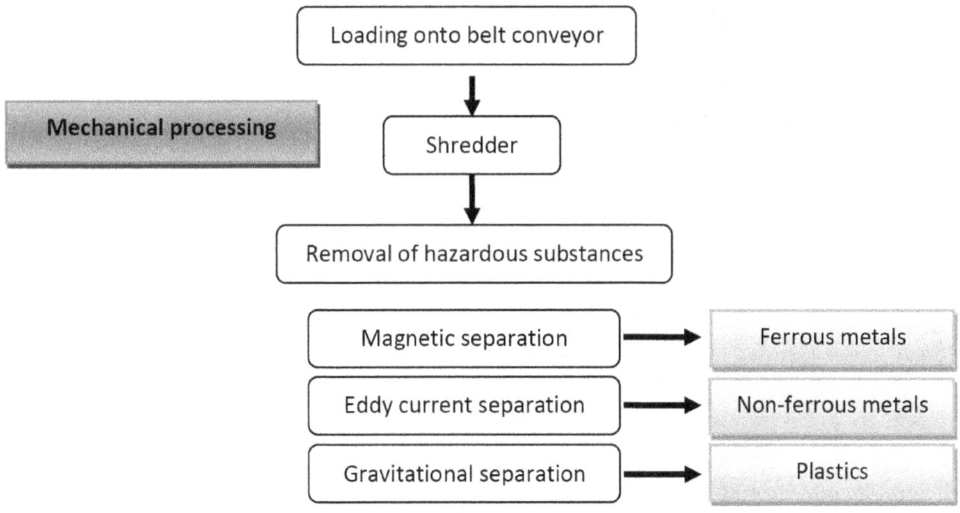

Figure 12. Disassembly scheme of small equipment.

Table 2. Main technologies for processing of the categories of e-waste from households.

Category of equipment	**Main technologies in e-waste treatment**	**Special processes**	**Type of output fractions**
Large household appliances (washing machines, dishwashers) excluding cooling equipment	Manual disassembly of large parts, crushing, shredding, various separation techniques	Removal of electrolyte condensers	Steel, plastics, concrete (counterweight), cables, non-ferrous metals
Cooling equipment (refrigerators, air conditioners, etc.)	Manual disassembly of large parts, crushing, shredding, various separation techniques including pneumatic separation of insulation	Removal of cooling agents and oils, mechanical process protection against explosion or fire by monitoring inflammable gasses (cooling agents) level.	Steel, plastics, cables, non-ferrous metals, cooling agents, oil, compressors, polyurethane foam
Small and medium households appliances	Mechanical disassembly, shredding, crushing, pulverizing, various separation techniques	Manual or semi-automatic removal of hazardous components, batteries and electrolytic condensers	Plastics, ferrous metals, non-ferrous metals, cables, minor electronic components, glass
Communication and information technologies equipment	Manual disassembly of high grade components, shredding, grinding pulverizing	Separation of fractions/components rich in precious and earth metals	Steel, plastics, copper, aluminum, other non-ferrous metals, printed circuit boards, precious and rare earth fractions, ceramics, cables
Lighting equipment	Shredding, crushing grinding, separation techniques	Separation of luminophorous powder and protecting against dusting and pollution	Glass, plastics, minor electronic components, diodes luminophorous powder, mercury containing powder

7. DISCUSSION AND CONCLUSIONS

Automated dismantling is required for high-performance WEEE processing. In this way, an increase in the efficiency of the process is achieved and consequently increasing the profitability of recycling these devices. Automated disassembly stations eliminate the performance of heavy work by employees and ensure the possibility of contact of people with hazardous substances that endanger health (Marconi et al. 2019).

The future application of a fully automated and robotized disassembly plant assumes all types of identification methods applied including barcodes, 2D codes, RFID tags, or image processing and recognition. Suitable programs supported by Artificial Intelligence (AI) should select the best disassembly method and robots should be the main operators of opening housings, cabinets, and removing hazardous substances (Alvarez-de-los-Mozos and Renteria 2017). Therefore, the capacity and the efficiency would be at the highest level and human operation would be limited to the minimum.

The methods and disassembly techniques presented in the previous sections allow selecting the most profitable and efficient method for each category of e-waste. Even partial materials' separation methods can benefit, especially in the developing countries where mostly manual disassembly is the method of treatment of e-waste. Currently, focus on Industry 4.0 and automation of various processes in the economy indicates the possible development of new methods to support the disassembly of e-waste (Poschmann et al. 2020).

As the Industry 4.0 issues are a broad area of expertise, therefore, the discussion section in this chapter limits on the three most significant subjects: identification and recognition of parts and materials in the waste equipment, automation, and robots supporting the disassembly, and changing of equipment within the time frame of the production.

Before the disassembly of e-waste equipment begins, the first step is to estimate the content of secondary raw materials and hazardous substances in the waste. It is possible by the identification of the material content of the item subject to disassembly. The best solution is to provide such information by a producer. This concept was proposed by Global System 1 (GS1 2010). The introduction and maintaining such a system by a large number of manufacturers worldwide is a very difficult task. A proposal for using 2D codes or radio frequency identification in the production stage to identify material contents that could be used after product end of life indicates some potential (Nowakowski 2018). Broad application by leading world manufacturers would benefit in the selection of the best disassembling strategies for equipment containing valuable metals or hazardous substances. Other proposals presented on case studies use an automatic object-recognition algorithm using labeling information for cameras (Hayashi et al. 2019) or focus directly on the identification of components containing critical raw metals (Charles et al. 2020).

Further development of preliminary recognition waste equipment was using deep learning techniques and algorithms. The case studies for television sets, refrigerators, washing machines indicated high accuracy of equipment image recognition (Nowakowski and Pamuła 2020) or screw detection for disassembly planning supported by robots (Foo et al. 2021). Sterkens proposes using a combination of X-ray technique supported by deep learning algorithms to identify batteries in waste equipment during disassembly (Sterkens et al. 2021). Locating hazardous substances included in batteries is a key factor for selecting these items and protect the mainstream of materials against pollution by hazardous substances.

After identification of the equipment for disassembling, several automated and robot-supported methods can be applied. A possible application of robots in the disassembly plant was discussed in a review (Bogue 2019) indicating areas where robots can support employees or operate independently. Selected specialized applications for robots can focus on selective components like printed circuit boards (Copani et al. 2019). Some disassembly plants require precise identification of equipment for disassembly for safety or environmental reasons. An example is cooling appliances' recycling and disassembling plant. In this case, it is necessary to identify a refrigerant applied in a disassembled unit. Although the information about this substance is provided (printed) on the foam, in real it can be different. To exclude other cooling equipment containing chlorofluorocarbons or hydrochlorofluorocarbons to enter the processing line, a robot equipped with a special sensor is used at the beginning of the disassembly line in a plant in Lublin, in Poland. The main advantages of automatic disassembly with the support of robots include increasing the efficiency of the process, reducing the share of human resources, and reduction of processing costs.

The main disadvantage of automatic dismantling is the high cost of infrastructure and the requirement to ensure a constant material stream of used devices of a given type. The final challenge for disassembly plants in this discussion is changing material contents and types of components in the waste equipment. It can change in the time frame of manufacturing. An example of that is refrigerators or general cooling appliances. In this case, we observe phasing out refrigerants. In this case, the disassembly process must take into consideration explosive properties of cyclo-pentane or iso-butane applied as refrigerants. Certain proportions of these currently used refrigerants can cause fire or even explosion. Refrigerants previously used were of environmental concern due to chloride or fluoride contents. Another issue in the cooling equipment is the application of a different type of insulation—Vacuum Insulation Panels. The panels are produced using microporous powders of silica (VIPA 2021) and are difficult or even impossible to process in shredder or crushers due to excessive dusting. Therefore, they have to be disassembled separately and excluded from shredding. Another innovation in refrigerating equipment is using more sophisticated control panels equipped even with Wi-Fi cards. In this case, additional components including LCD panels and printed circuit boards are a new type of waste components to be collected in the cooling equipment disassembly plant. Also, some new equipment enters disassembling and recycling facilities reaching the end of life. Photovoltaic panels were one of the equipments providing green energy from the sun, but now after expected life of 25 years of operation, large amounts of photovoltaic panels need to be disassembled and recycled. It requires working out new technologies and adaptation of disassembly lines (Chowdhury et al. 2020, Komoto et al. 2018). The dismantling of WEEE is a key stage ensuring obtaining clean fractions of raw materials for recycling. In this way, the basic materials used can be obtained in the construction of machines and devices: iron alloys, non-ferrous metals, plastics, glass, etc. The most efficient form of disassembling machines and devices is a mechanical process based on shredding, shredding larger components and subassemblies, and then sorting and separating using various techniques. The initial phase of the whole process—manual disassembly—is very often crucial to initiate the proper sequence of further processing. Tasks for workers who carry out manual demolition are aimed at the removal of hazardous substances. Then, the main components are disassembled and can be further processed, including mechanical.

This chapter discussed techniques commonly used worldwide for conducting disassembly of the various categories of e-waste. Taking into consideration the demands and concept of circular economy, it is possible to acquire high recycling rates for any category of the WEEE. At the same time, it is possible to protect the natural environment by removal of hazardous substances and components for special treatment. Also, various new technologies based on AI including deep learning, and heuristic algorithms are the main contributors to the implementation of the Industry 4.0 concept. Regardless of the selected method of disassembly, if it is conducted in conformity with the technical and environmental standards, it is beneficial for disassembling the plant as a source of income and for the natural environment by recycling materials and protection against hazardous substances pollution.

REFERENCES

Alvarez-de-los-Mozos, E. and Renteria, A. 2017. Collaborative Robots in e-waste Management. Procedia Manufacturing 11: 55–62.

Bilitewski, B., Härdtle, G. and Marek, K. 1997. Waste Management. Springer Link, ISBN: 978-3-662-03382-1

Bogue, R. 2019. Robots in recycling and disassembly. Industrial Robot: the international journal of robotics research and application 46: 461–466.

Charles, R.G., Douglas, P., Dowling, M., Liversage, G. and Davies, M.L. 2020. Towards Increased Recovery of Critical Raw Materials from WEEE—evaluation of CRMs at a component level and pre-processing methods for interface optimisation with recovery processes. Resources, Conservation and Recycling 161: 104923.

Chatterjee, S. and Kumar, K. 2009. Effective electronic waste management and recycling process involving formal and non-formal sectors. International Journal of Physical Sciences 4: 893–905.

Chowdhury, M.S., Rahman, K.S., Chowdhury, T., Nuthammachot, N., Techato, K., Akhtaruzzaman, M., Tiong, S.K., Sopian, K. and Amin, N. 2020. An overview of solar photovoltaic panels' end-of-life material recycling. Energy Strategy Reviews, 27, 100431.

Copani, G., Colledani, M., Brusaferri, A., Pievatolo, A., Amendola, E., Avella, M. and Fabrizio, M. 2019. Integrated technological solutions for zero waste recycling of printed circuit boards (PCBs), in: Factories of the Future. Springer, Cham, pp. 149–169.

Cucchiella, F., D'Adamo, I., Lenny Koh, S.C. and Rosa, P. 2015. Recycling of WEEEs: An economic assessment of present and future e-waste streams. Renewable and Sustainable Energy Reviews 51: 263–272.

Cui, J. and Zhang, L. 2008. Metallurgical recovery of metals from electronic waste: A review. Journal of Hazardous Materials 158: 228–256.

Das, A., Vidyadhar, A. and Mehrotra, S.P. 2009. A novel flowsheet for the recovery of metal values from waste printed circuit boards. Resources, Conservation and Recycling 53: 464–469.

Dyckhoff, H., Lackes, R. and Reese, J. 2013. Supply Chain Management and Reverse Logistics. Springer Berlin Heidelberg. ISBN: 978-3-540-24815-6.

European Commission, 2012. Directive 2012/19/EU of the European Parliament and of the Council of 4 July 2012 on waste electrical and electronic equipment (WEEE)Text with EEA relevance - LexUriServ.do.

European Union, 2011. Directive 2011/65/EU Of The European Parliament and of The Council of 8 June 2011 on the restriction of the use of certain hazardous substances in electrical and electronic equipment [WWW Document]. URL http://eur-lex.europa.eu/LexUriServ/LexUriServ.do?uri=OJ:L:2011:174:0088:0110:EN:PDF (accessed 10.28.11).

Foo, G., Kara, S. and Pagnucco, M. 2021. Screw detection for disassembly of electronic waste using reasoning and re-training of a deep learning model. Procedia CIRP 98: 666–671.

Goodship, V. and Stevels, A. 2012. Waste Electrical and Electronic Equipment (WEEE) Handbook. Elsevier Science.

GS1. 2010. Implementation Guide for the use of GS1 EPC global Standards in the Consumer Electronics Supply Chain. (https://www.gs1.org/standards/epc-rfid/guideline/implementation-guide-use-gs1-epcglobal-standards-consumer-electronics-supply)

Habich, U. 2007. Sensor-based sorting systems in waste processing. International Symposium MBT.

Hayashi, N., Koyanaka, S. and Oki, T. 2019. Constructing an automatic object-recognition algorithm using labeling information for efficient recycling of WEEE. Waste Management 88: 337–346.

Hester, R.E. and Harrison, R.M. 2009. Electronic Waste Management. Royal Society of Chemistry. ISBN - 978-0-85404-112-1

Jess, A. and Wasserscheid, P. 2019. Chemical Technology: From Principles to Products. John Wiley & Sons.

Jujun, R., Yiming, Q. and Zhenming, X. 2014. Environment-friendly technology for recovering nonferrous metals from e-waste: Eddy current separation. Resources, Conservation and Recycling 87: 109–116.

Kang, H.Y. and Schoenung, J.M. 2005. Electronic waste recycling: A review of US infrastructure and technology options. Resources, Conservation and Recycling 45: 368–400.

Kaya, M. 2016. Recovery of metals and nonmetals from electronic waste by physical and chemical recycling processes. Waste Management 57: 64–90.

Keri, C. 2019. Recycling cooling and freezing appliances, in: Waste Electrical and Electronic Equipment (WEEE) Handbook. Elsevier, pp. 357–370.

Komoto, K., Lee, J.-S., Zhang, J., Ravikumar, D., Sinha, P., Wade, A. and Heath, G.A. 2018. End-of-life management of photovoltaic panels: trends in PV module recycling technologies. National Renewable Energy Lab. (NREL), Golden, CO (United States).

Korhonen, J., Honkasalo, A. and Seppälä, J. 2018. Circular Economy: The Concept and its Limitations. Ecological Economics 143: 37–46.

Kuo, T.C. 2006. Enhancing disassembly and recycling planning using life-cycle analysis. Robotics and Computer-Integrated Manufacturing 22: 420–428.

Letcher, T. and Vallero, D. 2011. Waste: A Handbook for Management. Elsevier Science.

Marconi, M., Palmieri, G., Callegari, M. and Germani, M. 2019. Feasibility study and design of an automatic system for electronic components disassembly. Journal of Manufacturing Science and Engineering 141.

Morselli, L., Morselli, P.V., Passarini, F. and Vassura, I. 2009. Waste Recovery. Strategies, Techniques and Applications in Europe. Franco Angeli.

Ning, D. 2001. Cleaner Production, Eco-industry and Circular Economy. Research of Environmental Sciences 6: 1–4.

Nof, S.Y., Wilhelm, W.E. and Warnecke, H.J. 1997. Industrial Assembly. Chapman & Hall.

Nowakowski, P. 2020. Reconfigurable Recycling Systems of E-waste. *In*: Khan, A., Inamuddin, and Asiri, A.M. (Eds.). E-Waste Recycling and Management: Present Scenarios and Environmental Issues. Springer International Publishing, Cham, pp. 19–38.

Nowakowski, P. 2018. A novel, cost efficient identification method for disassembly planning of waste electrical and electronic equipment. Journal of Cleaner Production 172: 2695–2707.

Nowakowski, P. and Pamuła, T. 2020. Application of deep learning object classifier to improve e-waste collection planning. Waste Management 109: 1–9.

Ongondo, F. and Williams, I. 2011. Mobile phone collection, reuse and recycling in the UK. Waste Management 31: 1307–15.

Parajuly, K., Kuehr, R., Awasthi, A.K., Fitzpatrick, C., Lepawsky, J., Smith, E., Widmer, R. and Zeng, X. 2019. Future E-waste scenarios. StEP (Bonn), UNU ViE-SCYCLE (Bonn) & UNEP IETC (Osaka).

Pichtel, J. 2014. Waste Management Practices: Municipal, Hazardous, and Industrial, Second Edition. Taylor & Francis. ISBN – 978-146-658-5188.

Poschmann, H., Brueggemann, H. and Goldmann, D. 2020. Disassembly 4.0: A review on using robotics in disassembly tasks as a way of automation. Chemie Ingenieur Technik 92: 341–359.

Prasad, M.N.V., Vithanage, M. and Borthakur, A. 2019. Handbook of Electronic Waste Management: International Best Practices and Case Studies. Elsevier Science. ISBN: 978-012-817-0304.

Scharke, H. 2003. Comprehensive Information Chain for Automated Disassembly of Obsolete Technical Appliances. Gito-Verlag für industrielle Informationstechnik und Organisation mbH, Berlin.

Statista. 2020. Coronavirus impact: global device usage increase by country 2020 [WWW Document]. Statista. URL https://www.statista.com/statistics/1106607/device-usage-coronavirus-worldwide-by-country/ (accessed 8.13.21).

Sterkens, W., Diaz-Romero, D., Goedemé, T., Dewulf, W. and Peeters, J.R. 2021. Detection and recognition of batteries on X-Ray images of waste electrical and electronic equipment using deep learning. Resources, Conservation and Recycling 168: 105246.

Time, 2020. The WHO Just Declared Coronavirus COVID-19 a Pandemic [WWW Document]. Time. URL https://time.com/5791661/who-coronavirus-pandemic-declaration/ (accessed 8.21.20).

VIPA. 2021. What are Vacuum Insulation Panels? [WWW Document]. Vipa International. URL https://vipa-international.org/vacuum-insulation-panels (accessed 8.16.21).

Williams, E., Kahhat, R., Allenby, B., Kavazanjian, E., Kim, J. and Xu, M. 2008. Environmental, social, and economic implications of global reuse and recycling of personal computers. Environmental Science & Technology 42: 6446–6454.

Williams, J. A. S. 2006. A review of electronics demanufacturing processes. Resources, conservation and recycling, 47, 195–208.

Xavier, L.H. and Adenso-Díaz, B. 2015. Decision models in e-waste management and policy: A review. Decision Models in Engineering and Management 271–291.

Xue, M., Li, J. and Xu, Z. 2013. Management strategies on the industrialization road of state-of-the-art technologies for e-waste recycling: the case study of electrostatic separation—a review. Waste Management & Research 31: 130–140.

Yang, W., Dong, Q., Liu, S., Xie, H., Liu, L. and Li, J. 2012. Recycling and disposal methods for polyurethane foam wastes. Procedia Environmental Sciences 16: 167–175.

4

INSIGHTS INTO METALLURGICAL PROCESSES FOR SUSTAINABLE RECOVERY OF METALS FROM ELECTRONIC WASTE

Ramdayal Panda,[1] *Snigdha Mishra,*[1] *Amrita Preetam,*[1,2] *Prashant R. Jadhao*[1] and *Kamal K. Pant*[1,*]

1. INTRODUCTION

Advancement in digitalization and eagerness to ease life has increased the use of refrigerators, washing machines, mobile phones, computers, printers, and various electronic gadgets. High technology competition urges consumers to pace up with updated technology, thereby contributing to accumulation of e-waste. Mobile phones and computers are often having shorter life spans and contribute most to the rising pile of e-waste (Dave et al. 2020). Increasing users, advancement in technology and economic development are key reasons for the growth of e-waste. According to United Nation University (UNU), WEEE (Waste Electrical and Electronic Equipment) is growing at an alarming compounded annual growth rate (CAGR) of 3–5% and globally, 53.6 million tons of e-waste was generated in year 2019 and is estimated to reach 74.7 million tons by 2030 (Forti et al. 2020). Globally, there is an average increase of 2.5 million metric tons of WEEE (Mt) per year with 7.3 kg per capita. Forti et al. (2020) estimated that around 7–20% of e-waste is exported as metal scraps and for reuse.

Regulation of e-waste is one of the major challenges due to variable data on e-waste quantification. Developing nations are major importers of e-waste and due to lack of proper infrastructure and cost effective techniques, they handle e-waste via improper and unsafe methods, mainly hazardous cyanide, and concentrated acid usage. Open burning, and incineration are some informal recycling

[1] Chemical Engineering Department, Indian Institute of Technology Delhi, New Delhi, 110016, India.
[2] Center for Rural Development and Technology, Indian Institute of Technology Delhi, New Delhi, 110016, India.
* Corresponding author: kkpant@chemical.iitd.ac.in

practices which release secondary pollutants like dioxin, mercury, or lead. Several advanced as well as traditional strategies are employed to counter e-waste problems. Well planned e-waste management system is required incorporating proper collection, segregation, recycling, and reuse of e-waste at state and national level to avoid environmental and health hazards caused by recycling e-waste. A proper channel for e-waste management is shown in Fig. 1.

Non-metallic fractions, i.e., plastics, ceramics are low-cost materials and widely used in e-waste as it acts as thermal and/or electric insulator and can be processed into lightweight durable material. E-waste plastics are a mixture of different types of polymers like ABS, HIPS, PS, PP, PE, PVC, PET, PA, etc. (Maris et al. 2015, Martinho et al. 2012). The brominated flame retardants are the hazardous component of e-waste which has several environmental drawbacks; one of them is food contamination causing adverse human health effects (Leonetti et al. 2016).

E-wastes are rich in base metals (Cu, Fe, Sn, Al), precious metals (Ag, Au, Pd, Pt), with approximately 20–30% of copper in metallic composition (Behnamfard et al. 2013, Birloaga et al. 2013). Major metals used in the high technology industry are copper and gold, with 30% and 12% of their total consumption in society. Economically, 50% of economic returns are based on these two metals. Major pollutants in e-waste are heavy metals (As, Ba, Be, Cd, Cr, Pb), halogenated compounds (Polychlorinated biphenyls, Polybrominated biphenyls, Polychlorinated diphenyl ether, Chlorofluorocarbon. Polyvinyl chloride), plastics and ceramics. Special attention needs to be given to a metallic part of the e-waste which can be reutilized, and e-waste can act as a secondary source for metals like iron, gold, palladium, platinum, copper and various critical metals, which are present in a substantial amount in high-tech electronic gadgets. Gu et al. (2017) analysed Printed Circuit Board (PCB) of mobile phones and found that there is 203-mg/g copper, 48.39-mg/g nickel, 62.25-mg/g tin, and 1.11-mg/g gold. The amount of various elements used in manufacturing of electronic gadgets is shown in Fig. 2. Mining, processing of these metals is an energy-intensive process which not only affects the environment with harmful emissions but also pollutes natural resources. Less dependency on natural resources will reduce the burden on natural ores indirectly, conserve energy and reduce pollution. The economic value of various metals present in electronic waste motivates to consider it as a secondary source of metals.

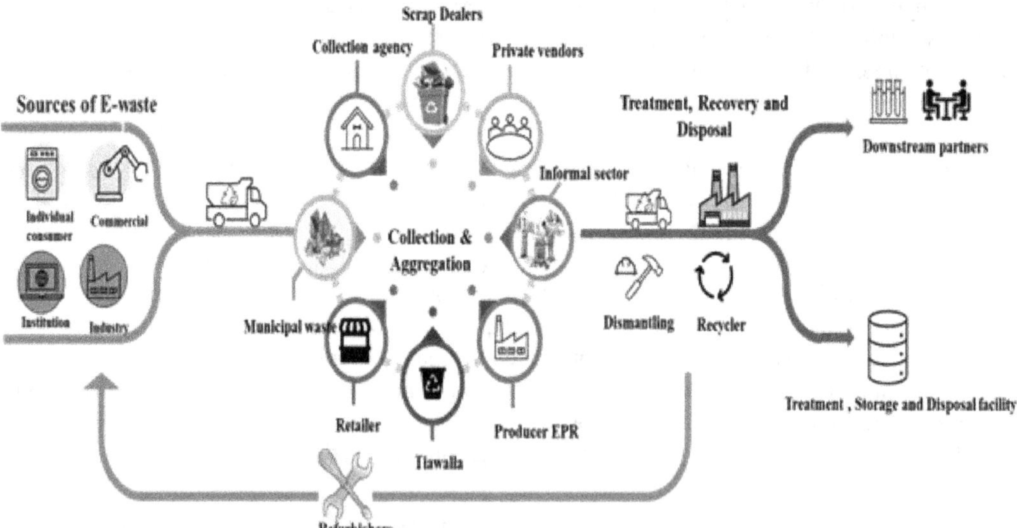

Figure 1. Steps in e-waste management (Magalini et al. 2019).

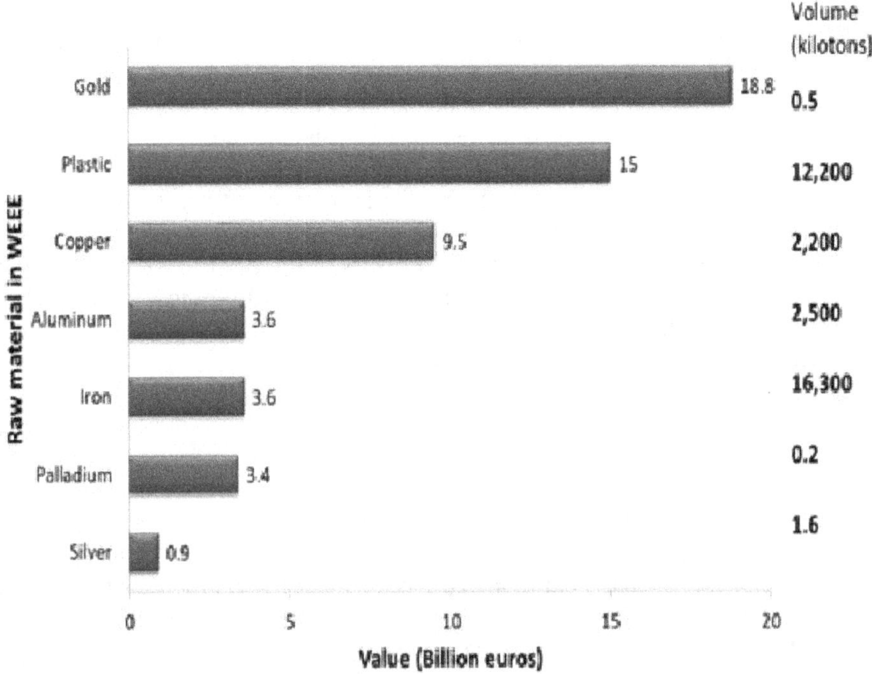

Figure 2. Volumetric value of metals present in e-waste (Data taken from Forti et al. 2020).

2. PYROMETALLURGICAL PROCESS

Pyrometallurgical techniques have been extensively used for recovery of metals from different mineral sources. Industrially, this is the most adopted technique for metal recovery due to simple process and mechanism. Smelting, incineration, roasting, and combustion are typical methods for metal recovery (Khaliq et al. 2014, Lee et al. 2007, Sum 1991, Williams and Drake 1983). The advantage of the pyrometallurgy process is that the same equipment can be used for different numbers of metal sources like e-waste, ores, etc. Efficiency of metal recovery is quite high in this technique, but recycling of polymeric fraction is not feasible; however, they can be used in place of coal as a source of energy. Emission due to presence of halogenated flame retardants cannot be avoided. Moreover, after the smelting process, one needs to look for other downstream techniques for selective recovery of metals. In the current scenario, copper and lead smelters are the most used pyrometallurgical process for recovery of metals depending upon feed composition (Khaliq et al. 2014, Anindya et al. 2013). Lead smelters generate toxic fumes; therefore, copper smelters are more environment friendly and preferred over lead smelters (Sohaili et al. 2012). Copper smelters are closed loop processes for PCBs recycling and both precious metals as well as base metals can be recovered using this facility. Precious metals can be recovered from slimes using traditional electrorefining processes (Anindya et al. 2013). Primary copper smelting which processes sulphide ores and secondary copper smelting which processes oxide ores, both processes can be used for recycling of PCBs. Copper matte and blister copper were produced in primary copper smelting route; similarly, black copper is produced using secondary smelting route.

2.1 Industrial processes

Pyrometallurgy uses high temperature (800°C–1200°C) for metal recovery and it includes incineration, high-temperature roasting, smelting, and calcination (Chauhan et al. 2020, Zhang and Xu 2016, Lee et al. 2007). Very few pyrometallurgical plants such as Ronnskar smelter in Sweden, Umicore in Belgium, Noranda copper smelter in Quebec, Canada, and Kosaka smelter in Japan are processing e-waste for metal recovery (Chauhan et al. 2018, Hagelüken 2006a, Maeda et al. 2000). The Ronnskar smelter in Sweden recovers Au, Ag, Cu, Ni, Pd, Se, and Zn from e-waste using pyrometallurgy (Mark and Lehner 2010). The major steps in the Ronnskar Smelters are drying, roasting, smelting, converting, and refining. Similarly, e-waste mixed with the lead concentrates is combusted in the Kaldo furnace in presence of oxygen and oil (Leirnes and Lundstrom 1983). The mixed copper alloy generated in the Kaldo furnace is then fed to the copper converter which treats this alloy. The treated alloy was then sent to the refining process for the recovery of Ag, Au, Cu, Ni, Pd, Se, and Zn. The vapor which contains volatile metals such as Pb, Sb, In, and Cd is sent to a different process for recovery of these metals. The gases generated are processed to produce sulphuric acid and SO_2 gas. The typical recovery of metals in this process exceeds 95%.

Umicore has prepared a large-scale smelter plant at Hoboken, Belgium (Fig. 3) for recycling of precious metals and rare earth metals from various metal sources like mineral ores, e-waste, spent catalyst, advanced materials, residues, etc. PCBs, cell phone and small IT devices (without batteries), CPU, IC, connectors and laptops (without battery and screen) are being recycled and refined to recover precious metals (Hagelüken 2006a,b, Brusselaers et al. 2005). Is a smelter, which uses submerged lance combustion technology, is used in this plant for heating. Oxygen rich air along with fuel are injected in a molten bath, whereas coke is injected as a reducing agent for reduction of metals. As electronic scrap contains polymers, it is being partly used as a reducing agent and energy source in place of coke. Precious metal operations and base metal operations are the key operations generally done here. Firstly, the metal source goes through the precious metal operation and the slag that was produced from precious metal operation processed through the base metal operation plant. Smelter, copper leaching and electro-winning, and precious metal refinery are the major steps being used in precious metal operation plants. Here precious metals were recovered in copper bullion and little in lead slag. The lead slag which contains the majority of the base metal is subsequently processed through the base metal operation plant. Lead blast furnace, lead refinery, and special metal plant are the major steps involved at base metal operation plant for recovery of base metals. Continuous measurement of SO_x, NO_x and periodical measurement of dust, metals, H_2SO_4, CO, C_{tot}, HCl, HF were done at the stack of the plant (Rajeshwari 2007). Similarly, continuous sampling for dioxins

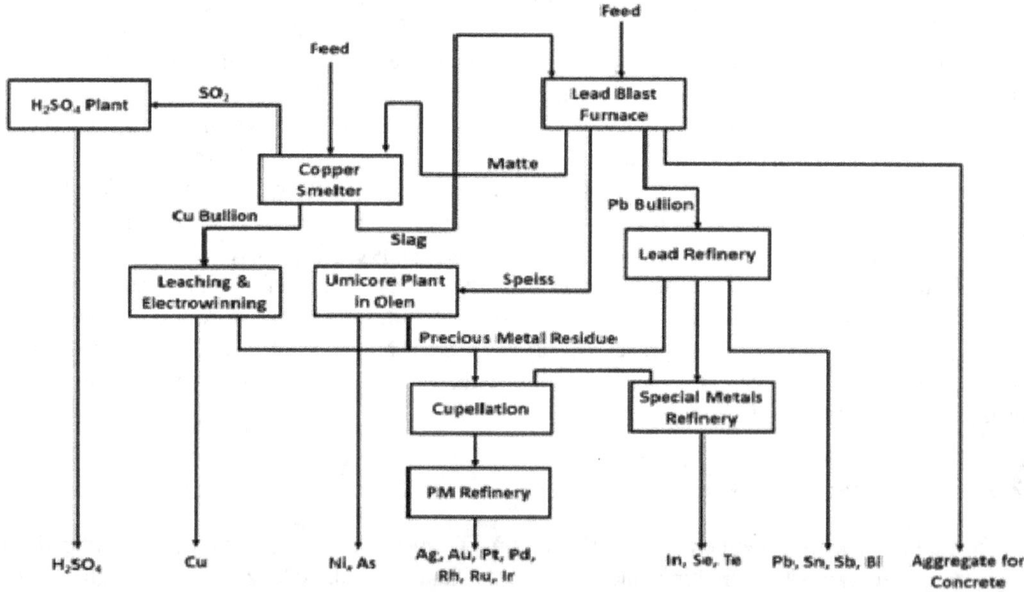

Figure 3. Flowsheet of Umicore-Hoboken integrated smelter and refinery plant (Redrawn from Hagelüken 2006a).

and furans detections were done to keep the environment clean. Umicore's plant at Belgium produces sulphuric acid as a by-product by absorbing these emitted SO_x. They mainly focus on recovery of gold, silver, palladium, platinum, rhodium, ruthenium, iridium, indium, selenium, copper, nickel, lead, tin, arsenic and bismuth. Another by-product is the slag, which is mainly used in construction work. The feed capacity of the plant is 250000 ton/annum, which is the largest facility for recovery of precious metals. This plant at Hoboken has capacity to produce 100 ton of Au, 2400 ton of silver and around 50 ton of other precious metals annually (Hagelüken 2006b).

Noranda is one of the biggest miners in Canada and has invested heavily to set up an electronic waste recycling plant at New Brunswick (Veldbuizen and Sippel 1994). The flowsheet of the electronic waste plant for metal recovery of Noranda smelting process is shown in Fig. 4. The plant is operated at 1250°C which is maintained by passing supercharged oxygen. A mixture of electronic scrap and copper concentrate is fed into the smelting chamber. Polymeric fractions of e-waste play a key role by providing a source of energy by combustion. During the oxidation process, precious metals are segregated into liquid copper and base metals like iron, lead and zinc are segregated into a slag by forming oxides. The liquid copper is thereafter fed into the anode furnace for refining of precious metals and casted in the form of anodes. The anodes are subsequently electro-refined for individual separation of precious metals. Around 100000 tons of e-waste is recycled in this facility annually and metals like copper, platinum, palladium, gold, silver, selenium, tellurium and nickel were recovered with high purity.

Ronnsker smelter at Skelleftehamn, Sweden having capacity of 100000 ton recycles a wide range of non-ferrous ores along with e-waste (Lehner 2003, Wernick and Themelis 2003, Theo 1998). This plant has two main processes, one is converter and other one is the furnace. High grade ores or electronic scrap whose copper content is high is directly fed into the converter whereas low grade waste and ores were fed to the Kaldo furnace. The Kaldo converter enriches low grade copper concentrate into high grade copper concentrate. Low grade copper concentrate along with lead concentrate are combusted in the Kaldo furnace in the presence of oxygen and oil. Drying, roasting, smelting, converting, and refining are the major operations carried out at Ronnskar smelter. Metals

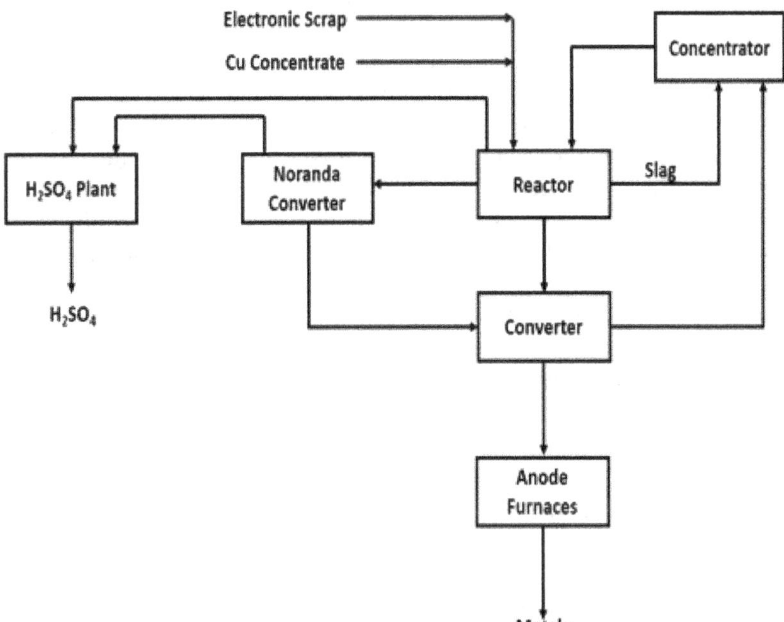

Figure 4. Flowsheet of the Noranda smelting process for electronic waste recycling (Redrawn from :Veldbuizen and Sippel 1994).

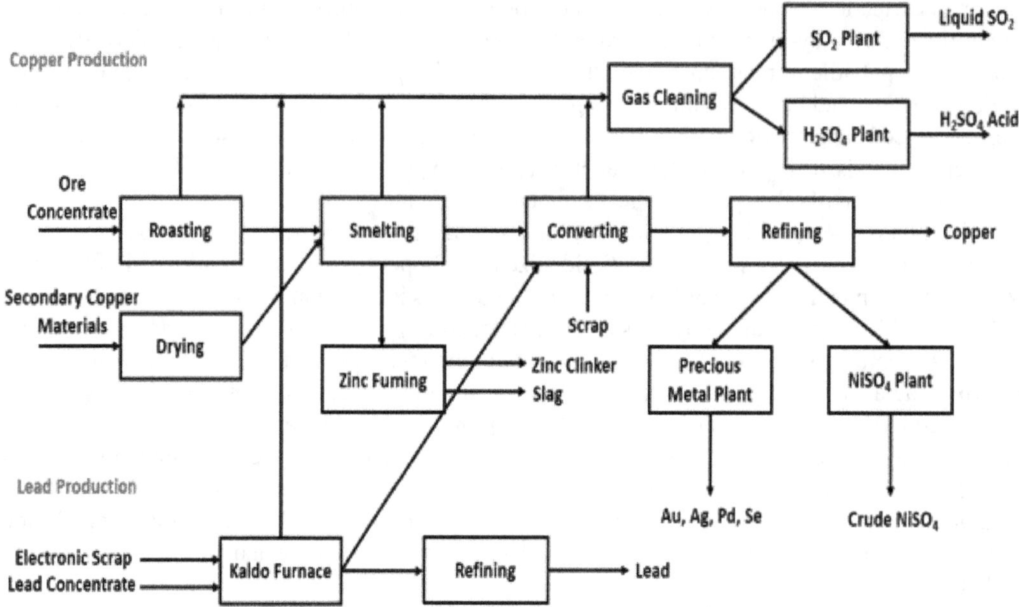

Figure 5. Flow sheet diagram of the Rönnskär smelters (Redrawn from: Mark and Lehner 2010).

like platinum, palladium, gold, silver, selenium, copper, nickel, tin, lead, indium and cadmium are recovered in this process. Sulfuric acid and SO_2 gas are produced as by-products by treating emitted gases. The flowsheet of the Rönnskär smelters for metal recovery is shown in Fig. 5.

Similarly, there is a facility at Akita Japan called the Kosaka smelter which recovers metal from batteries, ores, circuit boards, chips and copper scraps. In this facility, copper smelting and lead smelting can be done at the same place (Maeda et al. 2000). Off-gases coming from this plant are treated further to make sulfuric acid as a by-product. Metals like copper, silver and gold were recovered in this facility from different scraps (Maeda et al. 2000).

The pyrometallurgical process is efficient for the recovery of metals from different sources of metals but there are some limitations associated with this process in case of metal recovery from e-waste. E-waste contains halogenated flame retardants such as chlorinated and brominated biphenyl and processing of e-waste using pyrometallurgy leads to the generation of dioxins and furans (Hao et al. 2020). Therefore, there is a need to install a state-of-the-art facility to prevent the emission of toxic gases while processing e-waste, which makes this process cost-intensive (Jin et al. 2020, Yken et al. 2020). E-waste, especially metal bearing component PCB, contains ceramic which generates the slag in a furnace during the pyrometallurgy process which increases the possibility of losing precious metals to the slag (Chauhan et al. 2018). In addition to this, the recovery of iron and aluminium is also difficult as they land up in the slag phase. The use of high temperatures for the recovery of metals makes this process energy-intensive (Li et al. 2018). Therefore, the focus has been shifted from pyrometallurgy to hydrometallurgy to find a more economical and environment friendly approach for metal recovery from e-waste.

2.2 Laboratory scale processes

Laboratory scale pyrometallurgical methods for recovery of metals from e-waste are still unexplored and there lies a huge potential for development of novel technologies for recycling of e-waste. Molten salt oxidation, plasma technology and dry chlorination techniques have been explored by the researchers for recovery of metals from waste PCBs. These novel technologies working at low temperature conditions around 300°C are different from traditional industrial smelters, which require high temperature conditions above 1000°C for recovery of metals. Around 300°C polymeric fraction of e-waste cannot be utilized as an energy source; therefore, it is recommended to recycle polymeric fraction and remove BFRs.

2.2.1 Ammonium chloride roasting

Due to high reactivity, low roasting temperature, and easy extraction, chlorination roasting process has been extensively used for recovery of valuable metals from various mineral resources such as discarded lithium ion batteries, waste printed circuit boards, rare earth metals based magnets, electrolytic residue, and ores, etc. (He et al. 2021, Panda et al. 2021a, Lorenz and Bertau 2019, Xu et al. 2017, Zhang et al. 2012). Compared to other traditional solid chlorinating agents such as NaCl, $CaCl_2$, KCl, $FeCl_3$, and $AlCl_3$, NH_4Cl proves to be an excellent chlorination agent at low temperature range, leading to low energy consumption (Xiao et al. 2021, Lorenz and Bertau 2020, Panda et al. 2020, Xu et al. 2017). Ammonium chloride disintegrate into ammonia and hydrochloride vapor at higher temperature, from which hydrochloride vapor reacts with metals for highly water soluble metal chloride salts (Xiong et al. 2020, Cui et al. 2018).

Panda et al. (2020) studied the recovery of metals from PCBs of discarded mobile phones using ammonium chloride roasting-water leaching method. The illustrative flow diagram connecting different operations is shown in Fig. 6. Vital process variables such as roasting temperature, roasting time and ammonium chloride dosage were investigated during the roasting process. To evaluate the effectiveness of the roasting process, the roasted residue was water leached for extraction of valuable metals. PCBs herein contains approximately 28.2% Cu, 1.1% Ni, 0.3% Pb, 0.1% Zn, 230 ppm Au and 796 ppm Ag. Before the ammonium chloride roasting process, the PCBs powder was pyrolyzed at 400 °C to get rid of the polymeric fraction. The polymeric fraction during the pyrolysis degraded to form oil and combustible gases. The decrease in weight of the PCBs leads to enrichment of the

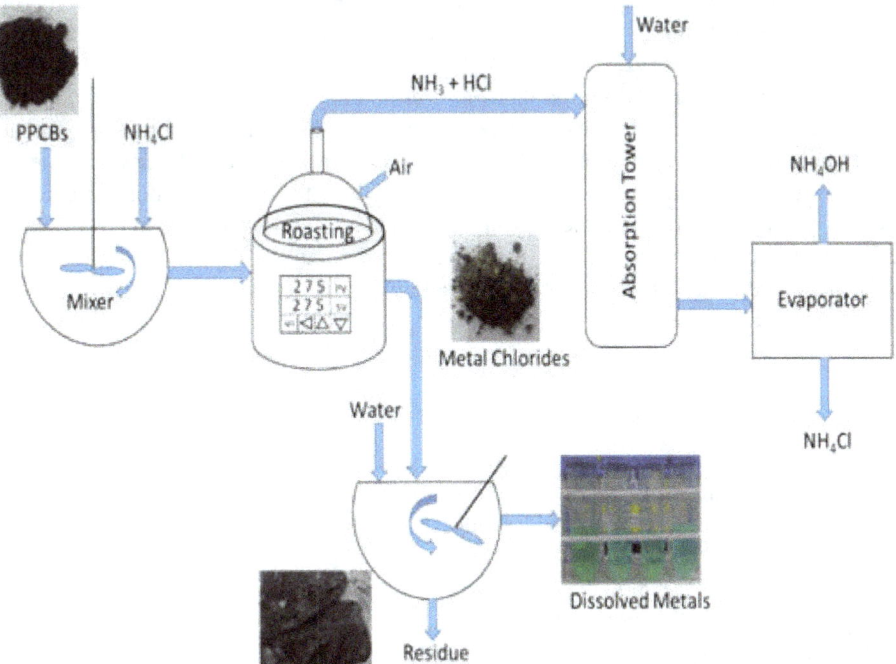

Figure 6. Illustrative flow diagram of the process. Reprinted with permission from (Panda et al. 2020).

metal fraction in the pyrolyzed residue. The final solid residue of the pyrolysis process contains 33.2% Cu, 1.5% Ni, 0.38% Pb, 0.22% Zn, 326 ppm Au and 1384 ppm Ag. Thereafter, the pyrolyzed PCBs, ammonium chloride mixture (1:2) was treated in the range of 200°C to 325°C in a roasting chamber. Metals from PCBs react with HCl vapor to form salts such as $CuCl_2$, $NiCl_2$, $ZnCl_2$, $PbCl_2$, AuCl, and AgCl. PCBs also contain refractory oxides containing oxides of silicon and aluminium, which does not form any complex or water soluble chloride while treated with ammonium chloride (Nadirov et al. 2013). Approximately, 92% Cu, 100% Ni, 100% Zn, 100% Pb, 43% Au and 42% Ag were extracted by water leaching method in this study. Following are the possible chemical reactions that may occur during the roasting process:

$$NH_4Cl(s) \rightleftharpoons HCl(g) + NH_3(g)$$

$$Cu(s) + 2HCl(g) + 0.5\,O_2(g) \rightleftharpoons CuCl_2(s) + H_2O(g)$$

$$Ni(s) + 2HCl(g) + 0.5\,O_2(g) \rightleftharpoons NiCl_2(s) + H_2O(g)$$

$$Pb(s) + 2HCl(g) + 0.5\,O_2(g) \rightleftharpoons PbCl_2(s) + H_2O(g)$$

$$Zn(s) + 2HCl(g) + 0.5\,O_2(g) \rightleftharpoons ZnCl_2(s) + H_2O(g)$$

$$Cu(s) + 2HCl(g) \rightleftharpoons CuCl_2(s) + H_2(g)$$

$$Ni(s) + 2HCl(g) \rightleftharpoons NiCl_2(s) + H_2(g)$$

$$Pb(s) + 2HCl(g) \rightleftharpoons PbCl_2(s) + H_2(g)$$

$$Zn(s) + 2HCl(g) \rightleftharpoons ZnCl_2(s) + H_2(g)$$

Similarly, in another study, Panda et al. (2021b) used ammonium chloride roasting-water leaching method; however, a chemical pre-treatment was used to remove the brominated epoxy resin (BER) from PCBs surface. In this study, dimethylformamide (DMF) was used for dissolution of BER followed by roasting with ammonium chloride (NH_4Cl) for recovery of metals. Separation of BER provides better metallic surface exposure and reduces the chance of generation of hazardous polybrominated dibenzo-p-dioxins and dibenzofurans (PBDD/Fs) during the chlorination-roasting process. Dimethyl formamide (DMF), dimethyl sulfoxide (DMSO), and N-methyl-2-purrolidone (NMP) are the most commonly used solvents for dissolution of BER (Verma et al. 2017, 2016, Wath et al. 2015, Zhu et al. 2013a,b). These organic solvents have hydrogen donor sites which interact with BER to form hydrogen bonding; hence, dissolution of BER from PCBs surface occurs. Afterwards, the roasting experiments were carried out in the range of 200°C to 300°C with varying ammonium chloride dose and time. It was found that roasting the maximum amount of metals were able to convert into water soluble form at 275°C, 300% dose and 180 min time. After roasting, water leaching was done at 80°C for 30 min to recover metals in solution form. Around 93% Cu, 100% Ni, 100% Zn, 100% Pb, 38% Au and 43% Ag were leached out at optimum roasting conditions.

2.2.2 Molten salts

Molten salt oxidation has been a popular method for gasification of coal, and eradication of various hazardous materials such as organic mixed oils, chlorinated organic solvents, radioactive organic materials, weapons, etc. (Yao et al. 2011, Yang et al. 2007, Hsu et al. 2000). These processes use molten carbonate salts with an air injection system to provide an oxidation environment for degradation of organics. The molten salt system works in the temperature range of 900–1000°C and traps generated gases such as carbons dioxide, hydrogen chloride and hydrogen fluoride. This method catalytically breaks down the hydrocarbon bonding in presence of molten alkaline salts. The inorganic materials don't react with the molten salt and get separated as final residue. The process flow sheet for separation of metals from e-waste using molten salts is shown in

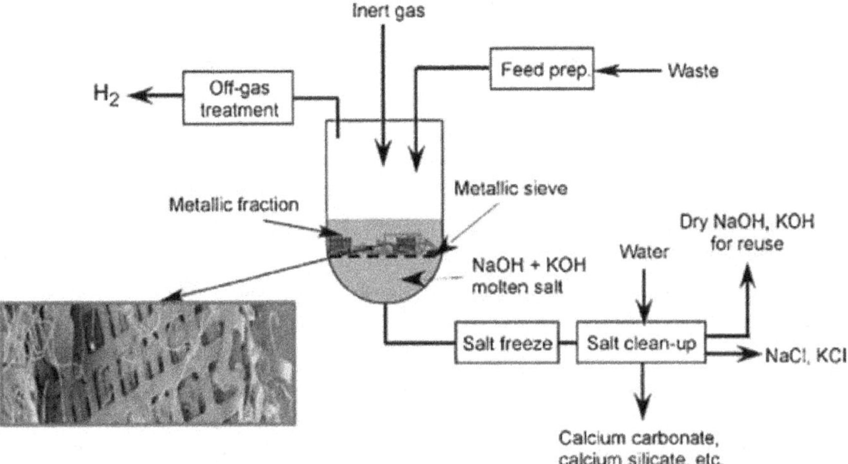

Figure 7. Process flowsheet to separate metals from PCBs using molten salts. Reprinted with permission from Stuhlpfarrer et al. (2016).

Fig. 7. Compared to incineration process, molten salt oxidation is a low cost enviro-friendly process possessing several advantages such as flameless oxidation, generates no dioxins/furans, high solubility, produces no effluents, and easy scale up (Flandinet et al. 2012). Stuhlpfarrer et al. (2016) synthesized an eutectic molten solution consisting of 41 wt.% NaOH and 59 wt.% KOH in a cylindrical nickel reactor. This combination of NaOH-KOH mixture melts at 170°C and acts as a catalyst. The temperature of the reactor system varied in the temperature range of 250–300°C for decomposition of polymeric fraction in an inert atmosphere. The study revealed, within 15 min of the decomposition process most of the polymers get cracked. The benefit of low working temperature range helps in conquering the formation of hazardous dioxins and furans. The study further reported that, around 99.8% of fluorides, 99.99% of bromides and 99.3% of chlorides were trapped by the eutectic molten solution and the gas was composed of CO and CO_2.

2.2.3 Plasma melting

The use of plasma technology for recovery of metals from various sources is growing day by day. In the past, many researchers used plasma technology to efficiently recover metals from e-waste (Wang et al. 2013, Rath et al. 2012, Mitrasinovic et al. 2011). Plasma is regarded as the fourth state matter, produced using highly ionized gas at high temperature and high energy density (Du et al. 2015). In general, plasma can be categorized into high temperature plasma and low temperature plasma. High temperature plasma can be generated above 10^6–10^7°C, whereas low temperature plasma at 10^1–10^5°C (Changming et al. 2018). However, low temperature plasma produces a limited amount of free electrons; therefore, most of the studies used high temperature plasma to maximize the amount of metal recovery. The high temperature plasma degrades the organic matter and converts inorganic fraction of the source into a ceramic slag (Huang et al. 2003). Additionally, the amount of generation of dioxins and furans is minimal compared to the incineration process. Rath et al. (2012) used a 35 kW DC extended plasma reactor for recovery of metals from e-waste, where all the metals were melted to form a slag with mixture of metals. Thereafter, a hydrometallurgical process was used to recover individual metals. Similarly, Szałatkiewicz in various studies shows that plasma technology can effectively be used for recovery of precious metals (Szałatkiewicz 2016, Szałatkiewicz et al. 2013). The experiment shows that 40% of the e-waste sample was degraded using plasma technology, whereas 20% and 40% were recovered as metals and slag. Overall, the plasma technology for recovery of metals from e-waste is highly economic and high energy value (Wang et al. 2013).

2.3 Limitations of pyrometallurgical processes

Current industrial pyrometallurgical processes which either use copper smelting route or lead smelting route are more economical and efficient for recovery of precious metals; nonetheless, they have some disadvantages (Cui and Zhang 2008, Hagelüken 2006b). The polymeric fraction of the e-waste cannot be recycled using pyrometallurgical processes; however, they can be used as a means of energy in place of traditional coke. The recovery of Fe and Al are difficult as they get segregated in the form of slag from smelters. Majority of the e-waste contains brominated flame retardants as fire suppressant, which produces hazardous dioxins and furans during pyrometallurgical recovery of e-waste. Hence, additional setup is required to reduce pollution. Large amount of ceramics present in e-waste generates huge amounts of slag, which may be responsible for the loss of some precious metals. Pyrometallurgical processes are energy intensive processes, and therefore are difficult to handle; the heterogeneous nature of e-waste also creates challenges to operate. The end product of pyrometallurgical processes is impure in nature; hence, it requires additional refining processes in the form of hydrometallurgical or electrorefining methods to purify the precious metals (Khaliq et al. 2014).

3. HYDROMETALLURGICAL PROCESSES

Electronic scrap contains a variety of metals that need to be leached to prevent health hazards and to recover precious metals. Base metals in e-waste are generally present in alloy state or zero oxidation state which are very stable in nature. Due to the stable nature of metals, it is hard to extract metals from e-waste in comparison to ores where metals are present in oxide or sulphide state. There are several novel techniques researchers have tried in recent years for successful recovery of metals. Acid leaching, halide leaching, cyanide leaching, thiourea leaching, thiosulfate leaching, ammonia-ammonium salt leaching, and chelation technology are some popular techniques on which extensive research is being carried out (Chauhan et al. 2018). A tabular comparison of parameters and amount of metal extraction from e-waste using various methods is shown in Table 1.

3.1 Mineral acid leaching

Acid leaching is one of the oldest hydrometallurgical methods used for dissolution of metals from different sources. For electronic waste, mineral acids such as H_2SO_4, HCl, HNO_3 and a mixture of HCl and HNO_3 (aqua regia) can be used to recover metals and aqua regia is best amongst all in terms of metal dissolution. However, due to the hazardous environmental effect of these acids, mentioned mineral acid leaching is not popular amongst miners. Jha et al. (2012) studied the effect of mineral acid such as HNO_3, H_2SO_4 and HCl on dissolution of lead and tin from solder materials of PCBs. Before dissolution experiments, the PCBs were treated with n-methyl-2-pyrolidone to separate metal sheets and brominated epoxy resin. For selective leaching of Pb and Sn, a two-step HNO_3 leaching and HCl leaching was used in series. Based on the two-step leaching process, 99.99% of Pb and 98.74% of Sn were recovered from solder materials of PCBs. The use of H_2SO_4 and HCl tends to precipitate Pb during the extraction process, which restricts their use for Pb dissolution. Due to strong oxidizing property of HNO_3, it swiftly dissolves Pb from solder material as given below (Vogel 1996, Puri et al. 1991):

$$Pb(s) + 2HNO_3(aq) \rightarrow PbO(s) + H_2O\,(aq) + NO_2(g)$$

$$PbO(s) + 2HNO_3(aq) \rightarrow Pb(NO_3)_2(aq) + H_2O(aq)$$

During the dissolution of Pb, Sn reacts with HNO_3 to form SnO_2, which precipitates on the surface of the metal. With increase in HNO_3 concentration, the SnO_2 forms metastannic acid. Furthermore, with addition of HCl, the metastannic acid dissolves into the solution as given below (Mecucci and Scott 2002):

$$Sn(s) + 4HNO_3(aq) \rightarrow H_2SnO_3(s) + 4NO_2(g) + H_2O(aq)$$

$$H_2SnO_3(s) + 6HCl(aq) \rightarrow H_2SnCl_6(aq) + 3H_2O(aq)$$

Waste multilayer ceramic capacitors (MLCCs) are one of the essential components of PCBs, which contain an abundant amount of precious and base metals. Liu et al. (2021) studied the extraction of these metals such as Ni, Cu, Ag, Pd, and Bi using HNO_3 solution. To find out the concentration of HNO_3 required for productive leaching of metals, Eh-pH diagrams for Pd and Ag were studied. Eh-pH diagram illustrates that HNO_3 solution with pH 9.2 has Pd solubility of 0.01 mol/L at the potential range of 0.85–1.85 V.

E-waste consists of a variety of metals either present in noble state or an alloy form. Numerous methods based on HCl and H_2SO_4 have been employed for extraction of target metals like Cu, Au, Ag, Pt, and Pd; nonetheless, the extensive use of these acids alone is not effective for leaching of valuable metals. Therefore, oxidants such as H_2O_2, Fe^{+3}, Cl_2, O_2 and bacteria are often used with HCl or H_2SO_4 for efficient extraction of target metals (Birloaga et al. 2013, Lee and Pandey 2012, Pant et al. 2012, Kim et al. 2011). However, HNO_3 itself being a strong oxidizing agent is more effective in leaching metals compared to HCl and H_2SO_4 (Dhawan et al. 2009). Bas et al. (2014) studied leaching of waste TV boards by using strong HNO_3. Parameters like acid concentration, time, and pulp density were varied to optimize the process. In this study, around 99.9% of Cu and 68% of Ag were dissolved at 5 M HNO_3. Jadhav and Hocheng (2015) studied leaching of metals from large computer PCBs by using various acids. The chemical coating or epoxy resin hinders the contact between solvent and metal; therefore, PCBs were initially treated with concentrated NaOH for removal of polymeric coating present on the top of the board. Several acids such as HCl, H_2SO_4, HNO_3, $C_2H_2O_2$, and $C_6H_8O_7$ were employed for the leaching process; however, HCl was found to be most efficient for metal recovery. It was reported that 100% of Cu, Ni, Zn, Pd, Fe, Au, and Ag were recovered in 1 M HCl in 22 h.

$$4Cu(s) + 4HCl(aq) + O_2(g) \rightarrow 4CuCl(aq) + 2H_2O(aq)$$

$$Zn(s) + 2HCl(aq) \rightarrow ZnCl_2\,(aq) + H_2\,(g)$$

$$Ni(s) + 2HCl(aq) \rightarrow NiCl_2\,(aq) + H_2\,(g)$$

$$Fe(s) + 2HCl(aq) \rightarrow FeCl_2\,(aq) + H_2\,(g)$$

$$2Al(s) + 2HCl(aq) \rightarrow 2AlCl_3\,(aq) + 3H_2\,(g)$$

$$2Ag(s) + 2HCl(aq) \rightarrow 2AgCl(aq) + H_2\,(g)$$

Similarly, Kumari et al. (2016) studied recovery of copper from waste printed circuit boards by using sulphuric acid leaching technique. Parameters like pressure, temperature, pulp density and concentration of sulphuric acid were varied to get the best result. Pyrolysis of PCBs as a pre-treatment technique was used to treat polymers. These pyrolyzed PCBs were further treated with 2.4 M H_2SO_4 at 150°C, 20 bar and 100 g/L pulp density for recovery of around 98% Cu. Chemical reactions of various metals with H_2SO_4 in presence of H_2O_2 is given below (Birloaga et al. 2013):

$$Cu(s) + H_2SO_4(aq) + H_2O_2(aq) \rightarrow CuSO_4\,(aq) + 2H_2O(aq)$$

$$Zn(s) + H_2SO_4\,(aq) + H_2O_2(aq) \rightarrow ZnSO_4\,(aq) + 2H_2O(aq)$$

$$Sn(s) + H_2SO_4\,(aq) + H_2O_2(aq) \rightarrow SnSO_4\,(aq) + 2H_2O(aq)$$

$$Fe(s) + H_2SO_4\,(aq) + H_2O_2(aq) \rightarrow FeSO_4\,(aq) + 2H_2O(aq)$$

$$Ni(s) + H_2SO_4\,(aq) + H_2O_2(aq) \rightarrow NiSO_4\,(aq) + 2H_2O(aq)$$

Table 1. Literatures on extraction of metals from electronic waste using different lixiviants

S. No.	Type of Waste	Leaching Method	Process Parameters	Extraction Efficiency	Reference
1.	Solder of PCBs	HNO_3 + HCl leaching	HNO_3: 0.2 M, S/L: 1/100 (g/mL), Temperature: 90ºC, Time: 120 min	Pb: 99.99%	(Jha et al. 2012)
			HCl: 3.5 M, S/L: 1/20, Temperature: 90ºC, Time: 120 min	Sn: 98.74%	
2.	PCBs	H_2SO_4 Leaching (Two step)	H_2SO_4: 2M, Temperature: 30ºC, S/L: 1:12, H_2O_2(30%): 20 mL, Time: 180 min	Cu: 75% Zn: 80% Sn: 57%	(Birloaga et al. 2013)
3.	PCBs	H_2SO_4 Leaching (Two step)	H_2SO_4: 2M, Temperature: 25ºC, S/L: 12:125, H_2O_2(35%): 25 mL, Time: 180 min	Cu: 99%	(Behnamfard et al. 2013)
4.	PCBs	H_2SO_4 Leaching (Three step)	H_2SO_4: 2M, Temperature: 30ºC, S/L: 1:12, H_2O_2(30%): 20 mL, Time: 180 min	Cu: 99% Al: 80% Fe: 77% Sn: 80% Ni: 87% Zn: 97% Ag: 60%	(Birloaga et al. 2014)
5.	PCBs	H_2SO_4 Leaching (Two step)	H_2SO_4: 1M, Temperature: 75ºC, S/L: 1:10, Time: 240 min	Cu: 100% Zn: 60% Al: 10%	(Silvas et al. 2015)
6.	Multilayer ceramic capacitors of PCBs	HNO_3 Leaching	HNO_3: 1 M, Temperature: 80ºC, Time: 6 h	Ni: 99.16% Ag: 99.09% Pd: 94.91% Cu: 99.61% Bi: 23.76%	(Liu et al. 2021)
7.	PCBs	Aqua regia leaching (HNO_3 + HCl)	S/L: 1/40, Time: 3 h	Ni: 100% Au: 100% Pd: 7.8% Ag: 7.2%	(Park and Fray 2009)
8.	PCBs	H_2SO_4 Leaching	H_2SO_4: 15 wt.%, Temperature: 23ºC, S/L: 1:10, H_2O_2 (30%): 10 mL, Time: 3 h	Cu: 96%	(Yang et al. 2011)
9.	TV Boards	HNO_3 Leaching	HNO_3: 3 M, S/L: 6/10 (g/mL), Temperature: 70ºC, Time: 120 min	Cu: 99.9%	(Bas et al. 2014)
			HNO_3: 5 M, S/L: 6/10 (g/mL), Temperature: 70ºC, Time: 120 min	Ag: 68%	
10.	PCBs	HCl Leaching	HCl: 2 M, S/L: 1/10, Temperature: 75ºC, Time: 240 min	Sn: 88% Pb: 99%	(Moosakazemi et al. 2019)

11.	PCBs	H_2SO_4 Leaching (High pressure and high temperature)	H_2SO_4: 2.4 M, Temperature: 150ºC, P_{O2}: 20 bar, S/L: 1:10, Time: 90 min	Cu: 31% Ni: 88% Fe: 100%	(Kumari et al. 2016)
12	PCBs	HCl Leaching	HCl: 1M Time: 22h	Cu: 100% Ni: 100% Au: 100% Ag: 100% Zn: 100% Pd: 100% Fe: 100%	(Jadhav and Hocheng 2015)
13.	PCBs	H_2SO_4 leaching (Two stage)	H_2SO_4: 2M, Temperature: 25 °C, H_2O_2(35%): 20 mL, Time: 3 h	Cu: 96%	(Huang et al. 2020)
14.	Electronic Scrap	Cyanide leaching	CN-: 0.1 M, S/L: 1:5, Temperature: 20ºC, pH: 9–11	Ag: 93% Au: 95%	(Quinet et al. 2005)
15.	PCBs	Cyanide Leaching	KCN: 6-8%, pH: 12.5, Temperature: 25ºC, Time: 2 h, S/L: 1/20	Au: 60–70%	(Petter et al. 2014)
16.	PCBs	Cyanide Leaching	NaCN: 4 g/L, pH: 10.5–11, Time: 15 days	Au: 46.6% Ag: 51.3%	(Montero et al. 2012)
17.	PCBs	Thiosulfate Leaching	Sodium thiosulfate: 0.1 M, Ammonium hydroxide: 0.2 M, Copper sulfate: 0.015 M to 0.03 M	Au: 15%	(Tripathi et al. 2012)
18.	PCBs	Thiosulfate Leaching	Ammonium sulfate: 0.1M, Copper sulfate: 40 mM, pH: 10–10.5, 250 rpm, 8 h	Au: 78.8%	(Petter et al. 2014)
19.	PCBs	Thiourea leaching	Thiourea: 24g/L, Fe^{3+}: 0.6%, H_2SO_4: 0.5M	Au: 90% Ag: 50%	(Jing-ying et al. 2012)
20.	PCBs	Thiourea leaching	Thiourea: 20g/L, Fe^{3+}: 6g/L, H_2SO_4: 0.1M	Au: 90% Ag: 75%	(Birloaga and Vegliò, 2016)
21.	Electronic scrap	Iodide leaching	I_2: 50g/L, NaOH: 1.5 M	Cu: 90% Au: 94-100%	(Serpe et al. 2015)
22.	PCBs	Chloride leaching	$NaClO_3$: 25 g/L NaCl: 75g/L	Cu: 90% Au: 90.2 %	(He and Xu 2015)
23.	PCBs	Chelation leaching	EDTA: 0.5 M, pH: 11, Solid to liquid ratio: 1:15, Time: 3h	Cu: 83.8%	(Jadhao et al. 2016)

Table 1. contd. ...

...*Table 1. contd.*

S. No	Type of Waste	Leaching Method	Process Parameters	Extraction Efficiency	Reference
24.	PCBs	Chelation leaching	EDTA: 0.5–0.6M, Temp.: 83–89°C, pH: 8–9, Time: 130–148 min S/L: 1/19–1/23 g/L	Cu: 90%	(Sharma et al. 2017)
25.	Mobile batteries	Chelation leaching	EDTA: 0.33–0.41 M, Temp.: 87–92°C, pH: 8.3–8.8, Time: 163–178 min S/L: 1/25–1/28 g/L	Co: 94%	(Sharma et al. 2017)
26.	Alloys	Ammonia-Ammonium chloride leaching	NH_4Cl: 2 kmol.m^{-3} NH_3: 5 kmol.m^{-3} $CuCl_2$:0.1 kmol.m^{-3} Pulp density: 1%	Cu: 98%	(Lim et al. 2013)
27.	Electronic scrap	Ammonia-Ammonium carbonate leaching	$(NH_4)_2CO_3$: 120 g/L, NH_3: 75.5 g/L Pulp density: 5%	Cu: 90%	(Sun et al. 2015)
28.	PCBs	Ammonia-Ammonium sulphate leaching	$(NH_4)_2SO_4$: 180 g/L, NH_3: 90 g/L, H_2O_2: 0.4 M L/S ratio: 20 mL/g, Temperature: 80°C	Cu: 100% Ni: 90%	(Jadhao et al. 2021)

3.2 Cyanide leaching

Cyanide based leaching techniques has been used in industries over a decade to recover precious metals like Au and Ag (Behnamfard et al. 2013). The popularity of this technique was quite high during the 20th century due to efficient recovery and low cost with respect to other leaching agents (Akcil et al. 2015). Although it has many advantages to metal recovery, it is highly toxic in nature (Ubaldini et al. 1998). This aspect was mainly neglected in the 20th century as the environment norms at that time were lenient. However, today's stringent government norms towards environment protection questions the scope of cyanide leaching in future. Generally, cyanide exists in two different forms during leaching, i.e., CN^- and HCN. The stability of the ions depends on the pH of the solution. HCN in comparison to CN^- is more toxic in nature due to its high volatilization nature. Stability of CN^- and HCN lies above 10.5 and less than 8.5 pH, respectively (Akcil et al. 2015). Though cyanide leaching is efficient for Au and Ag recovery, presence of other metals especially Cu in PCBs results in surplus consumption of cyanide, which may affect overall cost of the process. Formation of Cu-cyanide complexes may suppress the formation of Au-cyanide complexes (Arshadi and Mousavi 2014, Natarajan and Ting 2014). It may further hinder the extraction of precious metals by forming a protective layer of CuO or $Cu(OH)_2$ on particle surface (Montero et al. 2012). Additionally, the standard electrode potential of (E^0) of Au (–1.83V) is much less than base metal such as Ni (–0.67V) and Cu (–0.34V), which makes Au less reactive than base metals for cyanide reaction (Işıldar et al. 2016). The formation of residual weak acid dissociable cyanide complexes during the process demands a recovery or detoxification after precious metal recovery. To address these problems, a multistage process is required. Pre-treatment of PCB is required to avoid large consumption of cyanide due to presence of other metals, and recover base metals like Cu, Ni, Al, Sn, etc. Second and third stages can be used for precious metal recovery and detoxification, respectively (Li et al. 2018). Although cyanide itself is capable of metal recovery, having an oxidizing agent in the process can increase the rate as well as efficiency. Au reacts with cyanide by the Elsner's equation as follows (Campbell et al. 2001):

$$4Au + 8CN^- + O_2 + 2H_2O \rightarrow 4Au(CN)_2^- + 4OH^-$$

Quinet et al. (2005) studied a bench scale extraction process for recovery of precious metal from electronic scrap by using cyanide leaching. The electronic scrap was high on precious metal, containing 27.37% Cu, 0.52% Ag, 0.06% Au and 0.04% Pd. The extraction process was carried out in three stages. In the first stage, Cu and Ag were extracted using oxidative sulphuric acid. In the second stage, oxidative chlorine was used to recover Pd and in the final stage, cyanide was used to recover Au, Ag and Pd. Different techniques were applied to recover metals individually after the leaching process. Palladium was recovered from chlorine solution by cementation technique, whereas NaCl was used for silver recovery. Similarly, activated carbon was used as adsorbent to recover Au, Ag and Pd from cyanide leachate. In this three stage process, they were able to recover 93% of Ag, 95% of Au and 99% of Pd. Petter et al. (2014) recovered Au from PCBs of mobile phones by using KCN at pH: 12.5 and 25°C and the recovery of Au was 60–70%.

3.3 Thiosulfate leaching

Thiosulfate leaching of precious metals from electronic waste has been a promising technique with respect to cyanide leaching due to lesser environmental hazard, high selectivity, low cost etc. Generally, thiosulfate leaching is done in alkaline medium to avoid decomposition of thiosulfate which occurs at acidic solution. Tripathi et al. (2012) studied leaching of gold from mobile phone printed circuit boards by using thiosulfate. Process parameters like pH, pulp density, thiosulfate and copper sulphate concentration were varied to obtain desirable recovery results. The authors reported 78% gold recovery at optimum process conditions i.e. pH: 10–10.5, 0.1 M ammonium thiosulfate, 40 mM copper sulphate, and 8 h of process time. Petter et al. (2014) studied leaching of gold from mobile

PCBs using thiosulfate but they were not able to recover more than 15% of the gold at optimized condition. It was found that copper sulphate in thiosulfate leaching acts as a catalyst which enhances the rate of the leaching. The ammonium hydroxide was employed to maintain the pH.

$$Au + 5S_2O_3^{2-} + Cu\,(NH_3)_4^{2+} \rightarrow Au(S2O3)_2^{3-} + 4NH_3 + Cu(S_2O_3)_3^{5-}$$

$$2Cu(S_2O_3)_3^{5-} + 8NH_3 + 0.5\,O_2 + H_2O \rightarrow 2Cu(NH_3)_4^{2+} + 2OH^- + 6S_2O_3^{2-}$$

Tripathi et al. (2012) reported maximum 78.8% of gold dissolution when thiosulfate concentration of 0.1 M, and copper sulphate concentration of 40 mM were used at pH range of 10–10.5 at stirred condition for 8 h at room temperature. Similarly, Ha et al. (2014) carried out the kinetic study of dissolution of gold and found maximum dissolution rate of 2.395×10^{-5} mol.m^{-2}.S^{-1} at 72.71 mM of thiosulfate, 10 mM of copper ion, and 0.266 M of ammonia at pH range of 10–10.5.

3.4 Thiourea leaching

Thiourea leaching of precious metals is less toxic and eco-friendly in nature. Thiourea compared to thiosulfate is stable in acidic medium, especially less than 4.5 pH. Thiourea works better in pH range of 1–2, provided other parameters like concentration of thiourea, Fe^{3+} and leaching time were optimized (Li et al. 2018). Here Fe^{3+} works as an oxidant; however, excess use of Fe^{3+} can facilitate many side reactions. Jing-Ying et al. (2012) studied leaching of gold and silver from waste mobile phone printed circuit boards by using thiourea at acidic medium. They were able to recover 90% of Au and 50% of Ag at optimized conditions like 24 g/L of thiourea and 0.6% Fe^{3+} concentration in 2 h operation. 0.5 M sulfuric acid was used to maintain pH of the solution. Similarly, Birloaga and Vegilo (2016) studied multistep leaching of waste PCBs for recovery of Au, Ag and Cu. In the first step, Cu was recovered by using concentrated sulphuric acid along with hydrogen peroxide whereas in the second step, thiourea leaching technique was employed for recovery of Au and Ag. Around 90% of Au and 75% of Ag were leached at optimized conditions like 20 g/L of thiourea, 6 g/L of Fe^{3+} and 0.1 M H_2SO_4 for 1 h.

$$Au + 2SC(NH_2)_2 + Fe^{3+} \rightarrow Au(SC(NH_2)_2)_2^{+} + Fe^{2+}$$

$$Ag + 2SC(NH_2)_2 + Fe^{3+} \rightarrow Ag(SC(NH_2)_2)_2^{+} + Fe^{2+}$$

The concentration of sulphuric acid was 0.5 M to maintain pH of the solution (Jing-ying et al. 2012). The higher concentration of thiourea tends to oxidize in presence of ferric ions in acidic medium and form formamidine disulphide. This formed compound is not stable in acidic medium and decomposes to form cynamide and elemental sulphur, which prevents dissolution of gold and silver by forming a membrane on the surface of the pulp (Jing-ying et al. 2012). Therefore, excess of thiourea should be avoided while extracting Au and Ag from e-waste. Similarly, ferric ions concentration also plays an important role in recovery of Au and Ag. At low ferric ion concentration, Au and Ag will have low oxidation rate, whereas at high concentration, thiourea also decomposes, resulting in decrease of Au and Ag dissolution (Jing-ying et al. 2012). Thiourea tends to decompose above 4.5 pH. Hilson and Monhemius (2006) reported that pH range of 1–2 is generally considered ideal for extraction of Au and Ag using thiourea. Generally, a two-step approach is used for recovery of precious metals using thiourea method. In the first step, base metals are recovered and in the second step, acidic thiourea leaching helps in extracting Au and Ag (Birloaga and Vegliò 2016).

3.5 Halide leaching

Halide based leaching has been a promising technique for years for recovery of metals. Due to strong oxidizing properties, they are capable of oxidizing metal from different sources and subsequently to dissolve metal into the solution. However, due to their strong corrosive nature, they have some drawbacks in comparison to other hydrometallurgy techniques. Gold forms a stable complex with iodide. This property can be explored to recover precious metals. Serpe et al. (2015) were able to recover various metals by using iodide leaching from different mixtures of electronic scrap. Two stage process was used in this study: in first stage, base metals were extracted by using 3 M citric acid and in second stage, precious metals were leached by using iodide solution. Around 90% of copper and 94–100% of gold were selectively recovered in this study. He and Xu (2015) studied leaching of copper and gold from waste printed circuit boards by using chlorination technique. First pre-treatment of the crushed PCBs were done by ball milling and supercritical leaching. Sodium chlorate was provided to generate required chlorine. Two stage technique was employed for recovery of copper and gold separately. In another study, Ilyas et al. (2021) recovered Au from pre-processed PCBs using electro-generated Cl_2 in brine solution. The PCBs were pre-processed with HCl solution in presence of H_2O_2 for separation of base metals. Subsequently, the residual PCBs samples were fed into solution containing 2 M NaCl by maintaining S/L ratio of 1:20. Temperature of 25°C was maintained for 120 min for extraction of gold. *In-situ* Cl_2 gas was generated from an electrochemical cell made of polymethyl methacrylate. The cell was composed of HCl solution and graphite rods were used as electrodes for generation of Cl_2 gas. More than 99% of Au was leached, when the concentration of brine was maintained at 2 M NaCl and Cl_2 was charged at a rate of 0.62 mmol min^{-1}.

3.6 Chelation technology

Chelation technology is one of the novel technologies whose demand in the field of metallurgy is growing day by day. Chelation is denoted as the origination of stable metal-ligand complex (Verma and Hait 2019). Among various ligands, ethylene diamine tetra acetic acid (EDTA), diethylene triamine pentaacetic acid (DTPA), and nitrilotriacetic acid (NTA) are promising for metal recovery due to robust metal binding and sequestering capability (Chauhan et al. 2014). Earlier chelation technology was explored to recover metals from spent catalysis. It is quite eco-friendly in nature which attracted its usage for metal recovery from e-waste. In this technology a multi-dente chelator is used for selective recovery of metals from electronic waste in basic medium. Jadhao et al. (2016) studied leaching of copper from printed circuit boards by using EDTA (Ethylenediamine tetra acetic acid) as chelator. Around 83.8% of copper was recovered by using this technology. Due to easy recycling of EDTA, this technique has the potential to become a closed process. Parameters like pH, EDTA concentration, pulp density, etc., were varied to obtain desirable results. Similarly, Sharma et al. leached metals like copper and cobalt from PCBs and mobile batteries, respectively, using EDTA (Sharma et al. 2017). They were able to recover around 90% of copper from PCBs and 94% of cobalt from mobile batteries. Similarly, Verma and Hait (2019) studied extraction of Cu, Ni and Zn from waste PCBs using DTPA as ligand, where they were able to extract of 99% of Cu, 100% of Zn and 82% of Ni at optimum condition of 0.5 M DTPA, pH: 9, L/S: 50, T: 50°C, mixing: 450 rpm, 0.9 M H_2O_2. Furthermore, chemical precipitation method was employed to recover 98% Cu, 98% Ni and 95% Zn from the leached solution.

3.7 Ammonia-ammonia salt leaching

Ammonia-ammonium leaching technique is another promising method for recovery of metals from different metal sources. Base metals like Cu and Zn can be recovered by this process whereas precious metal cannot be recovered by this technique. Hydrogen peroxide has been employed as an oxidant to increase the leaching rate and capacity (Li et al. 2018). Lim et al. (2013) studied recovery of metals from alloys produced during smelting of WPCBs. Around 98% recovery of Cu was achieved by using ammonia and ammonium chloride solution. Optimum conditions reported in this study were 2 $kmol.m^{-3}$ NH_4Cl, 5 $kmol.m^{-3}$ NH_3, 0.1 $kmol.m^{-3}$ $CuCl_2$ at 200 rpm and pulp density of 1%. Copper in this study was acting as oxidant. Sun et al. studied recovery of copper from electronic scrap by using ammonia-ammonium carbonate solution (Sun et al. 2015). Parameters like pulp density, ammonia, ammonium carbonate and stirring speed were varied to optimize the process. Around 90% of copper was recovered at optimum condition by this technique. Recently, Jadhao et al. (2021) studied extraction of valuable metals such as Cu and Ni from PCBs using ammonia-ammonium sulphate solution. Around 100 % Cu and 90% Ni were extracted at optimum conditions of NH_3: 90 g/L, $(NH_4)_2SO_4$: 180 g/L, H_2O_2: 0.4 M, time: 4 h, L/S: 20 mL/g, temperature: 80°C. Thereafter, Cu was separated from leached solution using electrowinning process. Approximately 98.38% pure Cu was deposited on an aluminium cathode using 2.0 V. Overall, ammonia-ammonium salt leaching is an excellent method to extract metals from e-waste; however, further treatment process is required for recovery of precious metals.

3.8 Limitations of hydrometallurgical processes

Currently, various hydrometallurgical processes are being used for recovery of metals from e-waste due to its easy operation, low cost, and efficient extraction of metals. However, they still have some limitations for recovery of metals, which are discussed here (Cui and Zhang 2008, Hilson and Monhemius 2006, La Brooy et al. 1994). Generally, the majority of the hydrometallurgical processes are slow in nature; hence, they alter the recycling economy of the process. For effective extraction of precious metals, additional physical processing of e-waste is required to shorten the particle size. Along with the additional energy required for size reduction, approximately 20% of the precious metals get lost with dust generated through the mechanical treatment of e-waste. Majority of the solvents used for dissolution of metals are highly corrosive in nature; therefore, it requires additional investment to upgrade the reactor quality. Though cyanide leaching is very effective for recovery of precious metals, it poses higher threat to the environment. Similarly, thiourea and thiosulfate based leaching requires high operating cost due to high cost and exhaustion of reagents. Additionally, use of all these solvents generates large amounts of effluent, which needs to be neutralized before safe disposal (Khaliq et al. 2014).

4. CONCLUSION

Circular economy and environmental concerns have boosted the importance of the e-waste recycling sector. Additionally, 3Rs policy of reduce, recycle and reuse have given importance to reuse and recycling to reduce the dump of e-waste. Two main routes of extraction of metals, i.e., pyrometallurgy and hydrometallurgy are briefly discussed. Post treatment of feed is required in both the routes to avoid toxic gas emissions. However, each technology has certain limitations, and a single technology cannot serve a thorough approach to solve the e-waste problem. Major challenge is to develop a sustainable technique combining pre-treatment, physical and chemical route to segregate and purify individual metals with minimum environmental impact, low cost and maximum economy. Heterogeneous nature of e-waste and the presence of toxic materials in it poses an important challenge in recycling methods. Future research directions in processing e-waste should target development of new technology, optimizing current technologies, and finding sustainable alternatives. Dismantling of e-waste is a major step, particularly separating hazardous components from non-hazardous components to avoid contamination in the process. Physical and chemical techniques are applied to separate non-metallic fraction from metallic fraction. Mechanical treatment is required in almost all the processes including advanced hydrometallurgical processes. Chemical technique is generally separation of non-metallic fraction using solvent, supercritical fluids. Pyrometallurgy is a conventional route to extract metals. Any form of e-waste could be treated using pyrometallurgy, but the release of toxic gases and loss of metals is difficult to manage. However, low temperature roasting pyrometallurgical methods hold a promising future combating the disadvantages of pyrometallurgy. Hydrometallurgy has several advantages over pyrometallurgy in terms of control, exactitude and greenness. Acid leaching is primarily used, sulphuric acid being best for copper extraction along with nitric acid but it has below par selectivity in processing steps. Thiourea and thiosulfate are other greener options, but large reagent usage and slow kinetics limit its use. In hydrometallurgy route, metals are extracted in solution from which individual metal separation is a challenging task. There is an urgent need to develop sustainable technologies.

REFERENCES

Akcil, A., Erust, C., Gahan, C.S. ekha., Ozgun, M., Sahin, M. and Tuncuk, A. 2015. Precious metal recovery from waste printed circuit boards using cyanide and non-cyanide lixiviants—A review. Waste Management 45: 258–271.

Anindya, A., Swinbourne, D.R., Reuter, M.A. and Matusewicz, R.W. 2013. Distribution of elements between copper and FeOx–CaO–SiO2 slags during pyrometallurgical processing of WEEE. Mineral Processing and Extractive Metallurgy 122: 165–173.

Arshadi, M., Mousavi, S.M. 2014. Simultaneous recovery of Ni and Cu from computer-printed circuit boards using bioleaching: Statistical evaluation and optimization. Bioresource Technology 174: 233–242.

Bas, A.D., Deveci, H. and Yazici, E.Y. 2014. Treatment of manufacturing scrap TV boards by nitric acid leaching. Separation and Purification Technology 130: 151–159.

Behnamfard, A., Salarirad, M.M. and Veglio, F. 2013. Process development for recovery of copper and precious metals from waste printed circuit boards with emphasize on palladium and gold leaching and precipitation. Waste Management 33: 2354–2363.

Birloaga, I. and Vegliò, F. 2016. Study of multi-step hydrometallurgical methods to extract the valuable content of gold, silver and copper from waste printed circuit boards. Journal of Environmental Chemical Engineering 4: 20–29.

Birloaga, I., Coman, V., Kopacek, B. and Vegliò, F. 2014. An advanced study on the hydrometallurgical processing of waste computer printed circuit boards to extract their valuable content of metals. Waste Management 34: 2581–2586.

Birloaga, I., De Michelis, I., Ferella, F., Buzatu, M. and Vegliò, F. 2013. Study on the influence of various factors in the hydrometallurgical processing of waste printed circuit boards for copper and gold recovery. Waste Management 33: 935–941.

Brusselaers, J., Hageluken, C., Mark, F., Mayne, N. and Tange, L. 2005. An Eco-efficient Solution for Plastics-Metals-Mixtures from Electronic Waste: the Integrated Metals Smelter. 5th IDENTIPLAST 2005, the Biennial Conference on the Recycling and Recovery of Plastics Identifying the Opportunities for Plastics Recovery, Brussels.

Campbell, S.C., Olson, G.J., Clark, T.R. and McFeters, G. 2001. Biogenic production of cyanide and its application to gold recovery. Journal of Industrial Microbiology and Biotechnology 26: 134–139.

Changming, D., Chao, S., Gong, X., Ting, W. and Xiange, W. 2018. Plasma methods for metals recovery from metal–containing waste. Waste Management 77: 373–387.

Chauhan, G., Kaur, P., Pant, K.K., Nigam, K.D.P. 2020. Sustainable Metal Extraction from Waste Streams. WILEY-VCH Verlag GmbH.

Chauhan, G., Jadhao, P.R., Pant, K.K. and Nigam, K.D.P. 2018. Novel technologies and conventional processes for recovery of metals from waste electrical and electronic equipment: Challenges & opportunities—A review. Journal of Environmental Chemical Engineering 6: 1288–1304.

Chauhan, G., Pant, K.K. and Nigam, K.D.P. 2014. Chelation technology: a promising green approach for resource management and waste minimization. Environmental Sciences: Process & Impacts 17: 12–40.

Cui, F., Mu, W., Wang, S., Xin, H., Shen, H., Xu, Q., Zhai, Y. and Luo, S. 2018. Synchronous extractions of nickel, copper, and cobalt by selective chlorinating roasting and water leaching to low-grade nickel-copper matte. Separation and Purification Technology 195: 149–162.

Cui, J. and Zhang, L. 2008. Metallurgical recovery of metals from electronic waste: A review. Journal of Hazardous Material 158: 228–256.

Dave, S.R., Sodha, A.B. and Tipre, D.R. 2020. Microbial Processes for Treatment of e-Waste Printed Circuit Boards and Their Mechanisms for Metal(s) Solubilization. pp. 131–143. *In*: Microbes for Sustainable Development and Bioremediation, CRC Press. ISBN: 978-042-927-5876.

Dhawan, N., Wadhwa, W., Kumar, V. and Kumar, M. 2009. Recovery of metals from electronic scrap by hydrometallurgical route, EPD congress 2009 (CD-ROM). pp. 1107–1109. *In*: Howard, S. (ed.). P. Anyalebechi, L. Zhang (Section Eds.). TMS Knowledge Resource Center, TMS, Warrendale, PA.

Du, C., Wu, J., Ma, D., Liu, Y., Qiu, P., Qiu, R., Liao, S. and Gao, D. 2015. Gasification of corn cob using non-thermal arc plasma. International Journal of Hydrogen Energy 40: 12634–12649.

Flandinet, L., Tedjar, F., Ghetta, V. and Fouletier, J. 2012. Metals recovering from waste printed circuit boards (WPCBs) using molten salts. Journal of Hazardous Materials 213–214: 485–490.

Forti, V., Balde, C.P., Kuehr, R. and Bel, G. 2020. The Global E-waste Monitor 2020: Quantities, flows and the circular economy potential. United Nations University.

Gu, W., Bai, J., Dong, B., Zhuang, X., Zhao, J., Zhang, C., Wang, J. and Shih, K. 2017. Enhanced bioleaching efficiency of copper from waste printed circuit board driven by nitrogen-doped carbon nanotubes modified electrode. Chemical Engineering Journal 324: 122–129.

Ha, V.H., Lee, J.C., Huynh, T.H., Jeong, J. and Pandey, B.D. 2014. Optimizing the thiosulfate leaching of gold from printed circuit boards of discarded mobile phone. Hydrometallurgy 149: 118–126.

Hagelüken, C. 2006a. Recycling of Electronic Scrap at Umicore's Integrated Metals Smelter and Refinery. World of Metallurgy—ERZMETALL 59(3): 152–161

Hagelüken, C. 2006b. Improving metal returns and eco-efficiency in electronics recycling—A holistic approach for interface optimisation between pre-processing and integrated metals smelting and refining. IEEE International Symposium on Electronics and the Environment, pp. 218–223.

Hao, J., Wang, Y., Wu, Y. and Guo, F. 2020. Metal recovery from waste printed circuit boards: A review for current status and perspectives. Resources, Conservation and Recycling 157: 104787.

He, S., Wilson, B.P., Lundström, M. and Liu, Z. 2021. Hazard-free treatment of electrolytic manganese residue and recovery of manganese using low temperature roasting-water washing process. Journal of Hazardous Materials 402: 123561.

He, Y. and Xu, Z. 2015. Recycling gold and copper from waste printed circuit boards using chlorination process. RSC Advances 5: 8957–8964.

Hilson, G. and Monhemius, A.J. 2006. Alternatives to cyanide in the gold mining industry: what prospects for the future? Journal of Cleaner Production 14: 1158–1167.

Hsu, P.C., Foster, K.G., Ford, T.D., Wallman, P.H., Watkins, B.E., Pruneda, C.O. and Adamson, M.G. 2000. Treatment of solid wastes with molten salt oxidation. Waste Management 20: 363–368.

Huang, J., Shi, J., Liang, R., Liu, Z., Huang, J., Shi, J., Liang, R. and Liu, Z. 2003. A New Waste Disposal Technology-plasma arc Pyrolysis System. Plasma Science and Technology 5: 1743–1748.

Huang, Y.F., Pan, M.W. and Lo, S.L. 2020. Hydrometallurgical metal recovery from waste printed circuit boards pretreated by microwave pyrolysis. Resources, Conservation and Recycling 163: 105090.

Ilyas, S., Srivastava, R.R. and Kim, H. 2021. Gold recovery from secondary waste of PCBs by electro-Cl2 leaching in brine solution and solvo-chemical separation with tri-butyl phosphate. Journal of Cleaner Production 295: 126389.

Işıldar, A., van de Vossenberg, J., Rene, E.R., van Hullebusch, E.D. and Lens, P.N.L. 2016. Two-step bioleaching of copper and gold from discarded printed circuit boards (PCB). Waste Management 57: 149–157.

Jadhao, P.R., Pandey, A., Pant, K.K. and Nigam, K.D.P. 2021. Efficient recovery of Cu and Ni from WPCB via alkali leaching approach. Journal of Environment Management 296: 113154.

Jadhao, P., Chauhan, G., Pant, K.K. and Nigam, K.D.P. 2016. Greener approach for the extraction of copper metal from electronic waste. Waste Management 57: 102–112.

Jadhav, U. and Hocheng, H. 2015. Hydrometallurgical Recovery of Metals from Large Printed Circuit Board Pieces. Scientific Reports 51(5): 1–10.

Jha, M.K., Kumari, A., Choubey, P.K., Lee, J.C., Kumar, V. and Jeong, J. 2012. Leaching of lead from solder material of waste printed circuit boards (PCBs). Hydrometallurgy 121–124: 28–34.

Jin, H., Frost, K., Sousa, I., Ghaderi, H., Bevan, A., Zakotnik, M. and Handwerker, C. 2020. Life cycle assessment of emerging technologies on value recovery from hard disk drives. Resources, Conservation & Recycling 157: 104781.

Jing-Ying, L., Li, X.X. and Quan, L.W. 2012. Thiourea leaching gold and silver from the printed circuit boards of waste mobile phones. Waste Management 32: 1209–1212.

Khaliq, A., Rhamdhani, M.A., Brooks, G. and Masood, S. 2014. Metal Extraction Processes for Electronic Waste and Existing Industrial Routes: A Review and Australian Perspective. Resources 3: 152–179.

Kim, E.Y., Kim, M.S., Lee, J.C., Jeong, J. and Pandey, B.D. 2011. Leaching kinetics of copper from waste printed circuit boards by electro-generated chlorine in HCl solution. Hydrometallurgy 107: 124–132.

Kumari, A., Jha, M.K., Lee, J.C. and Singh, R.P. 2016. Clean process for recovery of metals and recycling of acid from the leach liquor of PCBs. Journal of Cleaner Production 112: 4826–4834.

La Brooy, S.R., Linge, H.G. and Walker, G.S. 1994. Review of gold extraction from ores. Minerals Engineering 7: 1213–1241.

Lee, J. and Pandey, B.D. 2012. Bio-processing of solid wastes and secondary resources for metal extraction—A review. Waste Management 32: 3–18.

Lee, J., Teak, H. and Yoo, J. 2007. Present status of the recycling of waste electrical and electronic equipment in Korea. Resources, Conservation & Recycling 50: 380–397.

Lehner, T. 2003. E&HS aspects on metal recovery from electronic scrap. IEEE International Symposium on Electronics and the Environment, Boston, USA, pp. 318–322.

Leirnes, J. and Lundstrom, M. 1983. Method for working-up metal-containing waste products, US Patent, US4415360 (C22B 1/00).

Leonetti, C., Butt, C.M., Hoffman, K., Hammel, S.C., Miranda, M.L. and Stapleton, H.M. 2016. Brominated flame retardants in placental tissues: associations with infant sex and thyroid hormone endpoints. Environmental Health 151(15): 1–10.

Li, H., Eksteen, J. and Oraby, E. 2018. Hydrometallurgical recovery of metals from waste printed circuit boards (WPCBs): Current status and perspectives—A review. Resources, Conservation & Recycling 139: 122–139.

Lim, Y., Kwon, O., Lee, J. and Yoo, K. 2013. The ammonia leaching of alloy produced from waste printed circuit boards smelting process. Geosystem Engineering 16: 216–224.

Liu, Y., Song, Q., Zhang, L. and Xu, Z. 2021. Separation of metals from Ni-Cu-Ag-Pd-Bi-Sn multi-metal system of e-waste by leaching and stepwise potential-controlled electrodeposition. Journal of Hazardous Materials 408: 124772.

Lorenz, T. and Bertau, M. 2020. Recycling of rare earth elements from SmCo5-Magnets via solid-state chlorination. Journal of Cleaner Production 246: 118980.

Lorenz, T. and Bertau, M. 2019. Recycling of rare earth elements from FeNdB-Magnets via solid-state chlorination. Journal of Cleaner Production 215: 131–143.

Maeda, Y., Inoue, H., Kawamura, S. and Ohike, H. 2000. Metal Recycling at Kosaka Smelter. pp. 22–25. *In*: Proceedings of the 4th International Symposium on Recycling of Metals and Engineered Materials, Pittsburgh, PA, USA.

Magalini, F., Khetriwal, D.S. and Kyriakopoulou, A. 2019. E-waste Policy handbook, Coffey International Development Ltd.

Maris, E., Botané, P., Wavrer, P. and Froelich, D. 2015. Characterizing plastics originating from WEEE: A case study in France. Mineral Engineering 76: 28–37.

Mark, F.E. and Lehner, T. 2010. Plastics Recovery from Waste Electrical & Electronic Equipment in Non-Ferrous Metal Processes. Technical Report, Association of Plastic Manufacturers in Europe, Brussels. https://www.informea.org/en/icl-ip-europe-apme-plastics-recovery-waste-electrical-electronic-equipment-non-ferrous-metal

Martinho, G., Pires, A., Saraiva, L. and Ribeiro, R. 2012. Composition of plastics from waste electrical and electronic equipment (WEEE) by direct sampling. Waste Management 32: 1213–1217.

Mecucci, A. and Scott, K. 2002. Leaching and electrochemical recovery of copper, lead and tin from scrap printed circuit boards. Journal of Chemical Technology and Biotechnology 77: 449–457.

Mitrasinovic, A., Pershin, L., Wen, J.Z. and Mostaghimi, J. 2011. Recovery of Cu and valuable metals from E-waste using thermal plasma treatment. JOM 63: 24–28.

Montero, R., Guevara, A. and Dela Torre, E. 2012. Recovery of gold, silver, copper and niobium from printed circuit boards using leaching column technique. International Journal of Earth Sciences and Engineering 2: 590.

Moosakazemi, F., Ghassa, S. and Mohammadi, M.R.T. 2019. Environmentally friendly hydrometallurgical recovery of tin and lead from waste printed circuit boards: Thermodynamic and kinetics studies. Journal of Cleaner Production 228: 185–196.

Nadirov, R.K., Syzdykova, L.I., Zhussupova, A.K. and Usserbaev, M.T. 2013. Recovery of value metals from copper smelter slag by ammonium chloride treatment. International Journal of Mineral Processing 124: 145–149.

Natarajan, G. and Ting, Y.P. 2014. Pretreatment of e-waste and mutation of alkali-tolerant cyanogenic bacteria promote gold biorecovery. Bioresource Technology 152: 80–85.

Panda, R., Pant, K.K. and Bhaskar, T. 2021a. Efficient extraction of metals from thermally treated waste printed circuit boards using solid state chlorination: Statistical modeling and optimization. Journal of Cleaner Production 313: 127950.

Panda, R., Pant, K.K., Bhaskar, T. and Naik, S.N. 2021b. Dissolution of brominated epoxy resin for environment friendly recovery of copper as cupric oxide nanoparticles from waste printed circuit boards using ammonium chloride roasting. Journal of Cleaner Production 291: 125928.

Panda, R., Jadhao, P.R., Pant, K.K., Naik, S.N. and Bhaskar, T. 2020. Eco-friendly recovery of metals from waste mobile printed circuit boards using low temperature roasting. Journal of Hazardous Materials 395: 122642.

Pant, D., Joshi, D., Upreti, M.K. and Kotnala, R.K. 2012. Chemical and biological extraction of metals present in E waste: A hybrid technology. Waste Management 32: 979–990.

Park, Y.J. and Fray, D.J. 2009. Recovery of high purity precious metals from printed circuit boards. Journal of Hazardous Materials 164: 1152–1158.

Petter, P.M.H., Veit, H.M. and Bernardes, A.M. 2014. Evaluation of gold and silver leaching from printed circuit board of cellphones. Waste Management 34: 475–482.

Puri, B.R., Sharma, L.R. and Kalia, K.C. 1991. Principles of Inorganic Chemistry, 21st Ed. Shoban Lal Nagin Chand & Co, Jalandhar, India.

Quinet, P., Proost, J. and Van Lierde, A. 2005. Recovery of precious metals from electronic scrap by hydrometallurgical processing routes. Mining, Metallurgy & Exploration 22: 17–22.

Rajeshwari, K.V. 2007. Tackling e-waste Towards Efficient Management Techniques, TERI Press, New Delhi. ISBN: 978-817-993-1103.

Rath, S.S., Nayak, P., Mukherjee, P.S., Roy Chaudhury, G. and Mishra, B.K. 2012. Treatment of electronic waste to recover metal values using thermal plasma coupled with acid leaching—A response surface modeling approach. Waste Management 32: 575–583.

Serpe, A., Rigoldi, A., Marras, C., Artizzu, F., Mercuri, M.L. and Deplano, P. 2015. Chameleon behaviour of iodine in recovering noble-metals from WEEE: towards sustainability and "zero" waste. Green Chemistry 17: 2208–2216.

Sharma, N., Chauhan, G., Kumar, A. and Sharma, S.K. 2017. Statistical Optimization of Heavy Metal (Cu^{2+} and Co^{2+}) Extraction from Printed Circuit Boards and Mobile Batteries Using Chelation Technology. Industrial & Engineering Chemistry Research 56: 6805–6819.

Silvas, F.P.C., Jiménez Correa, M.M., Caldas, M.P.K., de Moraes, V.T., Espinosa, D.C.R. and Tenório, J.A.S. 2015. Printed circuit board recycling: Physical processing and copper extraction by selective leaching. Waste Management 46: 503–510.

Sohaili, J., Kumari Muniyandi, S. and Suhaila Mohamad, S. 2012. A Review on Printed Circuit Boards Waste Recycling Technologies and Reuse of Recovered Nonmetallic Materials. International Journal of Scientific & Engineering Research 3: 1–7.

Stuhlpfarrer, P., Luidold, S. and Antrekowitsch, H. 2016. Recycling of waste printed circuit boards with simultaneous enrichment of special metals by using alkaline melts: A green and strategically advantageous solution. Journal of Hazardous Materials 307: 17–25.

Sum, E.Y.L. 1991. The recovery of metals from electronic scrap. JOM 43: 53–61.

Sun, Z.H.I., Xiao, Y., Sietsma, J., Agterhuis, H., Visser, G. and Yang, Y. 2015. Selective copper recovery from complex mixtures of end-of-life electronic products with ammonia-based solution. Hydrometallurgy 152: 91–99.

Szałatkiewicz, J. 2016. Metals Recovery from Artificial Ore in Case of Printed Circuit Boards, Using Plasmatron Plasma Reactor. Materials 9: 683.

Szałatkiewicz, J., Szewczyk, R., Budny, E., Missala, T. and Winiarski, W. 2013. Construction Aspects of Plasma Based Technology for Waste of Electrical and Electronic Equipment (WEEE) Management in Urban Areas. Procedia Engineering 57: 1100–1108.

Theo, L. 1998. Integrated recycling of non-ferrous metals at Boliden Ltd. Ronnskar smelter. IEEE International Symposium on Electronics and the Environment, Oak Brook, USA, pp. 42–47.

Tripathi, A., Kumar, M., Sau, D.C., Agrawal, A., Chakravarty, S. and Mankhand, R., T. 2012. Leaching of Gold from the Waste Mobile Phone Printed Circuit Boards (PCBs). with Ammonium Thiosulphate. International Journal of Metallurgical Engineering, 1, 17–21.

Ubaldini, S., Fornari, P., Massidda, R. and Abbruzzese, C. 1998. An innovative thiourea gold leaching process. Hydrometallurgy 48: 113–124.

Veldbuizen, H. and Sippel, B. 1994. Mining discarded electronics. Industry and Environment 17: 7.

Verma, A. and Hait, S. 2019. Chelating extraction of metals from e-waste using diethylene triamine pentaacetic acid. Process Safety and Environmental Protection 121: 1–11.

Verma, H.R., Singh, K.K. and Mankhand, T.R. 2017. Liberation of metal clads of waste printed circuit boards by removal of halogenated epoxy resin substrate using dimethylacetamide. Waste Management 60: 652–659.

Verma, H.R., Singh, K.K. and Mankhand, T.R. 2016. Dissolution and separation of brominated epoxy resin of waste printed circuit boards by using di-methyl formamide. Journal of Cleaner Productiom 139: 586–596.

Vogel, A.I. 1996. Vogel's Qualitative Inorganic Analysis, 7th edition, Revised by Svehla, G., 311 Pearson Education Ltd.

Wang, S.B., Cheng, C.M., Lan, W., Zhang, X.H., Liu, D.P. and Yang, S.Z. 2013. Experimental Study of Thermal Plasma Processing Waste Circuit Boards.Advanced Materials Research 652–654: 1553–1561.

Wath, S.B., Katariya, M.N., Singh, S.K., Kanade, G.S. and Vaidya, A.N. 2015. Separation of WPCBs by dissolution of brominated epoxy resins using DMSO and NMP: A comparative study. Chemcal Engineering Journal 280: 391–398.

Wernick, I.K. and Themelis, N.J. 2003. Recycling metals for the environment. Annual Review of Energy and the Environment 23: 465–497.

Williams, D.P. and Drake, P. 1983. Recovery of precious metals from electronic scrap. Precious Metals, Proceedings of the Sixth International Precious Metals Institute Conference, Newport Beach, California, pp. 555–565.

Xiao, J., Gao, R., Niu, B. and Xu, Z. 2021. Study of reaction characteristics and controlling mechanism of chlorinating conversion of cathode materials from spent lithium-ion batteries. Journal of Hazardous Materials 407: 124704.

Xiong, X., Li, G., Lu, X., Cheng, H., Xu, Q. and Li, S. 2020. Mechanistic studies on millerite chlorination with ammonium chloride. Physical Chemistry Chemical Physics 22: 4832–4839.

Xu, C., Cheng, H., Li, G., Lu, C., Lu, X., Zou, X. and Xu, Q. 2017. Extraction of metals from complex sulfide nickel concentrates by low-temperature chlorination roasting and water leaching. IInternational Journal of Minerals, Metallurgy and Materials 24: 377–385.

Yang, H., Liu, J. and Yang, J. 2011. Leaching copper from shredded particles of waste printed circuit boards. Journal of Hazardous Materials 187: 393–400.

Yang, H.C., Cho, Y.J., Eun, H.C. and Kim, E.H. 2007. Destruction of chlorinated organic solvents in a two-stage molten salt oxidation reactor system. Chemical Engineering Science 62: 5137–5143.

Yao, Z., Li, J. and Zhao, X. 2011. Molten salt oxidation: A versatile and promising technology for the destruction of organic-containing wastes. Chemosphere 84: 1167–1174.

Yken, J., Van, Cheng, K.Y., Boxall, N.J., Nikoloski, A.N., Moheimani, N., Valix, M., Sahajwalla, V. and Kaksonen, A.H. 2020. Potential of metals leaching from printed circuit boards with biological and chemical lixiviants. Hydrometallurgy 196: 105433.

Zhang, L. and Xu, Z. 2016. A review of current progress of recycling technologies for metals from waste electrical and electronic equipment. Journal of Cleaner Production 127: 19–36.

Zhang, M., Zhu, G., Zhao, Y. and Feng, X. 2012. A study of recovery of copper and cobalt from copper–cobalt oxide ores by ammonium salt roasting. Hydrometallurgy 129–130: 140–144.

Zhu, P., Chen, Y., Wang, L., Qian, G., Zhang, W.J., Zhou, M. and Zhou, J. 2013a. Dissolution of Brominated Epoxy Resins by Dimethyl Sulfoxide To Separate Waste Printed Circuit Boards. Environment Science & Technology 47: 2654–2660.

Zhu, P., Chen, Y., Wang, L.Y., Zhou, M. and Zhou, J. 2013b. The separation of waste printed circuit board by dissolving bromine epoxy resin using organic solvent. Waste Management 33: 484–488.

5

ENERGY RECOVERY FROM ELECTRONIC WASTE

A CRITICAL ASSESSMENT

R. Khanna,[1,*] *Y.V. Konyukhov*,[2] *R. Cayumil*,[3] *R. Saini*[4] and *P.S. Mukherjee*[5]

1. INTRODUCTION

Electronic waste is one of the fastest growing solid waste streams in the world with global generation estimated to reach 55.2 million tonnes by 2021 (Forti et al. 2020). Short life-spans, rapid technological advancements, almost continuous upgrades of electronic devices play a significant role in the generation of such waste. In 2017, there were over 5 billion mobile phone users worldwide (Statista 2017); almost three-quarters of the world population now uses mobile phones. Nearly fifteen billion devices were connected by 2015 and expected to increase to ~ 40 billion by 2020. With increasing availability and affordability, emerging economies are quickly catching up with developed nations (EIU 2015). Highest e-waste producing countries include (in million tonnes/annuum): China (7.2), USA (6.3), Japan (2.1), India (2.0), Germany (1.9), etc. However, through illegal trans-boundary movements as second hand equipment, a significant portion of the e-waste from developed economies gets transferred to the poor and/or developing nations (Global e-waste monitor 2017).

A large proportion of electronic waste continues to be consigned to landfills, which represents a major source of pollution as well as the loss of valuable material resources. With most recycling in developing nations still being carried out in the informal sector, unskilled employees, lack of education, rudimentary methods, e-waste management can cause significant environmental pollution,

[1] School of Materials Science and Engineering (Ret.), The University of New South Wales, NSW 2052, Sydney, Australia.

[2] Department of Functional Nanosystems and High-Temperature Materials, National University of Science and Technology "MISiS", Moscow, 119049, Russia.

[3] Facultad de Ingenieria, Santiago & Centro de Investigaci6n Para La Sustentabilidad, Universidad Andres Bello. Providencia, Santiago, 880 Chile.

[4] Department of Mechanical Engineering, ABES Engineering College, Ghaziabad, 201009, India

[5] Institute of Minerals and Materials Technology (Ret.), Council of Scientific and Industrial Research, Bhubaneshwar, Orissa 751013, India.

* Corresponding author: rita.khanna66@gmail.com

hazards and health related damages to areas near the recycling facilities. Indiscriminate disposal and processing of burgeoning volumes of e-waste has been known to cause considerable local area damage and environmental pollution (Rajarao et al. 2014).

Modern electronic devices use a wide variety of materials such as plastics, metals and refractory oxides. The production of smart phones can use up to 70 elements (Greenfield and Graedel 2013). A number of metallurgical techniques and industrial approaches are being used to recover precious and other metals from waste printed circuit boards (WPCBs). Typical concentrations of precious metals in mobile phones are estimated to be ~ gold (340 ppm), silver (3500 ppm), palladium (140 ppm) (Schluep et al. 2009); these metals account for up to 80% total material value of WPCBs (Hageluken and Corti 2010). Priya and Hait (2018) carried out an in-depth characterization of WPCBs from electronic devices such as mobile phones, computers, laptops, calculators, refrigerators etc. to determine relative concentrations of major, minor and precious metals. Maximum levels of Au (~ 316 ppm) and Ag (~ 636 ppm) were recorded for mobile and laptop WPCBs. However, these values depend on the type of WPCB, product brand, generation, class, etc., and could vary significantly across various boards. Some effort has also been expended on recovering rare earth elements (REEs) from WPCBs. Khanna et al. 2018) have reported high temperature pyrolysis investigations on WPCBs towards recovering REEs: La, Nd, Gd, Pr, Sm, Ce, Dy and Y; these were found to be concentrated in the slag phase of pyrolysis residues.

Several billion tonnes of waste plastics are expected to be produced from e-waste alone in coming decades. A number of electronic components such as instrument housing, printed circuit boards, covers, etc., contain valuable engineered plastics. Due mainly to difficulties with their separation, sorting and collection, a major proportion of these is generally reused in low-value applications (Guo et al. 2009). The U.S. Environmental Protection Agency has indicated recycling to be the preferred approach for these plastics (Kang and Schoenung 2005). However, the recycling rate of such waste plastics is still quite low as it sometimes costs more to recycle than the value of the recycled material/product itself.

Most research on e-waste has been focussed on material recovery, whereas investigations on energy recovery have been somewhat limited. E-waste is known to be a complex mixture of plastics, metals, ceramics and other trace impurities; WPCBs typically contain 40% metals, 30% organics/plastics and 30% ceramics. After removing batteries and capacitors, WPCBs from computers/TVs can contain up to 70% organics whereas WPCBs from mobile phones contain ~ 20% organics (Hall and Williams 2007). With metals and ceramics being highly stable, plastics/organic constituents are the main source of energy in e-waste. Thermal degradation, decomposition, gaseous release from the polymer fraction is expected to be the key contributor to energy during e-waste processing.

Challenges in processing e-waste include complexity, composition variance, intermingling and integration of low cost plastics with metals and ceramics. Waste processing should enhance material recovery, reduce final waste volumes as well as minimise toxic emissions. The catalytic influence of metals on polymer stability, thermal degradation behaviour of organics and the presence of ~ 15% fire-retarding brominated compounds are likely to impact the generation of harmful particulates and gaseous emissions during recycling of WPCBs (Herat and Pariatamby 2012). Thermal treatment of e-waste using incineration, combustion, smelting, pyrolysis, gasification, etc., can be used to degrade the organic fraction with marginal influence on non-volatile and metallic phases; the gaseous, liquid or solid byproducts could be used for energy recovery (Gurgul et al. 2017). During smelting, air enriched with up to 90 vol. % oxygen is blown into the furnace at 1200°C and the melt is mixed intensively by the process gases for achieving rapid equilibration (Gerardo et al. 2014). The combustion process requires high temperatures for ignition, turbulence for mixing all components with an oxidant, and the time to complete all oxidation reactions. Incineration is used for small scale volume reduction and processing of e-waste using rotary furnaces or fluidised bed reactors at ~ 850°C for 3 min under 6 vol.% oxygen; the formation of melts and slags is avoided during incineration (Woynarowska 2014). The scale of incineration plants is less flexible as compared to pyrolysis plants.

Pyrolysis is used for the thermal decomposition of organic materials at high temperatures in the presence of inert gases Ar, N_2, He and the absence of oxygen, producing gas, liquid products and a solid residue rich in carbon (char). Gaseous products are rich in CO, CO_2, CH_4, H_2, etc., making this gas valuable for further combustion. During gasification, organic fractions can be fully eliminated as combustible gas, volatiles and ash; typical gasification reagents include steam, oxygen (air, oxygen enriched air) or carbon dioxide (much less) (Kaya 2016). Relative proportions of pyrolysis biproducts depend on the operating conditions such as temperature, heating rate, and residence time and can be manipulated to a certain extent (Santella et al. 2018). In addition to recovering valuable chemicals, pyrolysis can be used to convert low energy density waste into high energy density fuels (Kwon et al. 2019). Carbon rich chars as solid residues from the process can be used as carbon and energy resource (Kumar et al. 2017a, Sahajwalla et al. 2015).

In this chapter, we present a critical assessment on recovering energy from e-waste with focus on technical challenges, opportunities and associated environmental impact. This chapter is organised as follows. Section 2 contains details on e-waste polymers, their thermal degradation behaviour, organic bi-products and gas-phase emissions. Results from the thermal treatment of e-waste including pyrolysis, combustion and open burning are reviewed in Section 3. Section 4 presents pollution results with focus on the release of fine particulates, generation of dioxins and furans during e-waste processing. Energy savings through material recovery and urban mining of e-waste are discussed in Section 5, followed by concluding remarks on prospects of energy recovery from e-waste in Section 6.

2. E-WASTE POLYMERS

2.1 E-waste polymers

Thermoset as well as thermoplastic resins are employed in different parts of electronic equipment. Thermoset resins such as epoxies and polyamides are used in printed circuit boards, electrical switch housing and electrical motor components, etc. (Hall and Williams 2007). Thermoplastics are used in frames, keyboards, connectors and other components (Wu et al. 2008). Several different types of plastics can be found in e-waste, e.g., polyethylene (PE); Polypropylene (PP); Polyamide (PA); Polycarbonate (PC); high-impact polystyrene (HIPS), acrylonitrile butadiene styrene (ABS) and poly methyl methacrylate (PMMA), polyphenylene ether (PPE), polyvinyl chloride (PVC). HIPS, ABS, PPE and PC can be found in TVs, whereas various blends are also present in computers (Kang and Schoenung 2005). Up to 12% of these plastics contain flame retardants such as tetra-bromo-bisphenol-A (TBBPA), polybrominated biphenyls (PBBs), polybrominated diphenyl ethers (PBDEs) (Cui and Roven 2011).

E-waste polymers are generally recycled through mechanical, thermal and/or chemical processes. Mechanical processing consists of the identification and separation of different types of polymers present in the electronic waste to make new polymeric materials. Sorting is the first step, as plastics may be found coated, mixed with additives, painted, or having flame retardants that have to be removed or separated, or the properties of recycled plastics may get altered by these substances. Sorting methods involve grinding, abrasion or the use of solvent to remove coatings. Size reduction is used to give uniformity to the particles produced and to liberate different materials. To remove other materials, processes such as magnetic separation, eddy current processing and air classification may be carried out. Identification of various polymers can be done using triboelectric separation techniques (Saini 2018).

Energy recovery from the incineration of waste plastics is viable in principle due to their high heating values being comparable to convention fuels (Panda et al. 2010). High heating values of plastics, often in excess of 40 MJ/kg, are attributed to high C and H contents and low ash content. Heating values for conventional fuels are: natural gas (48 MJ/kg), heating oil (43 MJ/kg), coal (28 MJ/kg), paper and wood (15–16 MJ/kg). Typical heating values of various polymers are: PE (41.8 MJ/kg), PP (30.9 MJ/kg); PVC (13.69MJ/kg); PA(36.76 MJ/kg); PS (38.7 MJ/kg); PET (21.8 MJ/kg); ABS (35.2 MJ/kg); PMMA (24.1 MJ/kg); PC/ABS (29.7 MJ/kg); HIPS (35.2 MJ/kg); PVC/

ABS (14.1 MJ/kg), etc. (Panda et al. 2010, Othman et al. 2008). Public distrust of incineration, however, limits waste plastic to energy technologies due to the production of highly toxic pollutants, dioxins and furans.

2.2 Thermal degradation behaviour

Pyrolytic decomposition of waste polymers into chars, gasoline-based fuels and synthetic gas is one of the main industrial approaches for treating waste polymers. These processes use hydrogen from polymer degradation towards hydrogenation of unsaturated intermediates generated during the thermal cleavage of polymer chains (Qureshi et al. 2020, Ratnasari et al. 2017). Polymer degradation, thermal decomposition, and volatile release can be estimated through mass reduction using thermogravimetric analysis (TGA) and recorded on-line using software. These experiments are performed typically in air, nitrogen and argon atmospheres (Gabbott 2008). Saini (2018) investigated the thermal degradation behaviour of mobile phones WPCB powders using thermo-gravimetric analysis (TGA-DTG); nitrogen, air, and oxygen were used as carrier gases (flow rate: 0.2 L/min). The results are shown in Fig. 1. For N_2 carrier gas, a 20% weight loss was observed in the temperature range of 300–400°C with a further gradual weight loss up to 6% down to 1200°C without much change in slope. A two stage mass loss, i.e., between 300°C to 400°C (10% weight loss) and 600°C to 700°C (~ 6% weight loss), was detected for the Ar gas. Under air, the weight loss between 280°C and 532 °C was about 17%. Net weight loss after heating up to 1200°C was higher (40%) for N_2 as compared to air or argon atmospheres (~ 30%). This study shows that up to 20–30% polymers were present in the waste WPCB powder investigated. This result is in good agreement with published literature (Ortuño et al. 2013, Molto et al. 2009). As per respective DTG plots, various weight losses in TGA plots indicate exothermic processes. While the DTG peak under air was fairly sharp, corresponding peaks under N_2 or Ar atmospheres were somewhat broad. The WPCB powder showed a tendency towards oxidation when heated in air; broad DTG peaks for N_2 and Ar indicate the breaking down of plastics and flame retardant molecules (Font et al. 2011).

Chandrasekaran et al. (2018) carried out TGA investigations on PC/PA and complex mixtures of PC/PA/ABS/ PMMA found in several e-waste streams. A small amount (~ 20 mg) of sample was placed on an alumina crucible pan hanging on a sensitive microbalance. The sample was then heated in the furnace to 700°C at a heating rate of 10°C/min under nitrogen flow. Initial temperatures for 1% weight loss ranged from 156 to 341°C for e-waste mixtures and PC/PA samples. For PC/PA samples, a rapid weight loss was observed in the temperature range 400–500°C with major degradation occurring between 450 and 525°C. Typical char yields at 600°C were found to range between 30–40 wt.%. These results are in good agreement with published literature wherein majority of polymer degradations were found to reach completion by 450–550°C (Apaydin et al. 2014).

Guo et al. (2020) carried out TGA studies on board resin, capacitor scarfskin, solder coating and wire jacket recovered from WPCBs in cathode ray tube television sets. TGA was performed under a helium flow of 30 mL/min within the temperature range 25–650°C at a heating rate of 10°C/min. Onsets of polymer degradation were recorded at 272°C, 272°C, 252°C, and 288°C for board resin, capacitor scarf skin, solder coating and wire jacket, respectively. Corresponding maximum rates of decomposition for board resin, capacitor scarf skin, solder coating and wire jacket were respectively determined as 309°C, 281/451°C, 351°C, and 293/443°C. Single peaks were observed for the board resin and solder coating indicating a one-stage process; double peaks observed for the capacitor scarf skin and wire jacket indicate two degradation stages. While the weight losses were found to range between 32.9% to 77.7%, residual chars were lowest for the board resin (22.3%) and highest for the solder coating (67.1%). HBr and phenol were the key degradation products from board resin, whereas HCl, toluene and benzene were released from capacitor scarfskin.

A large number of pyrolysis and thermal degradation studies on WPCBs have been focussed on the non-metallic fractions obtained after mechanical pulverisation, crushing and separation (Gao

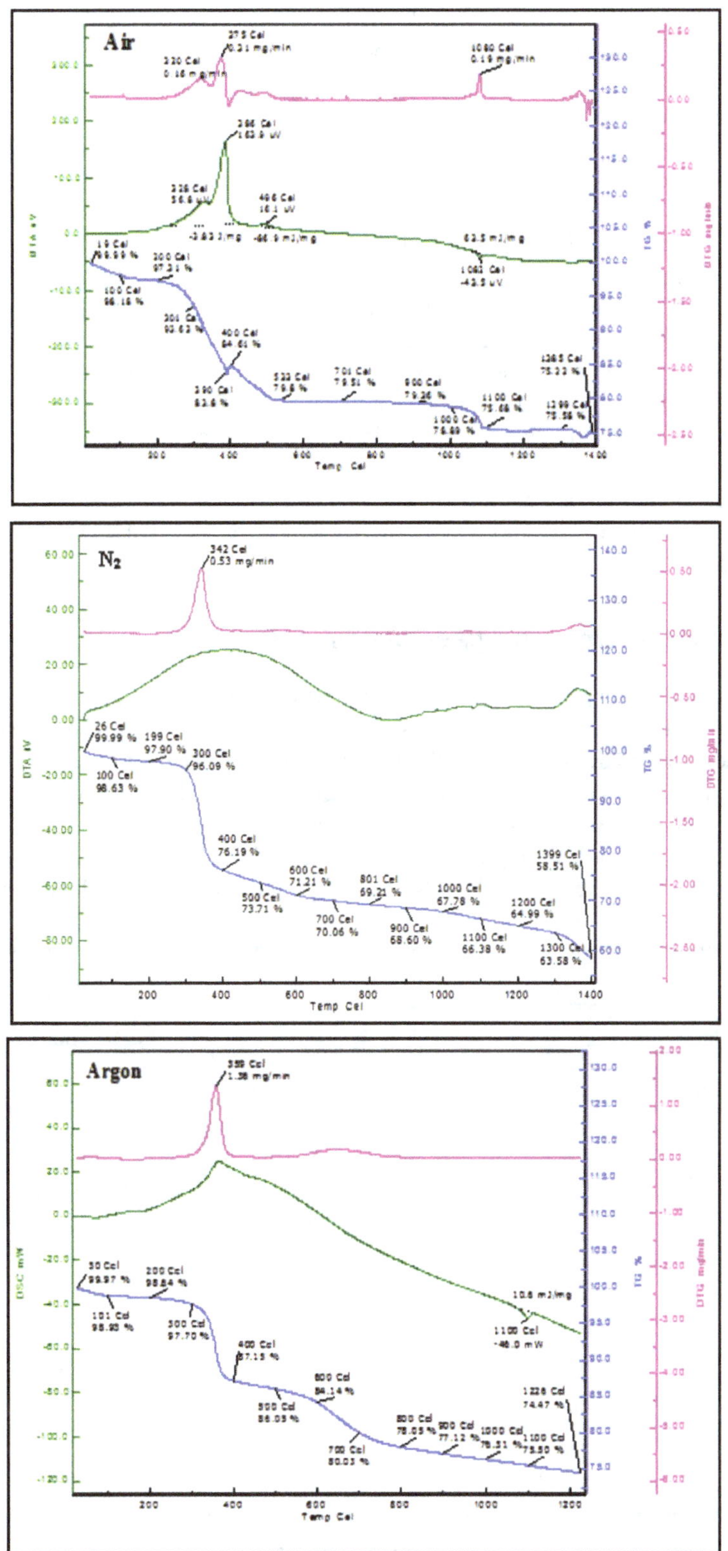

Figure 1. Thermogravimetric analysis (TGA-DTG) of WPCB powders under air, nitrogen and argon as carrier gases (Saini 2017).

et al. 2020, Ma and Kamo 2019, Kumar et al. 2018). However, it is very difficult to completely remove metals from these plastics using physical methods; up to 5% residual metals may still be present. Presence of metals can have a significant and catalytic influence on thermal degradation of organic materials (Chen et al. 2020). The presence of copper was found to promote the cleavage of Cu-Br bonds facilitating the decomposition of brominated epoxy resins (Gao et al. 2021). The presence of Fe was found to catalytically enhance the product yields and debromination performance (Ma and Kamo 2019); catalytic influences of CuO, Fe_2O_3, PbO, ZnO have also been reported (Oleszek et al. 2013).

To reduce local area contamination around e-waste recycling facilities, thermal treatment of WPCBs without prior mechanical treatments and pulverisation is gaining increasing attention (Khanna et al. 2018, Chen et al. 2021). The situation becomes even more challenging with recent trends towards the non-removal of electronic components prior to waste PCB processing (Khanna et al. 2020a). Liu et al. (2021) compared the pyrolysis characteristics, volatile emission, kinetics, and thermodynamic parameters between metal-free leftover pieces from WPCBs and as received PCBs. The TGA analysis showed that the characteristic pyrolysis temperatures were about 15°C higher for metal free boards thereby confirming the catalytic influence of metals on polymer degradation. Other challenges in thermal degradation of waste PCBs include the presence of complex polymer blends, the simultaneous presence of metals and ceramics, forever evolving WPCB compositions and relative proportions.

2.3 Organic bi-products and gaseous emissions

In a study on mobile phones, Saini (2018) characterised various gases evolving during thermal degradation of WPCBs under N_2 atmosphere using gas chromatography-mass spectroscopy (GC-MS) (Fig. 2). A broad peak was detected in the MS-chromatogram in the temperature range 30–600°C with the peak position located in the highest weight loss regime in TGA. This peak can be attributed to the evolution of large number of gaseous species during polymer degradation. The presence of phenol, esters, substituted phenol and bromophenol was detected; bromophenol is commonly used as a flame retardant in circuit boards (Ren et al. 2014). Some of the organics released included phosphonic acid,

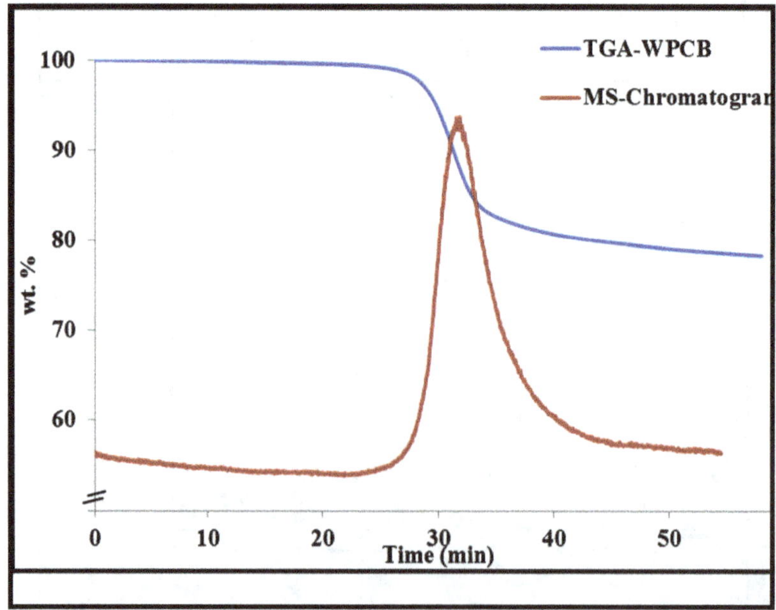

Figure 2. Thermogravimetric analysis of WPCB powders under nitrogen together with MS-chromatogram results on gaseous emissions (Saini 2017).

(p-hydroxyphenyl)-phenol, phenyl phosphoric acid, phenyl ester, 2-vinylfuran, phenyl-beta.-chloro propionate, 4-phenoxy-butanoic acid, (2-butenyloxy)-benzene, (1-methyl-2-propenyl)oxy-benzene, 2-phenoxy-ethanethiol, etc. (Hanxia et al. 2008). Negligible weight loss beyond 600°C indicates almost complete degradation of the polymers present.

During investigations on a variety of polymer based products from e-waste, Guo et al. (2020) reported on emissions of H_2O, CO_2, HBr, phenol, and methyl-phenols from board resins in the temperature range 272–382°C. The release of both H_2O and CO_2 can be attributed to the decomposition of the resin matrix or cellulose. The degradation of brominated flame retardants (BFR) started at about 270°C with the formation of free bromine radicals and led to the formation of HBr, phenols and methyl-phenols; the degradation of HBRs was complete by 310°C. The thermal degradation of capacitor scarfskin was found to take place in two distinct stages. With the key reaction involving the dehydrochlorination of the PVC, first degradation stage (peak temperature 330°C) saw the release of HCl, C_2H_3Cl, SO_2, benzene, and few other volatiles. Formations of benzene, toluene and HCl were seen to continue well into the second stage.

Sahajwalla et al. (2015) have used Fourier transform infrared spectrum (FTIR) spectroscopy to characterize volatile species during TGA of WPCBs under N_2 atmosphere in the temperature range 40–600°C. A small amount of CO_2 (2200–2400 cm^{-1}) and HBr (2400–2800 cm^{-1}) were released at low temperatures. Characteristic FTIR absorption bonds for different species are given in parenthesis. Significant level of volatiles were released after 300°C; these include: phenol (3600 cm^{-1}), bromo-phenol (3650 cm^{-1}), H_2O (3300 cm^{-1}), unsaturated C-H in aromatics (3000 cm^{-1}), CO (2000 cm^{-1}), CO_2 (2400 cm^{-1}), benzene groups (1600 and 1500 cm^{-1}), ketones and aldehydes (1750 cm^{-1}), substituted aromatic groups (between 600–900 cm^{-1}), ether groups (1250 and 1350 cm^{-1}), etc. Only small amounts of CO and CO_2 were detected after 460°C. These results indicate that a wide variety of aromatic volatiles are released during low temperature thermal degradation of WPCB polymers accompanied by weight loss.

3. THERMAL TREATMENT OF E-WASTE

3.1 Pyrolysis and combustion

Pyrolysis represents thermochemical decomposition at high temperatures in the absence of oxygen and in the presence of other gases (N_2, Ar, He, etc.). It may involve changes in chemical composition or thermal decomposition releasing pyrolytic vapours and residual chars. A fraction of vapours may be condensed into liquids and non-condensable gases, e.g., CO, CO_2, H_2, hydrocarbons remain in the gas phase (Xiang et al. 2018). Products of e-waste pyrolysis include gas, liquid biproducts, and carbon residues as chars and their relative proportions depend strongly on process operating conditions. Different types of pyrolysis include slow, fast, intermediate, and catalytic pyrolysis as per the operating conditions (Liu et al. 2015). Slow pyrolysis is typically carried out in the temperature range 300–800°C, slow heating rate 5–7°C/min and long residence time (> 1 hr) for pyrolytic vapours; chars are the main products. In fast pyrolysis, heating rates of 300–1000°C/min are used, and the resulting vapours are removed from the furnace within 0.5–2s; pyrolytic liquids are the main product. The heating rates for intermediate pyrolysis are between 200 to 300°C/min (Bridgwater 2012). Joo et al. (2021) have summarised pyrolysis results on a large variety of e-waste materials investigated by several research groups in a tabular form; data on pure polymers, polymer blends, epoxy and phenolic WPCBs, e-waste powders, computer discs, random access memories, central processing units, etc., was presented.

Zhou et al. (2010) carried out vacuum pyrolysis on two types of WPCBs at 240°C. Yields of pyrolysis gas were found to range between 2.7 to 4.3%. With CO, CO_2, CH_4 as key gaseous constituents, pyrolysis gas could be used for combustion and making the process self-sustained. The liquid yield as pyrolysis oil was found to range between 20 to 27.8%; the solid yield as chars and WPCB residuals ranged between 69.5 to 75%. Quan et al. (2012) carried out high temperature (700–800°C) pyrolysis to recover organic, metals and glass fibres from TV, PC and mobile phones WPCBs. Jie et al. (2010) carried out pyrolysis on WPCBs from computers in the temperature range 300–700°C for 30 minutes and recovered pyro-oils and gaseous products as fuel.

In the Haloclean process, e-waste is first heated to 350°C and then to 450°C under N_2 in a horizontal rotary kiln. Products were in the form of a gas/oil fraction and a solid residue from which metals could be recovered (Schöner et al. 2004). Zhou et al. (2010) conducted a three-step pyrolysis: pyrolysis of WPCBs at 600°C under vacuum, followed by vacuum centrifugal separation of the solder and the remelting of the solder. Flandinet et al. (2012) heated mixtures of WPCBs and molten salts (KOH and NaOH) at 300°C under Ar flow; solid products recovered were in the form of wires and foils and a calcium carbonate/silicate powder. de Marco et al. (2008) heat treated polyethylene (PE) wires containing Al and Cu from e-waste, table phones, mobile phones and WPCBs at 500°C under N_2 atmosphere. Gaseous products were mainly hydrocarbons along with large amounts of CO and CO_2, a liquid fraction of pyrolysis oil, and solid chars. During controlled pyrolysis of WPCBs in the temperature range 750–1550°C, Rajarao et al. (2014) observed that more than 90% of the lead present in WPCBs was lost during heating to 750°C itself; several other metals showed a similar trend. Metals such as Cu, Ag, Au, Pt, Pd, Ni, Se tended to concentrate in the solid phase, and high concentrations of volatile metals Pb, Sb, In and Zn were found in the vapour phase (Khanna et al. 2020b).

Combustion represents thermochemical decomposition at high temperatures in the presence of air and/or oxygen. Watanabe et al. (2008) carried out combustion of WPCBs in a pilot-scale incinerator equipped with a flue gas treatment system. While the formation of brominated, chlorinated and mixed halogenated dioxins and furans were observed, more than 90% were destroyed at the final exit. With most of the harmful dioxins/furans getting destroyed at high temperatures, the recycling of WPCBs should be carried at temperatures above 800°C (Mckay 2002). Ortuno et al. (2014) took ~ 100 gm of waste mobile phone WPCBs and crushed them to fine powders using a vibratory disc mill. A metal free fraction was isolated by treating WPCB powders with dilute aqueous solutions of HCl and H_2O_2 followed by washing and drying the residue. Pyrolysis (N_2) and combustion (air) studies were carried out at 600°C and 850°C. For WPCBs and their metal free fraction, light hydrocarbons methane, ethylene, benzene and toluene were most abundant; generally, higher yields were observed at 850°C. Semi-volatile compounds phenol, benzofuran and styrene were detected in all runs; however, these were found oxidised during combustion at 850°C. Emissions of bromophenol were found to be significantly higher than chlorophenols and chlorobenzenes; highest yields were detected during combustion at 600°C; polybrominated dibenzo-p-dioxins/furans (PBDD/F) emissions were found to be higher than polychlorinated dibenzo-p-dioxins/furans (PCDD/Fs) and dioxin-like polychlorinated biphenyls (DL-PCBs). There was a small influence of metals present as well.

A few pyro-metallurgical processes such as Noranda, Umicore and Rönnskar recycle e-waste under oxidising conditions. In the Noranda process, up to 14% e—waste is charged at 1250°C in up to 39% oxygen atmosphere; the feed material is composed of e-waste and mined copper concentrates. The Rönnskar process does not distinguish between different types of e-waste as long as its copper content is high. High grade copper waste is processed along with copper concentrates, producing a mixed copper alloy in the smelting step. Umicore process treats a feedstock of wastes from the non-ferrous industry, precious metal residues, and up to 10% e-waste; the charge is melted at 1200°C under an oxygen-enriched air atmosphere (Cayumil et al. 2014). While the Rönnskar process is focussed on the recovery of Cu and precious metals, both Noranda and Umicore processes recycle low Cu grade e-waste separating various metallic impurities such as iron, lead and zinc in the form of slags.

The energy from the burning of e-waste plastics gets consumed within the furnace and cannot be extracted as such. Both rotary as well as fluidized bed furnaces have been tested for incinerating e-waste in the laboratory (Woynarowska et al. 2013). WPCBs were heat treated at 840–850°C for 3 min under 6.2% O_2; weight losses up to 21 wt.% were recorded. Chemical processes use end-of-life plastics as feedstock for petrochemical processes, such as in coal-making or as reducing agent in smelting processes, such as steel, which use hydrocarbon sources to allow metal reduction to occur (Kang and Schoenung 2005). Due to high calorific value, plastics release significant amounts of heat energy during combustion (Fisher et al. 2005).

3.2 Open burning

The environmental impact and associated pollution caused by inappropriate processing of e-waste is of great concern especially in the poor and developing countries. Key factors include inadequate technology, processing costs, lack of safety procedures and legislation, as well as transboundary movement of waste electronics from developed nations (Kumar et al. 2017b). Open burning of e-waste is used for several purposes, e.g., component separation, solder recovery from WPCBs, melting plastic components as well as copper recovery from cables (Chan and Wong 2013). Typically practiced in open pits or small workshops, the working environment is below acceptable standards of work place safety and hygiene (Imran et al. 2017). Open burning is one of the most commonly used crude recycling practices, and can release harmful substances, contaminants in the atmosphere, soil and waterways (Cayumil et al. 2016). It is also known to release various metals (Pb, Sn, Ni, As, Hg, etc.), and organo-halogen compounds in the environment. The situation is extreme in Ghana where informal workers, often adolescents and children, work over 10 hr/day incessantly burning wires and cables producing thick black smoke which takes a long time to disperse (Wittsiepe et al. 2015). In south-west Hebron, open burning is used to extract copper from plastic covered wires (Davis and Garb 2015).

Fujimori et al. (2016) investigated the formation of PCDD/Fs, PCBs, PBDD/Fs and mono-brominated PCDD/Fs in soils sampled in open burning sites in Accra, Ghana. Predominant dioxin-related compounds were found to be PBDFs, PCDFs, PCDDs, and DL-PCBs. Chan et al. (2007) determined the concentrations of PBDEs in 30 samples taken from soil, biota and plants from an e-waste recycling site in China. PCBs and PCDD/Fs were also analysed. PBDEs were detected in all soil samples and the highest was up to 789 ng/g dry weight (dw). The toxic equivalency (TEQ) of PCBs and PCDD/Fs detected in e-waste recycling site was significantly high. Biota and plants were also highly contaminated thereby indicating serious environmental problems.

Being highly stable, these dioxins and furans are classified as persistent organic pollutants (POPs) which can travel over long distances and accumulate in the fatty tissue of living species and in food chains (Li et al. 2008, Darnerud et al. 2001). Chlorinated dioxins and furans are known to cause neurologic toxicity, hepatic, dermal, and gastrointestinal issues in humans, reproductive and immunologic toxicity in animals. Dioxin exposure can affect breast milk, placenta and hair, and may cause cancer, immune-toxicity, lethality and reproductive issues (Han et al. 2009). Heavy metals can also get released during open burning as a particulate matter or carbonaceous particles. These are generally carried away in the generated gaseous fraction. The burning of copper wires in the open is likely to produce ~ 100 times more dioxins than burning municipal waste (Gullett et al. 2007). The loss of hazardous/other metals as emissions is also a serious issue, especially during uncontrolled/open burning of e-waste, causing significant environmental pollution.

4. POLLUTANT EMISSIONS

The environmental impact and associated pollution of processing e-waste is of great concern in poor as well as developing regions of the world. Commonly used e-waste recycling techniques such as manual dismantling, chipping, open burning, copper, melting, burning wires, acid and cyanide salt leaching are major sources of pollution (Li et al. 2015, Dwivedy and Mittal 2012). High polymer contents of WPCBs also lead to the release of a range of organic pollutants which may either be present *in-situ* or produced during waste handling/processing. While environmental safeguards and extensive gas scrubbing/filters are integral procedures in developed nations, these are rarely used in informal sectors or developing nations (Chiang and Lin 2014). These activities can cause significant damage to the ecology of the region along with contaminating the local environment near recycling sites.

4.1 Emissions of fine particulates

A major environmental risk during e-waste processing is the release of potentially toxic elements (PTEs) as airborne fine particulates: PM10, PM2.5 and ultrafine particles. The term PTE collectively represents Pb, Zn, Cu, Hg, Ni, Cd, Mo, Cr, Se, As, etc., that can contaminate soils, water and the local environment (Xie et al. 2012). Manual disassembly, removal of electronic components, mechanical shredding, and crushing can release coarse as well as fine particles loaded with PTEs and flame retardants in the atmosphere (Fromme et al. 2016). These may redeposit near the emission sites or be transported over long distances depending on particle sizes, densities and morphologies (de Oliveira et al. 2012).

Ellamparuthy et al. (2017) heat treated WPCBs to 1500°C in a thermal plasma furnace and captured exhaust gases in gas-bags. Airborne particulates captured by the gas-bag filters were analysed in detail. With sizes between 3–200 μm, high concentrations of carbon were noted along with the presence of several metals: Cu, Pb, Sn, Zn, Al, Fe, Ni, Sb, As, Cd, and other elements as well. Extensive gas cleaning systems/filters are rarely used in developing nations or the informal sector due to technological limitations and cost factors.

Using alumina adsorbents, Saini et al. (2017) were able to capture some of the PTEs released during thermal treatment of waste mobile phone WPCB powders. The experimental assembly consisted of WPCB powders on a sample holder and a ceramic crucible with adsorbent powder located downstream in close proximity. Exhaust gases carrying particulates released during thermal treatment (600°C) were passed over the adsorbent powder; N_2 or Ar were used as carrier gases with a flow rate of 0.3L/min. Weight gains up to ~ 17% were observed, thereby indicating a significant capture of particulates in the adsorbent. Elements Cu, Pb, Si, Mg and C were identified as some of the species captured.

Khanna et al. (2020b) have reported an in-depth study to identify key factors influencing particulate generation during thermal processing of WPCBs at 600°C using adsorptive capture of released particulate matter. Using alumina, activated charcoal or silica gel as adsorbents located downstream, a significant proportion of the released particulates (5.3 to 37%) were captured. Adsorbed particulates were present as fines, spheres, clusters, oblongs, larger particles with sizes ranging between nano to over 10 μm. Of the 24 elements initially present in WPCBs, only 14 elements were detected in the captured particulates. Major emissions (up to 400 ppm) were observed for Cu, Pb, Zn and Sn; minor emissions (up to 10 ppm) were recorded for Ni, Mn, Cr, Cd and Ba along with trace levels of Bi, Be and Sb. Although both Al and Fe were present in significant amounts in the WPCBs, these were not detected in the emitted particulates. There was no well-defined correlation between initial elemental concentrations and their corresponding release as particulate matter. While the initial concentration of copper was an order of magnitude higher than Pb, the extent of Cu emissions was up to 24 times smaller. Gas-borne release of various elements was influenced by the melting points, vapour pressures, basic elemental characteristics, local bonding, mechanical strength, densities, initial concentrations, etc. PTE emissions during thermal processing of WPCBs were found to be significantly higher than permissible standards. These results indicate that thermal processing of e-waste should be carried out in closed and/or controlled environments to contain the release of harmful and toxic particulates.

4.2 Dioxin and furan generation

PCDFs and PCDDs are chlorinated, planar, aromatic compounds containing two benzene rings. Dioxins are produced as a by-product from the incomplete combustion of organic materials in the presence of chlorine. A dioxin molecule is bonded by two oxygen atoms, and a furan molecule by a single oxygen atom and a direct bond. There are over 75 different forms of PCDDs and 135 different forms of PCDFs, known as congeners, distinguished by the position and the number of chlorine atoms attached to the two benzene rings. Ritter and Bozzelli (1994) investigated the gas-phase transformation of polychlorinated biphenyls, chlorinated biphenyl ethers and chlorinated dibenzo furans. Hinton and Lane (1991) examined the synthesis of PCDDs in a tubular reactor at 300°C from pentachlorophenol over PCDD extracted fly ashes collected from various incinerators. Fly ash was found to act as a catalyst in producing dioxins; no dioxins were observed in the absence of fly ash. Stockholm Convention 2001 has passed resolutions towards a continual minimization and eventual elimination of dioxins, furans, hexachlorobenzene and polychlorinated biphenyls during combustion and incineration processes (Wielgosinski 2011).

Stewart and Lemieux (2003) investigated the incineration of personal computer motherboards, keyboards, cases, etc., using a pilot-scale rotary kiln incinerator. The concentrations of metals, halogens, volatile and semi-volatile organic products, PCDD/Fs in the flue gas were examined. While significant concentrations of Cu, Pb and Sb were detected, levels of PCDD/Fs were well within regulatory limits. Fujimori et al. (2016) investigated the interaction of metals (Cu, Pb, Zn) and bromine present in e-waste and their impact on the formation of dioxin-related compounds (DRCs) such as PCDD/Fs and DL-PCBs, PBDD/Fs and mono-brominated PCDD/Fs. PTEs Cu, Pb, Zn, and Br present in various e-wastes, wires/cables, plastics, and tires were found to have a strong catalytic influence on the generation of several DRCs.

Duan et al. (2012) created an inventory of PCDD/Fs and PBDD/Fs released during low temperature processing of WPCBs in terms of toxic equivalency quotient (TEQ/kg body weight). During heating of WPCBs to 250°C under air, up to 12 dioxin congeners were detected with a total content of 60,000 ng TEQ/kg. A rapid increase in PBDD/Fs to 160,000 ng TEQ/kg was observed at 275°C under air; however, much lower values (700 ng TEQ/kg) of PCDD/Fs were recorded. These emissions were attributed to the dismantling and thermal processing of WPCBs. Due to incomplete combustion, uncontrolled waste incinerators are among the worst culprits of dioxin release. Ma et al. (2011) investigated the levels of PBDEs and PCDD/Fs in human hair of workers at an e-waste recycling plant in China; observed concentrations of PBDEs (22.8–1020 ng/g dw) were three times higher and PCDD/F (126–5820 pg/g dw were up to 18 times higher than corresponding results from control sites.

Highly stable dioxins can only be destroyed by high temperatures (700–850°C), excess oxygen (~ 6%) or long residence times at high temperatures (> 2 sec). Once dioxins enter the body, these tend to be absorbed in the fat tissue, and get stored in the body for long periods of time (half-life ~ 7−11 years). Dioxins also tend to accumulate in the food chains in the environment; higher an animal in the food chain, higher the concentration of dioxins. These compounds can be found in dairy products, meat, fish and shellfish, some soils as well as sediments. Hazardous e-waste cannot be easily disposed without contaminating the environment and should be destroyed in special facilities through high temperature processing or incineration.

5. ENERGY SAVINGS THROUGH MATERIAL RECOVERY

An alternate and indirect source of energy from e-waste can be through recovering metals and saving energies consumed in the primary production of various e-waste metals. Energy is consumed at all stages in the production of primary metals including direct energy consumption in mining, milling, smelting, refining stages as well as indirect consumption in the production of reagents and electricity. Embodied energy during various metal production stages represents the total sum of direct and indirect energies. The embodied energy of common metals can vary widely typically from ~ 20 MJ/kg for lead and steel, ~ 50 MJ/kg for copper, ~ 211 MJ/kg for aluminium and ~ 160 MJ/kg for gold (Norgate and Rankin 2002). Based on data for 2019, estimated volumes and worth of key e-waste elements are summarized in Table 1. The potential worth of ~ 25 Mt of selected metals was ~ 57 billion USD with significant contributions from copper, gold and palladium (Ghimire and Ariya 2020).

Copper is one of the main metallic constituent (up to 20 wt.%) of WPCBs. The recovery of copper from e-waste has been investigated extensively using hydrometallurgical and pyro-metallurgical techniques (Schluep et al. 2009, Cui and Zhang 2008). During high temperature (800°C; 10 min) pyrolysis of WPCBs, Cayumil et al. (2018) were able to extract high purity (> 95%) copper foils in a single process step; impurity concentrations of Pb and Sn were as low as 0.3% and 3.1%. Low levels of impurities present in the recovered copper are likely to reduce its further refining/processing for commercial applications. All aspects of primary production of copper require energy in the form of electricity, diesel, natural gas, coal/coke, etc. In 1977, the energy consumption for producing electrode grade copper was ~ 85 million BTU/ton; corresponding energies for iron/steel, lead and zinc were 15 million BTU/ton, 24 million BTU/ton and 64 million BTU/ton, respectively (Rötzer and Schmidt 2020).

Total number of mobile phones, PC and laptops along with the number of lithium ion batteries produced annually have been tabulated in Table 2 (Hagelüken and Korti 2010). Concentrations (in tonnes) of Cu, Au, Ag, Pd and Co present in these WPCBs have also been provided. In addition, the total volumes of these metals produced through primary production and their fractional usage in electrical and electronic equipment (EEE) have also been listed; nearly 20% of Pd and Co metals produced were used in the EEE sector. Energies used for metal extraction from scrap recycling/secondary production are a small fraction of the corresponding energies for primary production, e.g., Al (5–10%), Cu (15%), steel (20–40%), Pb (35%), Ni (10%) and Zn (25–40%) (Kusik and Kenahan 1978). Secondary recovery of these metals from WPCBs can therefore result in significant savings in energy consumption in the mining and metal industries sectors.

Table 1. Estimated amounts (kilo-tonnes) of various metals present in e-waste globally (2019) along with their estimated worth (million USD). (Ghimire and Ariya 2020).

S. No.	Metal	Amount in e-waste (Kt)	Estimated worth (in Million USD)
1.	Aluminium	3046	6,062
2.	Copper	1808	10,960
3.	Cobalt	13	1,036
4.	Iron/steel	20, 466	24,645
5.	Gold	0.2	9,481
6.	Silver	1.2	579
7.	Palladium	0.1	3,532

Table 2. Total number of various electronic devices produced per annum (2009), concentrations of key valuable metals present and electronics share of the total mined product (Hagelüken and Korti 2010).

Electronic devices	Numbers produced (million units/a)	Concentration (tonnes)				
		Cu	Au	Ag	Pd	Co
Mobile Phones	1600	14,000	38	400	14	
PC/Laptop	350	175,000	78	350	28	
Li-ion batteries	1480					17,800
Urban Mine	(tonnes/a)	16 million	2,500	22,200	200	88,000
Electronics share	fraction	1%	5%	3%	21%	20%

6. PROSPECTS OF ENERGY RECOVERY FROM E-WASTE

As landfilling and incineration have become less accepted and more expensive, recycling complex hazardous electronic waste is no longer a choice but an essential requirement. Atmospheric pollution due to burning and dismantling activities at e-waste recycling sites appear to be the main cause of occupational and secondary exposure. Although the polymer content (the main source of energy) in WPCBs can range up to 30%, direct extraction of energy from WPCBs through polymer degradation is fraught with hazards in the form of harmful emissions of noxious gases, dioxins, furans, the release of volatile metals and fine particulates. However, indirect savings on energy consumption through metal recovery from e-waste and associated savings on the primary production of electrical and electronic metals appears to hold the key for recovering energy from e-waste. Recycling and environmentally sustainable management of electronic waste is expected to contribute significantly towards green technologies, circular economy and resource recovery.

ACKNOWLEDGEMENTS

Authors gratefully acknowledge Department of Science and Technology (DST) India for providing financial support for this project under the Australia-India Strategic Research Fund Round 6.

REFERENCES

Apaydin-Varol, E., Polat, S. and Putun, A.E. 2014. Pyrolysis kinetics and thermal decomposition behaviour of polycarbonate-a TGA-FTIR study. Thermal Science 18(3): 833–842.

Bridgwater, A.V. 2012. Review of fast pyrolysis of biomass and product upgrading. Biomass Bioenergy 38: 68–94.

Cayumil, R., Ikram-Ul-Haq, M., Khanna, R., Saini, R., Mukherjee, P.S., Mishra, B.K., Sahajwalla, V. 2018. High Temperature Investigations on optimising the recovery of copper from waste printed circuit boards. Waste Management 73: 556–565.

Cayumil, R., Khanna, R., Rajarao, R., Ikram-Ul-Haq, M., Mukherjee, P.S. and Sahajwalla, V. 2016. Environmental Impact of Processing Electronic Waste—Key Issues and Challenges. In edited by Florin-Constantin Mihai. E-Waste in Transition—From Pollution to Resource, 9-35, IntechOpen Ltd. Croatia.

Cayumil, R., Khanna, R., Ikram-Ul-Haq, M., Rajarao, R., Hill, A. and Sahajwalla, V. 2014. Generation of copper rich metallic phases from waste printed circuit boards. Waste Management, 34(10): 1783–1792.

Chan, J.K.Y. and Wong, M.H. 2013. A review of environmental fate, body burdens, and human health risk assessment of PCDD/Fs at two typical electronic waste recycling sites in China. Science of the Total Environment 463–464: 1111–1123.

Chan, J.K., Xing, G.H., Xu, Y., Liang, Y., Chen, L.X., Wu, S.C., Wong, C.K., Leung, C.K. and Wong, M.H. 2007. Body loadings and health risk assessment of polychlorinated dibenzo-p-dioxins and dibenzofurans at an intensive electronic waste recycling site in China. Environmental Science & Technology 41: 7668–7674.

Chen, T., Yu, J., Ma, C., Bikane, K. and Sun, L. 2020. Catalytic performance and debromination of Fe-Ni bimetallic MCM-41 catalyst for the two-stage pyrolysis of waste computer casing plastic. Chemosphere 248: 125964.

Chen, Y., Yang, J., Liang, S., Hu, J., Hou, H., Liu, B., Xiao, K., Yu, W. and Deng, H. 2021. New insights into the debromination mechanism of non-metallic fractions of waste printed circuit boards via alkaline-enhanced subcritical water route. Resources Conservation and Recycling 165: 105227.

Chandrasekaran, S.R., Avasarala, S., Murali, D., Rajagopalan, N. and Sharma, B.K. 2018. Materials and energy recovery from e-waste plastics. ACS Sustainable Chemical Engineering 6(4): 4594–4602.

Chiang, H.-L. and Lin, K.-H. 2014. Exhaust constituent emission factors of printed circuit board pyrolysis processes and its exhaust control. Journal of Hazardous Materials, 264, 545-551.

Cui, J. and Roven, H.J. 2011. Electronic Waste. pp. 281–296. *In*: Letcher, V. (ed.). Waste: A handbook for management. Academic Press, Boston.

Cui, J. and Zhang, L. 2008. Metallurgical recovery of metals from electronic waste: A review. Journal of Hazardous Materials 158: 228–256.

Darnerud, P. O., Eriksen, G. S., Jóhannesson, T., Larsen, P.B. and Viluksela, M. 2001. Polybrominated diphenyl ethers: occurrence, dietary exposure, and toxicology. Environmental Health Perspectives 109: 49–68.

Davis, J.-M. and Garb, Y. 2015. A model for partnering with the informal e-waste industry: rationale, principles and a case study. Resources Conservation and Recycling 105: 73–83.

de Marco, I., Caballero, B.M., Chomón, M.J., Laresgoiti, M.F., Torres, A., Fernández, G. and Arnaiz, S. 2008. Pyrolysis of electrical and electronic wastes. Journal of Analytical Applied Pyrolysis 82: 179–183.

de Oliveira, C.R., Bernardes, A.M. and Gerbase, A.E. 2012. Collection and recycling of electronic scrap: A worldwide overview and comparison with the Brazilian situation. Waste Management 32(8): 1592−1610.

Duan, H., Jinhui Li, J., Liu, Y., Yamazaki, N. and Jiang, W. 2012. Characterizing the emission of chlorinated/brominated dibenzo-*p*-dioxins and furans from low-temperature thermal processing of waste printed circuit board. *Environmental Pollution* 161: 185−191.

Dwivedy, M. and Mittal, R. 2012. An investigation into e-waste flows in India. Journal of Cleaner Production 37: 229−242.

EIU. 2015. Global e-waste systems: Insights for Australia from other developed countries. A report for ANZRP by the *Economist Intelligence Unit* (EIU) February 2015.

Ellamparuthy, G., Mukherjee, P.S., Khanna, R., Jayasankar, K., Cayumil, R., Ikram-Ul-Haq, M., Sahajwalla, V. and Mishra, B.K. 2017. Environmental impact of recycling electronic waste using thermal plasma: in-depth analysis of aerosol particulates captured in gas filters. Current Environmental Engineering, 4 (3), 169−176.

Fisher, M.M., Mark, F.E., Kingsbury T., Vehlow J. and Yamawaki, T. 2005. Energy recovery in the sustainable recycling of plastics from end-of-life electrical and electronic products. pp. 83–92. *In*: Electronics and the Environment, 2005. Proceedings of the 2005 IEEE International Symposium on, 16–19 May 2005.

Flandinet, L., Tedjar, F., Ghetta, V. and Fouletier, J. 2012. Metals recovering from waste printed circuit boards (WPCBs) using molten salts. Journal of Hazardous Materials 213–214: 485–490.

Font, R., Moltó, J., Egea, S. and Conesa, J.A. 2011. Thermogravimetric kinetic analysis and pollutant evolution during the pyrolysis and combustion of mobile phone case. Chemosphere 85: 516–524.

Forti, V., Baldé, C.P., Kuehr, R. and Bel, G. 2020. The Global E-waste Monitor 2020: Quantities, flows and the circular economy potential. United Nations University (UNU)/United Nations Institute for Training and Research (UNITAR)—co-hosted SCYCLE Programme, International Telecommunication Union (ITU) & International Solid Waste Association (ISWA), Bonn/Geneva/Rotterdam.

Fromme, H., Becher, G., Hilger, B. and Völkel, W. 2016. Brominated flame retardants–Exposure and risk assessment for the general population. International Journal of Hygiene Environmental Health 219(1): 1−23.

Fujimori, T., Itai, T., Goto, A., Asante, K.A., Otsuka, M., Takahashi, S. and Tanabe, S. 2016. Interplay of metals and bromine with dioxin-related compounds concentrated in e-waste open burning soil from Agbogbloshie in Accra, Ghana. Environmental Pollution 209: 155–163.

Gabbott, P. 2008. Thermogravimetric Analysis. *In*: Principles and Applications of Thermal Analysis. Oxford, UK: Blackwell Publishing, pp. 87–118.

Gao, R., Liu, R., Zhan, L., Guo, J., Zhang, J. and Xu, Z. 2021. Catalytic effect and mechanism of coexisting copper on conversion of organics during pyrolysis of waste printed circuit boards. Journal of Hazardous Materials 403: 123465.

Gao, R., Zhan, L., Guo, J. and Xu, Z. 2020. Research of the thermal decomposition mechanism and pyrolysis pathways from macromonomer to small molecule of waste printed circuit board. Journal of Hazardous Materials 383: 121234.

Ghimire, H. and Ariya, P.A. 2020. E-Wastes: Bridging the Knowledge Gaps in Global Production Budgets, Composition, Recycling and Sustainability Implications. Sustainable Chemistry 1: 154–182.

Greenfield, A. and Graedel, T.E. 2013. The omnivorous diet of modern technology. Resources Conservation and Recycling 74: 1–7.

Gerardo, R.F., Flores, A., Nikolic, S. and Mackey, P.J. 2014. SASMELT™ for the Recycling of E-Scrap and Copper in the U.S. Case Study Example of a New Compact Recycling Plant. Journal of The Minerals, Metals & Materials Society 66: 823–832.

Global E-Waste Monitor. 2017. Quantities, Flows, and Resources. accessed April 30, 2021.

Gullett, B.K., Linak, W.P., Touati, A., Wasson, S.J., Gatica, S. and King, C.J. 2007. Characterization of air emissions and residual ash from open burning of electronic wastes during simulated rudimentary recycling operations. Journal of Material Cycles and Waste Management 9: 69–79.

Guo, J., Xiaomei, L., Shufei, T., Ogunseitan, O.A. and Xu, Z. 2020. Thermal degradation and pollutant emission from waste printed circuit boards mounted with electronic components. Journal of Hazardous Materials 382: 121038.

Guo, J., Tang, Y. and Xu, Z. 2009. Wood Plastic Composite Produced by Nonmetals from Pulverized Waste Printed Circuit Boards. Environmental Science and Technology 44(1): 463–8.

Gurgul, A., Szczepaniak, W. and Zabłocka-Malicka, M. 2017. Incineration, pyrolysis and gasification of electronic waste. E3S Web of Conferences 22, 00060 (2017) DOI: 10.1051/e3sconf/20172200060. *ASEE17*

Hagelüken, C. and Art, S. 2006. Recycling of e-scrap in a global environment –chances and challenges: Umicore precious metals refining. Indo-European Training Workshop, May 2006, New Delhi.

Hagelüken, C. and Corti, C. 2010. Recycling of gold from electronics: Cost-effective use through 'Design for Recycling'. Gold Bulletin 43: 209–220.

Hall, W.J. and Williams, P.T. 2007. Separation and recovery of materials from scrap printed circuit boards. Resources Conservation & Recycling 51: 691–709.

Han, W.L., Feng, J.L., Gu, Z.P., Chen, D.H., Wu, M.H. and Fu, J.M. 2009. Polybrominated diphenyl ethers in the atmosphere of Taizhou, a major e-waste dismantling area in China. Bulletin of Environmental Contamination and Toxicology 83: 783–788.

Hanxia, L., Qunfang, Z., Yawei, W., Qinghua, Z., Zongwei, C. and Guibin, J. 2008. E-waste recycling induced polybrominated diphenyl ethers, polychlorinated biphenyls, polychlorinated dibenzo-*p*-dioxins and dibenzo-furans pollution in the ambient environment. Environmental International 34(1): 67–72.

Herat, S. and Pariatamby, A. 2012. E-waste: a problem or an opportunity? Review of issues, challenges and solutions in Asian countries. Waste Management & Research 30(11): 1–17.

Hinton, W.S. and Lane, A.M. 1991. Characterisation of municipal solid waste incinerator fly ash promoting the formation of polychlorinated dioxins. Chemosphere 22(5–6): 473–483.

Imran, M., Haydar, S., Kim, J., Awan, M.R., Bhatti, A.A. 2017. E-waste flows, resource recovery and improvement of legal framework in Pakistan. Resources Conservation and Recycling 125: 131–138.

Jie, G., Min, X., Sheng, W.C., Zhou, M., Li, S. 2010. The compounds study of waste PC main-board pyrolysis. Advanced Materials Research 113–114: 887–891.

Joo, J., Kwon, E.E. and Lee, J. 2021. Achievements in pyrolysis process in E-waste management sector. Environmental Pollution 287: 117621.

Kang, H.Y. and Schoenung J. 2005. Electronic waste recycling: A review of U.S. infrastructure and technology options. Resources Conservation & Recycling 45(4): 368–400.

Kaya, M. 2016. Recovery of Metals and Non-metals from Electronic Waste by Physical and Chemical Recycling Processes. Waste Management 57: 64–90.

Khanna, R., Mukherjee, P.S. and Park, M. 2020a. A critical assessment on resource recovery from electronic waste: Impact of mechanical pre-treatment. Journal of Cleaner Production 268: 122319.

Khanna, R., Saini, R., Park, M., Ellamparuthy, G., Biswal, S.K. and Mukherjee, P.S. 2020b. Factors influencing the release of potentially toxic elements (PTEs) during thermal processing of electronic waste. Waste management 105: 414–424.

Khanna, R., Ellamparuthy, G., Cayumil, R., Mishra, S.K. and Mukherjee, P.S. 2018. Concentration of rare earth elements during high temperature pyrolysis of waste printed circuit boards. Waste Management 57: 602–610.

Kumar, A., Holuszko, M.E. and Janke, T. 2018. Characterization of the non-metal fraction of the processed waste printed circuit boards. Waste Management, 75: 94–102.

Kumar, A., Choudhary, V., Khanna, R., Cayumil, R., Ikram-ul-Haq, M. and Sahajwalla, V. 2017a. Recycling polymeric waste from electronic and automotive sectors into value added products. Frontiers of Environmental Science & Engineering 11(5): 1–10

Kumar, A., Holuszko, M. and Espinosa, D. 2017b. E-waste: An overview on generation, collection, legislation and recycling practices. Resources. Conservation & Recycling, 122, 32–42.

Kusik, C.L. and Kenahan, C.B. 1978. Energy use patterns for metal recycling. Information Circular 8781. US Bureau of Mines: Washington DC.

Kwon, E.E., Kim, S. and Lee, J. 2019. Pyrolysis of waste feedstocks in CO_2 for effective energy recovery and waste treatment. Journal of CO_2 Utilisation 31: 173–180.

Li, J., Zeng, X., Chen, M., Ogunseitan, O.A. and Stevels, A. 2015. Control-alt-delete": Rebooting solutions for the e-waste problem. Environmental Science & Technology 49(12): 7095–7108.

Li, Y.M., GuiBin, J., YaWei, W., Pu, W. and Qing Hua, Z. 2008. Concentrations, profiles and gas-particle partitioning of PCDD/Fs, PCBs and PBDEs in the ambient air of an e-waste dismantling area, southeast China. Chinese Science Bulletin 53: 521–528.

Liu, J., Jiang, Q., Wang, H., Li, J. and Zhang, W. 2021. Catalytic effect and mechanism of *in-situ* metals on pyrolysis of FR4 printed circuit boards: Insights from kinetics and products. Chemosphere 280: 130804

Liu, W.-J., Jiang, H. and Yu, H.-Q. 2015. Development of biochar-based functional materials: toward a sustainable platform carbon material. Chemical Reviews 115: 12251–12285.

Ma, C. and Kamo, T. 2019. Enhanced debromination by Fe particles during the catalytic pyrolysis of non-metallic fractions of printed circuit boards over ZSM-5 and Ni/SiO_2-Al_2O_3 catalyst. Journal of Analytical Applied Pyrolysis 138: 170–177.

Ma, J., Cheng, J., Wang, W., Kunisue, T., Wu, M. and Kannan, K. 2011. Elevated concentrations of polychlorinated dibenzo-p-dioxins and polychlorinated dibenzofurans and polybrominated diphenyl ethers in hair from workers at an electronic waste recycling facility in Eastern China. Journal of Hazardous Materials 186: 1966–1971.

Molto, C.J., Font, R., Galvez, A. and Conesa, J.A. 2009. Pyrolysis and combustion of electronic wastes. Journal of Analytical Applied Pyrolysis 84: 68–78.

McKay, G. 2002. Dioxin characterisation, formation and minimisation during municipal solid waste (MSW) incineration: a review. Chemical Engineering Journal 86(3): 343–368.

Norgate, T.E. and Rankin, W.J. 2002. The role of metals in sustainable development. pp. 49–55. *In*: Green Processing. Australasian Institute of Mining and Metallurgy: Melbourne.

Oleszek, S., Grabda, M., Shibata, E. and Nakamura, T. 2013. Distribution of copper, silver and gold during thermal treatment with brominated flame retardants. Waste Management 33: 1835–1842.

Ortuño, N., Juan A. Conesa, J.A., Moltó, J. and Font, R. 2014. Pollutant emissions during pyrolysis and combustion of waste printed circuit boards, before and after metal removal. Science of The Total Environment 499: 27–35.

Ortuño, N., Moltó, J., Egea, S., Font, R. and Conesa, J.A. 2013. Thermogravimetric study of the decomposition of printed circuit boards from mobile phones. Journal of Analytical Applied Pyrolysis 103: 189–200.

Othman, N., Basri, N.E.A., Yunus, M.N.M and Sidek, L.M. 2008. Determination of physical and chemical characteristics of electronic plastic waste (Ep-Waste) resin using proximate and ultimate analysis method. ICCBT 2008 - D - (16): 169–180.

Panda, A.K., Singh, R.K. and Mishra, D.K. 2010. Thermolysis of waste plastics to liquid fuel. A suitable method for plastic waste management and manufacture of value added products—A world prospective. Renewable and Sustainable Energy Reviews 14: 233–248.

Priya, A., Hait and S. 2018. Comprehensive characterization of printed circuit boards of various end-of-life electrical and electronic equipment for beneficiation investigation. Waste Management 75: 103–123.

Qureshi, M.S., Oasmaa, A., Pihkola, H., Deviatkin, I., Tenhunen, A., Mannila, J. and Laine-Ylijoki, J. 2020. Pyrolysis of plastic waste: Opportunities and challenges. Journal of Analytical and Applied Pyrolysis 152: 104804

Quan, C., Li, A. and Gao, N. 2012. Research on pyrolysis of PCB waste with TG-FTIR and Py-GC/MS. Journal of Thermal Analysis and Calorimetry 110: 1463–1470.

Rajarao, R., Sahajwalla, V., Cayumil, R., Park, M. and Khanna, R. 2014. Novel Approach for Processing Hazardous Electronic Waste. *In*: L. Beesley (ed.). Procedia Environmental Sciences 21: 33–41.

Ratnasari, D.K., Nahil, M.A. and Williams, P.T. 2017. Catalytic pyrolysis of waste plastics using staged catalysis for production of gasoline range hydrocarbon oils. Journal of Analytical and Applied Pyrolysis 124: 631–637.

Ren, Z., Xiao, X., Chen, D., Bi, X., Huang, B., Liu, M., Hu, J., Peng, P.A., Sheng, G. and Fu, J. 2014. Halogenated organic pollutants in particulate matters emitted during recycling of waste printed circuit boards in a typical e-waste workshop of Southern China. Chemosphere 94: 143–150.

Ritter, E.R. and Bozzelli, J.W. 1994. Pathways to chlorinated dibenzodioxins and dibenzofurans from partial oxidation of chlorinated aromatics by OH radical: Thermodynamics and kinetic insights. Combustion Science & Technology 101: 153–69.

Rötzer, N. and Schmidt, M. 2020. Historical, current, and future energy demand from global copper production and impact on climate change. Resources 9: 44.

Sahajwalla, V., Cayumil, R., Khanna, R., Ikram-ul-Haq, M., Rajarao, R., Mukherjee, P.S. and Hill, A. 2015. Recycling Polymer rich Waste Printed Circuit Boards at High Temperatures: Recovery of Value-added Carbon Resources. Journal of sustainable metallurgy 1: 75–84.

Saini, R. 2018. Novel approaches for processing waste printed circuit boards. PhD Dissertation. Indian Institute of Technology, Roorkee. India.

Saini, R., Khanna, R., Dutta, R.K., Cayumil, R., Ikram-Ul-Haq, M., Agarwala, V., Ellamparuthy, G., Jayasankar, K., Mukherjee, P.S. and Sahajwalla, V. 2017. A novel approach for reducing toxic emissions during high temperature processing of electronic waste. Waste management 64: 182–189.

Santella, C., Cafiero, L., De Angelis, D., La Marca, F., Tuffi, R., Vecchio Ciprioti, S. 2016) Thermal and catalytic pyrolysis of a mixture of plastics from small waste electrical and electronic equipment (WEEE). Waste Management, 54, 143–152.

Schluep, M., Hagelüken, C., Magalini, R., Maurer C., Meskers C., Mueller E., Wang F. 2009) Recycling - From E-waste to resources. UNEP United Nations Environment Programme, Berlin, Germany.

Schöner, J., Hornung, A., Sagi, S., & Seifert, H. 2004) Post-treatment of pyrolysis residues of WEEE. Recovery of precious metals. In: International Conference on Incineration and Thermal Treatment Technologies, Phoenix, Arizona, May 10–14 2004.

Statista, (2017) https://www.statista.com › Technology & Telecommunications › Consumer electronics (accessed 24 March 2021).

Stewart, E. S., Lemieux, P. M. 2003) Emissions from the incineration of electronics industry waste. Proceedings of Electronics and the Environment, IEEE International Symposium, 271-275.

Watanabe, M., Kajiwara, N., Takigami, H., Noma, Y. and Kida, A. 2008. Formation and degradation behaviors of brominated organic compounds and PCDD/Fs during thermal treatment of waste printed circuit boards. Organohalogen Compounds 70: 78–81.

Wielgosinski, G. 2011. The reduction of dioxin emissions from the processes of heat and power generation. Journal of the Air & Waste Management Association 61(5): 511–526.

Wittsiepe, J., Fobil, J.N., Till, H., Burchard, G.D., Wilhelm, M. and Feldt, T. 2015. Levels of polychlorinated dibenzo-p-dioxins, dibenzofurans (PCDD/Fs) and biphenyls (PCBs) in blood of informal e-waste recycling workers from Agbogbloshie, Ghana, and controls. Environmental International 79: 65–73.

Woynarowska, A. 2014. Termiczna utylizacja odpadów elektronicznych w reaktorze fluidyzacyjnym (Politechnika Krakowska, Kraków, 2014.

Woynarowska, A., Baron, J., S. Kandefer, S. and Zukowski, W. 2013. Combustion of electronic waste in the reactor with bubbling fluidised bed. Przemysel Chemiczny 92: 997–1005.

Wu, B., Chan, Y.C., Middendorf, A., Gu, X. and Zhong, H. 2008. Assessment of toxicity potential of metallic elements in discarded electronics: a case study of mobile phones in China. Journal of Environmental Sciences 20: 1403–1408.

Xiang, Z., Liang, J., Morgan, H.M., Liu, Y., Mao, H. and Bu, Q. 2018. Thermal behavior and kinetic study for co-pyrolysis of lignocellulosic biomass with polyethylene over cobalt modified ZSM-5 catalyst by thermogravimetric analysis. Bioresource Technology 247: 804–811.

Xie, X., Pan, X. Z. and Sun, B. 2012. Visible and near-infrared diffuse reflectance spectroscopy for prediction of soil properties near a Copper smelter. Pedosphere 22: 351–366.

Zhou, Y., Wu, W. and Qiu, K. 2010. Recovery of materials from waste printed circuit boards by vacuum pyrolysis and vacuum centrifugal separation. Waste Management 30: 2299–2304.

6

ECO-DESIGN STRATEGIES FOR RECYCLING OF E-WASTE

Debarun Banerjee,[1] *Nidhi Kushwaha,*[1] *Ejaz Ahmad,*[2,*]
Shireen Quereshi[3] and *K.D.P. Nigam*[3,4]

1. INTRODUCTION

The rapid growth of society and urbanization have caused an exponential increase in electrical and electronic equipment (EEE) and accessories usage. As a result of frequent change and improvement in technologies, these EEE materials have a very short lifespan. Thus, they end-up as discarded waste electrical and electronic materials commonly referred to as e-waste. Indeed, unprecedented growth in e-waste generation has been witnessed since the last decade as shown in Fig. 1. Overall, a 25% rise in global e-waste generation has been recorded in just six years from 2014 (44.4 million metric tonnes) to 2020 (55.5 million metric tonnes) which is further projected to increase to 74.7 million metric tonnes by 2030 (Forti et al. 2020). It is noteworthy that approximately 80% of all e-waste generated worldwide is exported to developing nations as a part of a profitable business model (Zhang et al. 2012). However, most of these nations lack high-end technologies for e-waste disposal and rely on conventional technologies, thereby risking the environment and human health. For example, e-waste contains 3% pollutants that may move to soil and environment due to lack of suitable e-waste disposal and conversion technologies (Cherrington and Makenji 2019). Primarily, e-waste pollutants comprise hazardous heavy metals such as cadmium (Cd), lead (Pb), nickel (Ni), mercury (Hg), arsenic (As), and organic compounds such as polybrominated diphenyl ether (PBDE), polychlorinated biphenyl (PCB), brominated flame retardant (BFR), besides several other pollutants which are primarily responsible for water, and soil pollution (Guo et al. 2016, Huang et al. 2014, Zhang et al. 2012, Bhutta et al. 2011).

On the contrary, e-waste contains several high-value products also such as noble and rare earth metals including gold (Au), silver (Ag), platinum (Pt), palladium (Pd), gallium (Ga), lanthanum (La),

[1] Department of Fuel, Minerals and Metallurgical Engineering, Indian Institute of Technology (ISM), Dhanbad-826004, India.
[2] Department of Chemical Engineering, Indian Institute of Technology (ISM), Dhanbad-826004, India.
[3] Department of Chemical Engineering, Indian Institute of Technology Delhi-110016, India.
[4] Chemical Engineering School, The University of Adelaide, Australia.
* Corresponding author: ejaz@iitism.ac.in

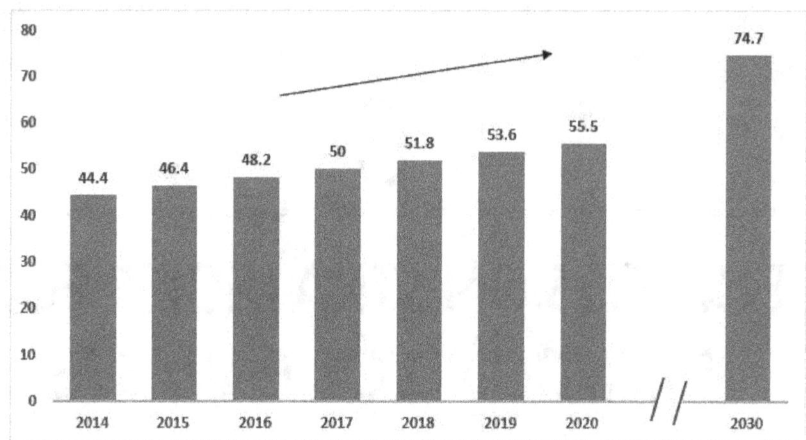

Figure 1. Rise in global e-waste generation (million metric tonnes) (Forti et al. 2020).

yttrium (Y), cerium (Ce), europium (Eu), terbium (Tb) and other metals (Akram et al. 2019, Tan and Li 2019). It is reported that up to 98% of noble metals can be recovered from e-waste which makes it an exciting feedstock for industries pertinent to noble metals production and usage (Huisman and Magalini 2007). Therefore, recovery of these metals from e-waste would have a significant societal, environmental and economic impact. Indeed, safe and efficient disposal of e-waste and recovery of metals from them will help to achieve the United Nations sustainable development goals for economic and environmental prosperity of the world (Forti et al. 2020).

It is worth mentioning here that landfilling and incineration are widely used processes for e-waste disposal. As a result of landfilling at unprotected sites, leaching of heavy metals and emission of greenhouse gases occur which eventually lead to soil, water and air contamination (Li et al. 2009, Spalvins et al. 2008). In addition, the plastic content of e-waste degrades at a negligible rate or may not degrade at all in dry conditions (Mundada et al. 2004). In another conventional process, incineration of e-waste is carried out at low (600–800°C) and high temperatures (> 1200°C) for its conversion (Ghimire and Ariya 2020). Notably, low-temperature incineration requires less energy input as gases produced from e-waste incineration can be recycled and reused for the heating purpose to sustain the reaction at a high rate. However, it emits extremely toxic chemicals including carbon monoxide, methane, toluene, phenol, polycyclic aromatic hydrocarbon (PAH), chlorinated and brominated complexes (Kaya 2016). In contrast, high-temperature incineration causes less emission of carbon monoxide and methane; however, NO_x emission is significantly increased at such a high temperature (Ning et al. 2017). Moreover, chlorinated and brominated complexes also get converted to highly toxic chlorine (Cl_2), hydrogen chloride (HCl), bromine (Br_2), and hydrogen bromide (HBr). Additionally, bottom ash resulted from incineration also contains Pb, Cd, Hg which are considered highly hazardous (Ni et al. 2012).

It is noteworthy that the nature and constituents of e-waste differentiate it from municipal solid waste (MSW) and industrial wastes (Garlapati 2016). Therefore, e-waste should be preferably processed separately than using conventional disposal processes of waste plastics such as landfilling, incineration, and dumping. It is well-known that developed nations export e-waste to maintain environmental balance and earn monetary benefit, whereas developing nations import e-waste to earn profit from it. Often, most e-waste is processed in the unorganized informal sector where metals and valuable products are recovered whereas other leftovers are dumped again. As a result, it creates substantial environmental hazard which directly opposes sustainable development goals set by United Nations. These drawbacks of conventional disposal processes of e-waste necessitate the development of environment-friendly recycling processes.

2. E-WASTE CLASSIFICATION

According to directive 2012/19/EU of the European Parliament and the council, e-waste can be broadly classified into ten categories that include large household appliances, small household appliances, information technology (IT) and telecommunication equipment, consumer equipment and photovoltaic panels, lighting equipment, electrical and electronic tools, toys, leisure and sports equipment, medical devices, monitoring and control instruments, and automatic dispensers as shown in Fig. 2 (Directive 2012/19/EU). The large household appliances e-waste include a discarded washing machine, cooker, dish washer, microwave, air conditioner whereas small household appliances include the discarded vacuum cleaner, coffee machine, iron, and toaster etc. Similarly, IT and telecommunications equipment include discarded computers, laptops, telephones, printers, typewriters, calculators, and fax machines. In contrast, consumer equipment and photovoltaic panels include radio, television, video and still camera photovoltaic panel e-waste. Lamps and bulbs come under the lighting equipment e-waste category whereas sewing, welding, soldering, drilling machine come under electrical and electronic tools e-waste. Similarly, toys, leisure and sports equipment include video games, sports equipment with electrical and electronic parts, and other leisure equipment e-waste whereas medical devices include cardiology, dialysis, radiotherapy equipment, ventilators, and freezers e-waste. The monitoring and control instruments include smoke detector, thermostat, weighing, and measuring equipment whereas automatic dispensers include e-waste of automatic dispensers for money, and cold drinks, etc.

Besides 2012/19/EU directives, there are several other classifications for e-waste. For example, Wang et al. (2012a) regrouped e-waste into 58 sub-categories based on primary category (from directive 2012/19/EU), collection category, and the average weight of the appliance. These sub-

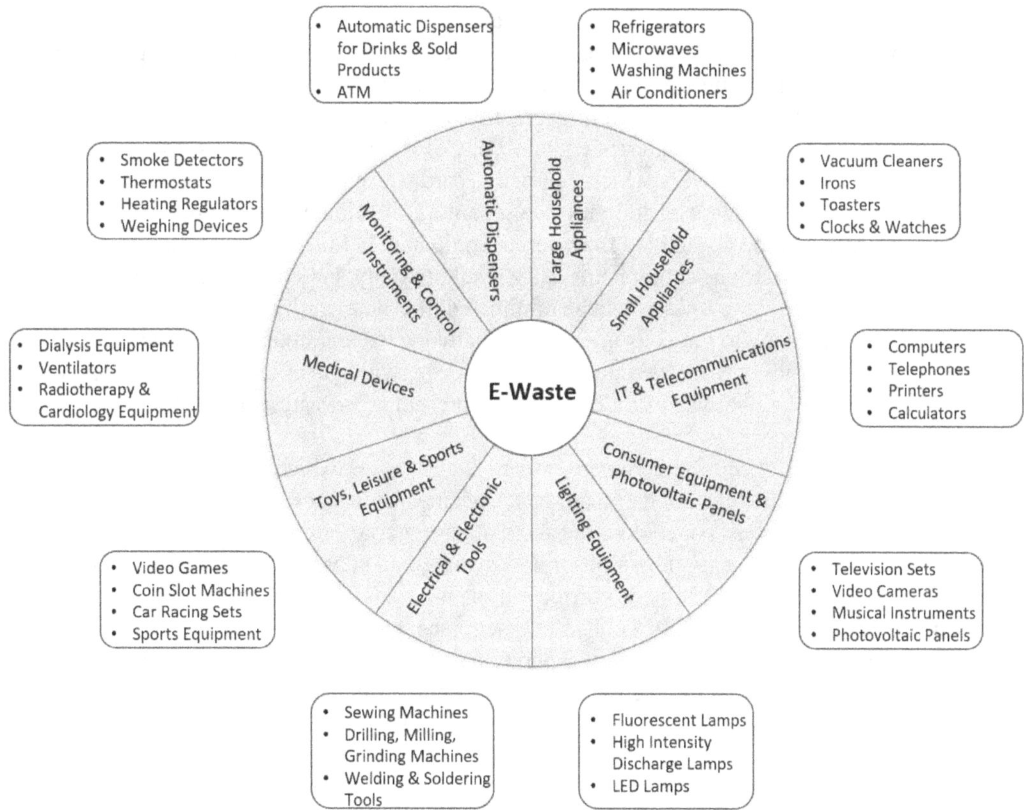

Figure 2. E-waste classification (Directive 2012/19/EU).

categories are uniquely represented using different codes provided by United Nations University and have been classified into successfully classified approximately 900 types of e-waste. It is based on compositional analysis, which is pivotal behind waste management as collection and treatment of e-waste varies according to its composition. In yet another classification for e-waste reported recently in 2018, it has been clustered into six groups based on the recycling process of the associated e-waste (DIRECTIVE 2012/19/EU) (Chagnes et al. 2016). These six groups are named as (i) temperature exchange equipment that includes refrigerators, air conditioners, freezers, and heat pumps, (ii) screens, monitors, and equipment containing screens having a surface > 100 cm^2, (iii) lamps, (iv) large equipment (any external dimension > 50 cm) such as washing machine, photovoltaic panels, cloth dryers, and other appliances, (v) small equipment (all external dimensions ≤ 50 cm) that includes vacuum cleaners, microwaves, clocks and watches, (vi) small IT and telecommunication equipment such as mobile phones, routers, personal computers, and pocket calculators. Notably, appliances in each of these groups have different composition, therefore different management processes need to be applied.

3. CONVENTIONAL APPROACH FOR E-WASTE RECYCLING AND THEIR SHORTFALLS

Conventional e-waste recycling consists of steps such as collection, separation of hazardous components, pre-processing, and end-processing. In this regard, the collection of e-waste is the first step which involves direct interaction between consumers and the technical e-waste recycling system. Thus, it acts as a bridge between societal and technical systems for e-waste recycling including environment, economy, technology, infrastructure and other aspects (Wang et al. 2012b). E-waste collected with MSW significantly enhances the probability of contamination due to toxic and hazardous substances present in e-waste. Therefore, it is advised to collect hazardous e-wastes including gas discharge lamps, cathode ray tubes, batteries, printed circuit boards, and external electric cables separately (DIRECTIVE 2012/19/EU). It is worth noting that different types of e-wastes contain different types of toxic materials. For example, batteries contain Pb, Cd, Hg, and other toxic substances whereas printed circuit boards contain BFR which can cause cardiovascular disease, renal problems, and cancer (Chen et al. 2012). In case of difficulties in separate collection of e-waste from MSW, at least these components should be dismantled from parent equipment before going into further processing to ensure removal of hazardous components. As a result, toxicity and contamination caused due to hazardous materials present in e-waste can be minimized. It is noteworthy that reusable components of e-waste can be fed back to the consumer supply network for maximum utilization of resources and energy-saving purposes whereas other parts of e-waste, which is at its end-of-life, can be sent for recycling. Especially, noble and rare earth metals present in e-wastes should be recovered to a maximum extent.

In general, e-waste collected undergo pre-processing processes, which include size reduction and sequential sorting for homogenization of materials. Both mechanical and manual processes can carry it out. Firstly, metals, plastics, paper, ceramics, and other materials are separated by manual sorting followed by size reduction using a hammer mill, crushing and subsequent screening via vibrating screens. Post this, material separation is carried out via the principle of magnetic, eddy current and density difference (Khaliq et al. 2014). The magnetic separator allows the separation of ferrous metals from bulk, whereas the eddy current separation process separates non-ferrous metals such as aluminum (Al), copper (Cu), and others. Choice of magnet in magnetic separator largely depends on energy consumption and efficacy of separation. For example, electromagnets usually possess greater magnetic power than permanent magnets but at the expense of higher energy consumption. Therefore, permanent magnets in series may be an excellent option to achieve targeted separation (Gramatyka et al. 2007). Similarly, plastics, which are the primary non-metal component of e-waste, are separated via a density separation process. Plastic materials usually have less density;thus, they can be separated from bulk under greater buoyancy force. Eventually, output fractions of pre-processing step are fed to end-processing units for treatment. For example, metal fractions are processed via a metallurgical route to further separate and recover individual metals. In this regard, pyrometallurgical, hydrometallurgical, and biometallurgical processes are widely reported for metal recovery from e-waste whereas plastic from e-waste is recovered using mechanical and thermo-chemical processes.

3.1 Pyrometallurgical process of metal recovery

In this process, non-metals such as plastics get converted into various chemicals and combustible gases using thermal energy, thereby leaving behind non-combustible metals as residue that can be recycled further. In general, the pyrometallurgical process is used to recover non-ferrous metals which could not be recuperated in pre-processing. Although the pyrometallurgical process does not necessarily require a pre-processing step, pre-processing of e-waste can certainly increase the extent of separation. The pyrometallurgical process is further divided into several categories and the process of metal recovery can be carried out using incineration, smelting, and other processes. Indeed, incineration is the most elementary pyrometallurgical process. At a temperature > 600°C, incineration of e-waste results in the burning of a combustible fraction such as plastics to increase metal concentration in residue. It is noteworthy that the standalone operation of incineration is insufficient for metal recovery from e-waste. Therefore, this process is carried out as a pre-treatment to remove combustible fraction and as a source of energy required for subsequent steps.

Smelting is also among the most practiced processes for metal recovery from e-waste. In a typical smelting operation, e-waste is fed into a molten bath at a very high temperature (~ 1250°C) and subsequently oxidized which removes combustible plastics and metallic residue in the form of slag and desired metals are obtained in the form of matte (Cui and Zhang 2008). Firstly, e-waste is agitated vigorously by supercharged air that contains significantly higher oxygen (up to 39%) than atmospheric air which allows complete combustion of e-waste plastics. The complete combustion of e-waste plastic components results in less pollutant emission than incomplete combustion. For example, emission of carbon monoxide is negligible for complete combustion whereas incomplete combustion results in significantly higher carbon monoxide emission. Moreover, ozone gas is formed at such a high temperature in the smelting process, which also aids in the oxidation of metals. As a result of smelting, the metallic fraction is transformed into a molten state comprising slag and matte (Wang et al. 2019). Notably, slag and matte are separated from each other under density differences (Desai et al. 2016). Matte settles down at the bottom of the smelter, whereas the slag phase floats over matte due to lower density. The metal residue slag contains oxides of Pb, zinc (Zn), and iron (Fe) that can be further mined after cooling for more significant metals recovery. The molten black Cu matte contains Cu, Ni, selenium (Se), tellurium (Te), and noble metals such as Au, Ag, Pt, and Pd which are fed to a converter for upgradation. The resultant blister Cu (98.5% pure) from the converter is obtained (Fig. 3) which is further passed into an anode furnace to obtain almost 99.5% pure anode Cu. Further electro-refining of the anode may yield ~ 100% pure Cu. Similar processes can be employed for Pb smelting as well. Despite yielding highly pure Cu, there exist economic challenges with the smelting process as it requires high energy input which increases operating cost. Moreover, this process requires high metal concentration in feed for adequate recovery and metal losses due to transfer in the slag are high. Also, the combustion of e-waste plastics emits toxic chemicals including halogenated compounds which are of great environmental concern.

Other pyrometallurgical processes in practice include pyrolysis and the sintering process. Pyrolysis is carried out at an elevated temperature (> 450°C) and an inert gas environment. Under such conditions, e-waste undergoes primary and secondary thermal decomposition which yields char consisting of metals and glasses that undergo further separation for metal recovery. Similarly, sintering

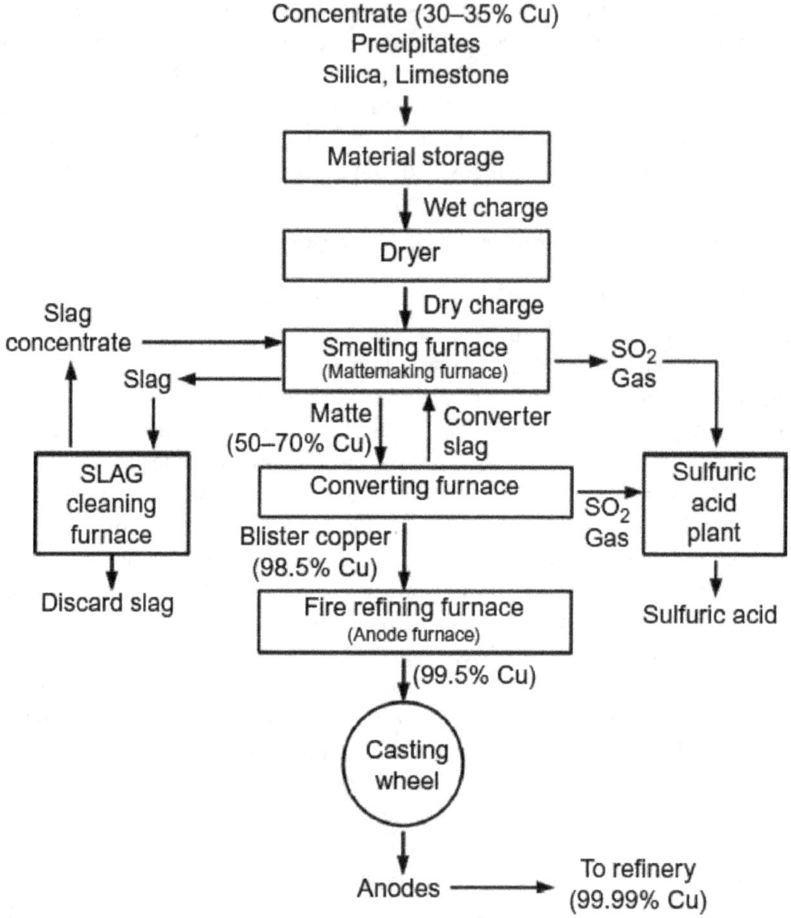

Figure 3. Flowchart of Cu smelting. Reproduced from Ref. (Moats and Davenport 2014) with permission.

is also performed to recover metals and glasses from e-waste by thermal treatment (1450°C) in a furnace (Karaahmet and Cicek 2019). In this process, e-waste feed is heated up to the melting point of the metal and glass fractions, which are separated and undergo subsequent quenching and solidification. However, these direct thermal treatment processes result in the emission of toxic compounds such as bromides and chlorides, which cause severe environmental contamination. Moreover, metals also tend to form alloys in the molten state which results in more incredible difficulty and cost in separation and purification of the metals. Therefore, although the pyrometallurgical process is very much famous for metal recovery from e-waste streams, its viability from an environmental and economic point of view is yet to be established.

3.2 Hydrometallurgical process of e-waste recycling

Based on the limitations of the pyrometallurgical process, several other processes have been explored as well for efficient and environmentally benign metal recovery from e-waste. In this regard, the hydrometallurgical process has gained considerable attention from the research community owing to its high control over the process, less environmental footprint and lower operating cost. In this process, metals from e-waste are recovered by leaching with the help of a suitable lixiviant (leachant) followed by subsequent purification and separation of the target metals. Most common metals recovered from

Table 1. Recoverable metals from different e-waste through hydrometallurgical route.

Source e-waste	Metals recovered
Printed Circuit Board	Au, Ag, Pt, Pd
Liquid Crystal Display	In
Lithium-ion Battery	Co
Ni-Metal Hydride Battery	La, Ce, Nd, Pr, Sm
Light Emitting Diode	Ga
Fluorescent Lamp	Eu, Y
Magnet	Fe, Nd, Pr, Dy

different types of e-waste using the leaching process are given in Table 1. The most commonly used leachants to recover these metals belong to the mineral acid family which includes HCl, sulfuric acid (H_2SO_4), and nitric acid (HNO_3) (Sethurajan et al. 2019). HCl is the most effective mineral acid when compared with H_2SO_4 and HNO_3 in terms of individual performance (Sakultung et al. 2007). One possible reason for the higher activity of HCl can be attributed to its higher dissociation constant (10^6) than H_2SO_4 (10^3) and HNO_3 (28) which allows a better dissolution of metals. Another possible reason could be activation energies of leaching reaction which is lowest for HCl (28.33kJ/mol) followed by H_2SO_4 (30.1-41.4 kJ/mol) and HNO_3 (52.3 kJ/mol) (Nayl et al. 2017, Shuva and Kurny 2013, Lee and Rhee 2003). However, more comprehensive practice is to mix HCl and HNO_3 in a 3:1 molar ratio to form aqua regia due to operational challenges.

Nevertheless, one persistent problem with conventional acids is their low selectivity towards a particular metal. Therefore, several other leach ants have also been explored. For example, cuprous chloride (CuCl) and ammonium sulfate $(NH_4)_2$ SO_4 are used for selective leaching of Ag from metals mixture. One possible reason could be attributed to the high solubility of Ag than other metals in these leachants at lower redox potential (Lister et al. 2014). It also indicates that a cascade leaching process can be designed to recover individual metals from a metal mixture having Ag. For example, Ag from the noble metal mixture can be selectively recovered using cuprous chloride or ammonium sulfate as leachant. Post these, each metal from the mixture can be recovered selectively one by one using different leachants. However, these inorganic leachants emit a considerable amount of pollutants such as NO_x, SO_x and Cl_2. Thus, several organic acids such as acetic acid, oxalic acid, succinic acid, tartaric acid, and citric acid have been explored as replacements of inorganic lixiviants (Chen et al. 2019, Jadhav et al. 2017, Li et al. 2015, Zeng et al. 2015, Saidan et al. 2012). In particular, organic acids are found very effective for indium (In) leaching from the liquid crystal display screen, cobalt (Co) leaching from a lithium-ion battery and rare earth elements such as neodymium (Nd), praseodymium (Pr), samarium (Sm), dysprosium (Dy) from batteries and magnets (Cui et al. 2019, Gergoric et al. 2018). Besides, several other reagents are used as leaching agents for metal recovery such as cyanides which are highly efficient leachants for Au recovery. However, cyanides rapidly decompose under sunlight and at a pH ≤ 10 to form highly volatile and toxic hydrogen cyanide (HCN). Indeed, the poisonous nature of cyanides limits its application as lixiviant.

Recently, thiosulfate, thiourea and halides other than chloride have been proposed as alternative leaching agents to extract noble metals, especially Au to overcome the challenges associated with earlier processes (Xu et al. 2017, Birloaga and Vegliò 2016, Gurung et al. 2013). In this process, metals are reduced by thiosulfate and thiourea to form intermediate complexes which are then required to undergo further processing to yield pure metals. These reactions are highly dependent on reactant concentration and reaction pH. Therefore, optimization of the process in terms of operating parameters is crucial for the efficient recovery of desired metals. Notably, thiourea has lone pair of electrons which allows the facile formation of complexes with metals and thus, enhanced leaching (Akcil et al. 2015, Birloaga et al. 2014). However, consumption of thiourea is high due to the rapid rate of oxidation which incurs a higher cost. On the contrary, thiosulfate is relatively economical than thiourea but its reaction rate and leaching performance are inferior. Additionally, iodides have been examined as leaching agent which is highly selective towards Au recovery (Konyratbekova et al. 2015). In contrast, despite having less corrosiveness and environmental footprint than cyanides and chlorides, iodides also suffer from drawbacks such as high consumption and thereby considerable operating cost (Ghosh et al. 2015). Similarly, organic leaching agent glycine has been found effective in Co leaching from lithium-ion batteries. In contrast, ionic liquid betainium bis (trifluoromethylsulfonyl) imide has been found effective for Eu and Y recovery from fluorescent lamps and Nd and Dy from scrap magnets (Nayaka et al. 2016, Dupont and Binnemans 2015a, b). Therefore, there exists a possibility to design a cascade-type system for individual metals recovery using different leachants in each step.

It is noteworthy that metals extracted in the leachate are not directly usable due to complex formation or dissolved in solution. Therefore, different processes are employed after leaching to recover metals from leachate. In this regard, chemical precipitation is a widely used process for metal recovery. In this process, metals are precipitated in the form of hydroxide, carbonate, sulfide or other salts in the presence of an appropriate reagent. Thus, solution pH has a profound effect on the extent of precipitation. However, the high amount of sludge formation limits this recovery process application at a large scale. Therefore, other processes such as adsorption and solvent extraction have emerged as an alternative to the chemical precipitation process. In adsorption, a suitable adsorbent generally with high surface area and porosity captures metal particles or adsorbates onto its surface pores and after specific treatment, adsorbate particles get desorbed from the adsorbent surface. Activated carbon, thioamide nanofiber, porphyrin polymer, and several other adsorbents have shown promising results in metal recovery from leachates (Hong et al. 2020, Li et al. 2013, Zhang et al. 2004). Similarly, solvent extraction or liquid-liquid extraction process requires a non-polar organic solvent which can selectively separate the target metal (s). The lixiviant used in the preceding leaching operation has an immense effect on the efficiency of the solvent extraction step. For example, HCl as leachant showed higher metal recovery than HNO_3, and H_2SO_4 was used as leaching agent (Krištofová et al. 2016). Different organic solvents such as tributyl phosphate, cyanex-272, cyanex-921 and others have also been used to extract metals frome-waste derived leachates (Correa et al. 2018, Jorjani and Shahbazi 2016, Alguacil et al. 1994). In addition, several other processes such as ionexchange, cementation, and electrowinning are also employed for metal recovery from leachate solution (Baniasadi et al. 2021, Mohebbi et al. 2019, Pereira et al. 2018, Innocenzi et al. 2017). Nevertheless, these recent developments and promising results need to be studied further for metal recovery from leachate.

3.3 Biometallurgical process of e-waste recycling

High energy associated with the pyrometallurgical process and contamination from various chemicals in the hydrometallurgical process is some of the limitations that make these processes environmentally unattractive. Therefore, the scientific community is also exploring biometallurgical recycling processes for metal recovery from e-waste. Biometallurgical processes also offer several advantages such as low operating cost, less environmental contamination, and less sludge generation (Marra et al. 2018). Moreover, these processes can possess high selectivity towards target metal(s) depending upon microbes used because of the interaction between metal and microorganisms. Biometallurgical metal recovery is carried out primarily via two routes which include bioleaching and biosorption. The operation of this process is similar to the hydrometallurgical process; however, the difference lies in the nature of the reagent. In the hydrometallurgical process, different chemicals are used for leaching and metal recovery whereas biometallurgical processes use microbes such as thermophilic, acidophilic, mesophilic, cyanogenic bacteria, etc., for recovery of metals.

Bioleaching can be further subdivided into two types, i.e., direct and indirect bioleaching. In the direct process, microbes directly act by oxidizing minerals and dissolve metals whereas the indirect approach involves generating some oxidizing agent by microbes that leaches desired metals.

However, in the actual scenario, both of these approaches intermingle and contribute to the overall leaching of metals. In general, metals such as Au, Ni, Zn, Al, Pb, and Cu can be recovered by the bioleaching process (Hong and Valix 2014, Yang et al. 2014). In contrast to bioleaching, biosorption is a physicochemical process where sorption between metal ions from e-waste and surficial charged species of microorganism occurs. A critical advantage of this process is its ability to utilize a wide variety of microbes including bacteria, yeasts, fungi, and algae either in living or dead form. This liberty is experienced because biosorption can proceed with or without microbe cell metabolism. For dead microbes, no cell metabolism occurs; thereby metal ions get adsorbed onto microbe cell walls, similar to the operation of a chemical adsorbent. From a mechanistic point of view, biosorption can proceed via different pathways such as chelation, micro precipitation, electrostatic interaction, and ion exchange process. Accordingly, different metals such as Cu, Au, Pt, and Pd have been successfully recovered from different e-waste streams by biosorption over the years (Das 2010). Despite showing huge promises, biometallurgical recycling processes are relatively slow and have difficulty controlling secondary reactions. Furthermore, these processes are yet to be tested in metal recovery from complex e-waste streams where multiple metal interactions are present. Overall, biometallurgical processes have great potential to be a game-changer for sustainable e-waste recycling in the future, subject to further research.

3.4 Recycling of plastics from e-waste

It is imperative to recycle non-metallic fractions or plastics along with metals. However, there is a fundamental difference between metals and plastics recycling. For example, metallurgical processes focus primarily on recovery of metals either in pure or compound form and plastic component is incinerated or not recovered. In contrast, plastics recycling processes possess the liberty to recover chemical components of plastics and convert them to a high-value product and energy (Sahajwalla and Gaikwad 2018). Recycling of plastics is carried out primarily via two routes, namely mechanical and chemical recycling. Mechanical recycling is employed to separate and reuse e-waste plastics in the form of some low-value products. In addition to manual sorting and separation processes discussed in the previous sections, several other mechanical recycling processes are also used to sort homogeneous waste plastics, which include flotation, air table, hydrocyclone, centrifugal, jig, optical and electrostatic sorting (Cherrington and Makenji 2019). All these sorting processes (except optical and electrostatic sorting) work based on the density difference between polymer particles. Optical sorting is a sensor-based operation that includes X-ray fluorescence, X-ray transmission, Infrared spectroscopy and other processes. In contrast, electrostatic sorting works based onvariation in electrical conductivity of different polymers (Maisel et al. 2020). The extra emphasis on sorting is particularly crucial to prevent dispersion and resulting contamination from BFR-containing waste plastics. Nevertheless, it is challenging to separate BFR-containing plastics via these sorting operations completely; thus, a manual sorting process for mechanical recycling becomes necessary (Ma et al. 2016).

3.4.1 Mechanical recycling

After sorting waste plastics from e-waste, different recycling processes are employed, such as agglomeration, extrusion, and injection molding. In agglomeration, flakes and films of polymers are subjected to densification by energy input. The energy can be provided in the form of pressure, temperature and/or rotation which creates a frictional heat in polymer particles that allow them to agglomerate together and undergo subsequent pelletizing. Instruments such as disc compactor, pot agglomerator are widely used and considered among highly efficient polymer densifiers. Similarly, the extrusion of e-waste plastics is another recycling process that melts the polymer into a continuous, viscous fluid via thermal and mechanical energy. The molten plastic is then passed onto an appropriate die which provides desired shape to it. Eventually, recycled e-waste plastic is converted to different products such as plastic sheets, pipes, railings, coatings, frames, and others based on the choice of die. Interestingly, the e-waste plastic extrusion process allows feed in pellet or granular form which is the output of the agglomeration process. Therefore, these two processes can be used to prepare high-value products from plastic flakes and films. Nevertheless, these processes are limited to produce only simple linear 2D plastic products. On the contrary, the injection molding process can produce comparatively complex 3D structures with discontinuity such as plastic combs, chess pieces, crates, and other items. In this process, the energetically excited polymer is injected into a mold whose movement is controlled via a clamp. Afterward, the polymer with the desired shape in the mold is cooled and ejected. However, injection molding has disadvantages of high capital and design cost. Therefore, further research is needed in this area for the optimization of this process.

3.4.2 Thermo-chemical recycling

As mentioned earlier, sorting is an essential step for mechanical recycling because recycled e-waste plastics are directly used to prepare ready-to-use products. However, only a narrow spectrum of e-waste can be recycled using the mechanical recycling process. Thus, different thermo-chemical recycling processes such as pyrolysis, co-pyrolysis, catalytic pyrolysis, and supercritical fluid treatment are employed to convert e-waste plastic into value-added products and fuel. Notably, pyrolysis is a thermo-chemical process that converts feed e-waste plastics into solid char, oil, and combustible gaseous products via a cracking reaction under an inert atmosphere. Polymers of e-waste plastic undergo a depolymerization reaction in which high molecular weight polymers convert into low molecular weight compounds and hydrocarbons. Pyrolysis process efficiency is affected by several parameters including temperature, heating rate, and vapor residence time which play a crucial role in end products distribution and quality. For example, higher temperature and heating rate yield less char, higher liquid and gaseous products. Also, pyrolysis is subdivided into different categories such as slow, conventional, and fast pyrolysis based on operating temperature (Ma et al. 2016). Besides operating temperature, composition and structure of e-waste plastic have an immense effect on end products' distribution. For example, high impact polystyrene (HIPS) yields more oil than acrylonitrile butadiene styrene (ABS) when pyrolyzed (Miskolczi et al. 2008). One possible reason can be attributed to more excellent thermal stability of ABS polymer chains than HIPS because of which the degree of volatilization of ABS molecules is less. Nevertheless, pyrolysis oil yield and energy can be improved by adding a fraction of pure polymers or biomass which is often referred to as co-pyrolysis. Pure polymer generally possesses a very high calorific value whereas biomass has high hydrogen content which results in a positive effect arising from synergy with e-waste plastics (Kushwaha et al. 2022). Furthermore, biomass addition causes maximum Br_2 transfer to the solid residue, thereby yielding high purity liquid and gaseous fuel having minor fractions of brominated compounds. Nevertheless, even the slight presence of Br_2 in the liquid product is major challenge as it can cause environmental contamination at the time of utilization of oil as fuel. Similarly, fast pyrolysis as an alternative to co-pyrolysis also allows the transfer of less than 0.2 wt.% Br_2 to product oil when ABS polymer was used as feed (Jung et al. 2012).

Interestingly, slow pyrolysis also causes the lowest Br_2 content (0.05 wt.%) in light oil products for HIPS as reactant (Miskolczi et al. 2008). The possible reason for such drastic variation in Br_2 content depending on the feedstock can be an exciting area of research. Noteworthy that despite less Br_2 content in liquid products, it cannot be used for commercial applications which necessitate the development of technologies capable of reducing Br_2 content further in the liquid product. In this context, the catalytic pyrolysis of e-waste plastic is suggested as an alternative process that uses a suitable catalyst to reduce operating temperature, thereby lowering the residence time and an upgrade in end-product quality. In general, catalytic pyrolysis has been carried out by different research groups primarily considering two aspects. The first one being catalytic cracking of e-waste plastic to high-value hydrocarbons, whereas the other aspect involves dehalogenation of e-waste plastics that primarily includes dechlorination and debromination. Accordingly, different types of catalysts such as zeolites, silica composites, metal catalysts and several other catalysts have been examined for catalytic cracking of the polymers (Ma et al. 2016, Muhammad et al. 2015). It is noteworthy that zeolites are the most popular catalyst used in catalytic cracking of e-waste plastics because of their high selectivity for end products and exceptional properties. However, attributes differ primarily from one type of zeolite to another; thus, choosing a suitable catalyst variant can be critical to achieving maximum selectivity for the desired product. Also, it is well known that Br_2 and Cl_2-containing chemicals are the reasons behind pollution caused by e-waste plastics. Therefore, catalysts such as alumina, activated carbon, silica, and metal salts have been widely used to degrade these pollutants from e-waste. Nonetheless, a practical approach would most likely involve both of these processes together in line with principles of sustainable development and green chemistry which is a part of eco-design strategies discussed in subsequent sections.

Besides pyrolysis, hydrothermal processes such as supercritical fluid treatment have also emerged as a viable process for disintegrating e-waste plastics. The hydrothermal process can process feedstocks with relatively high moisture content. Supercritical fluids can dissolve halogens by converting them to their respective acids; thus, end products are primarily halogen-free fuels. These acids are neutralized *in-situ* by the addition of a suitable base in the system. For example, sodium hydroxide (NaOH) and calcium hydroxide ($Ca(OH)_2$) are used as effective neutralizers of acids containing hazardous Br_2 and antimony (Onwudili and Williams 2009). Furthermore, supercritical fluids also possess several other advantages such as low viscosity, high thermal stability, and high diffusivity which are required for e-waste plastic processing. However, the use of supercritical fluids causes severe corrosion in the reactor, which eventually affects the longevity of the reactor. Moreover, supercritical fluids result in high energy consumption due to very harsh operating conditions.

4. MAJOR E-WASTE RECYCLING PLANTS IN INDIA AND WORLDWIDE

As mentioned earlier, developed nations export a very high quantity of e-waste to developing countries. Therefore, it is of importance for developing nations to upscale e-waste recycling capacity to meet the increasing quantity of generated e-waste. For example, consolidated data of the Central Pollution Control Board (CPCB), India, shows that 349,154.6 metric tons of e-waste were recycled per annum throughout India by 138 registered companies in 2014 (Ari 2016). Interestingly, the total e-waste recycling capacity of India has reached approximately 1.07 million tons per annum with an increase in the total registered number of companies to 400 in 2021 (List of Authorized E-Waste Dismantler/Recycler 2021). Among all these companies, E-Parisaraa Private Limited of Bengaluru, Karnataka, is the first e-waste recycling company in India which currently has an installed capacity of 8820 metric tons per annum. The e-waste recycling technology of this plant involves different steps such as manual dismantling, size reduction, density separation, and metal recovery (Ari 2016). Ash recyclers, located in Bengaluru, is another e-waste recycling company of similar age and working methodology as E-Parisaraa. However, the capacity of this plant is limited to only 120 metric tons per annum. On the contrary, Ariglonton Information System Pvt. Ltd. located in Ghaziabad, Uttar Pradesh, India, has an installed e-waste recycling capacity of 40000 metric tons per annum which is the highest among all companies in India. It is needless to say that India is still very far from meeting demands even after a significant escalation in e-waste recycling capacity. The worldwide scenario of e-waste recycling is also quite similar.

The consolidated data of the European Electronics Recyclers Association (EERA) shows a recycling capacity of around 2.2 million tons per annum, whereas the quantity of total e-waste generation across Europe is more than 9 million tons per annum (Recyclers—EERA). Hefty Metals (Bulgaria), Glencore (Switzerland), Müller-Guttenbrunn Group (Austria), E-Reciklaza (Serbia), Jacob Becker (Germany), Umicore (Belgium), Jacomij Electronics Recycling (The Netherlands), and several other e-waste recycling companies are members of EERA. Austria-based Andritz group, which is one of the largest technology companies worldwide, also carries out e-waste recycling as a part of their activities (RECYCLING: RECOVER RAW MATERIALS AND MAINTAIN VALUES 2020). The overall process involves size reduction via shredding before and after magnetic separation of the e-waste stream. The recycling process mainly divides entire metal components into ferrous and non-ferrous fractions. The company focuses on obtaining refurbished printed circuit boards, capacitors, transformers, batteries, and metals such as Cu, and Al. However, the recycling process does not aim towards precious and rare earth metal recovery to best of our knowledge.

The establishment of new facilities and scaling up of existing recycling plants is vital. In this regard, recently in 2019, Enviroserve established one of the world's most extensive integrated waste recycling facilities in Dubai with a total capacity of 100,000 metric tons per annum in which approximately 39000 metric tons per annum e-waste is processed (Sutton 2019). Thus, consistent and sustainable development of the e-waste recycling sector can overcome the issues associated with ever-increasing e-waste generation.

5. NEED FOR ECO-DESIGN STRATEGIES

Different conventional processes for e-waste recycling are discussed in the previous sections. It is well understood that primarily these processes emphasize on recovery of e-waste constituents and/or the reuse of products after necessary modification. Metals and plastics are the two most important constituents of e-waste. Metals are recovered via different metallurgical processes, whereas plastics undergo mechanical and thermo-chemical recycling. Mechanical recycling of plastics is a primary and secondary recycling process depending on the comparative value of the final product concerning the original item whereas thermo-chemical recycling strategies are tertiary processes. It is noteworthy that apart from these recovery processes, plastics also undergo a quaternary recycling process which is mainly carried out for energy recovery from plastics as they possess very high calorific value. The energy-producing processes include gasification, combustion and several other processes which are exothermic, and thereby release energy. Therefore, quaternary plastic recycling processes are considered a valuable part of waste to energy conversion. In contrast, these processes may cause severe environmental pollution especially due to Br_2-containing chemicals such as BFR, polybrominated dibenzodioxin (PBDD), polybrominated dibenzofuran (PBDF) and other toxic contaminants (Jadhao et al. 2020). Therefore, special attention is required to minimize environmental contamination. Nonetheless, this environmental aspect consideration faces stiff competition from an economic point of view. For example, co-pyrolysis of e-waste plastics with biomass can yield almost halogen-free oil but most of the Br_2 content of the feed is passed onto solid char product (Liu et al. 2013). Therefore, separation of Br_2 from char is required which incurs additional operating costs. On this account, it is crucial to design an environmentally benign and economically viable system with enhanced technologies for e-waste recycling. Significantly, the eco-design strategies become crucial for designing an overall reverse logistics model of e-waste recycling which is a function of environmental, technological, economic, and social factors. Interestingly, the forward logistics system deals with the supply chain of a commodity from raw material (point of origin) to its final version (point of consumption) for customers. In contrast, the reverse logistics system works in the opposite direction via the disintegration of the discarded product (Pokharel and Mutha 2009). Indeed, the focus should be turned to successfully implementing reverse logistics in e-waste management, governed by different eco-design strategies. Thus, subsequent sections discuss different aspects of eco-design strategies (Fig. 4) for e-waste recycling and trade-offs with challenges for sustainable development.

Figure 4. Different aspects of eco-design.

6. DIFFERENT ASPECTS OF ECO-DESIGN

6.1 Environmental aspect

As discussed earlier, the high environmental footprint of e-waste recycling processes is one of the significant challenges to overcome. Different BFR and halogenated compounds are responsible for resulting contamination from e-wastes. Therefore, the environmental aspect of eco-design strategies for e-waste recycling primarily focuses on removing BFR and halogenated substances before recycling. In this regard, solvothermal processes are widely used for the efficient removal of toxic materials. In a typical experimental procedure, a suitable solvent is mixed with e-waste and heated up to the desired temperature, followed by centrifugation. The mixture is divided mainly into 2 phases, i.e., solid and solvent. The solid phase contains plastic residue with a negligible amount of Br_2 and the solvent phase contains extracted Br_2. In particular, extracting agents such as methanol, ethanol, and isopropanol have been found very effective (Zhang and Zhang 2012). These solvents can reduce Br_2 content in e-waste plastic from 7.45 wt.% to 1.87 wt.% in a single pass. Notably, almost halogen-free plastic was obtained after secondary treatment was carried out keeping all the operating conditions constant. Notably, the affinity of solvents towards BFR and plastics plays a crucial role in the extraction process for selective dissolution of BFR over plastics (Das et al. 2021). In addition, operating conditions also play critical role because the extraction efficiency of solvents is usually higher at elevated temperatures and pressure. Nevertheless, breakdown of the polymeric matrix may occur under severe operating conditions due to excessive swelling of structure. Therefore, process optimization should be carried out carefully. Evangelopoulos et al. (2019) examined toluene as a potential solvent despite its toxicity and low polarity. It was aimed at a possible industrial approach as toluene can be produced from e-waste via catalytic pyrolysis and reused as a solvent. Similarly, several other solvents which include N-methyl-2-pyrrolidone, dichloromethane, ionic liquids, and deep eutectic solvents, have also been proposed as alternative solvents for extraction of BFR and metal recovery (Chandrasekaran et al. 2018, Yao et al. 2017).

It is noteworthy that conventional solvothermal processes suffer drawbacks such as high solvent consumption and longer residence time. Therefore, alternative processes such as microwave and ultrasonic wave-aided extraction processes can be used to remove BFR from e-waste plastic (Vilaplana et al. 2009, Morf et al. 2005). Moreover, extraction of BFR from e-waste plastic using supercritical fluid can also be examined as a potential green alternative to traditional processes (Ügdüler et al. 2020). Compared to the organic solvents, supercritical fluids possess relatively lower viscosity and greater diffusivity; thus, they selectively dissolve BFR from the polymer matrix. Especially, supercritical carbon dioxide (sc-CO_2) has been explored extensively as BFR extracting agent because of its low cost and ability to remain in the gaseous state at ambient conditions (Gripon et al. 2021). However, BFR removal efficiency using pure sc-CO_2 is still not up to the mark for industrial applications. Therefore, organic reagents are added as co-solvent for enhanced separation. Overall, only 44% Br_2 removal was reported in an ethanol-sc-CO_2 solvent at 40°C and 50 MPa (Gripon et al. 2021). In contrast, Zhang et al. measured more than 99% debromination efficiency using pure sc-CO_2 at 550°C and 1.5 MPa (Zhang and Zhang 2020). This considerable variation in BFR extraction efficiency certainly necessitates further research in process optimization for BFR removal from e-waste plastic using sc-CO_2. It is worth mentioning that majority of reports on BFR removal from e-waste have considered model polymers as feedstock or performed experiments at lab scale. Therefore, it is necessary to scale these processes to a larger size.

Fraunhofer Institute for process engineering and packaging IVV developed the CreaSolv® process to separate up to 99% BFR from e-waste plastics by solvothermal process (Schlummer et al. 2020). Moreover, this process was found very effective for a wide range of feedstocks including mixed plastics and virgin polymers present in e-waste. Recently, several small-scale pilot plants for e-waste plastics recycling using the CreaSolv® process have been commissioned which is undoubtedly a giant leap towards large-scale commercialization. Furthermore, another process Centrevap® is also reported which can efficiently remove different insoluble impurities such as dust, fillers, and non-polymeric substances from e-waste (Butturi et al. 2020, Freegard et al. 2006). However, the Centrevap® process does not cause satisfactory BFR removal from plastics as they are partially soluble in the solvent. Therefore, an integration of CreaSolv® and Centrevap® processes can be attempted to obtain BFR and insoluble impurities-free e-waste plastic for further recycling. It is evident that BFR possesses a tremendous environmental threat; therefore, BFR alternatives for application in EEE instruments and accessories should also be explored. For example, phosphates, polyphosphates, phosphonates, phosphinates, phosphate esters, metal hydroxides, and other alternatives can be viable green options as shown in Table 2 (BFRs in Electronics and Electrical Devices 2014, Nnorom and Osibanjo 2008). Nevertheless, these alternative chemicals are costlier and their flame retarding performance is also inferior to the commonly used BFR. In addition, existing reactors and process design also need to be modified to use these alternative chemicals. Therefore, extensive research action needs to be taken for closing these research gaps.

E-waste recycling processes require adding some chemicals at some point of processing which certainly increases the environmental footprint. Accordingly, it requires an additional purification step

Table 2. Green alternatives to conventional BFR (BFRs in Electronics and Electrical Devices 2014).

Use	BFR	Green Alternatives
Cabinet and Housing	Decabromo-diphenyl ether, Tetrabromo-bisphenol A, and other BFR	- Aluminum diethylphosphinate - Aluminum hydroxide - Melamine polyphosphate - Poly[phosphonate-co-carbonate] - Red phosphorous - Resorcinol bis-diphenylphosphate - Triphenyl phosphate - Aluminum hydroxide - Melamine polyphosphate
Printed Circuit Board	Tetrabromo-bisphenol A and other BFR	- Aluminum hydroxide - Melamine polyphosphate - Magnesium hydroxide

and separation of added chemicals, thereby increasing operating costs. Jadhao et al. (2020) proposed a chemical-free green e-waste recycling process which can be an excellent alternative to address these issues. The process utilizes an integrated concept of pyrolysis for e-waste plastic conversion to fuel, followed by ultrasonication of pyrolysis residue for metal separation. A 30 min ultrasonication resulted in approximately 90% total metal recovery including 100% noble metals (Au, Ag, Pt, and Pd). As the whole process was carried out without using any external chemical(s), the possibility of any contamination from the solvent is ruled out. Furthermore, 10% metal, which could not be recovered, was transferred to solid residue along with glass fibers and carbon. Notably, metallic fractions obtained from ultrasonication would again require metallurgical processing for the recovery of individual metals. Therefore, further research is required for the successful commercialization of this integrated technology at a larger scale.

Lately, ionic liquid as suitable and non-toxic organic salts have been explored for effective metal extraction from e-waste at ambient conditions. Ionic liquids possess low vapor pressure, high thermal stability, and control over the properties at ambient conditions, making these solvents a green alternative (AliAkbari et al. 2020, Ahrenberg et al. 2016). For example, Wstawski et al. (2021) examined hydrogen sulfate-containing ionic liquids as a solvent for recycling printed circuit boards and measured 30% metal recovery. Similarly, Carlesi et al. (2016) reported the use of ionic liquids as a catalyst to enhance the leaching activity of H_2SO_4 due to reduction in hydrophobic resistance to electron transfer between solid and solution. However, ionic liquids usually have high viscosity, and diluents must be added to the ionic liquid solution. Notably, the addition of diluents tends to decrease the extraction efficiency of ionic liquids. Therefore, exploration of highly efficient ionic liquid and diluent combination may be explored in the future. To overcome challenges associated with ionic liquid, deep eutectic solvents (DEES) which are another class of green solvents, have been examined for e-waste recycling lately (Zante and Boltoeva 2020). These solvents are made up of hydrogen bond acceptor and donor compounds. DEES have advantages such as low cost, facile synthesis, and purification processes (Entezari-zarandi and Larachi 2019). Moreover, biomass-derived oxygenates such as carboxylic acid, ester, and phenol constitute hydrogen bond donors in DEES, thereby increasing the eco-friendliness and sustainability of the process. Nevertheless, different halides such as chlorides and bromides, which are common hydrogen bond acceptors in DEES, may cause toxicity in the solvents which need to be studied further (Rodriguez et al. 2019). Therefore, the production of highly efficient non-toxic solvents poses great challenges for future research.

6.2 Technological aspect

Several processes have been developed over the years to minimize contamination from e-waste. Nonetheless, there are always some health hazards associated with e-waste handling. Therefore, it is essential to measure contaminants' exposure amount and assess the risk associated with it. The exposure quantification should be carried out in terms of air, water, and soil pollution including the levels of carbon monoxide, NO_x, SO_x particulate matter, heavy metals, PAH, and organic pollutants (Heacock et al. 2018). Notably, workers associated with e-waste recycling usually get exposed to contaminants through skin contact, inhalation and/or ingestion which can lead to severe life-threatening diseases. In some of the studies, metal exposure measurements were carried out at recycling sites of Asia, Africa, and Europe to measure metal concentration in blood, urine and plasma of the site workers well as office workers (Julander et al. 2014, Caravanos et al. 2011, Wang et al. 2011). As a result, a greater metal concentration was measured in site workers' samples than office employees. Furthermore, a linear trend was observed between the theoretical inhalable metal content of e-waste and metal concentration in workers' bodies. These observations show that a high metal concentration in the sample of site workers was because of e-waste exposure. Similar studies were carried out to measure other contaminants which showed the presence of five different PAH in urine samples of e-waste recycling workers (Feldt et al. 2014). In addition, other hazardous substances such as PBDE, PCB, PBDD, PCDF have also been detected in site workers' bodies and soil and water samples (Tue et al. 2016, Yang et al. 2013a, Chen et al. 2011). It was also observed that contaminants' concentration in the human body, soil and water samples varies from site to site possibly due to differences in safety measures practiced at each site. Overall, the difference in contaminants' concentration can also be correlated with the e-waste recycling process employed, protective gears used, and level of address to environmental aspects of eco-design strategies. In this context, several researchers carried out statistical analysis to determine the significance of different parameters on exposure measurement (Feldt et al. 2014). Interestingly, parameters like marital status, education level, religion were found to be statistically significant ($p \leq 0.05$). Furthermore, the chi-square test and 2-sided t-test were used to determine p-values of 14 health problems resulting from exposure at the recycling site. Concentrations of all 5 PAH in urine were found statistically significant which indeed connects health problems and exposure to PAH contaminants. A similar statistical approach has been taken up by Kumar et al. for the measurement of heavy metal exposure (Kumar and Fulekar 2019). The authors performed exposure quantification via principal component analysis (PCA) and subsequent cluster analysis. PCA transformed highly correlated variables into parameters with minimum correlation coefficient values. After cluster formation, the researchers determined values of various indexes such as pollution load index, geoaccumulation index and contamination factor. The results indicated that severe contamination is present due to heavy metals of e-waste, especially Cd, Cu, Pb, and chromium (Cr) present in e-waste. On the contrary, respective exposure indices revealed that Hg vapors were primary pollutant at fluorescent lamp recycling sites in France along with Pb and Y dust (Zimmermann et al. 2014). The statistical findings also include suggestive measures such as preventing Hg vapor emission to the atmosphere, minimization of accidental breakages, and incorporation of the proper ventilation system. Yang et al. (2013b) performed a similar statistical measurement of exposure to organic pollutants. Among the various contaminants, polybrominated biphenyls, dechlorane plus, PCB, and brominated diphenyl ether were found to cause the highest exposure. Furthermore, regression analysis showed that the properties of pollutants could cause significant variation in the degree of dispersion and thereby contamination. Pollutants with low vapor pressure and high molecular weight result in less dispersion and thus, contamination is limited to only site workers. On the contrary, organic pollutants with relatively high vapor pressure and low molecular weight such as polybrominated diphenyl ether, PCB, and brominated biphenyls are more dispersed and thereby result in contamination in local dwellers as well as recycling workers. On the contrary, earlier study revealed that contamination for site workers was much greater than office workers due to less dispersion of heavy metals. Thus, this area of research needs to be explored further. Notably, some organic pollutants show high dispersion, thereby causing greater threat to nearby people. Therefore, appropriate prevention processes should be employed at the e-waste recycling plants.

It is noteworthy that a degree of complexity is involved in computational modeling, considering different factors responsible for contaminant exposure. However, computational processes measure contaminants exposure amounts as well as help to introduce automation in e-waste recycling. For example, Maisel et al. (2020) used a computer vision system to determine the size distribution of plastics in e-waste. After clicking the picture of plastic flakes, photo filtration and noise removal were done followed by data storage in terms of contour and pixels. Afterward, the dimension and circle diameter of flakes were calculated based on which entire feed was divided into three categories, viz. large (20–50 mm), medium (10–20 mm), and small (6–15 mm) particle size. Later on, different size feeds were sent to different pre-processing technologies that can handle feed plastics of a specific size range. However, most pre-processing operations cannot shred and/or sort flakes of minimal size (< 5 mm). Therefore, it is also crucial to know the quantity of such fine plastics which should be handled very carefully. The efficiency of separation and sorting technologies significantly varies with feed particle size; therefore, controlling preceding size reduction to targeted size would undoubtedly maximize the efficacy of pre-processing processes. However, determining the number of particles in different size ranges is not very encouraging from an industrial perspective.

Therefore, Nowakowski and Pamula (2020) developed an image recognition system to facilitate systematic e-waste collection and subsequent classification. The proposed system works based on machine learning (ML); more specifically, a deep learning convolutional neural network is used to identify and classify e-waste. The authors claimed that close to 97% test accuracy is achieved for the ML system. Additionally, ML process such as faster region with the convolutional neural network was found highly effective in recognition and categorized different e-wastes from a single image. Furthermore, the developed algorithm was also able to measure the size of each e-waste, which would help to define the dimension of collection carriage. Needless to say, collection of e-waste is an integral part of the recycling system; therefore, it also requires optimization. Thus, researchers also carried out such work using an artificial immune system which performed better and faster than a conventional genetic algorithm (Nowakowski and Mrówczyńska 2018). Additionally, the model also provided information about economic feasibility and environmental footprint. Three scenarios for e-waste collection were considered, namely collection by mobile vans, in stationary bins and directly from consumers at the collection center, for which detailed analysis was carried out.

Integration of these processes in an application or server would thus enable to build convenient and self-sustainable system for e-waste collection and recycling. A concept for automated waste management is reported by Rahman et al. (2020) for implementation in e-waste recycling systems. Moreover, Acharekar et al. (2020) have developed an application, "RecyClick", consisting of modules for users, vendors and trade. Therefore, corresponding to these research studies, it is of utmost importance to commercialize such technologies as soon as possible for sustainable e-waste recycling.

It is worth noting that one of the significant drawbacks of the conventional e-waste recycling processes is low metal recovery which is caused by loss of heavy metals that result in environmental pollution. Therefore, to overcome these issues, different new innovative technologies are being examined as alternatives. In this context, metal recovery from e-waste by electrometallurgy is getting attention lately due to the high reaction rate, and less environmental footprint (Barragan et al. 2020). In this process, metal ions get reduced to the parent metal by electrons produced via anodic reaction in electrolytic cells, allowing metal recovery from e-waste. For example, electrometallurgy is employed to recover different metals such as Cu, Ni, Au, Eu from e-waste over the years (Kasper et al. 2018, O'Connor et al. 2018, Coman et al. 2013, Yang et al. 2012). The selective reduction of metals proceeds based on the difference in their reduction potential. Therefore, this process is not feasible for separating metals (Dy, Tb, Nd, Y) with reduction potential very close to each other (Kim et al. 2021). Furthermore, electrometallurgy has been explored as a follow-up process of the conventional hydrometallurgical process to enhance metal recovery from leachate. Therefore, its suitability as a standalone e-waste recycling process may be explored in the future. On the contrary, vacuum metallurgy is proposed as a green standalone process of e-waste recycling which causes negligible wastes with low energy consumption. As the name suggests, this process operates under vacuum conditions with significantly lower pressure (< 100 Pa) than atmospheric conditions (Zhan and Xu 2014). Due to low pressure, processes such as vacuum evaporation, sublimation, reduction, and pyrolysis, which are an integral part of vacuum metallurgy, can operate at a higher rate and result in more excellent selectivity than conventional processes. For example, vacuum metallurgy has been successfully applied to recover In, Zn, Hg, Cd, Pb, Ga, and As from different e-waste streams (Zhan et al. 2018, He et al. 2014, Lin and Qiu 2011, Zhan and Xu 2009). Moreover, this process can recover both metals and organics without toxic emission from BFR, thereby providing a green alternative to the conventional approaches. However, vacuum metallurgy requires costly pre-treatments, which incur a higher cost to the process. Moreover, a detailed mechanistic insight of the process is still unavailable which requires further attention of the research community.

6.3 Economic aspect

As discussed earlier, e-wastes contain numerous precious and rare earth metals. Therefore, the economy of the recycling process is certainly regulated by a fraction of the recovery of these valuable components. Earlier e-waste disposal strategies emphasized the linear economy concept; however, recent focus has been shifted to design waste management systems in line with the circular economy concept (Polverini and Miretti 2019). The linear economy concept does not focus on recycling; instead, it is based on the "take, make, use and dispose of" concept whereas the circular economy can be summarized by "make, use, reuse, remake and recycle" (Brydges 2021, Gonzalez et al. 2019). A circular economy assumes a business model to be developed as a closed-loop system. However, it is practically not possible to develop a fully closed system. Therefore, Geissdoerfer et al. (2020) provided a more realistic, flexible design of a business model basedon this concept (Fig. 5). Polverini and Miretti (2019) also proposed several requirements to connect eco-design and circular economy for enhancing product longevity, repairability and recyclability, reusability, and availability of spare components, which aid in the implementation of the circular economy concept in real. The authors suggested that cost analysis should also include the product's lifespan in terms of an "equivalent annual cost" and capture external factors by incorporating an "environmental damages fee". These strides would undoubtedly act as a stepping stone towards implementing the circular economy approach in practice. Additionally, it is crucial to perform an economic analysis of any new proposed technology to check its viability and perform necessary modifications if required. In this context, Ügdüleret al. (2020) performed an economic assessment of toxics removal from waste plastics by solvothermal process considering the ideal scenario. The authors showed that factors such as cost of feed plastic, percentage recovery of solvent, and extraction efficiency affect the overall economic feasibility of the process drastically. Also, solvent extraction was found to yield significantly higher profit than dissolution precipitation under different conditions. Notably, the collection of e-waste plays a critical role in the circular economy model of e-waste recycling. In this regard, a detailed economic assessment which includes fuel cost, employee wage, container replacement cost, maintenance cost, and insurance cost, which differ significantly from case to case for different collection strategies of

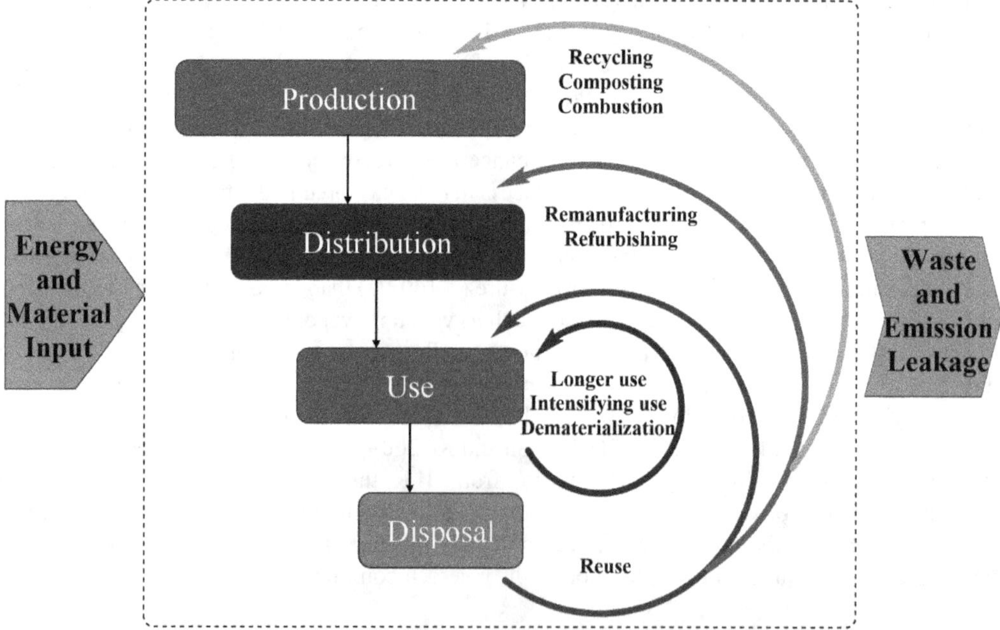

Figure 5. The Circular Economy Concept (Geissdoerfer et al. 2020).

e-wastes, was carried out (Nowakowski and Mrówczyńska 2018). Also, it is observed that e-waste collection by mobile vans would be more economically feasible for a densely populated residential zone. In contrast, e-waste collection via stationary containers placed at different spots would be a better option for a less populated area. Therefore, it is crucial to keep these points in consideration for the eco-designing of e-waste recycling.

It is worth mentioning that appropriate e-waste recycling strategies also aid in economic growth indirectly. It provides employment opportunities to people which is very important especially for developing and undeveloped countries (Heacock et al. 2018). However, the recycling of all e-wastes is not at the same economic level. For example, recycling noble and rare earth metal-containing e-waste is much more profitable than e-wastes with toxic and hazardous materials. Therefore, from a circular economy point of view, it is difficult to lay down a general economic framework for recycling all e-wastes. It also justifies the requirement for further research and development in converting non-precious components of e-waste into high-value products.

6.4 Social aspect

It is worth mentioning that one of the main bottlenecks for the eco-design of e-waste recycling is less consumer awareness. A survey conducted in Pune city of India revealed that only 3% of people are aware of e-waste collection and recycling facilities (Bhat and Patil 2014). On a similar note, a survey conducted in Jos metropolis of Nigeria showed that only 7% of residents sell e-waste to recycling companies (Miner et al. 2020). Such a lack of consumer awareness creates a void between e-waste generation and recycling quantities. Therefore, this is one of the reasons behind a collection of only a small percentage of e-waste for recycling purposes. For example, only 10% of e-wastes are systematically collected in Europe for recycling (Sarjaš 2018). Therefore, it is of utmost importance to increase the formal collection of e-waste for recycling by enhancing consumer awareness. Furthermore, vendors can also increase e-waste collection from consumers through incentive schemes such as exchange offer or cashback.

7. SUMMARY AND FUTURE OUTLOOK

Societal growth and e-waste generation are contemporary. Indeed, rapid changes in technology have reduced the life cycle of electrical and electronic equipment. As a result, a 25% surge in e-waste generation during the last six years is measured. Nevertheless, conventional technologies exist for metals and plastic recovery from e-waste; these processes have a considerably high environmental footprint. Therefore, the development of renewable and sustainable e-waste recycling processes is the need of the hour. In this regard, different eco-design strategies have been proposed which cover environmental, technological, economic, and social perspectives. The discussion constitutes consolidated information about various building blocks required for developing a holistic, eco-friendly e-waste recycling process. It is observed that the scientific community has focused on different aspects in a standalone manner and attempted to reach the optimum solution. However, all these standalone approaches must be integrated for the successful development of an eco-design process for e-waste recycling. For example, an integrated e-waste recycling process should include efficient automated e-waste collection and sort via ML, which is followed by exposure quantification and risk assessment by statistical processes. After toxicity measurement, efficient separation of the e-waste constituents (metals, plastics) and removal of toxic components such as BFR and halogenated substances are required to be carried out. Moreover, proper strategies should also be employed for purification and utilization of the waste materials generated in the separation step to make the process closed-loop in the circular economy principle. A commercial e-waste recycling process should include the end-processing of metals and plastics. Several drawbacks of conventional metallurgical recycling processes such as high energy requirement, low selectivity, and toxicity have encouraged the research community to develop green alternatives. In this regard, solvothermal metal extraction processes by ionic liquids and deep eutectic solvents, biometallurgy, electrometallurgy, and vacuum metallurgy have shown great potential and are thereby being paid a high attention by the researchers nowadays (Rao and Morrison 2020, Zhan and Xu 2014). On the contrary, the conventional plastics recycling process faces an uphill task from an energy and environmental point of view. Notably, emerging processes such as photo-reforming may show colossal potential for upcycling of plastics. Notably, no significant literature is available for photoreforming of e-waste plastics; however, this process has been employed for polymers such as polypropylene, and ABS which are an integral part of e-waste (Ouyang et al. 2021, Uekert et al. 2019). Therefore, this potential route may provide a considerable research opportunity for e-waste plastics recycling and production of hydrogen and other high-value chemicals under sunlight at ambient conditions with the help of a suitable photocatalyst. It is noteworthy that despite having various bottlenecks, the e-waste recycling sector is making progress rapidly. One prominent example of such progress is the recent 'Tokyo 2020 Medal Project' where the Tokyo 2020 Olympics and Paralympics medals were prepared by recycling 47 metric tons of e-waste, majorly from small electronics equipment (Paleja 2021). The total amount of recovered metals were 3500 kg silver, 2250 kg bronze, and 32 kg gold which is also an example of the holistic e-waste recycling process comprising of utilization of the end-product.

ACKNOWLEDGMENT

EA would like to thank the Indian Institute of Technology (ISM) Dhanbad, Delhi Research Implementation and Innovation (DRIIV) and Office of PSA to the Government of India for financial support of the solid waste management project.

LIST OF ABBREVIATIONS

E-waste	:	waste electrical and electronic materials
EEE	:	electrical and electronic equipment
Cd	:	cadmium
Pb	:	lead
Ni	:	nickel
Hg	:	mercury
As	:	arsenic
PBDE	:	polybrominated diphenyl ether
PCB	:	polychlorinated biphenyl
BFR	:	brominated flame retardant
Au	:	gold
Ag	:	silver
Pt	:	platinum
Pd	:	palladium
Ga	:	gallium
La	:	lanthanum
Y	:	yttrium
Ce	:	cerium
Eu	:	europium
Tb	:	terbium
PAH	:	polycyclic aromatic hydrocarbon
Cl_2	:	chlorine
HCl	:	hydrogen chloride
Br_2	:	bromine

HBr	:	hydrogen bromide
MSW	:	municipal solid waste
IT	:	information technology
Al	:	aluminum
Cu	:	copper
Zn	:	zinc
Fe	:	iron
Se	:	selenium
Te	:	tellurium
H_2SO_4	:	sulfuric acid
HNO_3	:	nitric acid
In	:	indium
Co	:	cobalt
Nd	:	neodymium
Pr	:	praseodymium
Sm	:	samarium
Dy	:	dysprosium
HIPS	:	high impact polystyrene
ABS	:	acrylonitrile butadiene styrene
CPCB	:	Central Pollution Control Board
EERA	:	European Electronics Recyclers Association
PBDD	:	polybrominated dibenzodioxin
PBDF	:	polybrominated dibenzofuran
sc-CO_2	:	supercritical carbon dioxide
DEES	:	deep eutectic solvent
PCA	:	principal component analysis
Cr	:	chromium
ML	:	machine learning

REFERENCES

Acharekar, K., Khedekar, P., Dsouza, J. and Vaidya, S. 2020. Machine Learning based Recy Click: Recycle at A Click. In: IEEE 2020 4th International Conference on Trends in Electronics and Informatics (ICOEI) - Tirunelveli, India, pp. 927–933.

Ahrenberg, M., Beck, M., Neise, C., Keßler, O., Kragl, U., Verevkin, S.P. and Schick, C. 2016. Vapor pressure of ionic liquids at low temperatures from AC-chip-calorimetry. Physical Chemistry Chemical Physics 18(31): 21381–21390.

Akcil, A., Erust, C., Gahan, C.S., Ozgun, M., Sahin, M. and Tuncuk, A. 2015. Precious metal recovery from waste printed circuit boards using cyanide and non-cyanide lixiviants—A review. Waste Management 45: 258–271.

Akram, R., Fahad, S., Hashmi, M.Z., Wahid, A., Adnan, M., Mubeen, M., Khan, N., Ishaq, M., Rehmani, A. and Awais, M. 2019. Trends of electronic waste pollution and its impact on the global environment and ecosystem. Environmental Science and Pollution Research 26: 16923–16938.

Alguaci, F.J., Caravaca, C., Cobo, A. and Martinez, S. 1994. The extraction of gold (I) from cyanide solutions by the phosphine oxide Cyanex 921. Hydrometallurgy 35(1): 41–52.

AliAkbari, R., Marfavi, Y., Kowsari, E. and Ramakrishna, S. 2020. Recent Studies on Ionic Liquids in Metal Recovery from E-Waste and Secondary Sources by Liquid-Liquid Extraction and Electrodeposition: a Review. Materials Circular Economy 2(1): Article No. 10.

Ari, V. 2016. A Review of Technology of Metal Recovery from Electronic Waste. pp. 121–158. *In*: F.-C. Mihai (ed.). E-Waste Transit From Pollution to Resources, Intech Open.

Baniasadi, M., Graves, J.E., Ray, D.A., De Silva, A.L., Renshaw, D. and Farnaud, S. 2021. Closed-Loop Recycling of Copper from Waste Printed Circuit Boards Using Bioleaching and Electrowinning Processes. Waste and Biomass Valorization 12(6): 3125–3136.

Barragan, J.A., Ponce de León, C., Alemán Castro, J.R., Peregrina-Lucano, A., Gómez-Zamudio, F. and Larios-Durán, E.R. 2020. Copper and Antimony Recovery from Electronic Waste by Hydrometallurgical and Electrochemical Techniques. ACS Omega 5(21): 12355–12363.

BFRs in Electronics and Electrical Devices. 2014. [accessed July, 2021]. https://www.cleanproduction.org/resources/entry/BFRs_in_electronics

Bhat, V. and Patil, Y. 2014. E-waste Consciousness and Disposal Practices among Residents of Pune City. Procedia Social and Behavioral Sciences 133: 491–498.

Bhutta, M.K.S., Omar, A. and Yang, X. 2011. Electronic Waste: A Growing Concern in Today's Environment. Economics Research International, 2011, Article ID:474230

Birloaga, I. and Vegliò, F. 2016. Study of multi-step hydrometallurgical methods to extract the valuable content of gold, silver and copper from waste printed circuit boards. Journal of Environmental Chemical Engineering 4(1): 20–29.

Birloaga, I, , Coman, V., Kopacek, B. and Vegliò, F. 2014. An advanced study on the hydrometallurgical processing of waste computer printed circuit boards to extract their valuable content of metals. Waste Management 34(12): 2581–2586.

Brydges, T. 2021. Closing the loop on take, make, waste : Investigating circular economy practices in the Swedish fashion industry. Journal of Cleaner Production 293: 126245.

Butturi, M.A., Marinelli, S., Gamberini, R. and Rimini, B. 2020. Ecotoxicity of plastics from informal waste electric and electronic treatment and recycling. Toxics 8(4): 1–19.

Caravanos, J., Clark, E., Fuller, R. and Lambertson, C. 2011. Assessing Worker and Environmental Chemical Exposure Risks at an e-Waste Recycling and Disposal Site in Accra, Ghana. Journal of Health and Pollution 1(1): 16–25.

Carlesi, C., Cortes, E., Dibernardi, G., Morales, J. and Muñoz, E. 2016. Ionic liquids as additives for acid leaching of copper from sulfidic ores. Hydrometallurgy 161: 29–33.

Chagnes, A., Cote, G., Ekberg, C., Nilsson, M. and Retegan, T. 2016.WEEE recycling : research, development, and policies, Elsvier, Amsterdem, Netherlands, pp. 219. ISBN: 978-012-803-3630.

Chandrasekaran, S.R., Avasarala, S., Murali, D., Rajagopalan, N. and Sharma, B.K. 2018. Materials and Energy Recovery from E-Waste Plastics. ACS Sustainable Chemical Engineering 6(4): 4594–4602.

Chen, A., Dietrich, K.N., Huo, X. and Ho, S.M. 2011. Developmental neurotoxicants in e-waste: An emerging health concern. Environmental Health Perspectives 119(4): 431–438.

Chen, X., Kang, D., Cao, L., Li, J., Zhou, T. and Ma, H. 2019. Separation and recovery of valuable metals from spent lithium ion batteries: Simultaneous recovery of Li and Co in a single step. Separation and Purification Technology 210: 690–697.

Chen, Y., Li, J., Chen, L., Chen, S. and Diao, W. 2012. Brominated Flame Retardants (BFRs) in Waste Electrical and Electronic Equipment (WEEE) Plastics and Printed Circuit Boards (PCBs). Procedia Environmental Sciences16: 552–559.

Cherrington, R. and Makenji, K. 2019. Mechanical methods of recycling plastics from WEEE. pp. 283–310. *In*: Waste Electrical and Electronic Equipment Handbook. 2nd Ed., Elsevier.

Coman, V., Robotin, B. and Ilea, P. 2013. Nickel recovery/removal from industrial wastes: A review. Resources, Conservation and Recycling 73: 229–238.

Correa, M.M.J., Silvas, F.P.C., Aliprandini, P., de Moraes, V.T., Dreisinger, D., Espinosa and D.C. R. 2018. Separation of copper from a leaching solution of printed circuit boards by using solvent extraction with D2EHPA. Brazilian Journal of Chemical Engineering 35(3): 919–930.

Cui, J. and Zhang, L. 2008. Metallurgical recovery of metals from electronic waste: A review. Journal of Hazardous Materials 158(2–3): 228–256.

Cui, J., Zhu, N., Luo, D., Li, Y., Wu, P., Dang, Z. and Hu, X. 2019. The Role of Oxalic Acid in the Leaching System for Recovering Indium from Waste Liquid Crystal Display Panels. ACS Sustainable Chemical Engineering 7(4): 3849–3857.

Das, N. 2010. Hydrometallurgy Recovery of precious metals through biosorption—A review. Hydrometallurgy 103(1–4): 180–189.

Das, P., Gabriel, J.C.P., Tay, C.Y. and Lee, J.M. 2021. Value-added products from thermochemical treatments of contaminated e-waste plastics. Chemosphere 269: 129409.

Desai, B., Tathavadkar, V. and Basu, S. 2016. Selenium partitioning between slag and matte during smelting. Russian Journal of Non-Ferrous Metals 57(4): 325–330.

DIRECTIVE 2012/19/EU OF The European Parliament and of the Council of 4 July 2012 on waste electrical and electronic equipment (WEEE). 2012. http://data.europa.eu/eli/dir/2012/19/oj (last accessed on: June 2021)

Dupont, D. and Binnemans, K. 2015a. Rare-earth recycling using a functionalized ionic liquid for the selective dissolution and revalorization of Y2O3:Eu3+ from lamp phosphor waste. Green Chemistry 17(2): 856–868.

Dupont, D. and Binnemans, K. 2015b. Recycling of rare earths from NdFeB magnets using a combined leaching/extraction system based on the acidity and thermomorphism of the ionic liquid [Hbet][Tf2N]. Green Chemistry 17(4): 2150–2163.

Entezari-zarandi, A. and Larachi, F. 2019. Selective dissolution of rare-earth element carbonates in deep eutectic. Journal of Rare Earths 37(5): 528–533.

Evangelopoulos, P., Arato, S., Persson, H., Kantarelis, E. and Yang, W. 2019. Reduction of brominated flame retardants (BFRs) in plastics from waste electrical and electronic equipment (WEEE) by solvent extraction and the influence on their thermal decomposition. Waste Management 94: 165–171.

Feldt, T., Fobil, J.N., Wittsiepe, J., Wilhelm, M., Till, H., Zoufaly, A., Burchard, G. and Göen, T. 2014. High levels of PAH-metabolites in urine of e-waste recycling workers from Agbogbloshie, Ghana. Science of Total Environment 466–467: 369–376.

Forti, V., Baldé, C.P., Kuehr, R. and Bel, G. 2020. The Global E-waste Monitor 2020: Quantities, flows and the circular economy potential. United Nations University (UNU)/United Nations Institute for Training and Research (UNITAR)—co-hosted SCYCLE Programme, International Telecommunication Union (ITU) and International Solid Waste Association (ISWA), Bonn/Geneva/Rotterdam.

Freegard, K., Tan, G. and Mortin, R. 2006. Develop a process to separate brominated flame retardants from WEEE polymers. The Waste and Resources Action Programme, Banbury, Oxon. https://polymerandfire.files.wordpress.com/2015/04/brominatedwithappendices-3712.pdf

Garlapati, V.K. 2016. E-waste in India and developed countries: Management, recycling, business and biotechnological initiatives. Renewable and Sustainable Energy Reviews 54: 874–881.

Geissdoerfer, M., Pieroni, M.P.P., Pigosso, D.C.A. and Soufani, K. 2020. Circular business models: A review. Journal of Cleaner Production 277: 123741.

Gergoric, M., Ravaux, C., Steenari, B.M., Espegren, F. and Retegan, T. 2018. Leaching and recovery of rare-earth elements from neodymium magnet waste using organic acids. Metals (Basel) 8(9): 1–17.

Ghimire, H. and Ariya, P.A. 2020. E-Wastes: Bridging the Knowledge Gaps in Global Production Budgets, Composition, Recycling and Sustainability Implications. Sustainable Chemistry, 1(2): 154–182.

Ghosh, B., Ghosh, M. K., Parhi, P., Mukherjee, P.S. and Mishra, B.K. 2015. Waste Printed Circuit Boards recycling: An extensive assessment of current status. Journal of Cleaner Production, 94: 5–19.

Gonzalez, E.D.R.S., Koh, L. and Leung, J. 2019. Towards a circular economy production system : trends and challenges for operations management. International Journal of Product Research 57(23): 7209–7218.

Gramatyka, P., Nowosielski, R. and Sakiewicz, P. 2007. WEEE Recycling of Waste Electrical and Electronic Equipment. Journal of Achievements in Materials and Manufacturing Engineering 20(1–2): 535–538.

Gripon, L., Belyamani, I., Legros, B., Seaudeau-pirouley, K., Lafranche, E. and Cauret, L. 2021. Brominated flame retardants extraction from waste electrical and electronic equipment-derived ABS using supercritical carbon dioxide. Waste Management 131: 313–322.

Guo, J., Zhang, R. and Xu, Z. 2016. Polybrominated diphenyl ethers (PBDEs) emitted from heating machine for waste printed wiring boards disassembling. Procedia Environmental Sciences 31: 849–854.

Gurung, M., Adhikari, B.B., Kawakita, H., Ohto, K., Inoue, K. and Alam, S. 2013. Recovery of gold and silver from spent mobile phones by means of acidothiourea leaching followed by adsorption using biosorbent prepared from persimmon tannin. Hydrometallurgy 133: 84–93.

He, Y., Ma, E. and Xu, Z. 2014. Recycling indium from waste liquid crystal display panel by vacuum carbon-reduction. Journal of Hazardous Material 268: 185–190.

Heacock, M., Trottier, B., Adhikary, S., Asante, K.A., Basu, N., Brune, M.N., Caravanos, J., Carpenter, D., Cazabon, D. and Chakraborty, P. 2018. Prevention-intervention strategies to reduce exposure to e-waste. Reviews on Environmental Health 33(2): 219–228.

Hong, Y., Thirion, D., Subramanian, S., Yoo, M., Choi, H., Kim, H.Y., Fraser Stoddart, J., Yavuz, C.T. 2020. Precious metal recovery from electronic waste by a porous porphyrin polymer. Proceedings of the National Academy of Sciences, USA, 117(28): 16174–16180.

Hong, Y. and Valix, M. 2014. Bioleaching of electronic waste using acidophilic sulfur oxidising bacteria. Journal of Cleaner Production 65: 465–472.

Huang, J., Nkrumah. P.N., Anim, D.O. and Mensah, E. 2014. E-Waste Disposal Effects on the Aquatic Environment: Accra, Ghana. Reviews of Environmental Contamination and Toxicology 229: 19–34.

Huisman, J. and Magalini, F. 2007. Where are WEEE now? Lessons from WEEE: Will EPR work for the US? IEEE International Symposium on Quality of Electronic Design, San Jose, USA, pp. 149–154.

Innocenzi, V., de Michelis, I. and Vegliò, F. 2017. Design and construction of an industrial mobile plant for WEEE treatment: Investigation on the treatment of fluorescent powders and economic evaluation compared to other e-wastes. Journal of Taiwan Institute of Chemical Engineering 80: 769–778.

Jadhao, P.R., Ahmad, E., Pant, K.K. and Nigam, K.D.P. 2020. Environmentally friendly approach for the recovery of metallic fraction from waste printed circuit boards using pyrolysis and ultrasonication. Waste Management 118: 150–160.

Jadhav, U., Su, C., Chakankar, M. and Hocheng, H. 2017. Acetic acid mediated leaching of metals from lead-free solders. Bioresources and Bioprocessing 4(1): Article No. 42.

Jorjani, E. and Shahbazi, M. 2016. The production of rare earth elements group via tributyl phosphate extraction and precipitation stripping using oxalic acid. Arabic Journal of Chemistry 9: S1532–S1539.

Julander, A., Lundgren, L., Skare, L., Grandér, M., Palm, B., Vahter, M. and Lidén, C. 2014. Formal recycling of e-waste leads to increased exposure to toxic metals: AN occupational exposure study from Sweden. Environment International 73: 243–251.

Jung, S.H., Kim, S.J. and Kim, J.S. 2012. Thermal degradation of acrylonitrile-butadiene-styrene (ABS) containing flame retardants using a fluidized bed reactor: The effects of Ca-based additives on halogen removal. Fuel Processing Technology 96: 265–270.

Karaahmet, O. and Cicek, B. 2019. Waste recycling of cathode ray tube glass through industrial production of transparent ceramic frits. Journal of Air Waste Management Association 69(10): 1258–1266.

Kasper, A.C., Veit, H.M., García-Gabaldón, M. and Herranz, V.P. 2018. Electrochemical study of gold recovery from ammoniacal thiosulfate, simulating the PCBs leaching of mobile phones. Electrochim Acta 259: 500–509.

Kaya, M. 2016. Recovery of metals and nonmetals from electronic waste by physical and chemical recycling processes. Waste Management 57: 64–90.

Khaliq, A., Rhamdhani, M.A., Brooks, G. and Masood, S. 2014. Metal extraction processes for electronic waste and existing industrial routes: A review and Australian perspective. Resources 3(1): 152–179.

Kim, K., Candeago, R., Rim, G., Raymond, D., Park, A.H.A. and Su, X. 2021. Electrochemical approaches for selective recovery of critical elements in hydrometallurgical processes of complex feedstocks. iScience 24(5): 102374.

Konyratbekova, S.S., Baikonurova, A. and Akcil, A. 2015. Non-cyanide leaching processes in gold hydrometallurgy and iodine-iodide applications: A review. Mineral Processing and Extractive Metallurgy Review 36(3): 198–212.

Krištofová, P., Rudnik, E. and Miškufová, A. 2016. Hydrometallurgical Methods of Indium Recovery From Obsolete Lcd and Led Panels. Metallurgy and Foundry Engineering 42(3): 157.

Kumar, P. and Fulekar, M.H. 2019. Multivariate and statistical approaches for the evaluation of heavy metals pollution at e-waste dumping sites. SN Applied Sciences General 1(11): 1–13.

Kushwaha, N., Banerjee, D., Ahmad, K.A., Shetti, N.P., Aminabhavi, T.M., Pant, K.K. and Ahmad, E. 2022. Catalytic production and application of bio-renewable butyl butyrate as jet fuel blend- A review. Journal of Environment Management 310: 114772.

Lee, C.K. and Rhee, K.I. 2003. Reductive leaching of cathodic active materials from lithium ion battery wastes. Hydrometallurgy 68(1–3): 5–10.

Li, L., Qu, W., Zhang, X., Lu, J., Chen, R., Wu, F. and Amine, K. 2015. Succinic acid-based leaching system: A sustainable process for recovery of valuable metals from spent Li-ion batteries. Journal of Power Sources 282: 544–551.

Li, X., Zhang, C., Zhao, R., Lu, X., Xu, X., Jia, X., Wang, C. and Li, L. 2013. Efficient adsorption of gold ions from aqueous systems with thioamide-group chelating nanofiber membranes. Chemical Engineering Journal 229: 420–428.

Li, Y., Richardson, J.B., Mark Bricka, R., Niu, X., Yang, H., Li, L. and Jimenez, A. 2009. Leaching of heavy metals from E-waste in simulated landfill columns. Waste Management 29(7): 2147–2150.

Lin, D. and Qiu, K. 2011. Recycling of waste lead storage battery by vacuum methods. Waste Management 31(7): 1547–1552.

List of Authorized E-Waste Dismantler/Recycler. 2021.m https://cpcb.nic.in/uploads/Projects/E-Waste/List_of_E-waste_Recycler.pdf (Last Accessed on September 2021).

Lister, T.E., Wang, P. and Anderko, A. 2014. Recovery of critical and value metals from mobile electronics enabled by electrochemical processing. Hydrometallurgy 149: 228–237.

Liu, W.J., Tian, K., Jiang, H., Zhang, X.S. and Yang, G.X. 2013. Preparation of liquid chemical feedstocks by co-pyrolysis of electronic waste and biomass without formation of polybrominated dibenzo-p-dioxins. Bioresource Technology 128: 1–7.

Ma, C., Yu, J., Wang, B., Song, Z., Xiang, J., Hu, S., Su, S. and Sun, L. 2016. Chemical recycling of brominated flame retarded plastics from e-waste for clean fuels production: A review. Renewable and Sustainable Energy Reviews 61: 433–450.

Maisel, F., Chancerel, P., Dimitrova, G., Emmerich, J., Nissen, N.F. and Schneider-Ramelow, M. 2020. Preparing WEEE plastics for recycling—How optimal particle sizes in pre-processing can improve the separation efficiency of high quality plastics. Resources, Conservation and Recycling 154: 104619.

Marra, A., Cesaro, A., Rene, E.R., Belgiorno, V. and Lens, P.N.L. 2018. Bioleaching of metals from WEEE shredding dust. Journal of Environmental Management 210: 180–190.

Miner, K.J., Rampedi, I.T., Ifegbesan, A.P. and Machete, F. 2020. Survey on household awareness and willingness to participate in e-waste management in jos, plateau state, Nigeria. Sustainability 12(3): 1047.

Miskolczi, N., Hall, W.J., Angyal, A., Bartha, L. and Williams, P.T. 2008. Production of oil with low organobromine content from the pyrolysis of flame retarded HIPS and ABS plastics.Journal of Analytical and Applied Pyrolysis 83(1): 115–123.

Moats, M.S. and Davenport, W.G. 2014. Copper Production. *In*: S. Seetharaman (ed.). Treatise Process Metallurgy, Vol. 3, Elsevier, ISBN: 978-008-096-9886.

Mohebbi, A., Mahani, A.A. and Izadi, A. 2019. Ion Exchange Resin Technology in Recovery of Precious and Noble Metals. pp. 193–258. *In*: Inamuddin, T.A. and Rangreez, A.M. Asiri (eds.). Applications of Ion Exchange Materials in Chemical and Food Industries, Springer International Publishing.

Morf, L.S., Tremp, J., Gloor, R., Huber, Y., Stengele, M. and Zennegg, M. 2005. Brominated flame retardants in waste electrical and electronic equipment: substance flows in a recycling plant. Environmental Science and Technology 39(22): 8691–8699.

Muhammad, C., Onwudili Jude, A. and Williams, P.T. 2015. Catalytic pyrolysis of waste plastic from electrical and electronic equipment. Journal of Analytical and Applied Pyrolysis 113: 332–339.

Mundada, M.N., Kumar, S. and Shekdar, A.V. 2004. E-waste: A new challenge for waste management in India. International Journal of Environmetnal Studies 61(3): 265–279.

Nayaka, G.P., Pai, K.V., Santhosh, G. and Manjanna, J. 2016 Recovery of cobalt as cobalt oxalate from spent lithium ion batteries by using glycine as leaching agent. Journal of Environmetnal Chemical Engineering 4(2): 2378–2383.

Nayl, A.A., Elkhashab, R.A., Badawy, S.M. and El-Khateeb, M.A. 2017. Acid leaching of mixed spent Li-ion batteries. Arabian Journal of Chemistry 10: S3632–S3639.

Ni, M., Xiao, H., Chi, Y., Yan, J., Buekens, A., Jin, Y. and Lu, S. 2012. Combustion and inorganic bromine emission of waste printed circuit boards in a high temperature furnace. Waste Management 32(3): 568–574.

Ning, C., Lin, C.S.K., Hui, D.C.W. and McKay, G. 2017. Waste Printed Circuit Board (PCB) Recycling Techniques. Topics in Current Chemistry 375(2): Article No. 43.

Nnorom, I.C. and Osibanjo, O. 2008. Sound management of brominated flame retarded (BFR) plastics from electronic wastes: State of the art and options in Nigeria. Resources, Conservation and Recycling 52(12): 1362–1372.

Nowakowski, P. and Pamuła, T. 2020. Application of deep learning object classifier to improve e-waste collection planning. Waste Management 109: 1–9.

Nowakowski, P. and Mrówczyńska, B. 2018. Towards sustainable WEEE collection and transportation methods in circular economy—Comparative study for rural and urban settlements. Resources, Conservation and Recycling 135: 93–107.

O'Connor, M.P., Coulthard, R.M. and Plata, D.L. 2018. Electrochemical deposition for the separation and recovery of metals using carbon nanotube-enabled filters. Environmental Science: Water Research and Technology 4(1): 58–66.

Onwudili, J.A. and Williams, P.T. 2009. Degradation of brominated flame-retarded plastics (Br-ABS and Br-HIPS) in supercritical water. Journal of Supercritical Fluids 49(3): 356–368.

Ouyang, Z., Yang, Y., Zhang, C., Zhu, S., Qin, L., Wang, W., He, D., Zhou, Y., Luo, H. and Qin, F. 2021. Recent advances in photocatalytic degradation of plastics and plastic-derived chemicals. Journal of Materials Chemistry A 9: 13402–13441.

Paleja, A. 2021. Tokyo's Olympic Medals Are Made of 47 Tons of Recycled E-Waste. Interesting Engineering, [accessed: August, 2021]. https://interestingengineering.com/tokyos-olympic-medals-are-made-of-47-tons-of-recycled-e-waste

Pereira, E.B., Suliman, A.L., Tanabe, E.H. and Bertuol, D.A. 2018. Recovery of indium from liquid crystal displays of discarded mobile phones using solvent extraction. Minerals Engineering 119: 67–72.

Pokharel, S. and Mutha, A. 2009. Perspectives in reverse logistics: A review. Resources, Conservation and Recycling 53(4): 175–182.

Polverini, D. and Miretti, U. 2019. An approach for the techno-economic assessment of circular economy requirements under the Ecodesign Directive. Resources, Conservation and Recycling 150: 104425.

Rahman, M.W., Islam, R., Hasan, A., Bithi, N.I., Hasan, M.M. and Rahman, M.M. 2020. Intelligent waste management system using deep learning with IoT. Journal of King Saud University —Computer and Information Sciences 34: 2072–2087.

Rao, M.D. and Morrison, C.A. 2020. Challenges and opportunities in the recovery of gold from electronic waste. RSC Advances 10: 4300–4309.

Recyclers - EERA. [accessed 2021 Aug 28]. https://www.eera-recyclers.com/recyclers RECYCLING: RECOVER RAW MATERIALS AND MAINTAIN VALUES. 2020. https://www.andritz.com/resource/blob/349008/99fb3c035a4a6eb7d8a10aa51da5ce85/recycling-image-brochure-en-data.pdf

Rodriguez, N.R., Machiels, L. and Binnemans, K. 2019. p - Toluenesulfonic Acid-Based Deep-Eutectic Solvents for Solubilizing Metal Oxides. ACS Sustainable Chemical Engineering 7: 3940–3948.

Sahajwalla, V. and Gaikwad, V. 2018. The present and future of e-waste plastics recycling. Current Opinions on Green Sustainable Chemistry 13: 102–107.

Saidan, M., Brown, B. and Valix, M. 2012. Leaching of electronic waste using biometabolised acids. Chinese Journal of Chemical Engineering 20(3): 530–534.

Sakultung, S., Pruksathorn, K. and Hunsom, M. 2007. Simultaneous recovery of valuable metals from spent mobile phone battery by an acid leaching process. Korean Journal of Chemical Engineering 24(2): 272–277.

Sarjaš, A. 2018. Recycling of electronic devices. pp. 1–15. *In*: Ecodesign Electron Devices. Ecosign Consortium, Project number: 562573-EPP-1-2015-1-SI-EPPKA2-SSA. http://www.ecosign-project.eu/wp-content/uploads/2018/09/ELECTRONICS_UNIT07_EN_Lecture.pdf

Schlummer, M., Fell, T., Maurer, A. and Altnau, G. 2020. The role of Chemistry in Plastics Recycling—A Comparison of Physical and Chemical Plastics Recycling. Kunststoffe International 5: 34–37.

Sethurajan, M., van Hullebusch, E.D., Fontana, D., Akcil, A., Deveci, H., Batinic, B., Leal, J.P., Gasche, T.A., Ali Kucuker, M. and Kuchta, K. 2019. Recent advances on hydrometallurgical recovery of critical and precious elements from end of life electronic wastes—a review. Critical Reviews in Environmental Science andTechnology 49(3): 212–275.

Shuva, M.A.H. and Kurny, A.S.W. 2013. Dissolution Kinetics of Cathode of Spent Lithium Ion Battery in Hydrochloric Acid Solutions. Journal of The Institution of Engineers (India): Series D 94(1): 13–16.

Spalvins, E., Dubey, B. and Townsend, T. 2008. Impact of electronic waste disposal on lead concentrations in landfill leachate. Environ Science and Technology 42(19): 7452–7458.

Sutton, M. 2019. Enviroserve opens world's largest e-waste facility. ITP.net [accessed: August, 2021]. https://www.itp.net/618785-enviroserve-opens-worlds-largest-e-waste-facility

Tan, Q. and Li, J. 2019. Rare earth metal recovery from typical e-waste. pp. 393–421. *In*: Waste Electrical and Electronica Equipment Handbook, Elsevier Ltd.

Tue, N.M., Goto, A., Takahashi, S., Itai, T., Asante, K.A., Kunisue, T. and Tanabe, S. 2016. Release of chlorinated, brominated and mixed halogenated dioxin-related compounds to soils from open burning of e-waste in Agbogbloshie (Accra, Ghana). Journal of Hazardous Materials 302: 151–157.

Uekert, T., Kasap, H. and Reisner, E. 2019. Photoreforming of Nonrecyclable Plastic Waste over a Carbon Nitride/Nickel Phosphide Catalyst. Journal of American Chemical Society 141: 15201–15210.

Ügdüler, S., Van Geem, K.M., Roosen, M., Delbeke, E.I.P. and de Meester, S. 2020. Challenges and opportunities of solvent-based additive extraction methods for plastic recycling. Waste Management 104: 148–182.

Vilaplana, F., Ribes-Greus, A. and Karlsson, S. 2009. Microwave-assisted extraction for qualitative and quantitative determination of brominated flame retardants in styrenic plastic fractions from waste electrical and electronic equipment (WEEE). Talanta 78(1): 33–39.

Wang, F., Huisman, J., Balde, K. and Stevels, A. 2012a. A systematic and compatible classification of WEEE. In: Electron Goes Green 2012, Berling, Germany.

Wang, F., Huisman, J., Meskers, C.E.M., Schluep, M., Stevels, A. and Hagelüken, C. 2012b. The Best-of-2-Worlds philosophy: Developing local dismantling and global infrastructure network for sustainable e-waste treatment in emerging economies. Waste Management, 32(11): 2134–2146.

Wang, H., Han, M., Yang, S., Chen, Y., Liu, Q. and Ke, S. 2011. Urinary heavy metal levels and relevant factors among people exposed to e-waste dismantling. Environment International, 37 (1), 80–85.

Wang, Q.M., Wang, S.S., Tian, M., Tang, D.X., Tian, Q.H. and Guo, X.Y. 2019. Relationship between copper content of slag and matte in the SKS copper smelting process. International Journal of Minerals, Metallurgy and Materials 26(3): 301–308.

Wstawski, S., Emmons-burzy, M., Rzelewska-piekut, M., Skrzypczak, A., Regel-rosocka, M. 2021. Studies on copper (II) leaching from e-waste with hydrogen sulfate ionic liquids: Effect of hydrogen peroxide. Hydrometallurgy 205: 105730.

Xu, B., Kong, W., Li, Q., Yang, Y., Jiang, T. and Liu, X. 2017. A review of thiosulfate leaching of gold: Focus on thiosulfate consumption and gold recovery from pregnant solution. Metals (Basel) 7(6): 222–247.

Yang, H., Huo, X., Yekeen, T.A., Zheng, Q., Zheng, M. and Xu, X. 2013. Effects of lead and cadmium exposure from electronic waste on child physical growth. Environmetnal Science and Pollution Research 20(7): 4441–4447.

Yang, J.G., Wu, Y.T. and Li, J. 2012. Recovery of ultrafine copper particles from metal components of waste printed circuit boards. Hydrometallurgy, 121–124: 1–6.

Yang, Q., Qiu, X., Li, R., Liu, S., Li, K., Wang, F., Zhu, P., Li, G. and Zhu, T. 2013. Exposure to typical persistent organic pollutants from an electronic waste recycling site in Northern China. Chemosphere 91(2): 205–211.

Yang, Y., Chen, S., Li, S., Chen, M., Chen, H. and Liu, B. 2014. Bioleaching waste printed circuit boards by Acidithiobacillus ferrooxidans and its kinetics aspect. Journal of Biotechnology 173: 24–30.

Yao, L., Zhao, J. and Lee, J.M. 2017. Small Size Rh Nanoparticles in Micelle Nanostructure by Ionic Liquid/CTAB for Acceptorless Dehydrogenation of Alcohols Only in Pure Water. ACS Sustainable Chemical Engineering 5(3): 2056–2060.

Zante, G. and Boltoeva, M. 2020. Review on Hydrometallurgical Recovery of Metals with Deep Eutectic Solvents. Sustainable Chemistry 1: 238–255.

Zeng, X., Li, J. and Shen, B. 2015. Novel approach to recover cobalt and lithium from spent lithium-ion battery using oxalic acid. Journal of Hazardous Materials 295: 112–118.

Zhan, L., Xia, F., Xia, Y. and Xie, B. 2018. Recycle Gallium and Arsenic from GaAs-Based E-Wastes via Pyrolysis-Vacuum Metallurgy Separation: Theory and Feasibility. ACS Sustainable Chemical Engineering 6(1): 1336–1342.

Zhan, L. and Xu, Z. 2014. State-of-the-art of recycling E-wastes by vacuum metallurgy separation. Environmental Science and Technology 48(24): 14092–14102.

Zhan, L. and Xu, Z. 2009. Separating and recycling metals from mixed metallic particles of crushed electronic wastes by vacuum metallurgy. Environmental Science and Technology 43(18): 7074–7078.

Zhang, C.C. and Zhang, F.S. 2020. Enhanced dehalogenation and coupled recovery of complex electronic display housing plastics by sub/supercritical CO2. Journal of Hazardous Materials 382: 121140.

Zhang, C.C. and Zhang, F.S. 2012. Removal of brominated flame retardant from electrical and electronic waste plastic by solvothermal technique. Journal of Hazardous Materials 221–222: 193–198.

Zhang, H., Ritchie, I.M. and La Brooy, S.R. 2004. The adsorption of gold thiourea complex onto activated carbon. Hydrometallurgy 72(3–4): 291–301.

Zhang, K., Schnoor, J.L. and Zeng, E.Y. 2012. E - Waste Recycling: Where Does It Go from Here? Environmetal Science and Technology 46: 10861–10867.

Zimmermann, F., Lecler, M.T., Clerc, F., Chollot, A., Silvente, E. and Grosjean, J. 2014. Occupational exposure in the fluorescent lamp recycling sector in France. Waste Management 34(7): 1257–1263.

7

RECOVERY AND CONVERSION OF E-WASTE PLASTIC VIA PHYSICAL AND CHEMICAL ROUTES

Prashant Ram Jadhao,[1] *Amrita Preetam,*[1] *Ramdayal Panda,*[1] *Snigdha Mishra,*[1] *K.K. Pant*[1,*] and *K.D.P. Nigam*[1,2]

1. INTRODUCTION

The rapid technological development in the field of electrical and electronic industry has decreased the life of electrical and electronic equipments (EEEs). The shortened lifespan of EEEs leads to the large generation of waste electrical and electronic equipments (WEEE) which is commonly termed as electronic waste (e-waste). E-waste is a complex mixture of metals, plastic, glass, ceramics, rubber, etc. (Jadhao et al. 2020). E-waste contains more than 69 elements of the periodic table including precious metals (Au, Ag, Pt. Pd, etc.), base metals (Cu, Pb, Ni, Sn, etc.), and critical metals (Co Pd, In, Bi, Sb, etc.) (Chauhan et al. 2018). Apart from this, plastic is also an important constituent of e-waste and approximately accounts for 25–30% of e-waste (Das et al. 2021). The e-waste plastic generated in 2019 was around 10.72 million metric tons (MMT) and it is expected to increase to 14.94 MMT by 2030 (Shi et al. 2021). Though metals and plastic are the major components of e-waste, the major research efforts are focused only on the recovery of metals and significant progress has also been made. The efforts for recycling the e-waste plastic are lagging compared to the metal recovery and most of the e-waste plastic is landfilled or burned after the metal recovery (Balde et al. 2017, Duan et al. 2011).

The plastic composition of e-waste differs subjected to the use of EEE. E-waste contains more than 15 different types of plastics such as Acrylonitrile Butadiene Styrene (ABS), High Impact Polystyrene (HIPS), Polycarbonate (PC), Polyethylene (PE), Polypropylene (PP), Polystyrene (PS), Polyurethane (PU), Polyamide (PA) and Polyphenylene Oxide (PPO) (Jadhao et al. 2022a). The plastic content of various EEEs is shown in Fig. 1 (Ma et al. 2016a). E-waste plastic is usually contaminated with toxic

[1] Chemical Engineering Department, Indian Institute of Technology Delhi, India.
[2] Chemical Engineering School, The University of Adelaide, Australia.
* Corresponding author: kkpant@chemical.iitd.ac.in

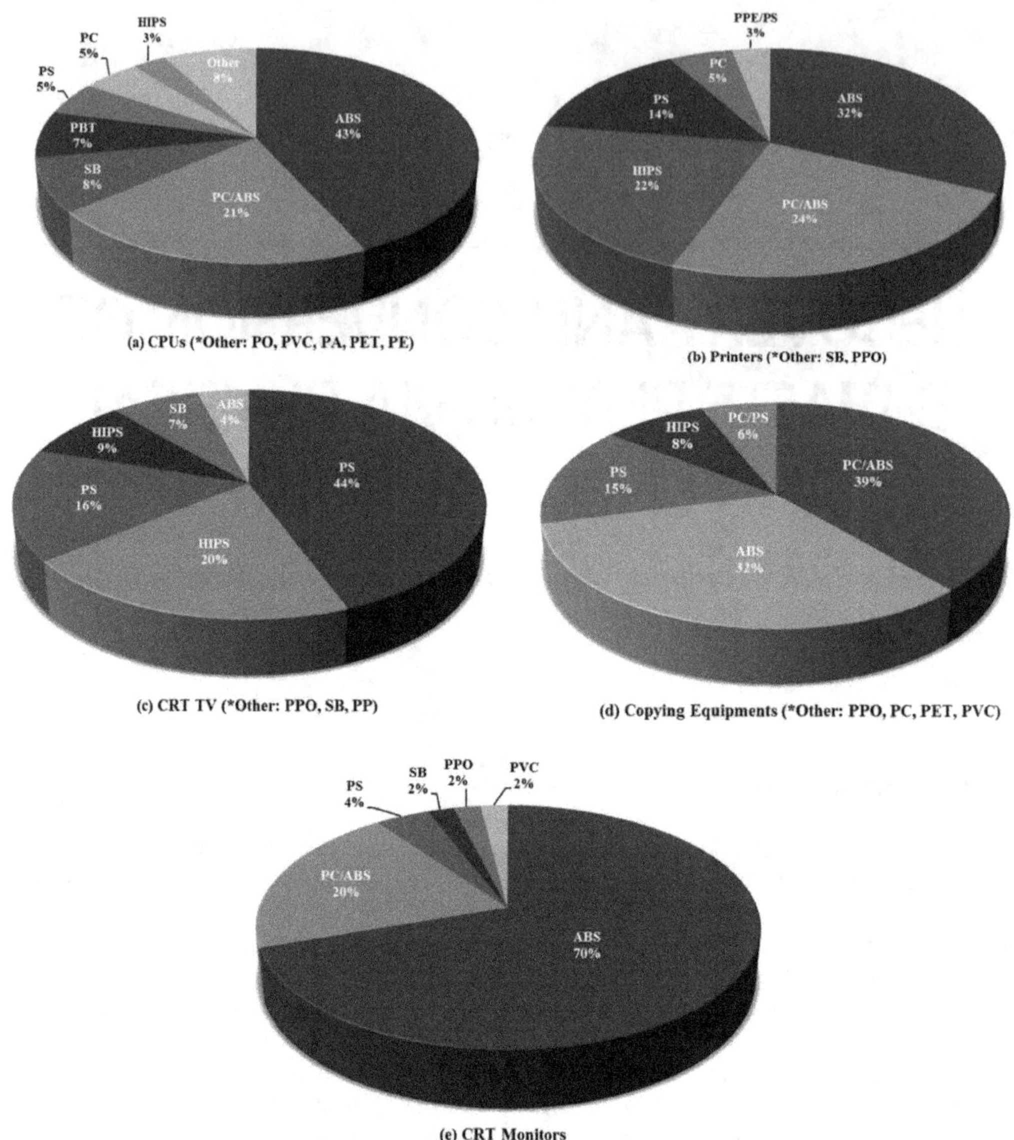

Figure 1. Plastic content of various EEEs (Ma et al. 2016a).

substances such as Pb, As, Sb, Cd, etc., and also consists of flame retardants (Debnath et al. 2018, Damrongsiri et al. 2016). The e-waste contains various flame retardants from the aspect of fire safety codes. Brominated flame retardants (BFRs) are the most commonly used flame retardants in EEE. Polybrominated diphenyl ethers (PBDEs), tetrabromobisphenol A (TBBPA), hexabromocyclodecane (BHCD), and decabromodipheny ether (deca-BDE) are the most commonly used BFRs in EEEs (Ma et al. 2016a). The BFRs are toxic and bio-accumulative that pose harmful effects on human health (Altarawnehet al. 2019). The improper processing of e-waste such as landfilling and open burning leads to environmental pollution and causes human health hazards. The open burning of e-waste results in the generation of toxic dioxins, furans, polyhalogenated aromatic hydrocarbons, polycyclic aromatic hydrocarbons (PAHs), and hydrogen halides (Das et al. 2021). The emission of these toxic chemicals into air leads to air pollution. Along with air pollution, soil and water pollution also increases due to e-waste. Tang et al. (2010) reported the presence of 371.8–1231.2 µg/kg PAH and 52–5789.5 µg/kg polychlorinated biphenyls in the soil of Wenling city of China which recycles e-waste. Similarly, Tue et al. (2017) also reported presence of 160–220 ng/g Cl-PAH and 19–40 ng/g Br-PAH in the soil near the e-waste recycling facility of Agbogbloshie city in Ghana where open burning is used to recycle e-waste. Several other studies have reported soil pollution due to e-waste recycling facilities (Zhang et al. 2019, Wu et al. 2015, Luo et al. 2011). Therefore, it has become critical to find out ways for sustainable e-waste recycling to mitigate environmental and human health hazards.

In this chapter, the physical and chemical routes for the recovery or recycling of plastic fraction of e-waste are discussed. The various physical separation processes such as gravity separation, magnetic separation, electrostatic separation, and flotation technique have been discussed. Along with physical separation processes, the different thermochemical processes such as pyrolysis, catalytic pyrolysis, gasification, and supercritical fluid treatments have also been covered in this chapter. All these processes can convert plastic into valuable products which can be used as fuel or feedstock for the recovery of mono- or oligomers.

2. PHYSICAL SEPARATION

The major focus of the physical separation process is the recovery of plastic fraction from e-waste without loss of metals. The physical separation process uses different physical properties such as size, shape, density, etc., for recovery of plastic fraction and these properties also influence the efficiency of the process. It is important to pre-treat the e-waste before employing physical separation. The pre-treatment of e-waste involves size reduction which can be carried out using shredding and crushing. This can be performed manually or using different equipments. The major physical separation processes employed for e-waste recycling are (1) gravity separation, (2) magnetic separation (3) electrostatic separation, and (4) flotation technique (Menad et al. 2016).

2.1 Gravity separation

The gravity separation process separates the metallic and non-metallic fractions based on their different densities, shape, and size (Kell 2009). Table 1 shows the values of densities for different components of e-waste. Other than gravity and size of particles, the medium employed for the separation of plastic fraction also plays a role for the separation of the metallic and non-metallic fractions. The different liquids such as tetra bromo ethane, acetone, etc. can facilitate the separation of metals and plastic fraction based on the density of particles (Kaya 2016). According to Richard et al. (2011), a cyclone type density separator will be helpful to achieve efficient recovery of plastic fraction from e-waste. The authors further stated that the process efficiency will depend upon the design of the cyclone type density separator and optimization of the process. Overall, the gravity separation process is of low efficiency as the shape and size of particles significantly affects the process operation and hampers the efficiency (Jadhao et al. 2020). Therefore, controlling the size of the feed will lessen the influence of size and separation will be more dependent on the density of particles which will increase the process efficiency.

Table 1. Difference in densities of different components of e-waste (Hsu et al. 2019).

Material	Specific gravity (g/cm^3)
Au, Pt, W	19.3–21.4
Pb, Ag, Mo	10.2–11.3
Cu, Ni, Fe, Zn	7.0–9.0
Al, Ti, Mg	1.7–4.5
Non-metallic fraction	1.8–2.0

2.2 Magnetic separation

The magnetic separation process is mainly used for separating ferrous metals from non-ferrous particles. The magnetic separation technique generally uses the low-intensity drum separator for the recovery of ferrous metals (Ghosh et al. 2015). However, in the last decade, the advances in magnetic separation techniques have led to changes in the design and operation of mechanical instruments employed (Kaya 2016). These advances have led to the use of high intensity magnetic separators which provide very high field strength and gradients. Viet et al. (2005) have employed magnetic separation for the recovery of ferrous metals, i.e., Fe and Ni in presence of 6000–6500 Gauss (G) magnetic fields. The authors reported the concentration of magnetic fraction consisting of around 43% Fe and 15.2% Ni. Apart from this, a substantial amount of copper was also reported in the magnetic fraction. Similarly, Yoo et al. (2009) have also investigated the application of magnetic separation for the recovery of ferrous metals from PCB. The authors employed two stage magnetic separation processes. A low magnetic field of 700 G was used in the first stage which separated around 83% Ni and Fe in the magnetic fraction. The Cu amount in the non-magnetic fraction was around 92%. In the second stage, 3000 G of magnetic field was employed which results in a higher amount of copper in the magnetic fraction. However, there are some drawbacks associated with magnetic separation process. The major problem with this separation process is the formation of clusters of particles that results in the impurity of non-ferrous particles to magnetic fraction which reduces the process efficiency (Ghosh et al. 2015).

2.3 Electrostatic separation

The electrostatic separation process employs the electrical conductivity of the particles for the separation of metals and non-metals. Mainly three types of electrostatic separation techniques have been used to recycle e-waste, i.e., corona electrostatic separation, triboelectric separation and eddy current separation (Wang et al. 2020). The electrostatic separation process can efficiently separate the conducting particles from semi-conductors and non-conductors if performed as a multi-step process (Wei and Realff 2003). The polarity of particles and charge attained by particles are the major factors which enable the separation using the electrostatic process.

Corona electrostatic separation process can separate the particles with big differences in conductivity. In the corona electrostatic separation, particle size is found to be an important aspect together with the pinning effect (Zhang et al. 1998). The effectiveness of corona electrostatic process decreases with the finer particle size (Li et al. 2007). Apart from the particle size, electrode system, rotor speed, and moisture content also influence the results in case of corona electrostatic separation. The use of an electrostatic field with high voltage results in the charging of the non-conducting particles and these particles stick to the drum as shown in Fig. 2 and therefore, the conducting particles can be separated using combined centrifugal and gravity force (Wang et al. 2020). Xue et al. (2012) stated that using higher voltage, the separation for conductor and semiconductor could be achieved with an efficiency of 82.5% and 88%, respectively, using the proposed integrated process.

Triboelectric separation can be employed for the separation of organic and inorganic material present in the non-metallic fraction of PCB (Yang et al. 2019). The organic material and fiberglass in PCB acquire the opposite charge which results in their separation (Wang et al. 2020). Zhang et al. (2016) stated that triboelectric separation can efficiently remove inorganics of non-metallic fraction of PCB, especially SiO_2 and Al_2O_3. Li et al. (2017) reported in their study that resins and fiberglass powder had different trajectories at the same conditions.

In the case of eddy current, different forces such as gravitational, centrifugal, frictional, and magnetic deflection act on the falling particles (Kaya 2016). However, the magnetic force majorly influences the deflection of particles and thus, it should be higher compared to other forces acting on the particles (Li et al. 2004). The eddy current is also useful for the separation of coarse particles

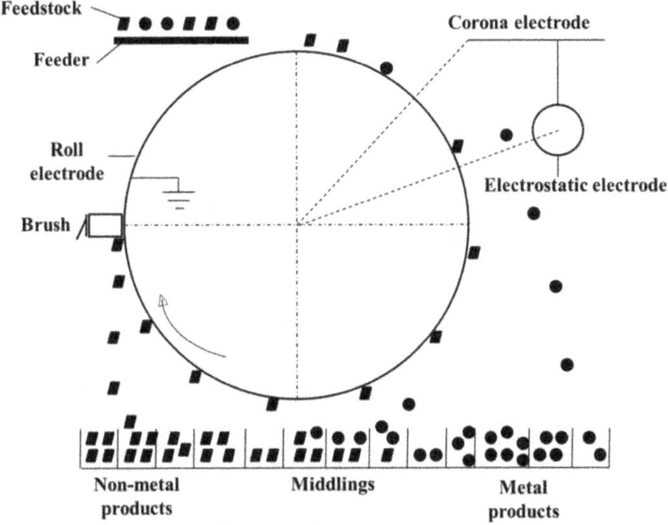

Figure 2. Separation process using corona electrostatic separator (Ruan and Xu 2016).

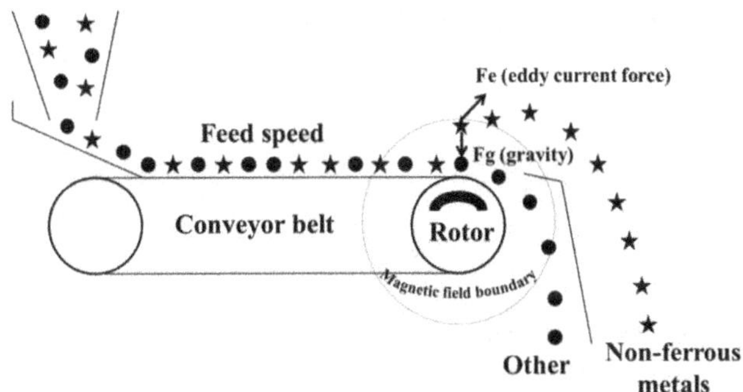

Figure 3. Schematic of eddy current separation process (Hsu et al. 2019).

along with the fine particles and therefore, it is preferred above corona separation. Figure 3 shows the schematic of the eddy current separation process. Wang et al. (2020) reported that eddy current separators used in industry have a magnet roller of around 300 mm diameter with a maximum revolution of 3000 rpm which have handling capacity of 2–2.6 tons of e-waste. The width of the belt ranges from 450 to 1250 mm with a maximum speed of 1 m/s. The separation efficiency of 95.54% was achieved using the 1.18 m/s feeding belt velocity, 3000 rpm rotating speed of magnetic roller, and 8.44 mm particle radius.

2.4 Flotation

Froth flotation separation technique uses difference in hydrophobicity of particles for the separation. Froth flotation is a common technique in mineral processing. One of the major challenges in physical separation processes is the recovery of fractions with a particle size ≤ 74 μm. These particles are generated during crushing and shredding operations (Kaya 2016). To overcome this problem, froth flotation can be a suitable technique (Ogunniyi and Vermaak 2009). Sarvar et al. (2015) employed jigging and flotation for the separation of metallic and non-metallic fractions from PCB. Jigging was employed for particles with size of 0.59 to 1.68 mm and flotation was employed for particles with a size less than 0.59 mm. The authors reported that the gasoline as a media resulted in effective recovery of non-metallic fraction. Around 85% and 95.6% recovery of metals was achieved using flotation and jigging, respectively. However, the authors reported the loss of 4.95% Cu and 24.5% Au during the process.

2.5 Perspective of physical separation processes

The physical separation processes are used for the separation of metals and non-metals. Different physical separation processes such as gravity separation, mechanical separation, electrostatic separation, and flotation separation were used for e-waste processing. Each of these processes has some drawbacks associated with them. The gravity separation process is greatly dependent upon the particle size and needs to monitor the particle size stringently. Moreover, the gravity separation process cannot separate the metals such as Au, Al, etc., which are present in a lamellar form in e-waste. The magnetic separation process is useful for the separation of only ferrous metals such as Fe, Ni, etc., and cannot separate the non-ferrous metals. This results in the transfer of metals into the non-metallic fraction and hampers the efficiency. The electrostatic separation processes are more effective than the magnetic and gravity separation processes. However, the middling product contains some fraction of metals and there is also a need to find the balance between the production capacity and efficiency of separation. The physical separation processes are relatively easy and convenient to use. The major challenge of physical separation process is the loss of metal during processing. The loss of metal is around 10–35% during the physical separation processes and this hampers the overall efficiency. Mainly precious metals are lost during the physical processing which provides the major economic incentives for e-waste recycling and therefore, more efficient physical processes are required for e-waste processing.

3. CHEMICAL RECYCLING

3.1 Pyrolysis

Pyrolysis is a route for conversion of organic substances to liquid, gas, and carbonaceous/char using thermal energy in the absence of oxygen. The products obtained through pyrolysis can be employed for the generation of energy or as feedstock to recover chemicals. The pyrolysis process degrades the polymers present in e-waste plastic into smaller molecules to generate range of hydrocarbons. The pyrolysis process has been widely employed for the conversion of e-waste plastic into fuel range products and recovery of valuable chemicals (Jadhao et al. 2022b, Ma et al. 2016b, Muhammad et al. 2015, Antonakou et al. 2014a, Hall and Williams 2006). The product distribution of pyrolysis process depends on different parameters such as temperature, process time, composition of feed, etc. The yield of the desired product can be increased by tweaking these parameters.

Temperature plays an important role to determine the product distribution; therefore, effect of temperature should be investigated to optimize the pyrolysis process for any feed. Jadhao et al. (2020) investigated the pyrolysis of PCB and studied the effect of temperature on pyrolysis. Also, TGA study was performed to determine the degradation temperature of e-waste plastic. The TGA study revealed that the thermal decomposition of PCB can be categorized into three zones, i.e., $<$ 250°C, 250–450°C, and $>$ 450°C. The authors reported the maximum weight loss of 35 wt.% in the second zone, i.e., 250–450°C. The authors also performed the experiments in the fixed bed reactor and found similar results. The 400°C of temperature was reported to be optimum for the degradation of e-waste plastic. At 400°C, the yield of solid and gaseous products was 60 wt.% and 35 wt.%, respectively. The gaseous product was a mixture of CH_4, H_2, CO, and CO_2. The heating value of the gaseous stream was around 28 MJ/kg. Furthermore, the authors reported the PCB used in the study was mainly composed of the HIPS, ABS, PC, polychlorinated and polybrominated biphenyls. The major components of PCB, i.e., HIPS and ABS decompose beyond 200°C and continue till 300°C. However, the halogenated compounds start degradation above 300°C and can be completely degraded at 400°C. The authors further separated the metallic fraction with an efficiency of 90%

using ultrasonication. The authors highlighted the benefits of pyrolysis as (1) efficiently converts e-waste plastic to valuable products (2) enhances the separation of metallic fraction and (3) metals can be recovered efficiently from separated metallic fraction.

The composition of feedstock is another factor that determines product composition and product distribution. The product can contain oxygenated hydrocarbons if the initial feed has oxygen containing compounds such as polyethylene terephthalate (PET), polycarbonate (PC), polylactic acid (PLA), polymethyl methacrylate (PMMA), etc. (Anuar Sharuddin et al. 2016). The brominated and chlorinated compounds present in the feedstock end up in the gas, liquid,and solid streams. The initial concentration of these compounds leads to the change in the product composition. Bhaskar et al. (2003) investigated the pyrolysis of Br-ABS and reported that below 500°C, the brominated compounds were found in liquid and solid streams, whereas Zhao et al. (2018) reported that above 500°C pyrolysis temperature bromine was present in the gas, liquid, and solid streams. The catalytic pyrolysis can be employed for the removal of halogenated compounds from product streams. Details have been discussed in the subsequent section. The presence of halogenated plastic such as PVC in e-waste leads to the generation of hydrogen halides such as HCl or HBr at high temperatures (Das et al. 2021). Similarly, BFRs present in e-waste plastic generates halogenated compounds during the pyrolysis which leads to the production of toxic chemicals such as PBDD/Fs while utilizing the liquid product as fuel (Anuar Sharuddin et al. 2016). Therefore, it is important to upgrade the liquid product before using it as fuel or eliminate the BFRs before pyrolysis. However, it is not easy and economical as often bromine is covalently linked with polymer chains.

At large, the gaseous product of e-waste plastic pyrolysis consists of H_2, C_1–C_4, CO, and CO_2 (Jadhao et al. 2020, Jung et al. 2012, Hall and Williams 2008). The share of CO and CO_2 in the gaseous stream depends upon the presence of oxygen containing compounds such as PC, PMMA, etc. Antonakou et al. (2014a) and Jadhao et al. (2020) reported that the generation of CO and CO_2 can be attributed to the presence of PC in the feed. Furthermore, the brominated compounds present in e-waste plastic reduced to HBr can be found in the gaseous product along with bromo methane and bromoethane (Das et al. 2021). The liquid product of e-waste plastic pyrolysis is mainly composed of phenolic compounds, furans such as 2 methyl-benzofuran, benzofuran, and brominated compounds such as bromomethane, and bromophenol (Ma et al. 2017, Antonakou et al. 2014b, Sun et al. 2011). The presence of BFRs in e-waste plastic leads to the generation of brominated compounds in the liquid product.

Caballero et al. (2016) studied the pyrolysis of landline and mobile phones. The pyrolysis was performed at 500°C for 30 min. The product distribution for landline phones was 58% liquid, 16.7% gas, and 28.3% solid, and for mobile phones was 54.5% liquid, 12.6% gas, and 32.4% solid. The liquid product obtained from both the wastes mainly consisted of toluene, styrene, ethyl-benzene, etc. The gaseous stream mainly consisted of hydrocarbons, CO, $CO_{2,}$ and H_2. Similarly, Chiang et al. (2010) investigated the PCB pyrolysis from 200 to 500°C for 30 min. The authors reported that 40%, 24%, and 38% yield of liquid, gaseous and solid products, respectively, at 500°C. The obtained liquid product is mainly composed of phenol, bromophenol, and methyl-phenol. The gaseous product is composed of CH_4,H_2, CO, and CO_2. Table 2 shows the summary of some more studies on pyrolysis.

Evangelopoulos et al. (2015) proposed a possible decomposition mechanism of brominated epoxy resin via pyrolysis as shown in Fig. 4. The thermal energy provided during the pyrolysis of e-waste leads to the dissociation of bonds which results in the production of free radicals. Free radical reactions cause the degradation of polymers by chain initiation, growth, and termination. The increase in temperature leads to the formation of lower molecular weight molecules. Phenol is the major product of PCB pyrolysis. The authors reported that with the increase in temperature, the production of phenol increases till 500°C and a further increase in temperature results in a lower amount of phenol (Yin et al. 2011). This can be attributed to the further degradation of phenolic compounds at higher temperatures, which generates lighter products such as CO_2, cyclopentadiene, and pentadiene (Brezinsky et al. 1998). Furthermore, cresol, 3,5-dimethylphenol, and 2,3-dimethylphenol are also formed during the thermal degradation process. The bisphenol A is formed due to the breakage of

Table 2. Summary of studies for the pyrolysis of e-waste.

E-waste material	Pyrolysis conditions	Pyrolysis product yield (%)			Reference
		Solid	Liquid	Gas	
PE wires	500°C, 30 min	32.9	44.1	23.0	De Marco et al. 2008
Table phones	500°C, 30 min	34.4	53.5	12.2	
Mobile phones	500°C, 30 min	30.3	57.4	12.3	
NMF of PCB computer	200°C, 30 min	96	NA	NA	Lin and Chiang 2014
NMF of PCB computer	300°C, 30 min	69	13	19	
NMF of PCB computer	400°C, 30 min	50	29	24	
NMF of PCB computer	500°C, 30 min	39	38	24	
Landline phones	500°C, 30 min	15	28.3	58	Caballero et al. 2016
Mobile phones	500°C, 30 min	15	32.9	54.5	
PCB	N_2, 10°C/min, 800°C	70.2	25.2	4.6	Lee et al. 2017
PCB	CO_2, 10°C/min, 800°C	67.4	9.6	23.0	
Waste electronics material	400°C, 0.250 kg/h, 5 h	29	52	19	Evangelopoulos et al. 2020
Waste electronics material	500°C, 0.250 kg/h, 5 h	27	49	24	
Waste electronics material	600°C, 0.250 kg/h, 5 h	17	39	44	
PCB	600°C, 20 min	78	20.5	1.5	Evangelopoulos et al. 2017

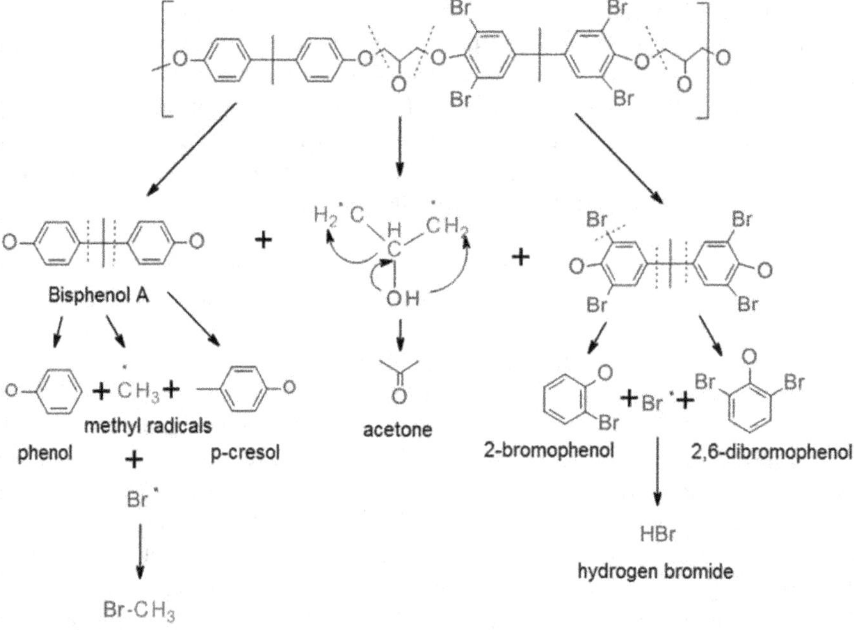

Figure 4. Brominated epoxy resin decomposition mechanism (Reproduced from Evangelopoulos et al. (2015)).

O-CH_2 bonds present in PCB. The amount of bisphenol A decreases with an increase in temperature and this happens due to further degradation at high temperatures which results in the formation of monomers. The other aromatic compounds such as styrene, benzene, toluene, and methylstyrene are also formed by breaking C-O bond at higher temperatures. Another aspect of e-waste pyrolysis is debromination and it occurs during the degradation by the formation of HBr (Barontini and Cozzani

2006). Moreover, the bromine also reacts with methyl radicals to form bromomethane, and some of the bromine is attached to aromatic phenolic rings to form bromophenol. The increase in temperature results in the formation of HBr at the expense of bromomethane and bromophenol (Barontini and Cozzani 2006).

Pyrolysis is a viable option for the chemical recycling of e-waste plastic. Pyrolysis of e-waste plastic converts it into products which can be employed for energy generation and as a feedstock for chemical recovery. The major problem associated with the pyrolysis of e-waste plastic is the presence of halogenated compounds in the product which hampers the utilization of the obtained product. Therefore, the removal of halogenated compounds from pyrolysis products will incur the extra cost and will increase the process cost. The efforts can also be made to simultaneously remove halogenated compounds during the pyrolysis of e-waste plastic using different catalysts or additives. The following section on catalytic pyrolysis will focus on this aspect.

3.2 Catalytic pyrolysis

The catalytic pyrolysisis used for the reforming of pyrolysis products to facilitate easy utilization of these products. The catalytic pyrolysis also helps to obtain the desired product via cracking and reforming reaction. The major role of catalyst in the pyrolysis of e-waste is to remove halogenated compounds from the products. In this regard, metal oxides can be employed as they can absorb halogens. The presence of metal oxides during the pyrolysis initiates the fission of C-Br covalent bonds over metal cations (Das et al. 2021). Metal oxides can remove Br from any brominated compounds. The iron and calcium based catalysts when employed during pyrolysis can convert halogenated compounds to their respective halides (HBr/HCl) (Sakata et al. 2003). The pyrolysis of e-waste plastic containing brominated compounds forms HBr at 400°C and if the iron oxide is used as a catalyst, it can absorb around 90% of HBr (Altarawneh et al. 2016). At higher temperatures, C-Br bond breaks which results in the adsorption of Br on the metal.

Zeolites have also been employed for the catalytic pyrolysis of e-waste plastic. Hall and Williams have investigated catalytic pyrolysis using Y-Zeolite and ZSM-5 (Hall and Williams 2008). The authors reported that both catalysts, i.e. Y-Zeolite and ZSM-5 can be employed for the debromination of pyrolysis products; however, Y-Zeolite was more effective for the removal of bromine than ZSM-5. Though both the catalysts were useful for debromination, their use leads to the reduction in the amount of liquid product and also reduces the formation of valuable styrene that can lessen the value of pyrolysis product. Y-Zeolite and ZSM-5 both catalysts only adsorb the brominated compounds without converting organic bromine into inorganic bromine. This raises the question of processing the spent catalyst. Another point is the reutilisation of catalyst and its effectiveness after reactivation. These points should be addressed to commercialize the use of these catalysts for e-waste pyrolysis. Ma et al. (2016b) investigated the effect of three zeolite catalysts, i.e., HY, Hβ, and HZSM-5 on the pyrolysis of Br-ABS. The authors also investigated their debromination efficiency. The authors reported that the addition of HY and Hβ reduces the oil yield from 63.6% to 45% and 44.3%, respectively, as compared to non-catalytic pyrolysis. The HZSM-5 has increased the gas yield by reducing the wax formation. The debromination efficiency of the studied catalysts was around 50% which means that 50% bromine could be removed from the oil product compared to non-catalytic pyrolysis. Chen et al. (2020) have investigated the efficiency of Fe-Ni/MCM-41to pyrolyze computer casing plastic along with debromination. The authors stated the synergistic effect of co-existence of Fe and Ni on oil composition which increases the formation of valuable single ring hydrocarbons. The catalysts also showed remarkable efficiency for the removal of bromine from the pyrolysis oil. The bromine content of oil was less than 4% in the presence of Fe-Ni/MCM-41 catalysts as compared to 10% during the non-catalytic pyrolysis. The higher content of Fe is more beneficial for debromination. The catalytic cracking of organobromines, reaction of metal oxides with HBr/$SbBr_3$, and deposition of organobromines on catalyst surface are the major steps in the debromination during catalytic pyrolysis using Fe-Ni/MCM-41.

Moreover, several other types of catalysts have also been employed for the debromination. Brebu et al. (2004) have investigated the catalytic pyrolysis of Br-ABS mixed with PS and PE in presence of FeOOH, Fe(Fe_3O_4)-C, and Ca($CaCO_3$)-C at 450°C. The authors reported that the FeOOH and Fe(Fe_3O_4)-C catalysts were more effective for bromine removal than Ca($CaCO_3$)-C. Also, the catalysts were more effective for bromine removal from PE/Br-ABS than PS/Br-ABS. A summary of some more studies on the catalytic pyrolysis of e-waste is shown in Table 3.

Zhang et al. (2021) has proposed a possible mechanism for the fixation of Br during the pyrolysis of WPCB as shown in Fig. 5. The authors considered TBBPA as the BFR of PCB. The Br attached with the phenyl ring was detached at a higher temperature to form bromine radicals. The bromine radicals formed reacts with hydrogen radicals and smaller molecules to form the HBr and organic bromine (Gao et al. 2020). In addition, Br could also react with carbon and forms C-Br bonds. The authors reported that the C-Br bond in pyrolysis residue of PCB (WPCB-PR) may be organic bromine. The alkali addition can be used for the Br fixation as shown in Fig. 5. The formed HBr reacts with alkali to form MBr (M: K and Na) and releases H_2O and CO_2 (Shen 2018). The authors reported that the KBr and NaBr form tiny crystals in WPCB-PR and this reduces the bromine content in liquid and gaseous streams.

Table 3. Summary of studies for the catalytic pyrolysis of e-waste.

Waste material	Catalysts	Pyrolysis conditions	Product Distribution (%)			Reference
			Solid	Liquid	Gas	
PCB	FCC	275°C, 1.5 h, Catalyst: PCB ratio – 10:90	58.0	22.4	19.5	Ng et al. 2014
	ZSM-5	275°C, 1.5 h, Catalyst: PCB ratio – 10:90	55.4	19.2	25.3	
	H-Y zeolite	275°C, 1.5 h, Catalyst: PCB ratio – 10:90	54.9	19.2	25.8	
	Dolomite	275°C, 1.5 h, Catalyst: PCB ratio – 10:90	55.1	14.4	30.4	
PCB	YZ	1 h, 500°C, Catalyst: PCB ratio – 10:50	81	4	15	Areeprasert and Khaobang 2018
	ZSM-5	1 h, 500°C, Catalyst: PCB ratio – 10:50	85	5	10	
	Fe/YZ	1 h, 500°C, Catalyst: PCB ratio – 10:50	73	3	24	
	Fe/ZSM-5	1 h, 500°C, Catalyst: PCB ratio – 10:50	86	6	7	
PCB	$CaCO_3$	1 h, 500°C, catalyst: PCB ratio – 1:1	70	17	12.3	Xie et al. 2017
	HZSM-5	1 h, 500°C, catalyst: PCB ratio – 1:1	62	18	20	
	CaO	1 h, 500°C, catalyst: PCB ratio – 1:1	78	15	6.44	
	Al_2O_3	1 h, 500°C, catalyst: PCB ratio – 1:1	71.84	19.2	8.96	
	FeOOH	1 h, 500°C, catalyst : PCB ratio – 1:1	65	22	12.8	
	$Ca(OH)_2$	1 h, 500°C, catalyst : PCB ratio – 1:1	66.2	25.3	8.4	
E-waste	HUSY	400°C, catalyst : feed ratio – 1:3, semi-batch	13.2	83	3	Santella et al. 2016
	HZSM-5	400°C, catalyst : feed ratio – 1:3, semi-batch	7.1	91	2	
PCB	Al_2O_3	400°C, catalyst : feed ratio – 1:2, 1 h	61.5	15	23.5	Wang et al. 2015
	Al_2O_3	500°C, catalyst : feed ratio – 1:2, 1 h	56.4	20.6	23	
	Al_2O_3	600°C, catalyst : feed ratio – 1:2, 1 h	56	19.3	24.6	

3.3 Gasification

Gasification is mainly employed for the generation of a major product as the gaseous stream. The gasification takes place in the presence of oxygen and yields the gaseous stream containing CO,

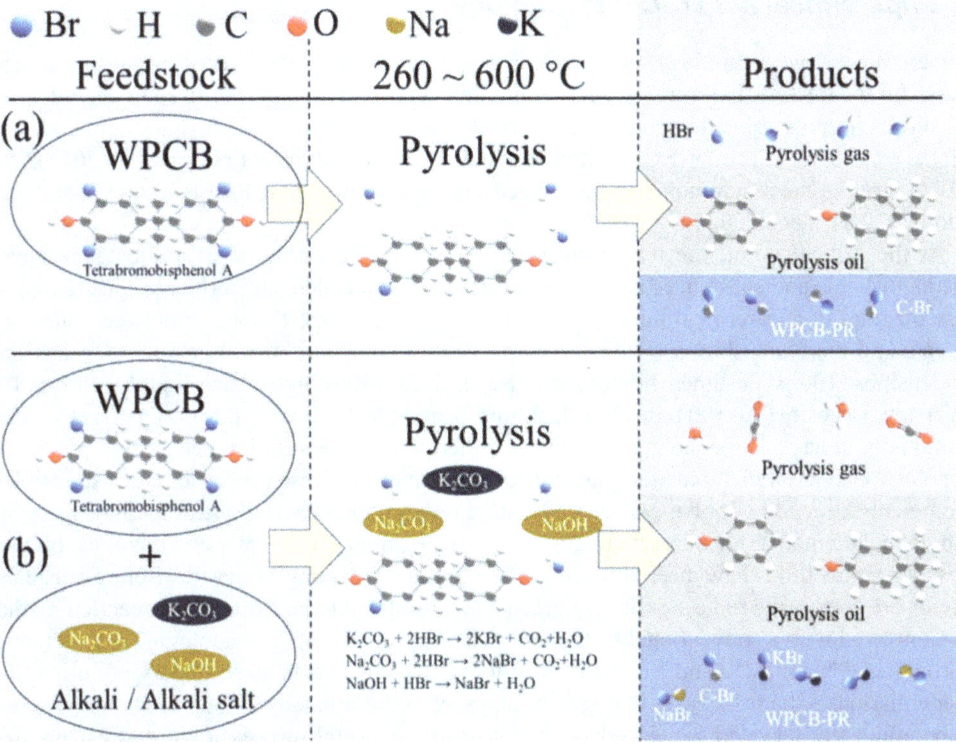

Figure 5. Possible mechanism of Br fixation during catalytic pyrolysis (Reproduced from Zhang et al. 2021).

H_2, CO_2, and CH_4 as major components (Devi et al. 2003). The gaseous product obtained from the gasification process can be employed for the generation of heat and electricity. Moreover, the syngas obtained from gasification can be employed for the synthesis of various chemicals such as methanol, dimethyl ether, etc. (Guan et al. 2016). The gasification process can be employed when the feedstock is a mixture of different waste streams. However, in the case of plastic waste processing using gasification, the major disadvantage is the generation of tar which requires the processing of product gas before utilizing it for the generation of energy or chemical production (Sansaniwal et al. 2017, Shahbaz et al. 2017, Molino et al. 2016, Heidenreich and Foscolo 2015). Very few studies have investigated e-waste gasification. Zhang et al. (2012) have studied the gasification of the epoxy circuit board at 600–700°C. The authors employed the ternary eutectic carbonate which is the mixture of lithium, sodium, and potassium carbonate (LNK-carbonate) during the gasification. CO_2 and H_2 were the major components of the gaseous product obtained from the gasification of the epoxy circuit board. The amount of CO and CH_4 was lesser in the gaseous product. The yield of the gas product was 60.8% and that of tar and phenol was 19.2% and 12.2%, respectively. The presence of LNK carbonate eliminates the formation of char. In the absence of LNK carbonate and using steam gasification, the formation of char increases to 17.4% and that of tar to 42.3% which reduces the yield of phenol and gas to 16.1% and 19.2%, respectively. Similarly, Salbidegoitia et al. (2015) studied the gasification of PCB in presence of nickel and LNK carbonate. The authors reported that in presence of LNK carbonate, the gasification completed at 675°C while the addition of nickel decreases the process temperature to 550°C and increases the H_2 yield by 3.2 times. Moreover, the presence of nickel eliminates the formation of tar and decreases the formation of char from 49% to 23% which ultimately increases the amount of carbon in the gaseous product. The increases in the H_2 production and carbon in the gaseous product can be attributed to the catalytic properties of nickel.

3.4 Supercritical fluid (SCF) technology

A supercritical fluid technology (SCF) is one of the emerging techniques used as an alternative process for the treatment of e-waste. Most commonly, SCF technology is utilized to degrade plastic and convert it into a valuable liquid product (crude oil). Several solvents, such as methanol, water, acetone, etc. can be utilized in SCF technology to degrade the polymers (Yildirir et al. 2015). Water and CO_2 are the most common and considered green solvents for SCF technology (Golzary and Abdoli 2020, Hsu et al. 2020).

At the critical point, the temperature, pressure, and vapor phase of a substance become indistinguishable and substances above the critical point are called supercritical fluids (SCF). The phase diagram of supercritical fluids is shown in Fig. 6. Its properties are in between pure liquid and gas, and therefore, they are called "compressible liquids" or "dense gases". SCF has liquid like densities (100–1000 times greater than gases), diffusivities higher than liquids (10^{-3} and 10^{-4} cm^2/s), low viscosity (10–100 times less than liquid), and good solvating power (Camel et al. 1993). Additionally, it has gas like compressibility and therefore possesses high penetrating power. SCF technology also has a high reaction rate and reaction kinetics. These properties are responsible for the efficient extraction of polymers from e-waste. Furthermore, these unique properties make SCFs a promising alternative for the recycling of e-waste. Beginning with water and alcohols, both sub- and supercritical fluids have been utilized in the chemical recycling of plastics from e-waste. SCF solvents are considered to be sustainably green and considered to be a promising reaction medium.

Compared with a conventional technique such as pyrolysis, SCF technologies lead in terms of environmental benefits (Li and Xu 2015). SCF technology generates no secondary pollution. It has lower emissions of harmful gases (NOx, SOx, soots, etc.). Additionally, in SCF technology, solvents can be efficiently recycled which reduces the cost of the overall process (Xiu and Zhang 2010). Another advantage of SCF technology over conventional techniques is that it can operate under low temperatures. Furthermore, SCF techniques are used for e-waste treatment with efficient results. Considerable work has been done for the degradation of e-waste using supercritical water (Calgaro et al. 2017). Supercritical fluids have emerged as an attractive substitute for high temperature thermal treatment and high metallic loss physical treatment in the field of e-waste recycling. Fig. 7 shows a general method of SCF for e-waste treatment (Hsu et al. 2019). Phenolic resin, epoxy resin, and polyurethane are the major polymers in the printed circuit board which have been widely delaminated and converted to liquid products by SCF technology. Significant work has been carried out using

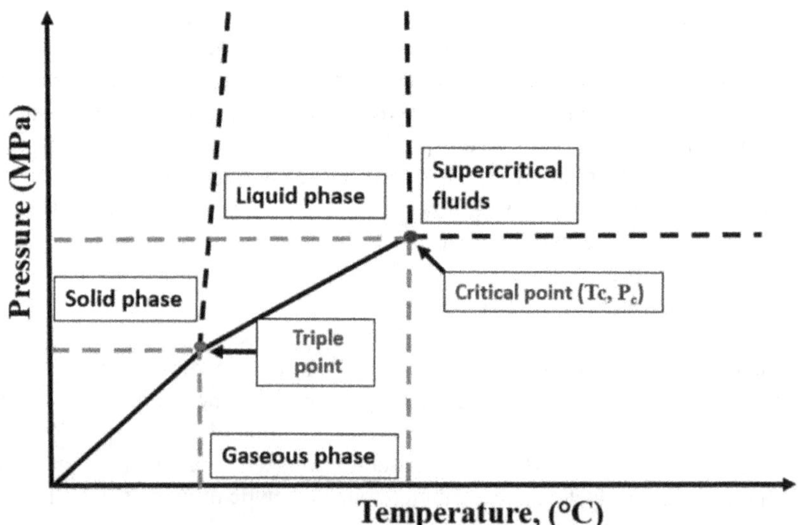

Figure 6. Phase diagram of supercritical fluids.

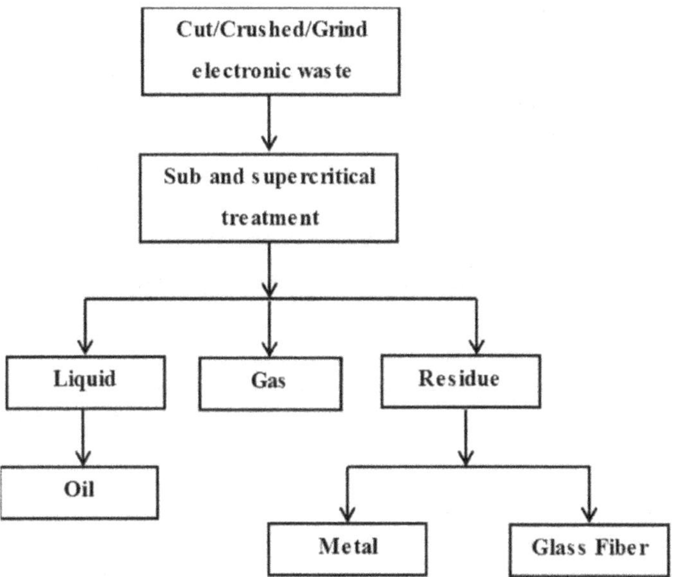

Figure 7. General method of SCF treatment of e-waste.

supercritical water for the detoxification of e-waste plastic. Li and Xu (2015) experimented to degrade the brominated epoxy resins (BERs) and recovery of metals at the same time from waste random access memory (RAM) by SCW treatment. The optimal condition was reported as temperature of 495°C, 33 MPa pressure, and time of 305 min. The major decomposition products were found as phenol, 2-methylphenol, 4-(1-methylethyl)-phenol, 4-methylphenol, 4-ethylphenol, 4-ethylphenol, 4-phenylphenol, and HBr.

Niu et al. (2017) utilizes a supercritical water oxidation process for the processing of epoxy resin and obtaining tantalum from capacitors in presence of H_2O_2. The authors employed 400°C temperature at 25 MPa for 90 min. Similarly, several other studies have reported the application of subcritical and supercritical water for e-waste recycling (Li and Xu 2019, Soler et al. 2017). However, the critical temperature (T_c) and critical pressure (T_p) of water are 374°C and 22.1 MPa, respectively, which are considerably high and limit the utilization of water as SCF. There are other solvents apart from water which can be used in the treatment of e-waste. Table 4 shows the critical condition of some typical solvents utilized in the SCF treatment (Li and Xu 2019).

The appropriate solvent can be chosen as per the need of the experiment and the critical point of the substance. Supercritical alcohols such as methanol, ethanol, and others have lower critical conditions than that of water, for example, the critical condition of methanol is Tc = 239°C and Pc = 8.09 MPa. Xiu and Zhang (2010) employed supercritical methanol for the degradation of BER. The authors reported the optimal conditions as 380°C temperature, 30 min time, and 8.03 MPa pressure. As a result, 67% conversion efficiency was found with 58% purity of phenol, 41% of HBr, and metal concentration up to 62%. Yildirir et al. (2015) investigated supercritical ethanol for the treatment of PCB. The major component of obtained liquid product was 62.4% phenol at 300°C temperature, pressure of 6.14 MPa, and 180 min of process time. Calgaro et al. (2017) used supercritical carbon dioxide modified with ethanol for liquefying polymers of waste printed circuit board. At 170°C and 7.5 MPa, 69.53% of organic is converted into oil. Supercritical isopropanol has the highest oil yield for oil recovery from brominated flame retardants plastics. Supercritical CO_2 with critical condition of Tc= 31°C and Pc = 7 MPa is a better and more promising option. It is also considered a more environmentally benign process for the delamination of PCB (Calgaro et al. 2015). Recently, some research groups have reported supercritical carbon dioxide as a greener and environment friendly process for easy delamination of plastics and further separating into copper foil, glass fiber, and polymers from e-waste (Peng and Park 2020, Hsu et al. 2019). Table 5 discusses some of the applications of supercritical fluid technology for the treatment of non-metallic fractions of e-waste.

The packaging of waste IC is a complex polymer in solid form. When the energy supplied by supercritical water is greater than the bond dissociation energy of resin molecules, the bond breaks and monomers are formed. Monomers are dissolved in supercritical water and hydrolyzed into bisphenol A and acetone, which further gets hydrolyzed into phenol and its derivatives. Finally, with various

Table 4. Typical solvents employed in SCF technology (Li and Xu 2019).

Solvents	Critical temperature(°C)	Critical Pressure (MPa)	Critical density(g/ml)
Water	374.15	22.05	0.322
Ethane	32.19	4.87	0.203
Ethylene	9.2	5.04	0.218
Propane	96.70	4.24	0.217
Ethanol	240.75	6.14	0.276
Xenon	16.55	5.87	0.118
Benzene	288.95	4.89	0.302
Methanol	239.45	8.09	0.272
Carbon dioxide	31.02	7.38	0.448
Acetone	234.95	4.70	0.278
Methane	–82.6	4.59	0.200
Toluene	318.55	4.11	0.292
Propylene	91.85	4.62	0.232
Nitrous Oxide	36.00	7.28	0.450
Ammonia	132.45	11.28	0.325

Table 5. Application of greener solvents utilised supercritical technology for the treatment of non-metallic fractions.

Solvents utilised	Feed	Optimum condition	Output	Remarks	References
$scCO_2$	Electronic display housing plastic	398°C, 90 min, 0.63 MPa	Debromination = 99.51% and Dechlorination = 99.12%	Debromination and dechlorination of BFRs and PVC are performed successfully in the presence of scCO2 solvent	Zhang and Zhang 2020
Water	Waste memory module	495°C, 33 Mpa, 305 min	Phenol and its derivatives	Metals (Cu and Au) are efficiently enriched and liquid product consists of phenol and its derivatives	Li and Xu 2015
Water	Printed circuit boards	519°C, 25 MPa, 10 min	Phenol and its derivatives	Effective degradation of brominated epoxy resins leads to the formation of phenol and its derivatives in the liquid product	Chien et al. 2000
Water	Shell of plastic	420°C, 22.5 MPa, 60 min	60% of oil produced	95.7% of debromination is achieved from brominated flame retardants	Wang and Zhang 2012
Water	Tantalum capacitor	400°C, 25 MPa, 90 min	Mold resins	Phenol and its derivatives are recovered in supercritical water oxidation condition	Niu et al. 2017
Isopropanol	Plastic shell	400°C, 18 MPa, 60 min	60% of oil product	Mixture of aromatics and benzene derivatives are recovered after the treatment of brominated flame retardants	Wang and Zhang 2012
Acetone	Epoxy resin	320°C, 1 MPa, 20 min	Carbon fibres	95.6% decomposition efficiency of epoxy resins	Okajima et al. 2017

complex side reactions, all the lower molecular weight compounds get hydrolyzed into H_2O and CO_2. It is investigated that the mechanism followed is of free radical mechanism. The possible decomposition behavior of epoxy resin of waste integrated circuits using supercritical water is depicted in Fig. 8.

Environmental problems and their impact can't be ignored, and SCF technology can be considered as a better alternative for e-waste treatment. SCF technology effectively balances the resources recovery, and detoxification without leading to any secondary pollution. SCF technology reports efficient dehalogenation of over 90%. Furthermore, the recovery of metals is also possible using SCF with a high recovery rate. However, some challenges such as salt precipitation, corrosion, and high investment cost of SCF technologies till now delayed its application at a large scale. But in recent times, problems like salt precipitation, and corrosion have reliable solutions. Future study would be required to explore a design process for the treatment of remaining polymer residue, solvent recycling, and detailed cost analysis that eventually incorporate the overall waste management and makes the process more sustainable.

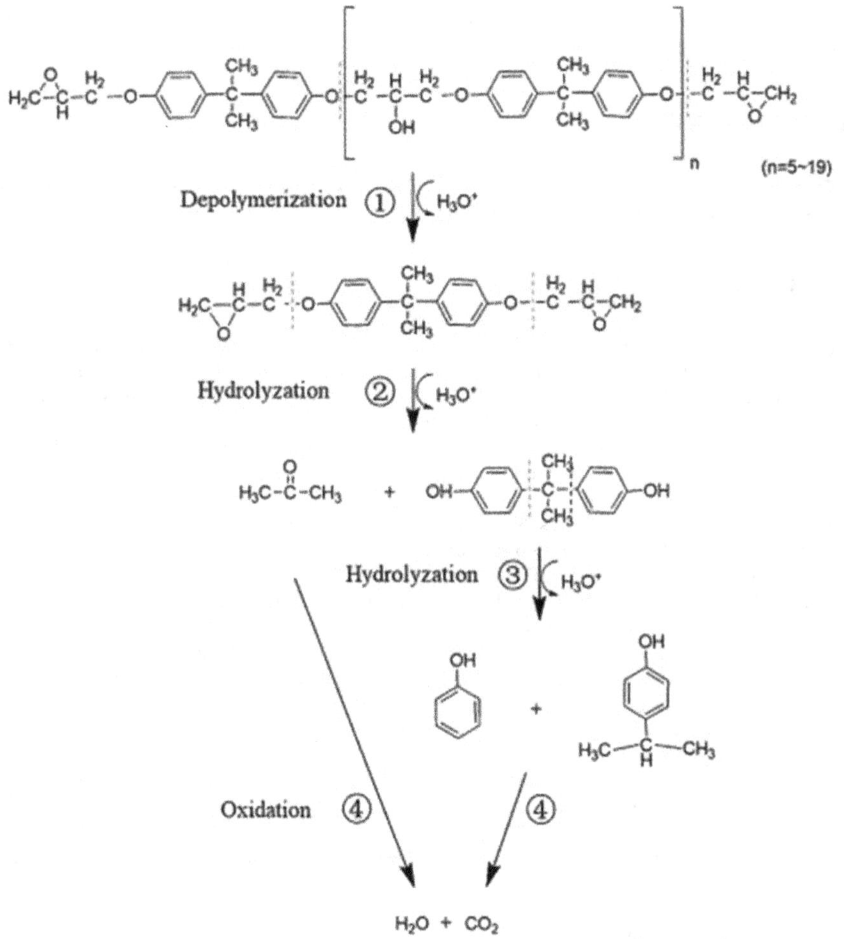

Figure 8. The proposed decomposition pathway of epoxy resin in SCW (Reproduced from Li et al. 2019).

CONCLUSIONS

The improper disposal of e-waste plastic causes soil, water, and air pollution which also hamper human health. The sustainable recycling of e-waste plastic should be given importance along with metal recovery. Presently, physical and chemical routes have been employed for the recovery or recycling of e-waste plastic. The different physical processes such as gravity separation, magnetic separation, electrostatic separation, and flotation have been investigated for the recovery of e-waste plastic. The physical separation process uses different physical properties such as size, shape, density, etc., for recovery of plastic fraction and these properties also influence the efficiency of the process. The currently employed physical processes for the recovery of plastic fraction leads to the loss of precious metals which is not desirable. The chemical processes provide an attractive option for the processing of e-waste plastic as their products can be used as fuels or feedstock for the generation of energy and valuable chemicals. The chemical processes which have been studied are pyrolysis, catalytic pyrolysis, gasification, and supercritical fluid technology. Out of these, pyrolysis and supercritical fluids can be more viable options for processing e-waste plastic. However, the presence of halogenated compounds in the products is the major challenge of the pyrolysis process. Supercritical fluid technology has not been extensively explored yet for e-waste processing. Therefore, presently, pyrolysis provides an attractive option to convert e-waste plastic into liquid and gaseous fuels without losing valuable metals. A more efficient and eco-friendly approach for the removal of halogenated compounds from pyrolysis products may pave a way for the industrial application of the pyrolysis process for processing e-waste plastic to generate a circular economy model for complete e-waste recycling.

ACKNOWLEDGMENT

The authors would like to thank the Office of the Principle Scientific Adviser to the Government of India for providing a research grant under DRIIV project.

REFERENCES

Altarawneh, M., Saeed, A., Al-Harahsheh, M. and Dlugogorski, B.Z. 2019. Thermal decomposition of brominated flame retardants (BFRs): Products and mechanisms. Progress in Energy and Combustion Science 70: 212–259.

Altarawneh, M., Ahmed, O.H., Jiang, Z.T. and Dlugogorski, B.Z. 2016. Thermal recycling of brominated flame retardants with Fe2O3. Journal of Physical Chemistry A 120: 6039–6047.

Antonakou, E.V., Kalogiannis, K.G., Stephanidis, S.D., Triantafyllidis, K.S., Lappas, A.A. and Achilias, D.S. 2014a. Pyrolysis and catalytic pyrolysis as a recycling method of waste CDs originating from polycarbonate and HIPS. Waste Management 34: 2487–2493.

Antonakou, E.V., Kalogiannis, K.G., Stefanidis, S.D., Karakoulia, S.A., Triantafyllidis, K.S., Lappas, A.A. and Achilias, D.S. 2014b. Catalytic and thermal pyrolysis of polycarbonate in a fixed-bed reactor: The effect of catalysts on products yields and composition. Polymer Degradation & Stability 110: 482–491.

Anuar Sharuddin, S.D., Abnisa, F., Wan Daud, W.M.A. and Aroua, M.K. 2016. A review on pyrolysis of plastic wastes. Energy Conservation & Management 115: 308–326.

Areeprasert, C. and Khaobang, C. 2018. Pyrolysis and catalytic reforming of ABS/PC and PCB using biochar and e-waste char as alternative green catalysts for oil and metal recovery. Fuel Processing and Technology 182: 26–36.

Baldé, C.P., Forti, V., Gray, V., Kuehr, R. and Stegmann, P. The Global E-waste Monitor – 2017, United Nations University (UNU), International Telecommunication Union (ITU) & International Solid Waste Association (ISWA), Bonn/Geneva/Vienna, 2017. ISBN 978-92-808-9053-2

Barontini, F. and Cozzani, V. 2006. Formation of hydrogen bromide and organobrominated compounds in the thermal degradation of electronic boards. Journal of Analytical and Applied Pyrolysis 77: 41–55.

Bhaskar, T., Murai, K., Matsui, T., Brebu, M.A., Uddin, M.A., Muto, A., Sakata, Y. and Murata, K. 2003. Studies on thermal degradation of acrylonitrile-butadiene-styrene copolymer (ABS-Br) containing brominated flame retardant. Journal of Analytical and Applied Pyrolysis 70: 369–381.

Brebu, M., Bhaskar, T., Murai, K., Muto, A., Sakata, Y. and Uddin, M.A. 2004. Thermal degradation of PE and PS mixed with ABS-Br and debromination of pyrolysis oil by Fe- and Ca-based catalysts. Polymer Degradation and Stability 84: 459–467.

Brezinsky, K., Pecullan, M. and Glassman, I. 1998. Pyrolysis and oxidation of phenol. Journal of Physical Chemistry A 102: 8614–8619.

Caballero, B.M., de Marco, I., Adrados, A., López-Urionabarrenechea, A., Solar, J. and Gastelu, N. 2016. Possibilities and limits of pyrolysis for recycling plastic rich waste streams rejected from phones recycling plants. Waste Management 57: 226–234.

Calgaro, C.O., Schlemmer, D.F., Bassaco, M.M., Dotto, G.L., Tanabe, E.H. and Bertuol, D.A. 2017. Supercritical Extraction of Polymers from Printed Circuit Boards using CO_2 and Ethanol. Journal of CO_2 Utilization 22: 307–316.

Calgaro, C., Schlemmer, D., da Silva, M., Maziero, E., Tanabe, E. and Bertuol, D. 2015. Fast copper extraction from printed circuit boards using supercritical carbon dioxide. Waste Management 45: 289–297.
Camel, V., Tambuté, A. and Caude, M. 1993. Analytical-Scale Supercritical Fluid Extraction: A Promising Technique for the Determination of Pollutants in Environmental Matrices. Journal of Chromatography A 642: 263–281.
Chauhan, G., Jadhao, P.R., Pant, K.K. and Nigam, K.D.P. 2018. Novel technologies and conventional processes for recovery of metals from waste electrical and electronic equipment: Challenges & opportunities—A review. Journal of Environmental Chemical Engineering 6: 1288–1304.
Chen, T., Yu, J., Ma, C., Bikane, K. and Sun, L. 2020. Catalytic performance and debromination of Fe–Ni bimetallic MCM-41 catalyst for the two-stage pyrolysis of waste computer casing plastic. Chemosphere 248: 125964.
Chiang, H.L., Lo, C.C. and Ma, S.Y. 2010. Characteristics of exhaust gas, liquid products, and residues of printed circuit boards using the pyrolysis process. Environmental Science & Pollution Research 17: 624–633.
Chien, Y.C., Wang, H.P., Lin, K.S. and Yang, Y.W. 2000. Oxidation of printed circuit board wastes in supercritical water. Water Research 34: 4279–4283.
Damrongsiri, S., Vassanadumrongdee, S. and Tanwattana, P. 2016. Heavy metal contamination characteristic of soil in WEEE (waste electrical and electronic equipment) dismantling community: a case study of Bangkok, Thailand. Environmental Science & Pollution Research 23: 17026–17034.
Das, P., Gabriel, J.C.P., Tay, C.Y. and Lee, J.M. 2021. Value-added products from thermochemical treatments of contaminated e-waste plastics. Chemosphere 269: 129409.
Debnath, B., Chowdhury, R. and Ghosh, S.K. 2018. Sustainability of metal recovery from E-waste, Front. Environmetnal Science & Engineering 12: 1–12.
De Marco, I., Caballero, B.M., Chomôn, M.J., Laresgoiti, M.F., Torres, A., Fernández, G., Arnaiz, S. 2008. Pyrolysis of electrical and electronic wastes. Journal of Analytical and Applied Pyrolysis 82: 179–183.
Devi, L., Ptasinski, K.J. and Janseen, F.J.J.G. 2003. A review of the primary measures for tar elimination in biomass gasification processes. Biomass and Bioenergy 24: 125–140.
Duan, H., Li, J., Liu, Y., Yamazaki, N. and Jiang, W. 2011. Characterization and inventory of PCDD/Fs and PBDD/Fs emissions from the incineration of waste printed circuit board. Environmetnal Science & Technology 45: 6322–6328.
Evangelopoulos, P., Persson, H., Kantarelis, E. and Yang, W. 2020. Performance analysis and fate of bromine in a single screw reactor for pyrolysis of waste electrical and electronic equipment (WEEE). Process Safety and Environmental Protection 143: 313–321.
Evangelopoulos, P., Kantarelis, E. and Yang, W. 2017. Experimental investigation of the influence of reaction atmosphere on the pyrolysis of printed circuit boards. Applied Energy 204: 1065–1073.
Evangelopoulos, P., Kantarelis, E. and Yang, W. 2015. Investigation of the thermal decomposition of printed circuit boards (PCBs) via thermogravimetric analysis (TGA) and analytical pyrolysis (Py-GC/MS). Journal of Analytical and Applied Pyrolysis 115: 337–343.
Gao, R., Zhan, L., Guo, J. and Xu, Z. 2020. Research of the thermal decomposition mechanism and pyrolysis pathways from macromonomer to small molecule of waste printed circuit board. Journal of Hazardous Material 383: 121234.
Ghosh, B., Ghosh, M.K., Parhi, P., Mukherjee, P.S. and Mishra, B.K. 2015. Waste printed circuit boards recycling: An extensive assessment of current status. Journal of Cleaner Production 94: 5–19.
Golzary, A. and Abdoli, M.A. 2020. Recycling of Copper from Waste Printed Circuit Boards by modified Supercritical Carbon Dioxide Combined with Supercritical Water Pretreatment. Journal of CO_2 Utilization 41: 101265.
Guan, G., Kaewpanha, M., Hao, X. and Abudula, A. 2016. Catalytic steam reforming of biomass tar: Prospects and challenges. Renewable and Sustainable Energy Reviews 58: 450–461.
Hall, W.J. and Williams, P.T. 2008. Removal of organobromine compounds from the pyrolysis oils of flame retarded plastics using zeolite catalysts. Journal of Analytical and Applied Pyrolysis 81: 139–147.
Hall, W.J. and Williams, P.T. 2006. Pyrolysis of brominated feedstock plastic in a fluidised bed reactor. Journal of Analytical and Applied Pyrolysis 77: 75–82.
Heidenreich, S. and Foscolo, P.U. 2015. New concepts in biomass gasification. Progress in Energy and Combustion Science 46: 72–95.
Hsu, E., Durning, C.J., West, A.C. and Park, A.H.A. 2020. Enhanced Extraction of Copper from Electronic waste via Induced Morphological Changes using Supercritical CO_2. Resources, Conservation & Recycling 168: 105296.
Hsu, E., Barmak, K., West, A.C. and Park, A.H.A. 2019. Advancements in the treatment and processing of electronic waste with sustainability: A review of metal extraction and recovery technologies. Green Chemistry 21: 919–936.
Jadhao, P.R., Ahmad, E., Pant, K.K. and Nigam, K.D.P. 2020. Environmentally friendly approach for the recovery of metallic fraction from waste printed circuit boards using pyrolysis and ultrasonication. Waste Management 118: 150–160.
Jadhao, P.R., Ahmad, E., Pant K.K. and Nigam, K.D.P. 2022a. Advancements in the field of electronic waste recycling: Critical assessment of chemical route for generation of energy and valuable products coupled with metal recovery. Separation & Purification Technology 289: 120773.
Jadhao, P.R., Vuppaladadiyam, A.K., Prakash, A. and Pant, K.K. 2022b. Co-pyrolysis characteristics and kinetics of electronic waste and macroalgae: A synergy study based on thermogravimetric analysis. Algal Research 61: 102601.
Jung, S.H., Kim, S.J. and Kim, J.S. 2012. Thermal degradation of acrylonitrile-butadiene-styrene (ABS) containing flame retardants using a fluidized bed reactor: The effects of Ca-based additives on halogen removal. Fuel Processing and Technology 96: 265–270.
Kaya, M. 2016. Recovery of metals and nonmetals from electronic waste by physical and chemical recycling processes. Waste Management 57: 64–90.

Kell, D. 2009. Recycling and recovery. *In*: Hester, R.E. and Harrison, R.M. (eds.). Electronic Waste Management, pp. 91–110.

Lee, J., Lee, T., Ok, Y.S., Oh, J.I. R.M. Kwon, E.E. 2017. Using CO2 to mitigate evolution of harmful chemical compounds during thermal degradation of printed circuit boards. Journal of CO_2 Utilization 20: 66–72.

Li, K. and Xu, Z. 2019. A Review of current progress of supercritical fluid technologies for e-waste treatment. Journal of Cleaner Production 227: 794–809.

Li, K., Zhang, L. and Xu, Z. 2019. Decomposition Behavior and Mechanism of Epoxy Resin from Waste Integrated Circuits under Supercritical Water Condition. Journal of Hazardous Material 374: 356–364.

Li, J., Jiang, Y. and Xu, Z. 2017. Eddy current separation technology for recycling printed circuit boards from crushed cell phones. Journal of Cleaner Production 141: 1316–1323.

Li, K. and Xu, Z. 2015. Application of supercritical water to decompose brominated epoxy resin and environmental friendly recovery of metals from waste memory module. Environmental Science &. Technology 49: 1761–1767.

Li, J., Xu, Z. and Zhou, Y. 2007. Application of corona discharge and electrostatic force to separate metals and nonmetals from crushed particles of waste printed circuit boards. Journal of Electrostatics 65: 233–238.

Li, J., Shrivastava, P., Gao, Z. and Zhang, H.C. 2004. Printed circuit board recycling: A state-of-the-art survey, IEEE Transactions on Electronics Packaging Manufacturing, 27, 33–42.

Lin, K.H. and Chiang, H.L. 2014. Liquid oil and residual characteristics of printed circuit board recycle by pyrolysis. Journal of Hazardous Material 271: 258–265.

Luo, C., Liu, C., Wang, Y., Liu, X., Li, F., Zhang, G. and Li, X. 2011. Heavy metal contamination in soils and vegetables near an e-waste processing site, south China. Journal of Hazardous Materials 186: 481–490.

Ma, C., Yu, J., Wang, B., Song, Z., Xiang, J., Hu, S., Su, S., Sun, L. (2017) Catalytic pyrolysis of flame retarded high impact polystyrene over various solid acid catalysts. Fuel Processing & Technology 155: 32–41.

Ma, C., Yu, J., Wang, B., Song, Z., Xiang, J., Hu, S., Su, S. and Sun, L. 2016a. Chemical recycling of brominated flame retarded plastics from e-waste for clean fuels production: A review. Renewable & Sustainable Energy Reviews, 61, 433–450.

Ma, C., Yu, J., Wang, B., Song, Z., Zhou, F., Xiang, J., Hu, S. and Sun, L. 2016b. Influence of zeolites and mesoporous catalysts on catalytic pyrolysis of brominated acrylonitrile-butadiene-styrene (Br-ABS). Energy and Fuels 30: 4635–4643.

Menad, N.E. 2016. Physical separation processes in waste electrical and electronic equipment recycling. *In*: Chagnes, A., Cote, G., Ekberg, C., Nilsson, M. and Retegan, T. (eds.). WEEE Recycling: Research, Development, and Policies, Elsevier Inc., pp. 53–74.

Molino, A., Chianese, S. and Musmarra, D. 2016. Biomass gasification technology: The state of the art overview. Journal of Energy Chemistry 25: 10–25.

Muhammad, C., Onwudili, J.A. and Williams, P.T. 2015. Catalytic pyrolysis of waste plastic from electrical and electronic equipment. Journal of Analytical & Applied Pyrolysis 113: 332–339.

Ng, C.H., Salmiaton, A. and Hizam, H. 2014. Catalytic pyrolysis and a pyrolysis kinetic study of shredded printed circuit board for fuel recovery. Bulletin of Chemical Reaction Engineering & Catalysis 9: 224–240.

Niu, B., Chen, Z. and Xu, Z. 2017. Recovery of Tantalum from Waste Tantalum Capacitors by Supercritical Water Treatment. ACS Sustainable Chemical Engineering 5: 4421–4428.

Ogunniyi, I.O. and Vermaak, M.K.G. 2009. Froth flotation for beneficiation of printed circuit boards comminution fines: An overview. Mineral Processing and Extractive Metallurgy Review 30: 101–121.

Okajima, I., Watanabe, K., Haramiishi, S., Nakamura, M., Shimamura, Y. and Sako, T. 2017. Recycling of Carbon Fiber Reinforced Plastic Containing Amine-Cured Epoxy Resin using Supercritical and Subcritical Fluids. Journal of Supercritical Fluids 119: 44–51.

Peng, P. and Park, A.H.A. 2020. Supercritical CO2-Induced Alteration of a Polymer-Metal Matrix and Selective Extraction of Valuable Metals from Waste Printed Circuit Boards. Green Chemistry 22: 7080–7092.

Richard, G.M., Mario, M., Javier, T. and Susana, T. 2011. Optimization of the recovery of plastics for recycling by density media separation cyclones. Resources, Conservation & Recycling 55: 472–482.

Ruan, J. and Xu, Z. 2016. Constructing environment-friendly return road of metals from e-waste: Combination of physical separation technologies. Renewable & Sustainable Energy Reviews 54: 745–760.

Sakata, Y., Bhaskar, T., Uddin, M.A., Muto, A. and Matsui, T. 2003. Development of a catalytic dehalogenation (Cl, Br) process for municipal waste plastic-derived oil. Journal of Material Cycles and Waste Management 5: 113–124.

Salbidegoitia, J.A., Fuentes-Ordóñez, E.G., González-Marcos, M.P., González-Velasco, J.R., Bhaskar, T. and Kamo, T. 2015. Steam gasification of printed circuit board from e-waste: Effect of coexisting nickel to hydrogen production. Fuel Processing & Technology 133: 69–74.

Sansaniwal, S.K., Pal, K., Rosen, M. A. and Tyagi, S.K. 2017. Recent advances in the development of biomass gasification technology: A comprehensive review. Renewable and Sustainable Energy Reviews 72: 363–384.

Santella, C., Cafiero, L., De Angelis, D., La Marca, F., Tuffi, R. and Vecchio Ciprioti, S. 2016. Thermal and catalytic pyrolysis of a mixture of plastics from small waste electrical and electronic equipment (WEEE). Waste Management 54: 143–152.

Sarvar, M., Salarirad, M.M. and Shabani, M.A. 2015. Characterization and mechanical separation of metals from computer Printed Circuit Boards (PCBs) based on mineral processing methods. Waste Management 45: 246–257.

Shahbaz, M., Yusup, S., Inayat, A., Patrick, D.O. and Ammar, M. 2017. The influence of catalysts in biomass steam gasification and catalytic potential of coal bottom ash in biomass steam gasification: A review. Renewable and Sustainable Energy Reviews 73: 468–476.

Shen, Y. 2018. Effect of chemical pretreatment on pyrolysis of non-metallic fraction recycled from waste printed circuit boards. Waste Management 76: 537–543.

Shi, P., Wan, Y., Grandjean, A., Lee, J.M. and Tay, C.M. 2021. Clarifying the in-situ cytotoxic potential of electronic waste plastics. Chemosphere 269: 128719.

Soler, A., Conesa, J.A. and Ortuño, N. 2017. Emissions of brominated compounds and polycyclic aromatic hydrocarbons during pyrolysis of e-waste debrominated in subcritical water. Chemosphere 186: 167–176.

Stevens, G.C. and Goosey, M. 2009. Materials used in manufacturing electrical and electronic products. pp. 40–74. *In*: Hester, R.E. and Harrison, R.M. (eds.). Electronica Waste Management.

Sun, J., Wang, W., Liu, Z. and Ma, C. 2011. Recycling of waste printed circuit boards by microwave-induced pyrolysis and featured mechanical processing. Industrial & Engineering Chemistry Research 50: 11763–11769.

Tang, X., Shen, C., Shi, D., Cheema, S.A., Khan, M.I., Zhang, C. and Chen, Y. 2010. Heavy metal and persistent organic compound contamination in soil from Wenling: An emerging e-waste recycling city in Taizhou area, China. Journal of Hazardous Materials 173: 653–660.

Tue, N.M., Goto, A., Takahashi, S., Itai, T., Asante, K.A., Nomiyama, K., Tanabe, S. and Kunisue, T. 2017. Soil contamination by halogenated polycyclic aromatic hydrocarbons from open burning of e-waste in Agbogbloshie (Accra, Ghana). Journal of Material Cycles and Waste Management 19: 1324–1332.

Veit, H.M., Diehl, T.R., Salami, A.P., Rodrigues, J.S., Bernardes, A.M. and Tenório, J.A.S. 2005. Utilization of magnetic and electrostatic separation in the recycling of printed circuit boards scrap. Waste Management 25: 67–74.

Wang, Q., Zhang, B., Yu, S., Xiong, J., Yao, Z., Hu, B. and Yan, J. 2020. Waste-printed circuit board recycling: Focusing on preparing polymer composites and geopolymers. ACS Omega 5: 17850–17856.

Wang, Y., Sun, S., Yang, F., Li, S., Wu, J., Liu, J., Zhong, S. and Zeng, J. 2015. The effects of activated Al2O3 on the recycling of light oil from the catalytic pyrolysis of waste printed circuit boards. Process Safety & Environmental Protection 98: 276–284.

Wang, Y. and Zhang, F.S. 2012. Degradation of brominated flame retardant in computer housing plastic by supercritical fluids. Journal of Hazardous Material 205–206: 156–163.

Wei, J. and Realff, M.J. 2003. Design and optimization of drum-type electrostatic separators for plastics recycling. AIChE Journal 49: 3138–3149.

Wu, Q., Leung, J.Y.S., Geng, X., Chen, S., Huang, X., Li, H., Huang, Z., Zhu, L., Chen, J. and Lu, Y. 2015. Heavy metal contamination of soil and water in the vicinity of an abandoned e-waste recycling site: Implications for dissemination of heavy metals. Science of the Total Environment 506–507: 217–225.

Xie, Y., Sun, S., Liu, J., Lin, W., Chen, N. and Ye, M. 2017. The effect of additives on migration and transformation of gaseous pollutants in the vacuum pyrolysis process of waste printed circuit boards. Waste Management Research 35: 190–199.

Xiu, F.R. and Zhang, F.S. 2010. Materials recovery from waste printed circuit boards by supercritical methanol. Journal of Hazard. Material 178: 628–634.

Xue, M., Yan, G., Li, J. and Xu, Z. 2012. Electrostatic separation for recycling conductors, semiconductors, and nonconductors from electronic waste. Environmetnal Science & Technology 46: 10556–10563.

Yang, J., Wang, H., Zhang, G., Bai, X., Zhao, X. and He, Y. 2019. Recycling organics from non-metallic fraction of waste printed circuit boards by a novel conical surface triboelectric separator. Resources, Conservation and Recycling 146: 264–269.

Yildirir, E., Onwudili, J.A. and Williams, P.T. 2015. Chemical Recycling of Printed Circuit Board Waste by Depolymerization in Sub- and Supercritical Solvents. Waste and Biomass Valorization 6: 959–965.

Yin, J., Li, G., He, W., Huang, J. and Xu, M. 2011. Hydrothermal decomposition of brominated epoxy resin in waste printed circuit boards. Journal of Analytical and Applied Pyrolysis 92: 131–136.

Yoo, J.M., Jeong, J., Yoo, K., Lee, J.C. and Kim, W. 2009. Enrichment of the metallic components from waste printed circuit boards by a mechanical separation process using a stamp mill. Waste Management 29: 1132–1137.

Zhang, T., Mao, X., Qu, J., Liu, Y., Siyal, A.A., Ao, W., Fu, J., Dai, J., Jiang, Z., Deng, Z., Song, Y., Wang, D. and Polina, C. 2021. Microwave-assisted catalytic pyrolysis of waste printed circuit boards, and migration and distribution of bromine. Journal of Hazardous Materials 402: 123749.

Zhang, C.C. and Zhang, F.S. 2020. Enhanced Dehalogenation and Coupled Recovery of Complex Electronic Display Housing Plastics by Sub/supercritical CO2. Journal of Hazardous Materials 382, 121140.

Zhang, T., Ruan, J., Zhang, B., Lu, S., Gao, C., Huang, L., Bai, X., Xie, L., Gui, M. and Qiu, R.L. 2019. Heavy metals in human urine, foods and drinking water from an e-waste dismantling area: Identification of exposure sources and metal-induced health risk. Ecotoxicology and Environment. Safety 169: 707–713.

Zhang, G., He, Y., Wang, H., Zhang, T., Yang, X., Wang, S. and Chen, W. 2017. Application of triboelectric separation to improve the usability of nonmetallic fractions of waste printed circuit boards : Removing inorganics. Journal of Cleaner Production 142: 1911–1917.

Zhang, G., Wang, H., Zhang, T., Yang, X., Xie, W. and He, Y. 2016. Removing inorganics from nonmetal fraction of waste printed circuit boards by triboelectric separation. Waste Management 49: 230–237.

Zhang, S., Yoshikawa, K., Nakagome, H. and Kamo, T. 2012. Steam gasification of epoxy circuit board in the presence of carbonates. Journal of Material Cycles and Waste Management 14: 294–300.

Zhang, S., Forssberg, E., Arvidson, B. and Moss, W. 1998. Aluminum recovery from electronic scrap by High-Force® eddy-current separators. Resources, Conservation & Recycling 23: 225–241.

Zhao, Y.B., Lv, X. D. and Ni, H.G. 2018. Solvent-based separation and recycling of waste plastics: A review. Chemosphere 209: 707–720.

8

CHEMICAL RECYCLING OF PLASTIC WASTE

Hossein Esfandian,[1] *Navid Motevalian,*[2]
Ehsan Shekarian,[3] *Reza Katal*[4,*] and *Sajjad Rimaz*[5,*]

1. INTRODUCTION

The word "plastics" refers to synthetic polymers that are produced from the light fraction of crude oil or natural gas. Plastic is so abundant in modern society that, on average, each individual in the United States uses 68 kg of plastic per year and each individual in the European Union uses 50 kg per year (Plastic Insight 2014). Overall, there are 26 different types of synthetic polymers that qualify as plastics, and seven of these polymers (Fig. 1) meet 80% of the global plastic demand. Polyethylene (PE), including low-density (LDPE) and high-density (HDPE) polyethylene,is the most highly consumed of these seven plastics, followed by polypropylene (PP), which constitutes about 16% of the global plastic production. Polypropylene has a wide range of applications in diverse industries including packaging, construction, and auto motives, among others (Bora et al. 2020).

Polyethylene terephthalate (PET), a semicrystalline thermoplastic polyester, is the third most consumed plastic and is widely used for bottles, films, sheets, and fibers, among other things. Because PET is lightweight, mechanically strong, resistant to chemicals, and highly aesthetic, it is widely used in the packaging and textile industries (Raheem et al. 2019). In 2015, approximately 72 million tons of PET were produced globally; of these, almost 48 million tons were amorphous and used for fiber applications, 20 million tons were used for packaging, and 4 million tons were used in films (Crippa and Morico 2019). Because the flexibility and convenience of plastics, everyday life is now pervaded by plastics in the form of packaging,sports and clothing equipment, electronic components,and biomedical devices,among other things (Day et al. 1999). Resin Identification Code (RIC) was introduced in 1988 by the Society of the Plastics Industry, dividing plastics to seven different groups. The purpose was to "provide a consistent national system to facilitate recycling of post-consumer plastics." RIC,

[1] Faculty of Engineering Technologies, Amol University of Special Modern Technologies, Amol, Iran.
[2] Faculty of Civil and Environmental Engineering, Shahid Beheshti University, Tehran, Iran.
[3] Department of Petroleum and Chemical Engineering, Science and Research Branch, Islamic Azad University, Tehran, Iran.
[4] Department of Civil and Environmental Engineering, National University of Singapore, Singapore.
[5] Department of Chemical and Biomolecular Engineering, National University of Singapore, Singapore.
* Corresponding authors: sajjad.rimaz@gmail.com; rezak@greenli-ion.com

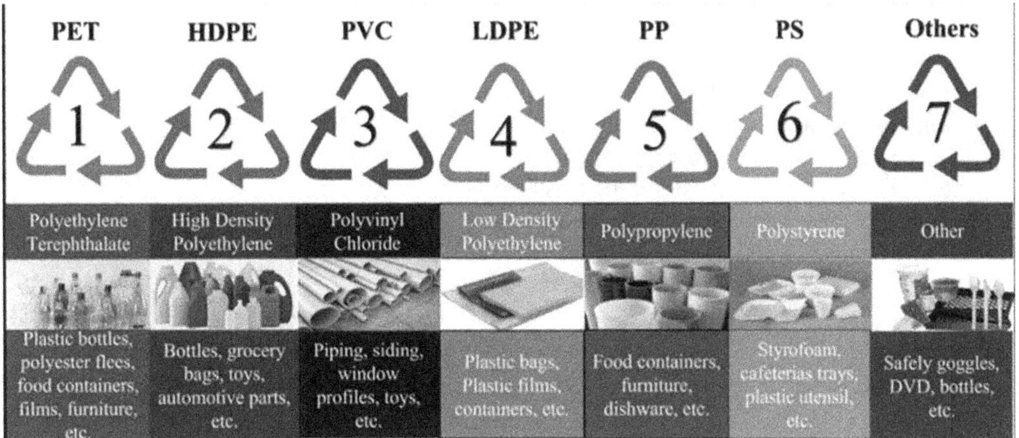

Figure 1. Different kinds of plastics, their applications and RIC numbers.

after minor revision, was recognized as the standard method for plastic classification. In Fig. 1, RIC code of each plastic is also introduced.

Despite the advantages and convenience of plastic, the widespread use of plastic has led to several concerns regarding appropriate plastic disposal (Jia et al. 2019, Lebreton et al. 2019, 2017), and several countries have enacted legislation to promote plastic recycling. For example, the European Union (2018) has targeted that fifty percent of plastic packaging need to be recycled by 2025 and that all plastic packaging needs to be recyclable by 2030. Additional recycling targets in the European Union include having 55% of plastic packaging and 65% of municipal solid waste (MSW) be recycled by 2035 (Solis and Silveira 2020, European Commission 2018). Estimates for the actual amount of plastics that are recycled vary among studies, but approximately 42% of all plastic packaging in the European Union and some 14–18% of all plastics globally were recycled in 2017 (OECD 2018). Martin et al. (2021) summarized the status of different plastic waste management in whole part of the world (Fig. 2). A large portion of plastics are either landfilled or lost, and as a result, plastics pollute the air we breathe and the water we drink. Incineration, a common method for recycling of wastes, is the worst form of plastic recycling; for example, burning one ton of plastics produces around 5 to 10 tons of CO_2 and other harmful gases, including dioxanes. Incineration facilities therefore require controlling units to capture these gases and prevent them from entering the environment (Vollmer et al. 2020).

As shown in Fig. 2, mechanical recycling is the currently most common recycling technology because it requires less resources and energy than other strategies for plastic management (Roosen et al. 2020, Garcia and Robertson 2017). However, challenges with mechanical recycling include immiscibility of various waste materials and the thermal-mechanical degradation of various polymers, which make it difficult to convert heterogeneous waste streams into recyclates that are of a sufficient quality for high-end recycling routes (i.e., recycling routes with a destination other than the common mixed-polymer bulk uses such as plant trays, garden benches, street pavement, or compost bins) (Huysveld et al. 2019, Ragaert et al. 2017). Another challenge facing mechanical recycling is the presence of definite components in waste material, including inks and odorous constituents (Vollmer et al. 2021). Definite components often contain halogens and metals, and because recycled materials could potentially release these pollutants, the adequate recycling of these compounds has become highly regulated. Pollutants in recycled material could otherwise pose a potential risk for the consumers' health, particularly for food contact materials (FCM) (Garcia et al. 2013, European Commission 2011).

On account of the disadvantages with mechanical recycling, recent efforts have explored chemical recycling technologies that can be used to recycle even heterogeneous and/or contaminated plastics. Chemical recycling methods can be used to convert plastics into synthesis gas (a mixture of hydrogen and carbon monoxide), which can then be used to produce several different fuels. Other routes of chemical recycling, including pyrolysis or solvolysis,can be used to convert plastics into monomers or oligomers that can then be purified and used again for plastic production (Weckhuysen 2020). Developing and applying the chemical recycling techniques for plastic wastes can considerably reduce the cost of the polymer and plastic production (virgin) because the products of chemical recycling can be immediately used to produce virgin plastic (Rahimi and García 2017). Chemical recycling of plastic wastes has the potential to save about 3.5 billion barrels of oil (USD$176 billion), which could help create a circular economy for plastic wastes (Fig. 3). The main purpose of a circular economy for plastic wastes is to keep the available materials for polymer production in a closed loop,thereby decreasing greenhouse gas emissions and minimizing the production of other waste products. A circular

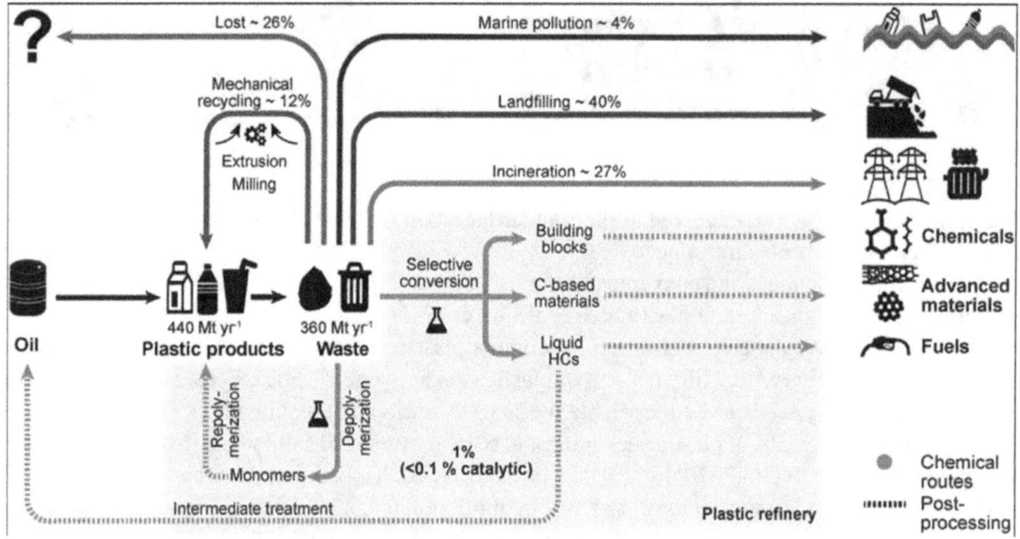

Figure 2. Different strategies for plastic management [Martin et al. (2021)]. Reproduced or/and reprinted with permission from Elsevier.

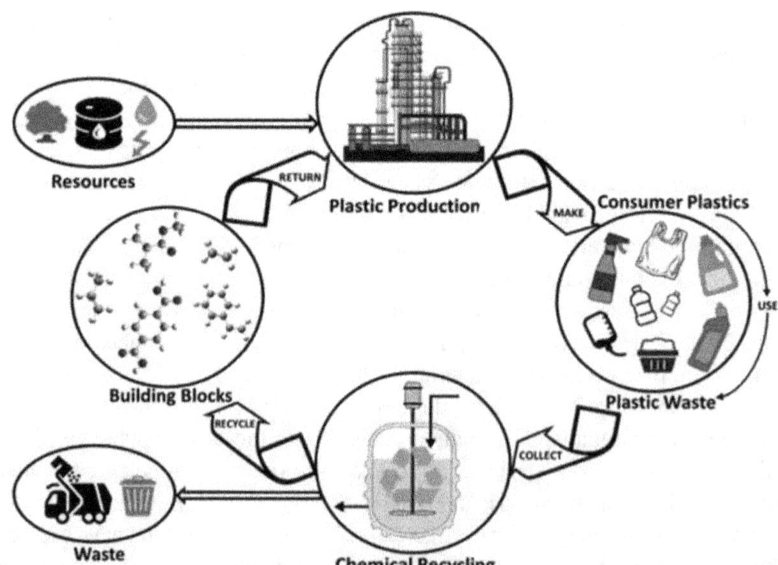

Figure 3. Role of chemical recycling in the circular economy (Dogu et al. 2021). Reproduced or/and reprinted with permission from Elsevier.

economy would also decrease our reliance on fuels and natural resources. However, achieving this circular economy will require the development of effective chemical recycling technologies.

This chapter summarizes the different methods for plastic recycling in the context of the environmental and economic problems that plastic wastes have created in the past few decades. We give particular attention to chemical methods because of their enormous advantages over other recycling techniques and their unique role in the circular economy for plastic wastes.

2. PLASTIC RECYCLING

There are four general classes of plastic solid waste (PSW) recycling procedures: primary, secondary, tertiary, and quaternary (Fig. 4). When a given material is brought through a recycling process, some features of the material are lost or altered, including tensile strength, dimensional accuracy, and wear features. Each PSW procedure generates different novel polymers, and there are advantages and disadvantages to every technique. Some of the different PSW techniques are described below.

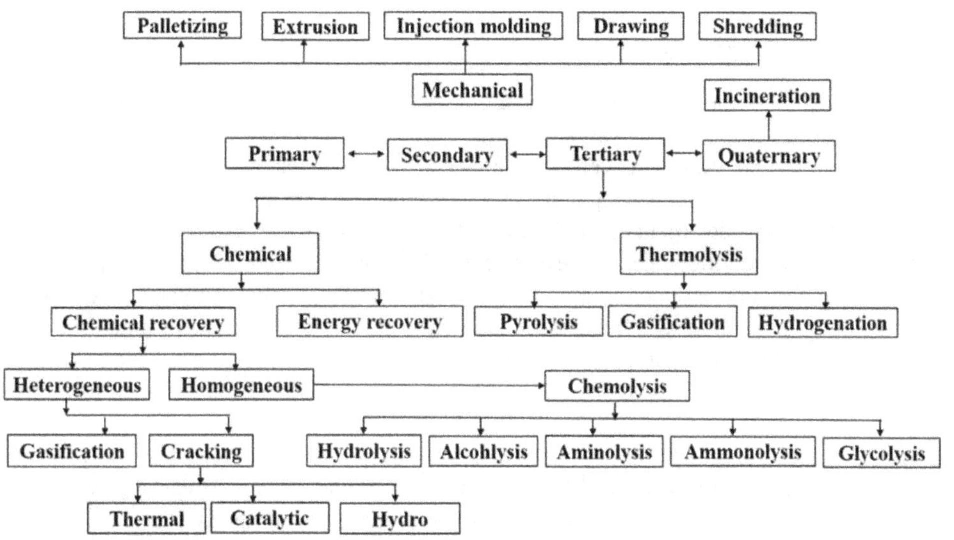

Figure 4. Different methods for PSW recycling (Solis and Silveira 2020). Reproduced or/and reprinted with permission from Elsevier; Copywrite 2020.

2.1 Primary recycling

Primary recycling is the most popular PSW process because it is simple and cost-effective. Primary recycling entails reusing products in their original structures without any chemical modifications. However, one drawback of primary recycling is that plastics can only be re-used a limited number of times before they start to break down or wear out (Grigore 2017, Francis 2016, Hopewell et al. 2009). Re-extrusion (Al-Salem 2009), also known as the closed-loop procedure (Sadat-Shojai and Bakhshandeh 2011), is a specific technique for recycling a single kind of uncontaminated polymer that still has features similar to virgin material (Kumar et al. 2011). In this procedure, scrap plastics with similar properties to the original products are used in place of the original products (Al-Salem 2009). Re-extrusion can only be performed with clean or semi-clean scrap plastics provided that any polluted parts are effectively sorted out and removed. Common re-extrusion plastics are polyethylene, polypropylene, polystyrene, nylon, and PVC. The conditions for extrusion of two main ones are listed in Table 1.

Table 1. The conditions for extrusion of most common plastics.

Plastic	Example	References
Polypropylene	Twin-screw extruder, twin-screw extruder had a length/diameter (L/D) ratio of 60:1; feed rate was controlled at 5 kg/h; the screw speed was 100 rpm, and the temperature of the screw was controlled at 230°C.	Phanthong et al. 2021
Polyethylene	Another twin-screw extruder. The conditions were a temperature of 200°C, screw rotation speed of 200 rpm, feed rate was 10 kg/h.	Okubo et al. 2021

MSW is normally not appropriate for primary recycling because it is highly polluted, impure, or contaminated (Yu et al. 2016). Clean scrap can occasionally be introduced into collected waste in order to achieve a waste product that is more similar to the virgin substance (Kumar et al. 2011). This technique is popular and easy in the manufacturing industry, as it results in the transformation of plastic waste into original-quality products (Sadat-Shojai and Bakhshandeh 2011). Injection molding and other mechanical recycling methods are also forms of primary recycling, though the quality of the material being recycled is different (Barlow 2008).

2.2 Secondary recycling

Only thermoplastic polymers can be used for secondary recycling because they can be melted and re-processed into end products. The alteration of the polymer during the process is not included in the mechanical recycling. Secondary recycling is a physical technique that involves shredding, washing, or cutting plastic waste into granules, pellets, or flakes. Washing includes two steps: a hot wash with 2% NaOH solution and a detergent at 80°C followed by a cold wash with water only. The washed and shredded plastics are then melted to create a novel product by extrusion. To achieve superior results, the reprocessed substance can also be mixed with virgin material. A major advantage of secondary recycling is that plastic waste can be processed directly into end products after it has been sorted, cleaned, and dried, thus reducing the amount of waste associated with plastic production.

One of the drawbacks of secondary recycling is that the final product is often of lower quality because trace acidic impurities and water (remained from washing step) cause chain scission reactions in the polymer. Strategies for preventing this reduction in the molecular weight of the final product include intensive drying,the use of chain extender compounds,and reprocessing with vacuum degassing. Although secondary recycling is relatively low-cost in comparison to primary, tertiary and quaternary, it requires a substantial initial investment (Grigore 2017, Francis 2016, Hopewell et al. 2009).

2.3 Tertiary recycling

Primary and secondary recycling methods are sometimes difficult to implement because they require waste materials to be identified and sorted. For example, contamination of polymer waste is a major challenge for primary recycling because most PSW is a combination of heterogeneous components (Kumar et al. 2011). Primary and secondary recycling methods also do not affect energy sustain ability. However, tertiary recycling of plastic polymers has positive effects on overall sustain ability of plastic products (Singh et al. 2017). Recyclers have been attracted to tertiary recycling techniques because they re-generate the raw materials that were used to produce the original plastics.

Tertiary recycling, also known as chemical recycling, specifically refers to the recovery of monomers from PSW via a depolymerization process. There are several techniques for tertiary recycling including pyrolysis, chemolysis, and gasification cracking. Thermolysis and chemolysis (solvolysis) are the depolymerization of PSW using either heat or chemicals, respectively. Pyrolysis refers to processing in the absence of air, and gasification refers to processing in controlled environments. Glycolysis, a subcategory of chemolysis, is the degradation of polymers using a glycol like ethylene glycol or diethylene. The use of methanol to degrade polymers, which results in a trans esterification reaction, is called methanolysis (Kumar et al. 2011). The various tertiary recycling techniques are shown in Fig. 5 and described in more detail in Section 3.

2.4 Quaternary recycling

All plastic polymers start losing some of their structural features as they go through multiple cycles of primary, secondary, and tertiary recycling, and eventually landfills become the only viable way to discard unusable plastic waste. However, depositing waste in landfills pollutes the Earth's surface. Quaternary recycling, or combustion,is the last option for MSW disposal if the other three recycling methods are no longer effective. Combustion is becoming a more popular method for MSW disposal as incineration facilities become increasingly efficient (Subramanian 2000). Quaternary recycling is also an effective energy recovery procedure because burning plastics that have been through other recycling releases energy that can potentially be harvested for other uses.

The energy content for heating oil is approximately 44.5 MJ/kg. This value is above 40 MJ/kg for HDPE, PP, LDPE, whereas it is less than 40 MJ/kg for PS and less than 30 MJ/kg for PET and PVC due to the presence of oxygen and chlorine (Dogu et al. 2021). Although these numbers initially make it seem like the incineration of plastics could be used as an alternative source of energy,the incineration process produces pollutants and residues that can be damaging to the environment (Gurgul et al. 2018, Hwang et al. 2017, Taylor et al. 2014, Roes et al. 2012). Incinerated plastic wastes can release toxic volatile organic compounds including dioxins, furans, and halogenated gasses. These waste products must be carefully treated so that they do not escape into the environment (Vinu et al. 2016). Even though quaternary recycling techniques can handle any kind of plastic waste and contamination, the associated production of toxic gases make it the least sustainable recycling method.

3. CHEMICAL RECYCLING OF PLASTICS

As mentioned earlier, plastic waste can be a promising source of fuels and chemicals. Most of the interest in plastic waste has been focused on mechanical recycling, energy recovery, and the production of valuable products like petrochemical feedstocks or monomers via chemical recycling. There are chemical recycling cases for some plastics, including Polyurethane (PUR), nylon, and PET, and there is growing interest in using these plastics as feedstocks because of their high hydrocarbon content and similarity to conventional petroleum fractions. Unlike biomass, plastic waste (especially polyolefin) does not contain significant amounts of oxygen and so can achieve higher carbon efficiency and gross margins than other feedstocks. Pyrolysis is an efficient recycling method for plastics without oxygen bonds or other heteroatoms, whereas solvolysis is a better option for plastics with heteroatoms. This section is mainly focused on chemical recycling techniques that include either monomer or feedstock recycling. It is worth mentioning that feedstock recycling is a part of chemical recycling that obtains its name from the primary product (petrochemical feedstock). This term is used to differentiate thermal process which is used for conversion of plastic waste into petrochemical plant feedstock, from chemical processes which purify the stream of plastic waste or break down waste product into monomers. These two forms of recycling are recognized as ideal approaches for preserving limited resources and protecting the environment by reducing the volume of non-degradable waste (Al-Salem et al. 2010, 2009).

3.1 Chemolysis

Chemolysis, also known as solvolysis, is one of the most well-known chemical methods for recycling plastic wastes. This method is specifically useful for recycling polymers with a carbonyl group (i.e., polymers that contain oxygen in their structure), including polyesters such as PET. A variety of solvents, including water, methanol, amine, and ethylene glycol, can be used for chemolysis (Pham and Cho 2021, Ügdüler et al. 2020, Raheem et al. 2019). Kosloski-Oh et al. (2020) reviewed different solvolysis methods for different kind of plastics. In the following paragraph, we briefly discuss the application of solvolysis methods for PET plastics.

PET can be entirely depolymerized into terephthalate (BHET), monomers, terephthalic acid (TPA), bis(hydroxylethylene), ethylene glycol (EG), and dimethyl terephthalate (DMT) using chemical recycling procedures (Das et al. 2021). The depolymerization process is the reverse reaction of the polymerformation pathway. PET can also be partly depolymerized to oligomers or other chemicalmaterials. There are various depolymerization paths including methanolysis, hydrolysis, glycolysis, aminolysis, and ammonolysis; the specific path is based on the chemical agent utilized for the PET chain scission. Fig. 5 shows various examples of PET chemolysis and the kind of products derived from PET depolymerization. The following sections discuss these examples in detail and summary of all of them is presented in Table 2.

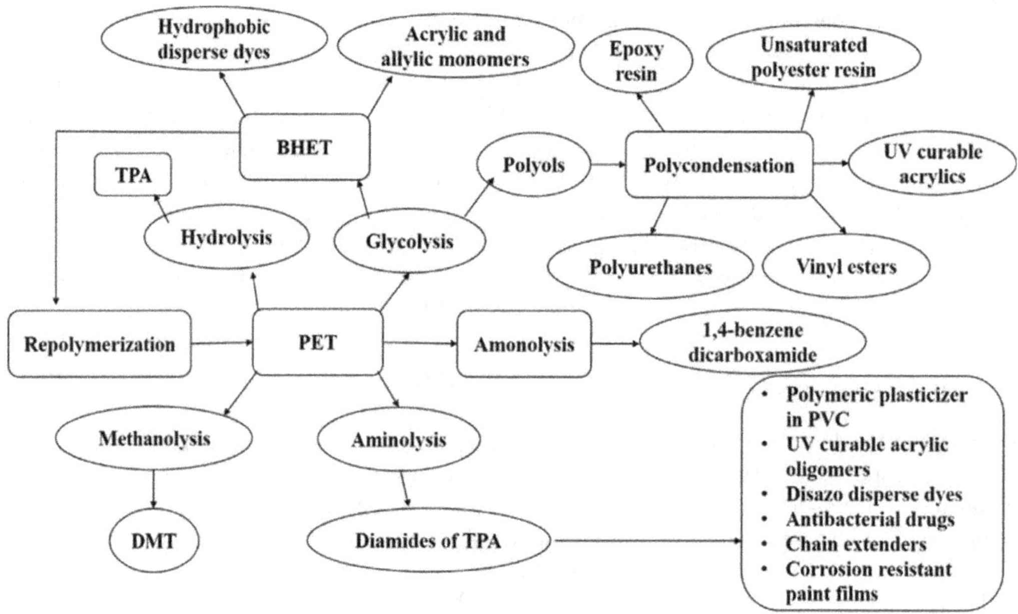

Figure 5. Different approaches for PET chemolysis and the resulting value-added products (Ragaert et al. 2017). Reproduced or/and reprinted with permission from Elsevier.

3.1.1 Alcoholysis

One of the most attractive approaches for PET degradation is alcoholysis, which involves utilizing alcohols as solvents. This process was developed over more than four decades and is currently the most common commercial PET recycling technique. Renowned chemical companies such as Shell Polyester, DuPont Goodyear, Zimmer, and Eastman Kodak all practice alcoholysis (George and Kurian 2014). Alcoholysis was also considered to be one of the most effective ways for recycling transparent PET scrap (i.e., PET without dyes or colorants) (Bedell et al. 2018, Welle 2011). According to previous studies, the products of PET alcoholysis can be used as raw materials for preparing waterborne polyurethane (WPU) (Zhou et al. 2017a,b, Fang et al. 2015). One study demonstrated that the products of PET alcoholysis products can enhance the particle size, thermal resistance, stability, and viscosity of WPU. The best performance of WPU is provided by formulations that comprise glycolyzed oligoesters within the hard segment sections of polyurethane. However,because the

Table 2. Summary of chemolysis method for plastic recycling (Raheem et al. 2019).

Technology	Reagents	Advantages	Disadvantages	Feed	Operation Condition	Products	Technology Statues	Ref.
Hydrolysis (Alkaline, Acidic, Neutral)	Alkali (NaOH) and acid (H_2SO_4) for acidification. Strong acids (Sulfuric, Phosphoric, and Nitric) Water or Steam	High product selectivity (i.e., TPA)	Highly alkaline or acidic condition is needed. Corrasion of equipment Yield in natural condition is low For neutral high temperature and pressure is needed In alkaline method, feed waste contaminants are carried over with the product, hence product purification is required	PET, PC	H_2SO_4: 25–100ºC; 1 bar.; 87 wt % Conc. HNO_3: 70–100ºC . Tempt. (200–250ºC); Press. (10–20 bar); time (> = 3 h); Tempt. (200–265ºC); Press. (10–40 bar); time (2 h); Alkaline metal acetate (Zinc acetate)	TPA and EG or alkaline metal salt of TPA and EG. Then, acidification to TPA and salt of used acid.	Alkaline hydrolysis is commercializing by DEMTEO	Raheem et al. 2019, Crippa and Morico (2019)
Glycolysis	Glycol (EG or PG or DEG or TEG, etc.)	Not all products can be separated from products	High product selectivity, e.g., bis (hydroxyethyl) terephthalate (BHET) selectivity	PET	Temperature (180–240ºC); high or low (1 bar) pressure; catalysts: Organicmetallic (e.g., Zn(OAc) 2); ionic liquids, etc.)	EG (BHET, BHET dimer, oligomers).PG (BHPT, BHET, hydroxypropyl e hydroxyethyl terephthalate).DEG (oligoesters diols).	Complies like CuRE and Garbo, and Ionica are using this technology.	Vollmer et al. 2020
Alcoholysis (transesterification process)	Alcohols (methanol, ethanol, etc.)	High selectivity to product, e.g., DMT in methanolysis	Catalyst deactivation (Catalyst is water sensitive)	PET	High temperature of (180–280ºC); pressure (20–40 bar); catalyst: Organometallic (metal acetate of Zn, Mg, Co and PbO); Ionic liquids, etc.	DMT and EG mixture of mainly DMT and glycols (EG), then alcohols, and phthalate derivatives.	Alcoholysis using methanol which results in DMT has been commercialized.	Pham and Cho 2021, Zhou et al. 2019

Table 2 contd. ...

... *Table 2 contd.*

Technology	Reagents	Advantages	Disadvantages	Feed	Operation Condition	Products	Technology Statues	Ref.
Aminolysis	Amines of methyl-, ethyl-, propyl-, butyl-, ethanol-, triethylene tetra-, hexamethylene di-, benzyl-; aniline;; piperidine; hydrazine monohydrate and some polyamines	Diamide product selectivity	Requires amine reagent	PET	Tempt. (20–100°C), Catalysts (Zinc acetate, Lead acetate, glacial acetic acid, and potassium sulfate.	EG and corresponding mono- and diamines of TPA like N,N[1]-Dialkyl terephthalamide (Alkyl amine); N,N[1] Diallyl terephthalamide (Allyl amine); BHETA (Ethanol amine); Terephthalohydroxamic acid (Hydroxyl amine hydrochloride); terephthalic dihydrazide (Hydrazine hydrate); and Bis (2- aminoethyl) terephthalamide(n ¼ 1) or α,Ψ(aminoligo ethylene) terephthalamide (n > = 2) (Ethylene diamine)	More	Raheem et al. 2019
Ammonolysis	Ammonia (NH_3)	High yield of the product	High pressure is required	PET	Temperature (70–180°C), Pressure (20 bar), time (1–7 h), catalyst: Zinc acetate	Terephthaldiamide (TPA di-amide) and ethylene glycol (EG)	-	Raheem et al. 2019

constituents are unstable, the non-purification alcoholysis products need to be matched with fresh chemical agents (e.g., polypropylene glycol, polyneopentyl glycol adipate, and polyethylene glycol) in the preparation of polyurethane to prevent separation (Fang et al. 2015). It is therefore necessary to analyze the key constituents of alcoholysis products from waste PET for use in more extensive applications.

Methanol is the most common alcohol used for alcoholysis,in a process known as methanolysis. Most PET is decomposed into dimethyl terephthalate (DMT) and EG when it undergoes methanolysis at high pressure and high temperatures. Because the product of PET methanolysis, i.e., DMT is not soluble in water, it can easily be purified and used for the production of PET or other products (Pudack et al. 2020). This process is carried out at temperatures between 180–280°C and pressures between 2–4 MPa (Scheirs 1998). Using methanol in supercritical condition can result in a high yield of product (~ 95%) in less time (almost one hour); however, this also requires a higher pressure (between 9–11 MPa) and therefore a higher operational cost (Genta et al. 2005). Another approach to decrease the reaction time is microwave heating. For example, Siddiqui et al. (2012) were able to depolymerize PET to DMT and EG in 0.5 h using microwave heating. They achieved an 80% yield at a temperature of 160°C using zinc acetate as catalyst. Other metal oxides,including lead acetate,can also be used as catalyst for this reaction; however, it is not feasible toquickly run the methanolysis reaction at low temperatures that are far from the melting point of PET (~ 260°C) (Pham and Cho 2021). A few studies have tried to further modify this process to make it more efficient and effective. For example, Phamand Cho (2021) demonstrated that complete decomposition of PET can be achieved using a combination of a highly active and selective catalyst and a cosolvent at ambient conditions. They specifically used potassium carbonate (K_2CO_3) as the catalyst and an appropriate aprotic polar solvent.

3.1.2 Hydrolysis

Hydrolysis is another method for PET degradation. During hydrolysis, the polyester chains in PET or other plastic waste react with water under different conditions to form EG and terephthalic acid (TPA). Acidic hydrolysis is typically performed using concentrated acids including phosphoric, nitric acid, or, most commonly, sulfuric acid (Karayannidis et al. 2002, Namboori et al. 1968). Alkaline hydrolysis is performed different concentrations of alkaline solutions of KOH or NaOH (Sinha et al. 2010, Scheirs 1998, Paszun and Spychaj 1997). The products of alkaline hydrolysis are EG and, depending on the alkaline solution, the equivalent terephthalate salt. Neutral hydrolysis of PET can also be conducted using water or steam with alkali metal acetates as catalysts (Padhan and Sreeram 2019, Turner et al. 2001). One drawback of hydrolysis is that the TPA produced after the reaction is less pure;however, hydrolysis can be performed in mild conditions and the reaction is highly tolerant of contamination in the feed stream (Ügdüler et al. 2020).

3.1.3 Aminolysis

Aminolysis, in which a kind of amine is used as solvent, is one of the less well-studied chemical techniques for PET degradation. Aminolysis is more thermodynamically favorable than alcoholysis and produces a spectrum of terephthalamides that are mostly used to form modifiers, additives, and blocks for high-performance materials (Fukushima et al. 2013). Aminolysis is conventionally performed using various amines including allyl amine, alkylamine, morpholine, hydrazine hydrate, polyamines, and triethanolamine (Spychaj et al. 2001).

Even though aminolysis reactions are more thermodynamically favorable than alcoholysis, effective depolymerization via aminolysis still requires microwave irradiation or catalysts (Pingale and Shukla 2009). It was recently reported that the depolymerization of PET is positively affected by a potent organic catalyst, 1,5,7-triazabicyclo [4.4.0] dec-5-ene (TBD), relative to the conventional metal catalysts (Fukushima et al. 2011). Sabot et al. (2007) found that TBD can be effectively used for aminolysis under solventless circumstances at room temperature, indicating the versatility and potency of TBD as a catalyst.

3.1.4 Ammonolysis

Ammonolysis is the process of degrading a plastic polymerusing anhydrous ammonia. End product of PET ammonolysis is terephthalamide (Blackmon et al. 1990). Terephthalamide can be transformed into para-xylylene diamine or l,4 bis (aminomethyl) cyclohexane and terephthalic acid nitrile. Blackmon et al. (1990) found that the ammonolysis of PET waste from post-consumer bottles yielded good results. They performed ammonolysis at a temperature range of 120–180°C and pressure of ~ 1 MPa for 1 to 7 h. After the reaction is completed, the resulting amide is filtered, washed with water, and dried at 80°C. The purity of the final product is almost 99%, with a yield of over 90%. A low-pressure technique for PET ammonolysis involves using ammonia asa degradation agent in an ethylene glycol setting. The procedure is catalyzed by 0.05 wt% of zinc acetate at 70°C and a PET:NH_3 ratio of 1:6. This procedure was used to make TPA amide with a yield of almost 87% (Spychaj 2002).

3.1.5 Glycolysis

Glycolysis uses lower quantities of reactants, lower temperatures, and lower pressures than methods like thermal degradation, supercritical methanolysis (Karayannidis et al. 2006, Yang et al. 2002, Kim et al. 2001, Sako et al. 2000, Chiu and Cheng 1999), or hydrolysis under basic or acidic circumstances, thus avoiding the pollution and corrosion problems associated with the latter methods (Karayannidis et al. 2006, Kao et al. 1998). Glycolysis can effectively be used to degrade high molecular weight PET to lower molecular weight or short-chain oligomeric, hydroxy-terminated products. These products then serve as raw materials for the preparation of unsaturated polyester resins, bis (hydroxyalkyl) terephthalates, or polyurethanes (especially rigid foams). The products of glycolysis specifically act as substrates for polybutylene terephthalate (PBT) and PET syntheses. In conventional glycolysis, PET waste is heated in combination with a glycol (e.g., diethylene glycol, ethylene glycol, propylene glycol, dipropylene glycol, or 1,4-butanediol) and a catalyst under higher or normal pressure at temperatures of 180–250°C. Typical catalysts include BHET, alkoxides, amines, or metal salts of acetic acid.

Glycolysis normally proceeds for 3–8 h (depending on the applied glycol) at approximately 200°C under reflux, provided that the PET: glycol weight ratio is within the range of 1:2 to 1:3. The reaction is performed under a constant nitrogen purge to prevent degradation of the resultant polyols (Spychaj 2002). There is a growing interest in using PET glycolysis to manufacture specialized products including unsaturated polyesters, vinyl esters, polyurethanes,polymer concretes, and epoxy resins (Rebeiz 1995, Rebeiz et al. 1992). Baliga et al. (1989) performed PET glycolysis using ethylene glycol (EG) and a variety of different catalysts. They reported that glycolyzed products contained 1–3 repeating units depending on the catalyst they used. Halacheva and Novakov (1995) assessed the chemical structure of the oligoesters produced from PET glycolysis using diethylene glycol (DEG) and found that products obtained at a higher excess of DEG contained secondary hydroxyl groups. Chen (2003), Chen et al. (2001), Mansour and Ikladius (2002), and Radenkov et al. (2003) have also investigated the glycolysis of recycled PET, and other studies have assessed the synthesis and applications of PET glycolysates (Colomines et al. 2005, Radenkov et al. 2003, Michel et al. 2002). For example, Farahat and Nikles (2002, 2001), Nikles and Farhat (2005) studied PET glycolysis with DEG and found a novel use for the resulting oligoester diols/polyols by transforming the hydroxyl terminals into acrylate/methacrylate groups. They examined these novel acrylated/methacrylated oligoesters as UV curable monomers, either alone or combined with other commercially available diacrylate/dimethacrylate monomers. Based on their mechanical features and curability by UV, these oligoesters have the potential to act as new binder systems for manufacturing solventless magnetic tape.

Atta et al. (2006, 2005) performed PET glycolysis with DEG and tetraethylene glycol (TEG). They prepared an epoxy resin from the reaction of epichlorohydrin and glycolyzed products. Novel dimethacrylate and diacrylate vinyl esters were also synthesized from the reaction of the terminal epoxy groups with acrylic acid and methacrylic. These vinyl esters were then used as cross-linking agents for an unsaturated polyester resin diluted with styrene. The resulting resins were assessed for their potential use in steel coating applications.

3.2 Pyrolysis

Pyrolysis is the process of converting plastic wastes to small molecules in the absence of oxygen. This process can be performed in presence of other additives such as H_2 and H_2O (Mark et al. 2020). Pyrolysis is particularly useful for polymers without heteroatoms and specifically for polyolefins. All plastic wastes are comprised primarily of carbon (80–90 wt%) and hydrogen (8–15 wt%). Some plastic wastes, including PS and HDPE, do not have any oxygen in their structure. Others plastic wastes, including PP and LDPE, contain a small amount of oxygen (< 0.7 wt%) (Mahari et al. 2021). Because of these unique atomic features, these kinds of plastics are more desirable than biomass as feedstock in pyrolysis units. The products of pyrolysis depend on the feedstock but broadly consist of gases, solids, and liquids (Fivga and Dimitriou 2018). For example, the products of PP pyrolysis are propylene gases, light oil, aromatics, heavy oil, and some solid char (Gracida-Alvarez et al. 2018).

Pyrolysis can generally be categorized into two general thermal and catalytic routes (Fig. 6) (Liu et al. 2021). Thermal pyrolysis must be performed at higher temperatures for a longer reaction time,

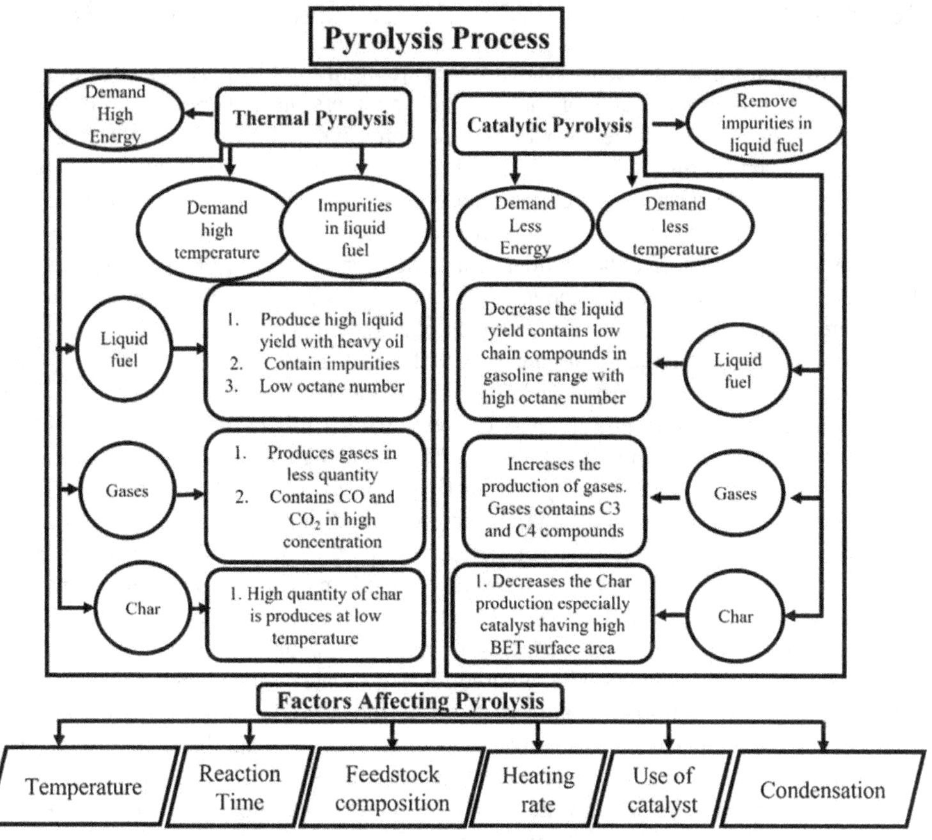

Figure 6. Flow chart of the two different pyrolysis processes (Miandad et al. 2016). Reproduced or/and reprinted with permission from Elsevier.

and the liquid products are of lower quality due to the presence of impurities. Catalytic pyrolysis is preferable to thermal pyrolysis because it can be performed at lower temperatures for less time while still yielding high-quality gas and liquid products. Various catalysts, including ZSM-5, Al_2O_3, NZ, and FCC, can be used for catalytic pyrolysis, though the specific properties of the chosen catalysts can have a large impact on the quality and composition of the end products. For example, catalysts with a high surface area and high acidity or catalysts with a microporous structure will result in more gaseous products and fewer liquid products. However, catalysts with a microporous structure will produce more liquid, oil, and char than gases. Accordingly, the performance of the catalyst, quality of the product, and efficiency of the process can also be enhanced by engineering the structure of the catalyst. These modifications include modifying the acidic and thermal properties of the catalyst via doping with other metal like Ni, Mo, and Co (Miandad et al. 2016).

It is worth mentioning that the main purpose of pyrolysis is the production of high-quality liquid oil. Obtaining high-quality product requires both an appropriate catalyst and an appropriate reactor. The liquid products produced from the pyrolysis of plastic wastes performed in conventional fixed-bed, tubular, or fluidized-bed reactors mainly consist of alkanes, alkenes, and aromatics. The main drawbacks of using these conventional reactors are a longer reaction time and the requirement for a large amount of carrier gases to maintain an inert pyrolysis environment. Microwave heating can help decrease the reaction time by providing fast heating in a more efficient way. Operating under vacuum can help lower the reaction temperature by reducing the boiling points of volatile compounds (Mahari et al. 2021). Overall, efficient recycling of plastic wastes via pyrolysis can be achieved with an effective catalyst and a microwave reactor that operates under vacuum.

4. RECYCLING OF PLASTIC WASTES UNDER MILD CONDITIONS

Because the cross-linked structures of plastic wastes are highly stable, plastic wastes can generally only be converted to valuable fuels or their initial monomers under harsh conditions. However, maintaining these conditions is energy- and resource-intensive and the recycling process therefore produces pollution on its own. The development of new technologies for plastic recycling under mild conditions would be a breakthrough for the recycling and up cycling of different kind of plastic wastes. Potential strategies that should be further investigated include the solar-driven conversion of plastic wastes into valuable bulk chemicals, the use of superoxides, organocatalyzed reductive depolymerization, the enzyme-driven conversion of plastic wastes (specifically PET to EG and TPA), or the low-temperature-driven catalytic conversion of polyolefins into other fuels.

Photocatalysis is a technique that can be used for CO_2 reduction, water splitting, and other environmental applications (Pelaezet al. 2012). Photocatalytic methods can also be used for recycling plastic wastes at low temperature and pressure. Uekret et al. (2019, 2018) used photocatalytic methods to convert some plastics, including PET/PLA,to hydrogen and organic carbonaceous fuels at ambient conditions in the presence of CdS/CdO_x QDs or $CN_x|Ni_2P$ as catalyst. Jiao et al. (2020) similarly converted various plastic wastes to C_2 fuels at ambient conditions via the photoinduced cleavage of C-C bonds. They specifically reported that plastics formed from PP, PE, or PVC, which include single-use bags, disposable food containers, and food wrap films, can be successfully converted into CH_3 COOH. For example, they found that PE can be totally converted into CO_2 after 40 h, and the CO_2 can later be converted into CH_3 COOH. Despite these outstanding results, these processes remain slow, and so further modification will be required for large-scale applications.

Another promising room-temperature method for the degradation of plastic waste involves using superoxide. The superoxide radical O_2^- is an oxygen species that has both anionic and free radical properties. This superoxide can be stably generated from the reaction of sodium or potassium hydroxide with hydrogen peroxide in water at ambient conditions (Stoin et al. 2013). The resulting superoxide can then be used to degrade plastic waste; this process can optionally be performed in the presence of a phase transfer catalyst. The oxidation reaction products include carboxylic acids and alcohols (Asson and Stoin 2021).

Organocatalyzed reductive depolymerization at mild conditions is also a promising method for recycling of plastic wastes at room temperature. This method is particularly useful for recycling polyesters and polycarbonate that have oxygen in their structure. A variety of functional chemicals can be obtained from organocatalyzed reductive depolymerization. For example, Feghali and Cantat (2015) applied hydrosilanes as a reductant and used a metal-free catalyst to depolymerize polyethers, polyesters, and polycarbonates to alcohols and phenols at room temperature. The main advantages of organocatalyzed reductive depolymerization is that the reaction is tolerant of contamination and active at room temperature. This process can also be used to depolymerize mixtures of polymers.

A final method for recycling plastic wastes at mild conditions is biological degradation using bacteria and enzymes. This method has attracted a lot of attention over the past few years. For example, a novel bacterium called *Ideonella sakaiensis* 201-F6 was isolated from the dilutions of a microbial consortium synthesized on discarded PET films. *I. sakaiensis* 201-F6 is capable of recycling PET to TPA and EG in 6 weeks at 30°C (Yoshida et al. 2016). This process is currently very slow and therefore not viable for large-scale recycling processes; however (Tournier et al. 2020), developed a better enzyme that is capable of depolymerizing PET to TPA in less than 10 hours. The TPA produced from biological degradation can be used for the production of virgin PET.

5. SUMMARY AND CONCLUSION

Advances in polymer science and engineering have helped with the development of different kind of plastics, and these new plastics have widespread applications in food packaging, construction materials, and electronic devices, among other things. However, the durability of plastic materials and their slow rate of degradation have unfortunately made them a threat to human health and the environment. There are a small number of recycling techniques for plastic materials, each with their own advantages and disadvantages. As an example, mechanical recycling can be used to convert plastic waste to products of the same quality; however, the presence of different additives in plastic materials (e.g., dyes and softeners) has made mechanical recycling challenging. Chemical recycling can be used to convert plastics to their chemical components, thus circumventing the problems associated with mechanical recycling. For example, some polymers (including PET and PS) can be chemically converted back to their original monomers and then re-used to produce virgin plastics. Other plastics (including PP and PE) can be converted to valuable chemicals using methods like pyrolysis. Current recycling methods can be energy-intensive and so it is important to develop new technologies for converting plastic waste to valuable products under mild conditions. Promising new technologies that require further investigation include cultivating bacteria that can tolerate hydrolysis conditions, developing new catalytic pyrolysis methods, and developing new photothermo catalytic methods in which plastics are broken down with thermal energy and then converted to valuable products under sunlight.

Each method for plastic recycling has its own advantages and disadvantages and combination of these methods can help decreasing this waste from our environment.

REFERENCES

Al-Salem, S., Lettieri, P. and Baeyens, J. 2010. The valorization of plastic solid waste (PSW) by primary to quaternary routes: From re-use to energy and chemicals. Progress in Energy Combustion Science 36(1): 103–129.

Al-Salem, S., Lettieri, P. and Baeyens, J. 2009. Recycling and recovery routes of plastic solid waste (PSW): A review. Waste management 29(10): 2625–2643.

Asson, Y. and Stoin, U. 2021. Process for treating plastic waste, US10961368B2.

Atta, A.M., Abdel-Raouf, M.E., Elsaeed, S.M. and Abdel-Azim, A.-A.A. 2006. Curable resins based on recycled poly (ethylene terephthalate) for coating applications.Progress in organic coatings 55(1): 50–59.

Atta, A.M., Elnagdy, S.I., Abdel-Raouf, M.E., Elsaeed, S.M. and Abdel-Azim, A.-A.A. 2005. Compressive properties and curing behaviour of unsaturated polyester resins in the presence of vinyl ester resins derived from recycled poly (ethylene terephthalate). Journal of Polymer Research 12(5): 373–383.

Baliga, S. and Wong, W.T. 1989. Depolymerization of poly (ethylene terephthalate) recycled from post-consumer soft-drink bottles. Journal of Polymer Science Part A: Polymer Chemistry 27(6): 2071–2082.

Barlow, C. 2008. Intelligent recycling (presentation). Department of engineering. Institute for Manufacturing, University of Cambridge, 10.

Bedell, M., Brown, M., Kiziltas, A., Mielewski, D., Mukerjee, S. and Tabor, R. 2018. A case for closed-loop recycling of post-consumer PET for automotive foams. Waste Management, 71: 97–108.

Blackmon, K.P., Fox, D.W. and Shafer, S.J. 1990. Process for converting pet scrap to diamine monomers, US4973746A.

Bora, R.R., Wang, R. and You, F. 2020. Waste Polypropylene Plastic Recycling toward Climate Change Mitigation and Circular Economy: Energy, Environmental, and Technoeconomic Perspectives. ACS Sustainable Chemistry & Engineering 8(43): 16350–16363.

Chen, C.H. 2003. Study of glycolysis of poly(ethylene terephthalate) recycled from postconsumer soft-drink bottles. III. Further investigation. Journal of Applied Polymer Science 87(12): 2004-2010.

Chen, C.H., Chen, C.Y., Lo, Y.W., Mao, C.F. and Liao, W.T. 2001. Studies of glycolysis of poly (ethylene terephthalate) recycled from postconsumer soft-drink bottles. II. Factorial experimental design. Journal of applied polymer science 80(7): 956–962.

Chiu, S. and Cheng, W. 1999. Thermal degradation and catalytic cracking of poly(ethylene terephthalate). Polymer Degradation Stability 63(3): 407–412.

Colomines, G., Robin, J.-J. and Tersac, G. 2005. Study of the glycolysis of PET by oligoesters. Polymer 46(10): 3230–3247.

Crippa, M. and Morico, B. 2019. PET depolymerization: a novel process for plastic waste chemical recycling. Studies in Surface Science and Catalysis (179): 215–229.

Das, S.K., Eshkalak, S.K., Chinnappan, A., Ghosh, R., Jayathilaka, W., Baskar, C. and Ramakrishna, S. 2021. Plastic Recycling of Polyethylene Terephthalate (PET) and Polyhydroxybutyrate (PHB)—a Comprehensive Review. Materials Circular Economy 3(1): 1–22.

Day, M., Cooney, J., Touchette-Barrette, C. and Sheehan, S. 1999. Pyrolysis of mixed plastics used in the electronics industry. Journal of Analytical Applied Pyrolysis (52): 199–224.

Dogu, O., Pelucchi, M., Van de Vijver, R., Van Steenberge, P.H., D'hooge, D.R., Cuoci, A., Mehl, M., Frassoldati, A., Faravelli, T. and Van Geem, K.M. 2021. The chemistry of chemical recycling of solid plastic waste via pyrolysis and gasification: State-of-the-art, challenges, and future directions. Progress in Energy and Combustion Science 84: 1–69.

European Commission. 2018. A European Strategy for Plastics. European Commission. 10.1021/acs.est.7b02368.

European Commission. 2011. European Commission. Commission Regulation (EU) No 2011/10 on Plastic Materials and Articles Intended to Come into Contact with Food. http://data.europa.eu/eli/reg/2011/10/oj.

Fang, C., Lei, W., Zhou, X., Yu, Q. and Cheng, Y. 2015. Preparation and characterization of waterborne polyurethane containing PET waste/PPG as soft segment. Journal of Applied Polymer Science 132(45): 42757.

Farahat, M.S. and Nikles, D.E. 2002. On the UV Curability and mechanical properties of novel binder systems derived from poly(ethylene terephthalate)(PET) waste for solventless magnetic tape manufacturing, 2. Methacrylated oligoesters. Macromolecular Materials Engineering 287(5): 353–362.

Farahat, M.S. and Nikles, D.E. 2001. On the UV curability and mechanical properties of novel binder systems derived from poly(ethylene terephthalate) (PET) waste for solventless magnetic tape manufacturing, 1. Acrylated oligoesters. Macromolecular Materials Engineering 286(11): 695–704.

Feghali, E. and Cantat, T. 2015. Room temperature organocatalyzed reductive depolymerization of waste polyethers, polyesters, and polycarbonates. ChemSusChem 8(6): 980–984.

Fivga, A. and Dimitriou, I. 2018. Pyrolysis of plastic waste for production of heavy fuel substitute: A techno-economic assessment. Energy 149: 865–874.

Francis, R. 2016. Recycling of polymers: methods, characterization and applications. John Wiley & Sons. ISBN: 978-3-527-33848-1.

Fukushima, K., Lecuyer, J.M., Wei, D.S., Horn, H.W., Jones, G.O., Al-Megren, H.A., Alabdulrahman, A.M., Alsewailem, F.D., McNeil, M.A. and Rice, J.E. 2013. Advanced chemical recycling of poly (ethylene terephthalate) through organocatalytic aminolysis. Polymer Chemistry 4(5): 1610–1616.

Fukushima, K., Coulembier, O., Lecuyer, J.M., Almegren, H.A., Alabdulrahman, A.M., Alsewailem, F.D., Mcneil, M.A., Dubois, P., Waymouth, R.M. and Horn, H.W. 2011. Organocatalytic depolymerization of poly(ethylene terephthalate). Journal of Polymer Science Part A: Polymer Chemistry 49(5): 1273–1281.

Garcia, J.M. and Robertson, M.L. 2017. The future of plastics recycling. Science 358(6365): 870–872.

Garcia, P.S., Scuracchio, C.H. and Cruz, S.A. 2013. Effect of residual contaminants and of different types of extrusion processes on the rheological properties of the post-consumer polypropylene. Polymer Testing 32(7): 1237–1243.

Genta, M., Iwaya, T., Sasaki, M., Goto, M. and Hirose, T. 2005. Depolymerization mechanism of poly (ethylene terephthalate) in supercritical methanol. Industrial & Engineering Chemistry Research 44(11): 3894–3900.

George, N. and Kurian, T. 2014. Recent developments in the chemical recycling of postconsumer poly(ethylene terephthalate) waste. Industrial & Engineering Chemistry Research 53(37): 14185–14198.

Gracida-Alvarez, U.R., Mitchell, M.K., Sacramento-Rivero, J.C. and Shonnard, D.R. 2018. Effect of temperature and vapor residence time on the micropyrolysis products of waste high density polyethylene. Industrial & Engineering Chemistry Research 57(6): 1912–1923.

Grigore, M.E. 2017. Methods of recycling, properties and applications of recycled thermoplastic polymers. Recycling 2(4): 24–35.

Gurgul, A., Szczepaniak, W. and Zabłocka-Malicka, M. 2018. Incineration and pyrolysis vs. steam gasification of electronic waste. Science of the Total Environment 624: 1119–1124.

Halacheva, N. and Novakov, P. 1995. Preparation of oligoester diols by alcoholytic destruction of poly(ethylene terephthalate). Polymer 36(4): 867–874.

Hopewell, J., Dvorak, R. and Kosior, E. 2009. Plastics recycling: challenges and opportunities. Philosophical Transactions of the Royal Society B: Biological Sciences 364(1526): 2115–2126.

Huysveld, S., Hubo, S., Ragaert, K. and Dewulf, J. 2019. Advancing circular economy benefit indicators and application on open-loop recycling of mixed and contaminated plastic waste fractions. Journal of Cleaner Production 211: 1–13.

Hwang, K.L., Choi, S.M., Kim, M.K., Heo, J.B. and Zoh, K.D. 2017. Emission of greenhouse gases from waste incineration in Korea. Journal of Environmental Management 196: 710–718.

Jia, L., Evans, S. and van der Linden, S. 2019. Motivating actions to mitigate plastic pollution. Nature communications 10(1): 1–3.

Jiao, X., Zheng, K., Chen, Q., Li, X., Li, Y., Shao, W., Xu, J., Zhu, J., Pan, Y. and Sun, Y. 2020. Photocatalytic Conversion of Waste Plastics into C2 Fuels under Simulated Natural Environment Conditions. Angewandte Chemie International Edition 59(36): 15497–15501.

Kao, C.Y., Cheng, W.H. and Wan, B.Z. 1998. Investigation of alkaline hydrolysis of polyethylene terephthalate by differential scanning calorimetry and thermogravimetric analysis. Journal of Applied Polymer Science 70(10): 1939–1945.

Karayannidis, G.P., Nikolaidis, A.K., Sideridou, I.D., Bikiaris, D.N. and Achilias, D.S. 2006. Chemical recycling of PET by glycolysis: polymerization and characterization of the dimethacrylated glycolysate. Macromolecular Materials Engineering 291(11): 1338–1347.

Karayannidis, G., Chatziavgoustis, A. and Achilias, D. 2002. Poly(ethylene terephthalate) recycling and recovery of pure terephthalic acid by alkaline hydrolysis. Advances in Polymer Technology: Journal of the Polymer Processing Institute 21(4): 250–259.

Kim, B.K., Hwang, G.C., Bae, S.Y., Yi, S.C. and Kumazawa, H. 2001. Depolymerization of polyethyleneterephthalate in supercritical methanol. Journal of Applied Polymer Science 81(9): 2102–2108.

Kosloski-Oh, S.C., Wood, Z.A., Manjarrez, Y., de los Rios, J.P. and Fieser, M.E. 2020. Catalytic methods for chemical recycling or upcycling of commercial polymers. Materials Horizons 8: 1084–1129.

Kumar, S., Panda, A.K. and Singh, R.K. 2011. A review on tertiary recycling of high-density polyethylene to fuel. Resources, Conservation Recycling 55(11): 893–910.

Lebreton, L.C., Van Der Zwet, J., Damsteeg, J.-W., Slat, B., Andrady, A. and Reisser, J. 2017. River plastic emissions to the world's oceans. Nature communications 8(1): 1–10.

Lebreton, L. and Andrady, A. 2019. Future scenarios of global plastic waste generation and disposal. Palgrave Communications 5(1): 1–11.

Liu, S., Kots, P.A., Vance, B.C., Danielson, A. and Vlachos, D.G. 2021. Plastic waste to fuels by hydrocracking at mild conditions. Science Advances 7(17): 1–9.

Mahari, W.A.W., Azwar, E., Foong, S.Y., Ahmed, A., Peng, W., Tabatabaei, M., Aghbashlo, M., Park, Y.-K., Sonne, C. and Lam, S.S. 2021. Valorization of municipal wastes using co-pyrolysis for green energy production, energy security, and environmental sustainability: A review. Chemical Engineering Journal 421: 129749.

Mansour, S. and Ikladious, N. 2002. Depolymerization of poly(ethylene terephthalate) wastes using 1, 4-butanediol and triethylene glycol. Polymer Testing 21(5): 497–505.

Mark, L.O., Cendejas, M.C. and Hermans, I. 2020. Cover Feature: The Use of Heterogeneous Catalysis in the Chemical Valorization of Plastic Waste. ChemSusChem 13(22): 5773–5773.

Martín, A.J., Mondelli, C., Jaydev, S.D. and Pérez-Ramírez, J. 2021. Catalytic processing of plastic waste on the rise. Chem 7(6): 1487–1533.

Miandad, R., Barakat, M., Aburiazaiza, A.S., Rehan, M. and Nizami, A. 2016. Catalytic pyrolysis of plastic waste: A review. Process Safety and Environmental Protection 102: 822–838.

Michel, A., Cassagnau, P. and Dannoux, M. 2002. Synthesis of Oligoester α, ω-diols by Alcoholysis of PET through the Reactive Extrusion Process. The Canadian Journal of Chemical Engineering 80(6): 1075–1082.

Namboori, C.G. and Haith, M.S. 1968. Steric effects in the basic hydrolysis of poly (ethylene terephthalate). Journal of Applied Polymer Science 12(9): 1999–2005.

Nikles, D.E. and Farahat, M.S. 2005. New motivation for the depolymerization products derived from poly(ethylene terephthalate) (PET) waste: A review. Macromolecular Materials Engineering 290(1): 13–30.

OECD. 2018. Improving Plastics Management: Trends, policy responses, and the role of international co-operation and trade. Environ. Policy Pap. 12: 20. https://www.oecd.org/environment/waste/policy-highlights-improving-plastics-management.pdf

Okubo, H., Kaneyasu, H., Kimura, T., Phanthong, P. and Yao, S. 2021. Effects of a Twin-Screw Extruder Equipped with a Molten Resin Reservoir on the Mechanical Properties and Microstructure of Recycled Waste Plastic Polyethylene Pellet Moldings. Polymers 13: 1058.

Padhan, R.K. and Sreeram, A. 2019. Chemical depolymerization of PET bottles via combined chemolysis methods. pp. 135–147. *In*: Recycling of Polyethylene Terephthalate Bottles. Elsevier. https://doi.org/10.1016/B978-0-12-811361-5.00007-9.

Paszun, D. and Spychaj, T. 1997. Chemical recycling of poly(ethylene terephthalate). Industrial Engineering Chemistry Research 36(4): 1373–1383.

Pelaez, M., Nolan, N.T., Pillai, S.C., Seery, M.K., Falaras, P., Kontos, A.G., Dunlop, P.S., Hamilton, J.W., Byrne, J.A. and O'shea, K. 2012. A review on the visible light active titanium dioxide photocatalysts for environmental applications. Applied Catalysis B: Environmental 125: 331–349.

Phanthong, P., Miyoshi, Y. and Yao, S. 2021. Development of Tensile Properties and Crystalline Conformation of Recycled Polypropylene by Re-Extrusion Using a Twin-Screw Extruder with an Additional Molten Resin Reservoir Unit. Applied Sciences 11: 1707.

Pham, D.D. and Cho, J. 2021. Low-energy catalytic methanolysis of poly (ethyleneterephthalate). Green Chemistry 23(1): 511–525.

Pingale, N. and Shukla, S. 2009. Microwave-assisted aminolytic depolymerization of PET waste. European Polymer Journal 45(9): 2695–2700.

Plastic Insight. 2014. World Per-Capita Consumption Of PE, PP & PVC Resins.https://www.plasticsinsight.com/world-per-capita-consumption-pe-pp-pvc-resins-2014/.

Pudack, C., Stepanski, M. and Fässler, P. 2020. PET Recycling–Contributions of crystallization to sustainability. Chemie Ingenieur Technik 92(4): 452–458.

Radenkov, P., Radenkov, M., Grancharov, G. and Troev, K. 2003. Direct usage of products of poly(ethylene terephthalate) glycolysis for manufacturing of glass-fibre-reinforced plastics. European Polymer Journal 39(6): 1223–1228.

Ragaert, K., Delva, L. and Van Geem, K. 2017. Mechanical and chemical recycling of solid plastic waste. Waste Management 69: 24–58.

Raheem, A.B., Noor, Z.Z., Hassan, A., Abd Hamid, M.K., Samsudin, S.A. and Sabeen, A.H. 2019. Current developments in chemical recycling of post-consumer polyethylene terephthalate wastes for new materials production: A review. Journal of Cleaner Production 225: 1052–1064.

Rahimi, A. and García, J.M. 2017. Chemical recycling of waste plastics for new materials production. Nature Reviews Chemistry 1(6): 1–11.

Rebeiz, K. 1995. Time-temperature properties of polymer concrete using recycled PET. Cement Concrete Composites 17(2): 119–124.

Rebeiz, K., Fowler, D. and Paul, D. 1992. Polymer concrete and polymer mortar using resins based on recycled poly(ethylene terephthalate). Journal of Applied Polymer Science 44: 1649–1655.

Roes, L., Patel, M.K., Worrell, E. and Ludwig, C. 2012. Preliminary evaluation of risks related to waste incineration of polymer nanocomposites. Science of the Total Environment 417: 76–86.

Roosen, M., Mys, N., Kusenberg, M., Billen, P., Dumoulin, A., Dewulf, J., Van Geem, K.M., Ragaert, K. and De Meester, S. 2020. Detailed Analysis of the Composition of Selected Plastic Packaging Waste Products and Its Implications for Mechanical and Thermochemical Recycling. Environmental Science Technology 54(20): 13282–13293.

Turner, J.A., Royall, D.J., Hugall, D.S., Jones, G.H. and Woodcock, D.C. 2001. Process for the Production of Terephthalic Acid. Patent, US6307099B1.

Sabot, C., Kumar, K.A., Meunier, S. and Mioskowski, C. 2007. A convenient aminolysis of esters catalyzed by 1, 5, 7-triazabicyclo [4.4. 0] dec-5-ene (TBD) under solvent-free conditions. Tetrahedron Letters 48(22): 3863–3866.

Sadat-Shojai, M. and Bakhshandeh, G.-R. 2011. Recycling of PVC wastes. Polymer Degradation Stability 96(4): 404–415.

Sako, T., Okajima, I., Sugeta, T., Otake, K., Yoda, S., Takebayashi, Y. and Kamizawa, C. 2000. Recovery of constituent monomers from polyethylene terephthalate with supercritical methanol. Polymer Journal 32(2): 178–181.

Scheirs, J. 1998. Recycling of PET In: Polymer recycling: science, technology applications. Wiley Series in Polymer Science, Wiley, Chichester, UK.

Siddiqui, M.N., Redhwi, H.H. and Achilias, D.S. 2012. Recycling of poly(ethylene terephthalate) waste through methanolic pyrolysis in a microwave reactor. Journal of Analytical and Applied Pyrolysis 98: 214–220.

Singh, N., Hui, D., Singh, R., Ahuja, I., Feo, L. and Fraternali, F. 2017. Recycling of plastic solid waste: A state of art review and future applications. Composites Part B: Engineering, 115, 409–422.

Sinha, V., Patel, M.R. and Patel, J.V. 2010. PET waste management by chemical recycling: a review. Journal of Polymers the Environment 18(1): 8–25.

Solis, M. and Silveira, S. 2020. Technologies for chemical recycling of household plastics–A technical review and TRL assessment. Waste Management 105: 128–138.

Spychaj, T. 2002. Chemical recycling of PET: methods and products. *In*: S. Fakirov (Ed.) Handbook of Thermoplastic Polyesters: Homopolymers, Copolymers, Blends, and Composites 1252–1290. https://doi.org/10.1002/3527601961.ch27.

Spychaj, T., Fabrycy, E., Spychaj, S. and Kacperski, M. 2001. Aminolysis and aminoglycolysis of waste poly(ethylene terephthalate). Journal of Material Cycles Waste Management 3(1): 24–31.

Stoin, U., Shames, A.I., Malka, I., Bar, I. and Sasson, Y. 2013. *In situ* generation of superoxide anion radical in aqueous medium under ambient conditions. ChemPhysChem 14(18): 4158–4164.

Subramanian, P. 2000. Plastics recycling and waste management in the US. Resources, Conservation Recycling 28(3–4): 253–263.

Taylor, P., Yamada, T., Striebich, R., Graham, J. and Giraud, R. 2014. Investigation of waste incineration of fluorotelomer-based polymers as a potential source of PFOA in the environment. Chemosphere 110: 17–22.

Tournier, V., Topham, C., Gilles, A., David, B., Folgoas, C., Moya-Leclair, E., Kamionka, E., Desrousseaux, M. L., Texier, H. and Gavalda, S. 2020. An engineered PET depolymerase to break down and recycle plastic bottles. Nature 580: 216–219.

Uekert, T., Kasap, H. and Reisner, E. 2019. Photoreforming of nonrecyclable plastic waste over a carbon nitride/nickel phosphide catalyst. Journal of the American Chemical Society 141(38): 15201–15210.

Uekert, T., Kuehnel, M.F., Wakerley, D.W. and Reisner, E. 2018. Plastic waste as a feedstock for solar-driven H2 generation. Energy & Environmental Science 11(10): 2853–2857.

Ügdüler, S., Van Geem, K.M., Denolf, R., Roosen, M., Mys, N., Ragaert, K. and De Meester, S. 2020) Towards closed-loop recycling of multilayer and coloured PET plastic waste by alkaline hydrolysis. Green Chemistry 22(16): 5376–5394.

Vinu, R., Ojha, D. and Nair, V. 2016. Polymer pyrolysis for resource recovery, Reference Module in Chemistry, Molecular Sciences and Chemical Engineering. 2016. http://doi.org/ 10.1016/B978-0-12-409547-2.11641-5.

Vollmer, I., Jenks, M.J., González, R.M., Meirer, F. and Weckhuysen, B.M. 2021. Plastic Waste Conversion over a Refinery Waste Catalyst. Angewandte Chemie International Edition, 60: 16101–16108..

Vollmer, I., Jenks, M.J., Roelands, M.C., White, R.J., van Harmelen, T., de Wild, P., van Der Laan, G.P., Meirer, F., Keurentjes, J.T. and Weckhuysen, B. M. 2020) Beyond mechanical recycling: Giving new life to plastic waste. Angewandte Chemie International Edition 59(36): 15402–15423.

Weckhuysen, B.M. 2020. Creating value from plastic waste. Science 370: 400–401.

Welle, F. 2011. Twenty years of PET bottle to bottle recycling—an overview. Resources, Conservation Recycling 55(11): 865–875.

Yang, Y., Lu, Y., Xiang, H., Xu, Y. and Li, Y. 2002. Study on methanolytic depolymerization of PET with supercritical methanol for chemical recycling. Polymer Degradation Stability 75(1): 185–191.

Yoshida, S., Hiraga, K., Takehana, T., Taniguchi, I., Yamaji, H., Maeda, Y., Toyohara, K., Miyamoto, K., Kimura, Y. and Oda, K. 2016. A bacterium that degrades and assimilates poly(ethylene terephthalate). Science 351(6278): 1196–1199.

Yu, J., Sun, L., Ma, C., Qiao, Y. and Yao, H. 2016. Thermal degradation of PVC: A review. Waste management 48: 300–314.

Zheng, J. and Suh, S. 2019. Strategies to reduce the global carbon footprint of plastics. Nature Climate Change 9(5): 374–378.

Zhou, L., Lu, X., Ju, Z., Liu, B., Yao, H., Xu, J., Zhou, Q., Hu, Y. and Zhang, S. 2019. Alcoholysis of polyethylene terephthalate to produce dioctyl terephthalate using choline chloride-based deep eutectic solvents as efficient catalysts, Green Chemistry 21: 897–906.

Zhou, X., Fang, C., Lei, W., Su, J., Li, L. and Li, Y. 2017a. Thermal and Crystalline Properties of Waterborne Polyurethane by *in situ* water reaction process and the potential application as biomaterial. Progress in Organic Coatings 104: 1–10.

Zhou, X., Fang, C., Yu, Q., Yang, R., Xie, L., Cheng, Y. and Li, Y. 2017b. Synthesis and characterization of waterborne polyurethane dispersion from glycolyzed products of waste polyethylene terephthalate used as soft and hard segment. International Journal of Adhesion Adhesives 74: 49–56.

9

DEGRADATION MECHANISM AND VALORIZATION OF WASTE PLASTIC TO ENERGY AND CHEMICALS

Perminder Jit Kaur,[1,*] *Uttam Kumar Mandal*[2] and *Vinay Shah*[2]

1. INTRODUCTION

The advent of modern processing technology has resulted in the development of materials with unique properties suitable for any particular task. Based on construction material, products can be divided into metal, non-metals, and polymer-based materials. Each group of materials has certain associated advantages and limitations. While metal-based materials are strong, and resistant to chemicals, their processing is energy-intensive and arduous. Non-metals are proven to be robust, corrosion and heat resistant. Howbeit, the corrosion and heat resistance of non-metals is jeopardised by deterrents like resistance to bending and brittleness. The third group of materials called polymers are long-chain molecules with flexible properties. The most commonly used polymers include plastics, which are highly tough and fatigue-resistant. Remarkable features, like low cost, low weight, high durability, and high chemical and water resistance, proffers its wide popularity.

Global plastic production was about 335 million tonnes in 2016, with Europe sharing 18% of this production. In Europe, around 60000 plastic production companies have employed about 1.6 million people with an annual turnover of 360 billion euros in 2018. As shown in Fig. 1, Europe employs plastic mainly for the packaging industry (40% of total plastic produced in the country), including the consumer and industrial sectors. Building and construction (20%) is the second most plastic consuming sector. Plastic like polyvinyl chloride is used for window and door frames and piping. Other segments like automotive parts (10%), electronic equipment (6%), non-packaging housing waste and sports (4%), and agriculture (3%)depend on plastic in various ways (Plastics Europe 2020). As shown in Table 1, plastic is a multi-utility material and can be used to manufacture items, such as bottles, jars, cable insulations, and wires. The increase in per capita usage of plastic from 1 kg in 1978 to 46 kilograms in 2017 shows the increasing importance of plastic in everyday life (Jiang et al. 2020).

[1] Department of Science and Technology, Centre for Policy Research, Indian Insttiute of Science, Bangalore, India.
[2] University School of Chemical Technology, Guru Gobind Singh Indraprastha University, New Delhi-India.
* Corresponding author: perminder.dua@gmail.com

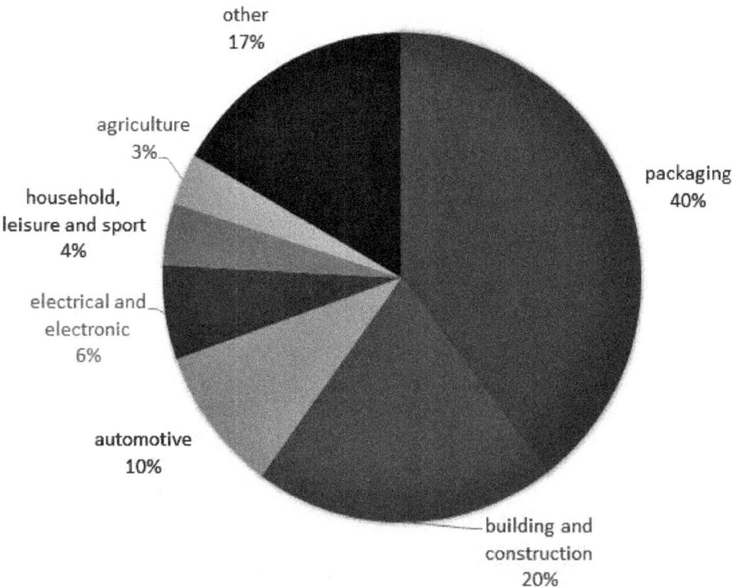

Figure 1. Segment-wise demand of plastic in Europe (Plastics Europe 2020).

The service life of plastic products varies from less than a year to more than 50 years, depending on its intended application. While in the packaging, use and throw concept results in frequent replacement and less than a year service life, the construction industry requires a service life of several decades.

Since the packaging industry comprises a significant consumption of plastic, there are serious concerns related to accumulating a large amount of waste material in the ecosystem, giving rise to white pollution or plastic pollution (Dogu et al. 2021). Inappropriate waste management techniques affect the environment negatively and make it onerous for the global attainment of sustainable development goals. Approximately 4900 Mt of plastic in various forms was reported to be accumulated in the ecosystem in 2017, expected to rise to 12000 Mt by 2050 (Geyer et al. 2017).

Further, due to the current global covid-19 pandemic outbreak, there is a sudden increase in the medical use of plastic bottles, which has further aggravated the plastic management problem. It is expected that the global plastic packaging market will grow at the compounded annual growth rate of 5.5% to USD 1012.6 billion by the end of 2021 (Prajapati et al. 2021). All the available data points out that this enormous amount of waste plastic needs special attention to be managed scientifically and systematically. The present chapter thus presents a thorough understanding of the chemistry, degradation mechanism, and valorization techniques of plastic, which has a strong potential to contribute to the circular economy construct.

2. CHEMISTRY OF PLASTIC MATERIAL

Based on the moulding method, plastic can be of 3 types: thermoset, thermoplastic, and elastomer. Thermoset plasticsare hard and brittle with narrow cross-linked monomer chains. Polyurethane, polyester, vinyl ester, silicon, and melamine resin are thermoset plastics. They cannot be heated and reshaped. On the other hand, plastic-like polyethene (PE), polyamides (PA), polypropylene (PP), polystyrene (PS), polyarylsulfone (PSU), and polyvinyl-chloride (PVC) are known as thermoplastic. They can be heated and moulded into different shapes reversibly. They are more popularly used due to their ability to be reheated and reshaped multipletimes. Elastomers like butadiene resin, styrene-butadiene resin, and polyurethane resin contain a net-like link between the monomer chains. They possess a soft elasticity and are usually non-meltable. Plastic like PE and PP are also known as

Table 1. Specific properties and applications of various types of plastic.

Type of plastic	Raw material (Monomer)	Specific properties	Applications	Reference
PE	Ethene	specific gravity<1, Molecular mass: 130–300 kg mol^{-1}, hyrdophobic	Packaging and disposable materials	Anderson 1982, Gopinath et al. 2020, He and Benson 2013
LDPE	Ethene	Density 0.91–0.93 x 10^3 kg/m^3, high branching, low crystallinity (55%), opaque, electrical insulating	Plastic bag, film, food and beverage packaging, squeeze bottles, pipes	Anderson 1982, Hariadi et al. 2021, He and Benson 2013
HDPE	Ethene	Density 0.94–0.96 x 10^3 kg m^{-3}, low branching, high crystallinity (85-95%), high mechanical strength	Jugs, containers, water pipes	Anderson 1982, Hariadi et al. 2021, He and Benson 2013
UHMWPE	Ethene	Molecular weight 2–6 million, low density, high strength	Joint replacement, artificial bones	Anderson 1982, Hariadi et al. 2021
Polypropylene (PP)	Propylene or ethylene and butylene	Molecular mass: 75–200 kg mol^{-1}	Fibres, films, pipes, bottles, injection moulding	Gilbert 2017
PS	Ethylene and benzene or styrene	Molecular mass: 50–200 kg mol^{-1} - Adjustable properties can be hard as well soft - Heat resilience - Durable, strong and light	Electronics, protective packaging, food containers, lids, pots, bottles, trays, construction, toys, disposable cups, and cutleries	Gilbert 2017, Sharuddin et al. 2016
PA 66	Hexamethylene diamine and adipic acid	Molecular mass: 12–20 kg mol^{-1}, tough, abrasion and corrosion-resistant	Conveyor belts, gears, parts of machines	Gilbert 2017
PVC	Vinyl chloride	Molecular mass: 50–120 kg mol^{-1}	Electrical insulations, window frames, shoes, medical devices, blood bags, interior parts of vehicles, packaging, credit cards	Gilbert 2017, Sharuddin et al. 2016
Poly(ethylene terephthalate (PET)	Ethylene glycol, terephthalic acid/ dimethyl terephthalate	Molecular mass: 30–80 kg mol^{-1} strong, durable, with density 1.41 g cm^{-3}, melting temperature 255–265°C	Packaging, electronics, automotive parts, lighting products, sports goods, photographic applications, X-ray sheets	Webb et al. 2013

Polyolefins for their oil-like texture. They constitute almost half of total global plastic production, around 63.4% contribution in the packaging industry (Rabnawaz et al. 2017). The percentage share of HDPE, PP, PS,PVC and PET is around 35%, 23%, 10%, 13%, 7%, respectively, of the total plastic consumption. Among packaging materials, HDPE, LDPE, PP are the most prominently used polymers.

Theunique electrical properties like high dielectric strength and very low electrical conductivity make PE suitable to make electrical insulating applications. Based on structure and density, polyethene can be of three types. Low-density polyethene (LDPE) with a density of 0.91 to 0.93 x 103 kg m^3 exhibits high branching, high flexibility,and low crystallinity. High-density polyethene (HDPE) possess a density in the range of 0.94–0.96 x 10^3 kg m^{-3}, is a low branching, high crystallinity, water-resistance and high mechanical strength polymer (Anderson 1982). Medium-density ultra-high molecular weight polyethene (UHMWPE) with molecular weight in the range of 2–6 million has a low density and high strength (He and Benson 2013). UHMWPE is a suitable joint and bone replacement material as it has high biocompatibility, wear resistance, and mechanical strength (Macuvele et al. 2017).

Polypropylene (PP) is a low density, high hardness, and abrasion-resistant plastic. It is obtained as a by-product during the manufacture of ethylene. For the large scale preparation of PP, the reaction of ethylene with butylene using olefins conversion technology is used. Polystyrene (PS) is produced from the polymerization process of styrene produced from ethylene and benzene. It is inexpensive, with a low softening temperature and elastic properties. The use of a suitable polymerization technique can make it glass-like, hard and transparent. It can also be made soft, insulating and expandable like foam (Gilbert 2017).

3. ENVIRONMENTAL, SOCIAL AND ECONOMIC EFFECTS OF PLASTIC POLLUTION

Most of the plastic produced is not biodegradable and last in the environment for thousands of years. Municipal Solid Waste (MSW) on land filling generates a toxic chemical solution called leachate. The leachate composition depends on the type and age of waste, environmental conditions and type of landfill technology (Kaur et al. 2020). Plastic products can form microplastic in leachate and, if remain untreated, can store pathogens and contaminants (Silva et al. 2021).When the waste containing plastic debris is burnt in open grounds (incineration),it generates toxins and harmful gases like carbon monoxide, chlorine, amines, nitrides, styrene, benzene, and butadiene (Jha and Kannan 2020). Plastic, on prolonged direct exposure to the sunlight,gets fragmented into smaller particles of size in the range of few micrometers to millimeters, known as microplastic. The myriad of plastic waste accumulated on land is the largest source of soil pollution. Digestion of plastic by various animals leads to blockage of their whole digestive system. Polyethene also reacts and form complexes with digestive salts, which results in irritation in the rumen (Derraik 2002). Moreover, evidence of the occurrence of polyethene bags in some terrestrial animals' stomachs is forcing environmentalists to circumvent the present situation.

At the same time, a fraction of plastic litter from sewage and shores also makes the water polluted. Researchers have investigated the toxic effects of these microplastic on marine and freshwater systems (Geyer et al. 2017). The annual leakage of 8 million tonnes of plastic into the marine is an environmental disquietude (Fadeeva and Van Berkel 2021). Researchers have studied the ocean waters between California and Hawaii. Great Pacific Garbage Patch was contaminated with a plastic waste of 79 k tonnes, consisting of 1.8 trillion plastic pieces of microplastic on an area of 1.6 million km^2 (Lebreton et al. 2018). Such an enormous quantity of plastic in water bodies leads to direct and frequent exposure of around 700 aquatic animals through either ingestion, entanglement or smothering of plastic. Accumulation of plastic debris in the marine is known as the greatest threat to biodiversity. This debris is known to be fatal not only for turtles, fish and marine mammals but also for birds (Puskic et al. 2020). The severity of impact depends on species and the type of waste ingested. Some birds and animals can digest hard items and prey, while others are susceptible to the chemicals present in these plastics. Accumulation of plastic in the guts of these animals have various sublethal effects (Gall and Thompson 2015). The plastic disposed of on the terrestrial system also becomes an integral part of the soil and affects the flora and fauna of that community (Huang et al. 2021).

There are serious sustainability issues associated with the imprudent use of plastic materials. Consequently, United Nations Environment Programme (UNEP) also directed all countries to transfer to 6R economy, i.e., Reduce, Redesign, Remove, Reuse, Recycle and Recover. United Nation's sustainable development goal (SDG) emphasized adopting technologies such that minimum waste in any form is disposed of in landfills and marine resources (UNEP 2016). SDG 6, 11, and 14 address the need to reduce litter in water bodies, cities, and marine systems, respectively. SDG emphasized importance of the development of such technologies that sustainably consume and produce materials. In the United States, environmental regulations like Microbead-Free Waters Act (2015) are implemented to reduce plastic microbeads' usage in the manufacture and sale of cosmetic products. European Union has also banned non-recyclable plastic products like straws, bags and cotton-bud sticks (European Union: Directive (EU) 2019/904). It has launched the strategy for a plastic-based circular economy. The commission has prepared guidelines to recycle all plastic products by 2030. G20 nations have realized the need to restrict plastic waste to enter the aquatic system;therefore,actions are being planned for the final goal of zero plastic waste disposal in marines by 2050.

3. MECHANISM OF PLASTIC DEGRADATION

A remarkable change in molecular weight, chemical, optical, and mechanical properties of plastic is known as 'plastic degradation'.Plastic degradation can follow eitherchain-end or random degradation reactions. Chain-end degradation or depolymerization reactions are similar to cutting off a long piece of paper into small pieces from its end. The method is a suitable method to recover monomers from their polymer units (Fig. 2). The molecular weight reduces very slowly in a stepwise manner with the liberation of one monomer unit in each step.

The other way to degrade plastic is by following the random plastic degradation mechanism. The plastic undergoes degradation at random locations in random chain degradation reactions rather than at endpoints (Fig. 3). Unlike the previous case where monomers are recovered, here, no recovery of monomers is obtained. Various low molecular weight compounds of different compositions can be recovered in this type of degradation. The total molecular weight of plastic reduces drastically at each step with no monomer recovery.

Kiran and Gillham (1976) showed that plastic might undergo inter and intramolecular radical transfer processes producing different products under thermal conditions. Two primary radicals are generated in the free-radical transfer mechanism for plastic degradation due to main chain scission.

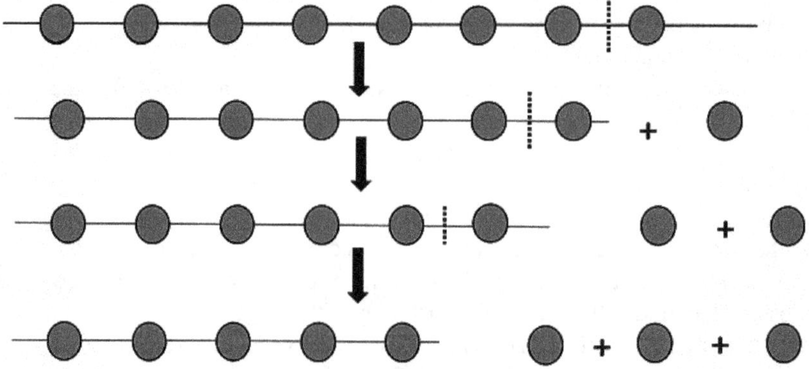

Figure 2. Schematic representation of chain-end plastic degradation mechanism where each circle represents a monomer unit.

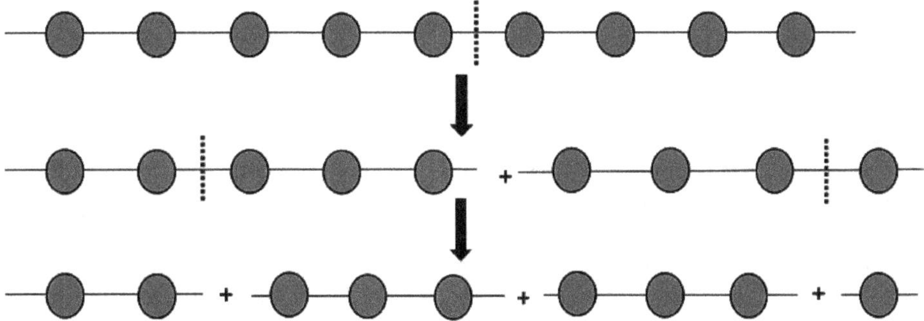

Figure 3. Schematic representation of random plastic degradation mechanism where each circle represents a monomer unit.

$$-CH_2-(CH_2)_n-CH_2-CH_2-CH_2- \rightarrow -CH_2-(CH_2)_n-CH_2\cdot + \cdot CH_2-CH_2- \quad (1)$$

These free radicals can further process undergoing inter and intramolecular transfer, forming a monomer and free radical.

$$-CH_2-(CH_2)_n-CH_2\cdot \rightarrow -CH_2-(CH_2)_{n-2}-CH_2\cdot + CH_2=CH_2 \quad (2)$$

There can be an intermolecular transfer of free radicals, deactivation of primary free radicals, and the formation of secondary free radicals.

$$\begin{aligned} &-CH_2-(CH_2)_n-CH_2\cdot + -CH_2-CH_2-CH_2-CH_2-CH_2-R \rightarrow \\ &\quad -CH_2-(CH_2)_n-CH_3 + -CH_2-CH_2-CH\cdot CH_2-CH_2-R \end{aligned} \quad (3)$$

This secondary free radical undergoes a beta-scission reaction leading to the formation of an alkene of low molecular weight

$$-CH_2-CH_2-CH\cdot CH_2-CH_2-R \rightarrow CH_3-CH_2-CH=CH_2 + \cdot CH_2-R \quad (4)$$

$\cdot CH_2$-R on getting hydrogen from neighbouring molecules to form an alkane CH_3-R. Equations(3) and (4) show that end product of plastic degradation can be saturated or unsaturated hydrocarbon depending on the-R group.

5. FACTORS AFFECTING PLASTIC DEGRADATION

Various physical factors like heat, light, mechanical stress or chemicals such as oxygen, ozone, acids or alkalies may be used to degrade plastic. Depending on the type of agent causing deterioration in plastic, it can be classified as photo-oxidative, ozone-induced, mechanochemical, catalytic, biological, and thermal degradation. The most common agents that carry out plastic degradation are UV radiation and mechanical abrasion. When light reacts with the plastic, it removes hydrogen atoms and forms peroxides in a photo-oxidation reaction. The process ends when one of the oxygen cleaves its polymer. While one end of the chain has a methyl group, the other end contains a carbonyl group (Biber et al. 2019). Researchers reported that the average molecular weight of polymers has a significant effect on their tensile strength. Continuity in polymer chain length is directly related to its tensile strength and makes it difficult to break. Cross-linking in the polymer chain makes it susceptible to cohesion and thus less stable (Sawai et al. 2006).

Ozone is a powerful oxidising agent. Even a minute concentration of ozone present in the air causes rapid ageing of polymeric materials like polyethene, forming carboxylic and carbonyl compounds. A cross-linking network is formed due to solid-phase oxidation of polystyrene powder on ozonation. The formation of three-dimensional structures is a common characteristic of solid-phase oxidation processes (Kefeli et al. 1971). In polyamides, the C-N bond undergoes stretching,leading to the formation of the degraded polymer (Ozen et al. 2003). Surface oxidation of PP happens due to ozonation. The effect is severe if the membrane has a large surface area. Vulcanised rubber undergoes unsaturation and surface cracks on exposure to ozone.

In the mechanochemical degradation, the mechanical force is applied in combination with chemical reagents to change the properties of plastic (Kim et al. 2000). For instance, mechanical degradation of PS occurs under turbulent flow. Likewise, various oxides like CaO, Fe_2O_3, SiO_2 and Al_2O_3 reduce PVC's molecular weight (Inoue et al. 2004).

Bioremediation is an efficient process used to remove contaminants from a source using biological agents (Kaur et al. 2020). This causes shortening of chain and reduces polymer molecular weight by microorganisms like fungi, bacteria, and enzymes. However, the efficiency of the process depends on the type of contaminant to be treated, its bio-availability, and the type of microorganism used for the degradation process (Kaur et al. 2020). The basic principle of thermal and photo-oxidation processes is quite similar.

Both methods are based on the depolymerisation of plastic using heat and UV-rays. The depolymerization reaction begins from any weak link or surface defects present in plastic. However, the sequence for initiation of the process is different in these two processes. Also, while photo-oxidation is mainly a surface process, thermal degradation is a bulk phenomenon. Thermal processes do not require the removal of contaminants from feedstock like a conventional plastic recycling plant. Depending on the type of reactor used, chemical composition of plastic, and reaction conditions maintained in the reactor, pyrolysis yields can be varied (Kaur et al. 2021). Among chemical constituents, the most influential parameters are reported to be volatile matter and ash content. The plastic richer in the high volatile matter and lower in ash content produces a higher amount of liquid fuel. Higher ash content favours the formation of gaseous fuel and bio-char (Sharuddin et al. 2016).

There could be several factors responsible for the degradation of plastics depending upon the choice of the process. Some of the major factors have been discussed in the following sections.

5.1 Type of polymer

The degradation of plastic depends strongly on its main chain and the side groups' constituents. More number of substitution groups decreases the stability of the polymer backbone. The bond dissociation energies of C-C bond in CH_3-CH_3, CH_3-CH_2-CH_3, $(CH_3)_3$-C-CH_3 and C_6H_5-CH_2-CH_3 are 88, 85, 80, 70 kcal/mol, resepctively, that indicates the lower stability of backbones with larger substituent groups. As a result, the strong carbon-carbon bond present in polyolefins (PP and PE) limits their ability to biodegrade. PE with the lowest number of side groups is a highly stable polymer (Xu et al. 2019). Polythene is a carbon-carbon chain based on a highly inert polyolefin plastic. Though this property makes it attractive for several applications, it acts as a barrier to its natural decomposition. PE contains no UV-visible chromophores and is photo-oxidative resistant. However, some structural defects present on its surface may include double bonds that can quickly react to form unstable hydroperoxides, which can be further converted to stable carbonyl groups. Due to reactive branching, LDPE can undergo this type of reaction more easily. The thermal oxidative response of PE requires exposure to sunlight and a temperature higher than 100°C. The product of both photo-oxidative and thermal degradation are similar. In the absence of air, anaerobic degradation of plastic requires a temperature as high as 350°C, which is impossible under natural conditions. Thus, anaerobic degradation of plastic waste does not happen naturally in landfills (Chamas et al. 2020).

LDPE is a highly branched hydrocarbon chain based material with high ductility, low crystallinity, low tensile strength and hardness. As there are branching in its structure, it is efficiently heated and moulded into various shapes. Its water-resistance properties are suitable for applying household applications to manufacturing plastic bags, packaging material, etc. The wide range of applications leads to the second-highest contribution to the total MSW. HDPE is a long linear chained, highly crystalline and strong polymer suitable for making bottles, containers, and toys. HDPE is the third most significant type of plastic found in MSW. Various studies have attempted to convert HDPE wastes to energy under reaction conditions. However, the degradation temperature of HDPE is very high, making its degradation an energy-intensive process. PP is a chemical and heat resistant polymer as its melting point is more than 160°C. A significant fraction of plastic waste in municipal solid waste (MSW) consists of PP. It has low density, high hardness and high rigidity and is a preferable material for an extensive range of applications from flower pots to automotive parts. Polypropylene (PP) with a methyl side group is less stable than PE. Polyisobutylene with two methyl side groups is highly unstable.

PS with a long hydrocarbon chain is prepared from the polymerisation of styrene monomers. The structure contains a phenyl group attached to alternative carbon atoms of the hydrocarbon chain. The presence of a phenyl side group helps it degrade with ease as compared to PE. High strength and lightweight make it an attractive choice for various applications, constituting a large MSW portion (Sharuddin et al. 2016).

The plastics containing esters or amides can be easily converted to oligomers or monomers. For instance, PET can be recycled to its constituent materials like terephthalic acid and ethylene glycol. Synthetic polyamide can produce diamine/diacids or ω-amino carboxylic acids, with the highest amount of residues. PVC contains vinyl chloride as the monomer unit, with chlorine and carbon as 57% and 43%, respectively. The presence of a significant amount of chlorine makes it fire resistant and contributes to its lower rate of degradation (Vijayakumar and Jilse 2018). The degradation of PVC produces hydrochloric acid and polyethene. Some surface defects in PVC helps in its degradation at a temperature lower than its processing temperature (100°C). With time, the chain length of produced polyethene increases, thus shifting the energy of electronic absorption. The colour of initial of the product also changes from white to yellow, orange, brown and black. Additional problems associated with corrosion of equipment due to produced HCl,degraded quality, and high brittleness adds to the complications of the process. The HCl release also led to the closed own of some of the commercial installations of plastic pyrolysis plants. The low amount of PVC in MSW and associated complications due to the release of the chlorinated compound makes the pyrolysis of PVC not a preferable choice. Additional steps of chlorine removal from produced oil are required, which leads to extra cost. Also, modified pyrolysis processes like stepwise pyrolysis, catalytic pyrolysis and pyrolysis are reported to be expensive (Sharuddin et al. 2016).

5.2 Types of additives

During the plastic manufacturing process, various chemical compounds are added to it. Calcium carbonate minerals, talc, wood floor, aluminium hydroxide, Wollastonite, mica, glass fibres and zinc oxide are used as filler in the plastic. Nanomaterials like nanoclays, poly-brominated diphenyl ethers, hexabromocyclododecanes, carboxylic acids, and organosilicon are added to make it fireproof (Gilbert 2017). Lubricants, fillers, pigments, processing aids, and impact modifiers added to PVC further hinder the degradation process. Stabilizers are added to plastic to strengthen the stability of plastic on exposure to outdoor conditions. Photostabilizers such as 2-(hydroxy phenol) benzotriazole are added to absorb UV radiations and transmit the absorbed energy to the environment in some non-toxic form. Plasticisers (maybe up to 70% of the weight of plastic) like phthalates and adipates are added to give space between plastic molecules to make them flexible. Curing agents like 4,4-diamino diphenyl methane are added to protect the plastic from microbial attack. Azo compounds and pigments are added to enhance the visual look and give colours to plastic. Antioxidants are added to preserve plastic against oxidative degradation and thus the formation of peroxides and hydroperoxides.

They tend to react with reactive radicals like P· PO·, POO·, HO· encountered to form stable products. Generally, Di-t-butyl p-cresol (TBC), Phenyl beta-naphthyl amine (PBNA) are used as antioxidants. Phenyl groups get unstable at elevated temperatures by producing free radicals. Though they are added to promote the properties of plastic, these compounds make them more challenging to degrade. They contain different proportions of carbon and can act as a source of energy for various microorganisms (Gu 2021). Even during the service life of the product, they become a cause of product failure. Depending on their relative proportions in the plastic, these compounds hinder the degradation and add to the toxic properties of plastic (Sanchez 2021).

5.3 Environment

Environment plays a prominent role to degrade plastic. Researchers have reported that exposure to air and water affects both tensile properties and chemical bonds of plastic. The lack of heat formation causes the same plastic to degrade at a slower rate in water than that in the air. Also, in water, surface biofouling of plastic may occur, leading to a lower rate of degradation (Pegram and Andrady 1989). Studies have shown that PE, PS, PP lost their extensibility in illuminated air. Moreover, light also influenced the degradation rate, which was higher in shader air than in water. All types of plastic, including PS, PE, polyester fibres and production beads in marine water, show their poor degradation in the aquatic system. Comparative study on specific surface degradation rates (SSDR) of different types of plastic has shown significantly different degradation rates under different environment and humidity conditions. While SSDR of HDPE and PLA were similar in water, PLA degraded at a rate 20 times faster on land. The study addressed the need to standardize reporting rates and procedures for plastic degradation (Chamas et al. 2020).

6. THERMAL TREATMENT TECHNIQUES

Plastic recycling can be done broadly in three different ways using material, chemical or energy recovery options. The material recovery process focuses on the recovery of monomers and other raw materials to produce the same plastic. This process's low cost and eco-friendly nature make it an attractive option to recycle clean and single-use plastic. In the chemical recovery process, plastic is broken down into its chemical constituents to form new plastic. In the thermal recovery process, heat is supplied to plastic in specially designed incinerators to produce energy and valuable chemicals ranging from oligomers to a mixture of other hydrocarbon compounds. Chemical and thermo-chemical means are used to alter the chemical structure of plastic to aid in its recycling.

Thermal treatment of plastic can be done either bypyrolysis or gasification techniques. Pyrolysis is an eco-friendly process of heating material in specially designed reactors in the absence/limited availabiltiy of oxygen to allow for the breakdown of plastic into smaller molecules to form gaseous products and liquid products composed of paraffins, olefins and naphthenes. Some aromatic compounds and a wide range of other heteroatoms containing hydrocarbons are also formed during the process (Demirbas 2005). Gasification is another thermal treatment process where feed material is treated using a gasifying agent under controlled conditions. These processes have an additional advantage of tolerance for resins and organic contaminants (Siddiqui 2009). Thermal degradation leads to the breaking down of complex polymeric molecules into small molecules of low molecular weight. Catalysts like red mud, kaolin, zeolite, MgO, Al_2O_3, ZSM-5 can also be added to the reactor to enhance the yield and selectivity of the product.

Based on the feedstock properties and introduction method into the reactor, different classes of reactors can be used to carry out pyrolysis and gasification of plastic. They can be either fixed bed or moving bed reactors. A fixed bed reactor with a stationary bed is a simple batch reactor with no moving parts and less maintenance cost. Further, a moving bed reactor can include pneumatic mode like bubbling, spouted or fluidized bed reactor or mechanical mode like a rotary kiln, rake or stirred reactors. A fluidized bed reactor is the most commonly used moving bed, continuous reactor using gasifying mediums like water and oxygen. A modified fluidized bed with a rotating drum for the uniform distribution of heat is called a rotating drum reactor. Factors affecting the efficiency of these reactors are the type of feedstock, reactor and reaction conditions used (Kaur et al. 2021). Both pyrolysis and gasification processes have been investigated at lab-scale, pilot-scale, and semi-commercial scales using various reactors under different reaction conditions to optimise the conversion of waste plastic into fuel (Kaminsky et al. 2004, Kaur et al. 2021).

A fixed bed reactor is a simple batch reactor using a fixed bed of plastic with or without catalyst in the presence or absence of air. The experiment setup for studying pyrolysis of mixed plastic waste using PP (45%), PE (35%), HDPE (25%) at laboratory scale in a fixed bed reactor showed that heating at temperature 500°C for 45 minutes produced pyrolysis oil (32.8%), gaseous product (65.75%) and solid residue (1.46%). Pyrolysis oil was composed of aromatics (11.49%), paraffin (82.11%) and naphthenes (6.4%). The high heating value (HHV) of pyrolysis oil (45.9 MJ/kg) was observed higher than coal (43 MJ/kg) and crude oil (44 MJ/kg),which suggests its potential as an alternative fuel (Papuga et al. 2015).

A fluidized bed reactor, opposed to a fixed batch reactor, is a continuous reactor and very popular for commercial pyrolysis processes. Experimental setup for laboratory scale studies of pyrolysis of waste tires, biomass and plastic waste has been investigated by researchers using a fluidized bed reactor. The fluidized bed reactor consists of a screw feeder, cyclone for removing solid particles from exhaust gases, and a condenser for cooling and recovery of bio-oil from pyrolysis gases (Lewandowski et al. 2019). In one of the studies, a lab-scalefluidized bed reactor consisting of a continuous feeding system and a cooling system was carried out in a silica-sand bed of reactor vessel of 90 mm diameter. It was observed that at temperature 540°C, a significant product of white wax and liquid oil (94%) was obtained. On further increasing the temperature to 737°C, a marked increase in gaseous components (43.75%) was obtained (Shui-e et al. 2007). However, pyrolysis in a fluidised bed reactors is accompanied by clump formation and reactor blockage, which results in loss of energy and thus higher energy consumption. The reactor is expensive, has a higher working cost, and uses additional equipment to separate fluidized gas and gaseous products (Singh et al. 2019). It utilises sweeping gas like nitrogen or argon to fluidise the bed, which decreases the calorific value of product gas.

Thermoplastic constitutes a significant portion of total waste plastic and takes a long time to degrade. Thus, most of the researchers have selected them for investigations (Siddiqui 2009). The studies on pyrolysis of 60–120 gm of LLDPE show that 425–450°C is suitable for its conversion to alkanes of chain length 13.6 and 70.8 mol% unsaturationequivalent quality to synthetic lubricant (Mccaffrey et al. 1995). Pyrolysis of polyalkenes like HDPE, LDPE and PP starts at around 304°C and finishes at 577°C, and yields liquid oil rich in alkenes, alkanes, and alkadienes. The gaseous products consisting of hydrogen, alkane, and alkene are also obtained, accompanied by anegligible generation of char. Pyrolysis of PP and PE results in higher paraffin and olefins, while PS produces a higher amount of liquid products (Demirbas 2005). A kinetic study was carried out under moderate reaction conditions, between 350 and 450°C. The initial charge of polymer to the reactor used for this study was relatively large, 60–120 g, compared with many previous investigations into the degradation of polyethene. The reaction products were characterized by their suitability for the production of synthetic lubricants. By varying pressure, temperature and residence time, products of different compositions can be obtained. To obtain light olefins, high temperature and short residence time are desirable. At medium temperature, higher production of wax is obtained (Lopez et al. 2017).

As shown in Table 2, most of the research is dedicated to the production of liquid fuel. In addition to this, wax, monomers, additive, aromatic and several chemical compounds can also be recovered during the process (Qureshi et al. 2020). Gaseous products with an energy capacity of 25–45 MJ/kg can also be recovered profitably during the process. Liquid products with high calorific value and blendable with conventional petroleum fuels can be obtained using suitable reaction conditions in these specially designed reactor vessels.

Table 2. Literature studies on thermal degradation of plastic waste

Raw material	Reactor	Temperature (°C)	Gas yield (%)	Oil yield (%)	Comments	
HDPE	Batch pyrolyser	450	13	84	Thermal cracking produced long H-C chain. Catalytic cracking using Zeolite Y and ZSM-5 resulted in formation of both aromatics and branched hydrocarbons	Seo et al. 2003
PP	Pyrolysis (800 gm)	500–700	82%		PP mixed with sand and alumina, Al2O3 catalyst	Kodera et al. 2006
HDPE, PP, PS, PET, PVC (2-5 mm granules)	Batch autoclave reactor (30–40 gm capacity)	500	93 HDPE 95 PP 71 PS 15 PET	7 HDPE 5 PP 2 PS 32 PET	Heating rate of 5°C min−1 for 60 min. Pyrolysis of PVC not completed due to HCl formation and corrosion	Williams and Slaney 2007
PET/PP/PS	Pyrolysis reactor (25 cm3)	430–440	53.3	-	PET, PP and PS in 1:1:2 proportion tested to yield maximum gaseous yield. PS tested alone degraded completely	Siddiqui and Redhwi, 2009
Mixed plastic waste	Batch pyrolyser (100 gm capacity)	460	-	72	Optimum reaction time: 15–30 min	López et al. 2011
Mixed plastic waste	Batch pyrolyser (100 gm capacity)	500	25.6	40.9	HHV of gaseous and liquid product: 47.1, 40.9 MJ/kg, respectively	Adrados et al. 2012
PS	Metal coil reactor (20 gm capacity)	-	88	9–10	Microwave pulse applied at 2 min pulse for 10 mins	Hussain et al. 2012
Expanded PS (thermocol)	Pyrolyser (20 gm)	500	2.5	93	Kaoline clay and plastic in 1:10 further improved the liquid yield to 94.5%	Panda et al. 2012
PS	Autoclave pyrolyser	500	-	94.5	Clay to plastic ratio(1:10)	Panda et al. 2012
Mixed plastic waste	Batch pyrolyser (10 mm particle size)				Full decomposition required 6652 s at 500 C, 5299 s at 550 C and 4187 s at 600 C	Ateş et al. 2013
HDPE	Pyrolyser (400 gm)	440	-	74 wax and oil	Waste plastic oil appeared appropriate as a blend for diesel fuel	Sharma et al. 2014
Mixed plastic waste	1st stage: Pyrolysis (1.6–2.6 kg capacity) 2nd stage: catalytic reforming Pyrolysis gas form 1st reactor sent to catalytic reactor with 100 gm catalyst	450	< 40	< 50	Thermal pyrolysis produced more liquid fraction. Zeolite-Y produced more gaseous fraction	Syamsiro et al. 2014

Mixed plastic waste	2-stage fixed bed pyrolyser	500	10	80	1st stage: plastic pyrolysis heating the plastic at 10°C/min, held at 500C for 30 mins 2nd stage	Muhammad et al. 2015
HDPE	Batch pyrolyser (2Lt capacity)	460	-	86.2	Y-zeolite and MgCO3 further reduced the temperature to 430 and 450, respectively, to yield about the same liquid fraction	Kunwar et al. 2016
Mixed (PP, HDPE, LDPE)	Fixed batch reactor (500 gm capacity)	525	70	28.8		Papuga et al. (2016)
PP, PET, PS, HDPE, LDPE, PA, PVC	Semi-batch reactor (800 gm capacity)	400–550	-	80–90	200 ml/min for 10–20 min. Oil yield of HDPE and PP > PS With PET, no liquid oil obtained	Singh et al. 2019
LDPE	Retort pyrolyser (6000 gm capacity)	250		99.7	Time of pyrolysis affects the oil yield significantly with optimum time: 110 min	Hariadi et al. 2021

7. PRESENT STATUS AND FUTURE TRENDS

The gargantuan waste plastic has forced countries to look at waste plastic as an alternative source of energy. Many countries, specially European nations, have shifted towards plastic as an energy option to treat waste plastic. The data for the current trend in Asia with 131 million tonnes of waste plastic in 2015 (Liang et al. 2021). Among Asian countries, China generates the maximum amount of waste plastic (49.71 Mt), accounting for 31% of total globally generated plastic, followed by India (17.66 Mt), Japan (11.19 Mt) and Turkey (6.28 Mt) (Liang et al. 2021, Plastics Europe 2020). China, India and Indonesia contribute the most to marine plastic pollution (Schmidt et al. 2021). According to EPA (2018), in the United States, 35.7 million tonnes of plastic was generated and constituted 12.2% of total MSW generated. The overall recycling rate of plastic was only 8.7%. In the U.K., the implementation of Packaging and Packaging Waste Directives has resulted in about 31% of waste plastic being recycled (Elliott and Elliott 2020). In Nigeria, less than 12% of 2.5 million tonnes of generated plastic is recycled, with the rest 80% being directed towards landfills and dumpsites (Aisien et al. 2021). As per reports from EPA(2018), about 35 million tonnes of waste plastic was generated in the United States, out of which about 16% is used to recover energy (EPA 2018). In Europe, approximately 58 million tonnes (2019) of waste plastic was generated in 2019, out of which an average of 42% was used to recover energy. Eurpoean countries like Finland, Austria, Germany, Ireland are recovering 45.9%, 28.8%, 27.9% and 27.4%, rescpectively. In Europe, the percentage of plastic recycled and landfilled was about 35% and 23%, respectively (PlasticsEurope 2019). It is estimated that if all the plastic waste is utilized to produce energy, it can generate energy equivalent to 3.5 barrels of oil annually (Bhattacharya et al. 2018).

There are many latest research investigations with the potential to be scaled-up. In a few studies, attempts have been made to evaluate the profitability and efficiency of lab-scale experiments on a commercial scale. One of the studies (Fahim et al. 2021) in the literature has proposed a design of a commercial plant with the feeding capacity of 20,000 tonnes of waste plastic daily in a fixed bed reactor. The plastic to catalyst ratio (ZSM-5) was 10:1 at the temperature of 550°C and the heating rate of 15°C/min. The conversion efficiency for HDPE and mixed plastic waste was 94% and 85.6–89.5%. The scheme proposed in the studies involves pre-treatment of plastic using cleaning, drying, size reduction and washing to remove toxins. Feed was mixed with a catalyst and fed to this fixed bed reactor. Products of pyrolysis process comprised of heavy oil, which can be used as fuels for ships. Bio-oil produced can be transported to refineries to produce gasoline and diesel oil (Fahim et al. 2021). One Australian company plans to convert waste plastic to energy based on liquefaction of waste plastic. Catalytic pyrolysis, followed by distillation, results in the generation of fuel suitable to run a diesel engine (http://www.ozmo- tech.com.au/). In India, low temperature degasifies catalytic process is used to produce liquid hydrocarbon fuel from plastic waste daily. The quality of fuel is reported to be comparable to petrol (Dnaindia report 2021). A similar plant in Germany by WASTX Plastic is processing 250 kg of waste plastic daily (Iamexpat Report 2021). The successful operations of these plants show the commercial potential of the present technology to produce fuel, which can replace petrol and diesel. Installation of more such plants will decrease our dependence on fossil fuels and reduce the myriad of plastic waste accumulated in the environment.

8. CONCLUSION AND FUTURE DIRECTION

The unique properties of plastic make them versatile materials for various applications ranging from domestic products to high-end engineered products. The increasing usage of plastic has led to the accumulation of large amounts of waste plastic in terrestrial and aquatic systems. Disposal of plastic as landfills has various health and environmental hazards. Uncontrolled, open burning of plastic produces toxic gases. Thermal degradation of plastic under controlled conditions can result in gaseous and liquid fuel with high calorific value and various valuable chemicals. Though fixed bed reactors showed the potential to operate efficiently at batch scale, fluidized bed reactors hold more significant potential for the commercial-scale production of fuel. The studies have shown that different types of plastic degrade differently, producing different proportions of various products. To get the highest pyrolytic oil yield, HDPE and PP are preferable to plastic. Under optimum conditions, more than 90% recovery of pyrolysis oil is possible.

In mixed type of plastic waste, generally found in MSW, the presence of PVC in the waste plastic leads to corrosion of reaction vessels and the generation of chlorine gas, which is undesirable. Thus, additional steps are required for the segregation of waste at source and removal of PVC. The development of suitable technology for converting PVC to eco-friendly products, non-toxic to the reaction vessel, can help utilise PVC as a source of valuable products.

However, scientific understanding of reaction mechanisms, reaction pathways, and the exact degradation rate of individual plastic is disproportionately thin. Most of the literature omits critical information, such as the chemistry of the polymer and the role of various additives on pyrolysis products. High variability exists in plastics concerning their molecular weight distribution and degree of polymerisation. The effects of these factors on the product of pyrolysis need to be documented. Many laboratory scale investigations have been attempted to optimize reaction conditions using batch reactors of small capacity. There is negligible information on the energy balance and heat recovery potential of these processes in the literature. These deficiencies and spareness on large scale studies limit the ability to transform technology to a commercial scale. Some of the companies are running successfully at pilot scale, which show the potential of energy generation waste plastic. Using computational techniques like modelling and simulation can help in the smoother transition of these technologies to commercial scale. Installation of commercial setups will pave the way for the inclusion of waste plastic in circular economy construct.

REFERENCES

Adrados, A., de Marco, I., Caballero, B.M., López, A., Laresgoiti, M.F. and Torres, A. 2012. Pyrolysis of plastic packaging waste: A comparison of plastic residuals from material recovery facilities with simulated plastic waste. Waste Management 32(5): 826–832.

Aisien, E.T., Otuya, I.C. and Aisien, F.A. 2021. Thermal and catalytic pyrolysis of waste polypropylene plastic using spent FCC catalyst. Environmental Technology and Innovation 22: 101455.

Anderson, J.C. 1982. High density and ultra-high molecular weight polyethenes: their wear properties and bearing applications. Tribology International 15(1): 43–47.

Ateş, F., Miskolczi, N. and Borsodi, N. 2013. Comparision of real waste (MSW and MPW) pyrolysis in batch reactor over different catalysts. Part I: Product yields, gas and pyrolysis oil properties. Bioresource Technology 133: 443–454.

Bhattacharya, R.R.N.S., Chandrasekhar, K., Deepthi, Roy, P. and Khan, A. 2018. A Report on Challenges and Opportunities: Plastic Waste Management in India, June. (T.E.R.I.). http://www.teriin.org/sites/default/files/2018-06/plastic-waste-management_0.pdf

Biber, N.F.A., Foggo, A. and Thompson, R.C. 2019. Characterising the deterioration of different plastics in air and seawater. Marine Pollution Bulle 141: 595–602.

Chamas, A., Moon, H., Zheng, J., Qiu, Y., Tabassum, T., Hee Jang, J., Abu-Omar, M., Scott, S.L. and Suh, S. 2020. Degradation Rates of Plastics in the Environment. ACS Sustainable Chemistry and Engineering 8: 3494–3511.

Demirbas, A. 2005. Recovery of Chemicals and Gasoline-Range Fuels from Plastic Wastes via Pyrolysis Recovery of Chemicals and Gasoline-Range Fuels from Plastic Wastes via Pyrolysis. Energy Sources 27: 1313–1319.

Derraik, J.G. 2002. The pollution of the marine environment by plastic debris: a review. Marine Pollution Bulletin 44: 842–852.

Dna India report. 2021. https://www.dnaindia.com/mumbai/report-now-in-nagpur-fuel-from-plastic-waste-6967(Last Accessed on: August, 2021).

Dogu, O., Pelucchi, M., Vijver, R. Van de, Steenberge, P.H.M. Van, Dagmar, R. D., Cuoci, A., Mehl, M., Frassoldati, A., Faravelli, T., Geem and K.M. Van. 2021. The chemistry of chemical recycling of solid plastic waste via pyrolysis and gasification: State-of-the-art, challenges, and future directions 84: 100901.

Elliott, T. and Elliott, L. 2020. A Plastic Futures, Plastic consumption and waste management in the UK. In report for WWF, Prepared by Eunomia Research and Consulting Ltd. https://doi.org/10.7551/mitpress/10711.003.0009 (Last Accessed on: August, 2021)

EPA. 2018. https://www.epa.gov/facts-and-figures-about-materials-waste-and-recycling/plastics-material-specific-data#:~:text=In%202018%2C%20plastics%20generation%20was,measure%20the%20recycling%20of%20plastic (Last Accessed on : August, 2021).

European Union: Directive (EU) 2019/904 of the European Parliament and of the Council of 5 June 2019 on the reduction of the impact of certain plastic products on the environment, Official Journal of the European Union, Brussels, Belgium (2019). https://eur-lex.europa.eu/legal-content/EN/TXT/PDF/?uri=CELEX:32019L0904 (Last Accessed on: September 2021)

Fadeeva, Z., Van Berkel, R. (2021) Unlocking circular economy for prevention of marine plastic pollution: An exploration of G20 policy and initiatives'. Journal of Environmental Management 277: 111457.

Fahim, I., Mohsen, O. and Elkayaly, D. 2021. Production of Fuel from Plastic Waste : A Feasible Business. Polymers 13: 915–923.

Gall, S.C. and Thompson, R.C. 2015. The impact of debris on marine life. Marine Pollution Bulletin 92(1–2): 170–179.

Geyer, R., Jambeck, J.R. and Law, K.L. 2017. Production, use, and fate of all plastics ever made - Supplementary Information. Science Advances 3(7): 19–24.

Gilbert, M. 2017. Plastic Mterials: Introduction and Historical development. *In*: Gilbert , M. (ed.). Brydson's Plastics Materials. Elsevier Inc. ISBN: 978-0-323-35824-8. https://doi.org/10.1016/c2014-0-02399-4

Gopinath, K.P., Nagarajan, V.M., Krishnan, A. and Malolan, R. 2020. A critical review on the influence of energy, environmental and economic factors on various processes used to handle and recycle plastic wastes: Development of a comprehensive index. Journal of Cleaner Production 274: 123031.

Gu, J.D. 2021. Biodegradability of plastics: the issues, recent advances, and future perspectives. Environmental Science and Pollution Research 28(2): 1278–1282.

Hariadi, D., Saleh, S.M., Anwar Yamin, R. and Aprilia, S. 2021. Utilization of LDPE plastic waste on the quality of pyrolysis oil as an asphalt solvent alternative. Thermal Science and Engineering Progress 23: 100872.

He, W. and Benson, R. 2013. Polymeric Biomaterials. *In*: S. Ebnesajjad (ed.). Handbook of Biopolymers and Biodegradable Plastics: Properties, Processing and Applications. Elsevier Inc., ISBN: 978-1-4557-2834-3. https://doi.org/10.1016/B978-1-4557-2834-3.00005-7

Huang, D., Xu, Y., Lei, F., Yu, X., Ouyang, Z., Chen, Y., Jia, H. and Guo, X. 2021. Degradation of polyethylene plastic in soil and effects on microbial community composition. Journal of Hazardous Materials 416: 126173.

Hussain, Z., Khan, K.M., Perveen, S., Hussain, K. and Voelter, W. 2012. The conversion of waste polystyrene into useful hydrocarbons by microwave-metal interaction pyrolysis. Fuel Processing Technology 94(1): 145–150.

Iamexpat report. 2021. https://www.iamexpat.de/lifestyle/lifestyle-news/dresden-company-transforms-plastic-waste-fuel (Last Accessed on : September 2021)

Inoue, T., Miyazaki, M. and Kamitani, M. 2004. Mechanochemical dechlorination of polyvinyl chloride by co-grinding with various metal oxides. Advanced Powder Technology 15(2): 215–225.

Jiang, X., Wang, T., Jiang, M., Xu, M., Yu, Y., Guo, B., Chen, D., Hu, S., Jiang, J., Zhang, Y. and Zhu, B. 2020. Assessment of Plastic Stocks and Flows in China: 1978–2017. Resources, Conservation and Recycling 161: 104969.

Jha, K.K. and Kannan, T.T.M. 2020. Recycling of plastic waste into fuel by pyrolysis - a review. Materials Today: Proceedings 37: 3718–3720.

Kaminsky, W., Predel, M. and Sadiki, A. 2004. Feedstock recycling of polymers by pyrolysis in a fluidized bed. Polymer Degradation and Stability 85: 1045–1050.

Kaur, P.J., Chandra, U., Hussain, C.M. and Kaushik, G. 2020a. Landfarming: A Green Remediation Technique. *In*: C.M. Hussain (ed.). The Handbook of Environmental Remediation: Classic and Modern Techniques, Royal Society of Chemistry, pp. 357–378.

Kaur, P.J., Chauhan, G., Kaushki, G. and Hussain, C.M. 2020b. Green techniques for remediation of soil using composting. *In*: C.M. Hussain (ed.). The Handbook of Environmental Remediation: Classic and Modern Techniques, Royal Society of Chemistry, pp. 254–267.

Kaur, P.J., Kaushik, G., Hussain, C.M. and Dutta, V. 2021. Management of waste tyres: properties, life cycle assessment and energy generation. Environmental Sustainability 4: 261–271

Kefeli, A.A., Razumovskii, S.D. and Zaikov, G.Y. 1971. Polymer Science U.S.S.R. 13: 904–911.

Kim, C.A., Kim, J.T., Lee, K., Choi, H.J. and Jhon, M.S. 2000. Mechanical degradation of dilute polymer solutions under turbulent flow. Polymer 41(21): 7611–7615.

Kiran, E. and Gillham, J.K. 1976. Pyrolysis-molecular weight chromatography: A new on-line system for analysis of polymers. II. Thermal decomposition of polyolefins: Polyethylene, polypropylene, polyisobutylene. Journal of Applied Polymer Science 20(8): 2045–2068.

Kodera, Y., Ishihara, Y. and Kuroki, T. 2006. Novel process for recycling waste plastics to fuel gas using a moving-bed reactor. Energy and Fuels 20(1): 155–158.

Kunwar, B., Moser, B.R., Chandrasekaran, S.R., Rajagopalan, N. and Sharma, B.K. 2016. Catalytic and thermal depolymerization of low value post-consumer high density polyethylene plastic. Energy 111: 884–892.

Lebreton, L., Slat, B., Ferrari, F., Sainte-Rose, B., Aitken, J., Marthouse, R., Hajbane, S., Cunsolo, S., Schwarz, A., Levivier, A., Noble, K., Debeljak, P., Maral, H., Schoeneich-Argent, R., Brambini, R. and Reisser, J. 2018. Evidence that the Great Pacific Garbage Patch is rapidly accumulating plastic. Scientific Reports 8(1): 1–16.

Lewandowski, W.M., Januszewicz, K. and Kosakowski, W. 2019. Efficiency and proportions of waste tyre pyrolysis products depending on the reactor type—A review. Journal of Analytical and Applied Pyrolysis 140: 25–53.

Liang, Y., Tan, Q., Song, Q. and Li, J. 2021. An analysis of the plastic waste trade and management in Asia. Waste Management 119: 242–253.

Lopez, G., Artetxe, M., Amutio, M., Bilbao, J. and Olazar, M. 2017. Thermochemical routes for the valorisation of waste polyole fi nic plastics to produce fuels and chemicals . A review. Renewable and Sustainable Energy Reviews 73: 346–368.

López, A., de Marco, I., Caballero, B.M., Laresgoiti, M.F. and Adrados, A. 2011. Influence of time and temperature on pyrolysis of plastic wastes in a semi-batch reactor. Chemical Engineering Journal 173(1): 62–71.

Macuvele, D.L.P., Nones, J., Matsinhe, J.V., Lima, M.M., Soares, C., Fiori, M.A. and Riella, H.G. 2017. Advances in ultra high molecular weight polyethylene/hydroxyapatite composites for biomedical applications: A brief review. Materials Science and Engineering C 76: 1248–1262.

Mccaffrey, W.C., Kamal, M.R. and Cooper, D.G. 1995. Thermolysis of polyethylene. Polymer Degradation and Stability 47: 133–159.

Muhammad, C., Onwudili, J.A. and Williams, P.T. 2015. Thermal degradation of real-world waste plastics and simulated mixed plastics in a two-stage pyrolysis-catalysis reactor for fuel production. Energy and Fuels 29(4): 2601–2609.

Ozen, B.B.F., Mauer, L.J. and Floros, J.D. 2003. Effects of Ozone Exposure on the Structural, Mechanical and Barrier Properties of Select Plastic Packaging Films. Packaging Technology and Science 15: 301–311.

Panda, A.K., Singh, R.K. and Mishra, D.K. 2012. Thermo-catalytic degradation of thermocol waste to value added liquid products. Asian Journal of Chemistry 24(12): 5539–5542.

Papuga, S.V., Gvero, P.M. and Vuki, L.M. 2015. Temperature and time influence on the waste plastics pyrolysis in the fixed bed reactor. Thermal Science, 20(2): 731–741.

Pegram, J.E. and Andrady, A.L. 1989. Outdoor Weathering of Selected Polymeric Materials under Marine Exposure Conditions. Polymer Degradation and Stability 26: 333–345.

Plastics Europe .2020. Plastic—the Facts 2020https://plasticseurope.org/knowledge-hub/plastics-the-facts-2020/ (Last Accessed on : October 2021)

PlasticsEurope. 2019. Plastics—the Facts 2019, An analysis of European plastics production, demand and waste data, pp. 14–35. https://www.plasticseurope.org/en/resources/market-data (Last Accessed on : August, 2021)

Prajapati, R., Kohli, K., Maity, S.K. and Sharma, B.K. 2021. Potential Chemicals from Plastic Wastes. Molecules 26(3175): 1–23.

Puskic, P.S., Lavers, J.L. and Bond, A.L. 2020. A critical review of harm associated with plastic ingestion on vertebrates. Science of the Total Environment 743: 140666.

Qureshi, M.S., Oasmaa, A., Pihkola, H., Deviatkin, I., Tenhunen, A., Mannila, J., Minkkinen, H., Pohjakallio, M. and Laine-Ylijoki, J. 2020. Pyrolysis of plastic waste: Opportunities and challenges. Journal of Analytical and Applied Pyrolysis 152: 104804.

Rabnawaz, M., Wyman, I., Auras, R. and Cheng, S. 2017. A roadmap towards green packaging: The current status and future outlook for polyesters in the packaging industry. Green Chemistry 19(20): 4737–4753.

Sanchez, C. 2021. Microbial capability for the degradation of chemical additives present in petroleum-based plastic products: A review on current status and perspectives. Journal of Hazardous Materials 402: 123534.

Sawai, D., Nagai, K., Kubota, M., Ohama, T. and Kanamoto, T. 2006. Maximum Tensile Properties of Oriented Polyethylene, Achieved by Uniaxial Drawing of Solution-Grown Crystal Mats: Effects of Molecular Weight and Molecular Weight Distribution. Journal of Polymer Science: Part B: Polymer Physics 44: 153–161.

Schmidt, H.P., Kammann, C., Hagemann, N., Leifeld, J., Bucheli, T.D., Sánchez Monedero, M.A. and Cayuela, M.L. 2021. Biochar in agriculture—A systematic review of 26 global meta-analyses. *In*: GCB Bioenergy 13: 1708–1730. John Wiley and Sons Inc.

Seo, Y.H., Lee, K.H. and Shin, D.H. 2003. Investigation of catalytic degradation of high-density polyethylene by hydrocarbon group type analysis. Journal of Analytical and Applied Pyrolysis, 70(2): 383–398.

Sharma, B.K., Moser, B.R., Vermillion, K.E., Doll, K.M. and Rajagopalan, N. 2014. Production, characterization and fuel properties of alternative diesel fuel from pyrolysis of waste plastic grocery bags. Fuel Processing Technology 122: 79–90.

Sharuddin, S.D.A., Abnisa, F., Wan Daud, W.M.A. and Aroua, M.K. 2016. A review on pyrolysis of plastic wastes. Energy Conversion and Management 115: 308–326.

Shui-e, Y.I.N., Peng, D. and Ru-shan, B.I.E. 2007. Basic Study on Plastic Pyrolysis in Fluidized Bed with Continuous-feeding. International Conference on Power Engineering, Hanghou, China 2: 123–124.

Siddiqui, M.N. 2009. Conversion of hazardous plastic wastes into useful chemical products. Journal of Hazardous Materials 167(1–3): 728–735.

Siddiqui, M.N. and Redhwi, H.H. 2009. Pyrolysis of mixed plastics for the recovery of useful products. Fuel Processing Technology 90(4): 545–552.

Silva, A.L.P., Prata, J.C., Duarte, A.C., Soares, A.M.V.M., Barceló, D. and Rocha-Santos, T. 2021. Microplastics in landfill leachates: The need for reconnaissance studies and remediation technologies. Case Studies in Chemical and Environmental Engineering 3: 100072.

Singh, R.K., Ruj, B., Sadhukhan, A.K. and Gupta, P. 2019. Thermal degradation of waste plastics under non-sweeping atmosphere: Part 1: Effect of temperature, product optimisation, and degradation mechanism. Journal of Environmental Management 239: 395–406.

Syamsiro, M., Saptoadi, H., Norsujianto, T., Noviasri, P., Cheng, S., Alimuddin, Z. and Yoshikawa, K. 2014. Fuel oil production from municipal plastic wastes in sequential pyrolysis and catalytic reforming reactors. Energy Procedia 47: 180–188.

UNEP. 2016. Marine Plastic Debris and Microplastics—Global Lessons and Research to Inspire Action and Guide Policy Change. United Nations Environment Programme, Nairobi. ISBN No: 978-92-807-3580-6. https://wedocs.unep.org/rest/bitstreams/11700/retrieve (Last Accessed on: July 2021)

Vijayakumar, A. and Jilse, S. 2018. Pyrolysis process to produce fuel from different types of plastic—a review. IOP Conf. Series: Materials Science and Engineering 396: 012062.

Webb, H.K., Arnott, J., Crawford, R.J. and Ivanova, E.P. 2013. Plastic Degradation and Its Environmental Implications with Special Reference to Poly(ethylene terephthalate). Polymers 5: 1–18.

Williams, P.T. and Slaney, E. 2007. Analysis of products from the pyrolysis and liquefaction of single plastics and waste plastic mixtures. Resources, Conservation and Recycling 51(4): 754–769.

Xu, J., Cui, Z., Nie, K., Cao, H., Jiang, M., Xu, H., Tan, T. and Liu, L. 2019. A quantum mechanism study of the C-C bond cleavage to predict the bio-catalytic polyethylene degradation. Frontiers in Microbiology 10: Article 489.

10

CATALYTIC PYROLYSIS OF WASTE PLASTICS FOR THE PRODUCTION OF LIQUID FUELS

Hanan El-Sayed

1. INTRODUCTION

Modern lifestyle has adapted the use of plastic in many applications due to its light weight, low cost, durability and easy processing. Plastics are widely used as essential raw materials in many industries such as electronics, automotive, healthcare, agriculture, transportation, packaging and much more. To put this in perspective, the global production of plastics has increased from about 2 million tons in 1950 to 380 million tons in 2015 (Geyer et al. 2017). Recovery or recycling of waste plastics is not sufficient, and the millions of tons end up in either oceans, landfills or incinerated which impose severe environmental and health risks. Vast amount of waste plastics (79%) ends up disposed in landfill or dumped into the oceans, incineration represents about 12% of waste management methods while recycling accounts for only 9% due to the high processing cost. Many nations have recently banned waste plastic incineration due to the adverse health and environmental risks associated with burning the waste plastics (Murthy et al. 2020).

Waste-to-energy technologies have been adapted for many years to deliver liquid fuels as a renewable source of energy to replace fossil fuels. Plastics are derived from petrochemical resources; therefore, they have high calorific values and low moisture content that make plastics an excellent candidate for conversion into useful forms of energy. There are different thermochemical routes that are being used for converting plastics into high end products. These include hydrothermal liquefaction (HTL), gasification, hydrotreating and pyrolysis. Pyrolysis process offers an environmental friendly solution to waste plastics by *catalytic* or *non-catalytic* degradation of long chain molecules into smaller ones which can subsequently be utilized to produce liquid fuels with high end market value.

Pyrolysis is a well-known thermochemical technology that is defined as the thermal decomposition of complex organic compounds into simpler ones in an oxygen free environment at high temperatures

Senior Process Engineer, Greenfield Global Inc. Chatham ON and Research Associate and Sessional Instructor, Department of Civil and Environmental Engineering, University of Windsor, Windsor ON.
Email: Hanan.El-Sayed@uwindsor.ca

and pressures. The origin of the word is the two Greek words: "pyro" which means fire and "lysis" which means decomposition into integral constituents. The very old well-established technology goes back to more than 5500 years ago, but it has been growing since and extensively advancing over the years (Abbas-Abadi et al. 2014). Pyrolysis technology may be divided into four main categories, i.e., slow, intermediate, fast and flash on the basis of reaction heating rate, temperature and residence time (Erdogan 2020). Each technology type has its own characterizing operating conditions which greatly influence the yield and distribution of different products. The products formed as a result of the pyrolysis process include solid (char), liquid (pyrolysis oil or bio-oil) and gas. Liquid, solid and gas products are often used as energy sources for power generation or to supply the heat required for the process itself to improve overall process heat integration and carbon footprint. Fast and flash pyrolysis are characterized by high temperatures, high heating rates and low residence times which result in higher bio-oil yields than char and gas products (Nanda and Berruti 2021). On the other hand, slow pyrolysis features longer residence times, moderate temperatures and slow heating rates which favors the production of more char than bio-oil. Pyrolysis oil can further be upgraded to higher quality liquid fuel or other chemical products via different chemical processes or using catalyst in the second step after catalytic pyrolysis. The product calorific value is highly dependent on the carbon content (fixed carbon) of the type of plastic waste.

This chapter will focus on evaluating the influence of employing different catalyst types in the pyrolysis of individual or mixed plastic wastes on the quality of the products particularly the distribution of liquid product. The chapter will also present a summary of the raw material, plastic categories and the properties of each. In addition, the different types of pyrolysis reactors are presented comparing the advantages and drawbacks of each type when used in catalytic pyrolysis. The effect of main operating parameters on the performance of the catalytic pyrolysis of plastics will be also reviewed. The chapter is also stating the most important catalytic pyrolysis commercial plants worldwide and emphasizing the key challenges facing the scale up of the different technologies. Conclusions will be given highlighting a summary of findings of the most recent experimental studies and addressing the gaps for future research.

2. WASTE PLASTIC CATEGORIES

Plastics are high molecular compounds known sometimes as polymers that are composed of various elements such as carbon, oxygen, hydrogen, nitrogen, sulfur and chlorine. The polymeric nature of plastics makes it take hundreds of years to degrade naturally. Petroleum crude accounts for about 99% of the raw material to manufacture plastics (Nanda and Berruti 2021).

Plastics may be divided into six categories based on their frequent use: polyethylene terephthalate (PET), high-density polyethylene (HDPE), low-density polyethylene (LDPE), polyvinyl chloride (PVC), polypropylene (PP), and polystyrene (PS) (Murthy et al. 2020). Polyethylene are thermoplastic polymers composed of multiple units of ethylene monomer. They are divided to high and low density and their chemical formula is $(\text{-}CH_2\text{-}CH_2\text{-})_n$. PE is the leader with respect to volumes consumed and global demand among all types of plastic. HDPE, LDPE and PP are called polyolefins because the building unit of the polymer or the monomer unit is an olefin (ethylene and propylene, respectively). But despite the large usage of polyolefins, they are the least recycled types of plastic due to their complex composition in post-consumer wastes (Bonifazi et al. 2018). Packaging is one of the most uses of plastic in our daily lives with PET as the main type used for this purpose. PVC is a high-performance type of plastic and is used in the manufacturing of pipes for water and wastewater and in many other architectural and construction applications. PVC is also used in electrical insulation due to its high fire-resistance. PP has higher hardness and rigidity and is mainly used for chairs, desks, and some car parts. PS is relatively low-cost plastic and is used in medical and electronic supplies as well as toys. It is important to understand the composition of the different plastic categories to determine how this might affect the distribution and the properties of the products corresponding to the pyrolysis of each individual type or their mixtures. Table 1 shows the proximate and elemental analysis of different types of plastics.

2.1 Polyethylene terephthalate (PET)

Due to vast applications of PET, its accumulation as a waste became a noticeable issue. With the high costs associated with sorting and recycling, pyrolysis of PET as the feedstock was investigated by many researchers. The chemical formula of PET is $(C_{10}H_8O_4)_n$. It is characterized by its high resistance to heat and solvents (Nanda and Berruti 2021). Figure 1 presents the chemical structure of various types of plastics discussed in this chapter. As explained earlier, the long chain polymer of each plastic is composed of repeated smaller building units. In case of PET, the building units are ethylene glycol and terephthalic. It is worth noting that more than half of world's synthetic fiber is made from PET which is known as polyester when used for this purpose (PETRA 2015). PET is considered a fully recyclable plastic and it is reported as being the most recycled type of plastics in US and the world (PETRA 2015). The properties of PET include light weight, high strength, and versatility.

2.2 High-density polyethylene (HDPE)

The high density of the polyethylene is attributed to its long chain linear structure. It is also characterized by its high tensile strength, high crystallinity, and high rigidity. Due to its vast usage, it contributes to about 18% of the MSW (Nanda and Berruti 2021). The demand for HDPE has a global annual increase of about 3.3%, going from 11.9 million tons in 1990 to 43.9 million tons in 2017 (Amjad and Fatemi 2021).

2.3 Low-density polyethylene (LDPE)

The low density of polyethylene arises from the low branching in the polymer chain. LDPE is characterized by its high opacity, tensile strength, rigidity, and chemical resistance. It is also flexible even at a low temperature due to its degree of crystallinity which is averaged at 55% (Sen and Raut 2015). LDPE is used in many industrial applications such as textile and paper coatings, trays and plastic bags for food and non-food items. It is estimated that each year about 500 billion to one trillion plastic bags are consumed (Roy et al. 2008). Many countries have adopted the use of reusable bags for groceries and stores have started to ask if customers want to purchase plastic bags and charge for them if they accept to purchase. Some municipalities started to ban all plastic bags and plastic wrapping from recycling boxes or carts due to the damage they might cause to the sorting equipment.

2.4 Polyvinyl chloride (PVC)

The building monomer of PVC is vinyl chloride with a chlorine content of approximately 56%. Due to its low thermal stability, some additives such as plasticizers and stabilizers must be added to improve PVC properties. Any processing/degradation of PVC releases HCl which is a major problem that leads to equipment corrosion. De-chlorination is a possible solution to such problem but imposes high cost and higher energy consumption (Rahimi and Garcia 2017). PVC is the second largest global demand after PE at a yearly demand exceeding 35 million tons (Sadat-Shojai and Bakhshandeh 2011).

2.5 Polypropylene (PP) and polyethylene (PE)

Polypropylene is thermoplastic polymer formed by repetitive propylene monomers (CH_2=CH-CH_3). It has a high fusion temperature (160°C–170°C) and high affinity to crystallization (about 90%) (Sharuddin et al. 2016). Polyethylene is a non-polar low-cost plastic and has high chemical and

Table 1. Proximate and Ultimate Analysis of Different Types of Plastics (Rajendran et al. 2020, Al-Salem 2019, Sharuddin et al. 2016).

Plastic Category	High Heating Value (HHV), MJ/kg	Fixed Carbon (wt%)	Volatile Matter (wt%)	Hydrogen (wt%)	Oxygen (wt%)	Carbon (wt%)	Nitrogen (wt%)	Ash (wt%)	Moisture (wt%)
PET	30.2	7.77–13.17	86.83–91.75	3.90	20.35	75.21	0.54	0.00–0.02	0.46–0.61
HDPE	46.4–49.4	0.01–0.30	98.57–99.81	14.90	0.74	83.90	0.05	0.18–1.40	0.00
LDPE	46.4	0.00	99.60–99.70	10.12–13.00	0.90–18.55	69.67–85.70	0.09	0–0.4	0.30
PVC	35.2	5.19–6.30	93.70–94.82	6.14	37.34	55.98	0.54	0.00	0.74–0.80
PP	45.3–46.4	0.16–1.22	95.08–97.85	13.7–14.22	0.20–7.46	77.52–86.1	0.10	1.99–3.85	0.15–0.18
PS	41.9	0.12–0.20	99.50–99.63	7.82–7.90	0.00–8.88	83.10–92.70	0.21	0.00	0.25–0.30

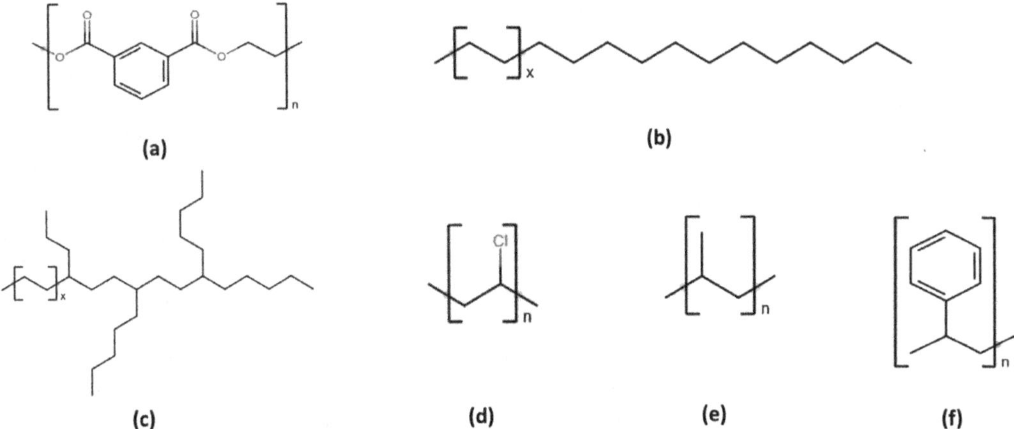

Figure 1. Chemical structure of (a) Polyethylene terephthalate (b) High density polyethylene (c) Low density polyethylene (d) Polyvinyl chloride (e) Polypropylene (f) Polystyrene.

fatigue resistances. Its uses are mainly in automotive industry. It represents the highest share of the MSW at about 24.3% (Nanda and Berruti 2021).

2.6 Polystyrene (PS)

Polystyrene is hydrocarbon polymer made of the styrene monomer building block. Its chemical formula is $(C_8H_8)_n$ and is characterized by its light weight which makes it good for applications such as packaging and electronics. On the other hand, it has low resistance to chemicals and poor barrier to gases and moisture (Nanda and Berruti 2021). The major problem with PS is its higher degree of reactivity when compared to other plastic types. It could slowly degrade under environmental conditions to form microplastics that could potentially leak into water bodies or soil and harm the ecosystem (Kwon et al. 2014).

3. TYPES OF PYROLYSIS CATALYST

Catalyst is typically used in most chemical processes to accelerate the chemical reactions involved in the process. This occurs due to increase in the reaction rate by lowering the activation energy. Use of catalyst results in reducing the reaction residence time as well as the process energy consumption by reducing the temperatures at which the reactions take place according to the well-known Arrhenius equation (Koch 1977). A catalyst would be considered effective if it has good activity, selectivity, and stability at reaction conditions. All these factors are a function of catalyst porosity and surface area (Butler et al. 2011).

Two approaches are used to introduce catalyst to pyrolysis process: (i) mixing with the waste plastic feedstock in liquid/solid form before reaction (homogeneous), and (ii) mixing within the vapor phase of the reactor volatile stream which is the most recent approach (heterogeneous). Many researchers have shown that the ratio of catalyst and waste plastic in the feed affects the process conversion efficiency, which increases with increasing the amount of catalyst until certain limit, then decreases after (Hazrat et al. 2015). It was reported in many studies that direct contact between the catalyst and the plastic feedstock will negatively impact catalyst recovery due to two possible reasons: first, impurities in the waste plastic such as sulfur, chlorine and nitrogen may deposit on the surface of the catalyst; secondly, the sticky nature of the plastic may block the catalyst pores during thermal degradation (Miandad et al. 2016).

Most widely used catalysts in pyrolysis process of waste plastics are: (1) zeolite based catalysts such as ZSM-5, HZSM-5, β-zeolite or Y-zeolite, and US-Y, (2) alumina, (3) silica, (4) basic catalysts such as $BaCO_3$, (5) bimetallic catalysts Al-Zn composites, (6) FCC catalyst (silica-alumina), (7) MCM-41, and (8) others (Zhang et al. 2019, Miandad et al. 2016, Sharuddin et al. 2016).

A summary of literature highlighting the effect of type of catalyst on the yield of products obtained via pyrolysis of single or mixture of plastics is presented in Table 2. The summary is categorized based on catalyst types, the reactor type and operating conditions.

Zeolite based catalysts have higher acid strength which promote higher conversion when used in pyrolytic reactors than non-zeolite-based catalysts. Moreover, zeolite-based catalysts were observed to increase the aromatic content in the produced liquid fuel (Sebestyén et al. 2017, Hazrat et al. 2015). It was also reported that liquid products yielded from zeolite based catalytic pyrolysis of waste plastics is composed of low carbon number compounds such as gasoline range hydrocarbons and less heavy fraction molecules (Kasar et al. 2020, Artetxe et al. 2013). On the other hand, it is worth mentioning that zeolite-based catalysis increases the gas yield at the expense of liquid yield products, particularly C_3 and C_4 gases (Rehan et al. 2017, Lopez-Urionabarrenechea et al. 2012).

Lopez et al. (2011a) compared the performance of the pyrolysis process in the presence of two different catalysts: ZSM-5 and Red Mud for a mixture of plastics. The plastic mixture used in this study was composed of 3 wt.% PVC, 4 wt.% PET, 18 wt.% PS, 35 wt.% PP and 40 wt.% PE. The experiments were performed in a semi-batch reactor at 440°C and at 500°C. The results support the ones obtained from another study (Lopez et al. 2011b) conducted by the same research group using same catalyst that ZSM-5 increases the yield of gases and the portion of aromatic content in the liquid. Moreover, it was concluded in the given study that it is possible to completely recover ZSM-5's initial activity after regeneration by heating at 550°C in oxygen atmosphere when compared with fresh catalyst. Red Mud showed similar behavior but at higher pyrolysis temperatures than for ZSM-5. This fact was attributed to the presence of Fe_2O_3 and TiO_2 as Red Mud components which led to the occurrence of additional reaction pathways. Red Mud was also used as a catalyst in another study (Adrados et al. 2012), utilizing simulated mixture of plastics (PE, PP, PS, PVC and PET) as well as a mixture of waste plastic from a real sorting plant. The pyrolysis experiments were performed in a semi-batch reactor at 500°C (This temperature which is the 500°C optimum temperature as suggested in literature) for 30 min at 20°C/min heating rate at a plastic to catalyst ratio of 10:1. The results showed higher liquid and gas yields in simulated plastics versus the real waste. It was attributed to the fact that real plastic waste obtained from an industrial recovery facility contained higher percentages of inorganics which couldn't be converted to either liquids or gases. The liquid product contained more than 70 wt.% aromatics which was attributed to the cracking and aromatization effect of Al_2O_3 component of the Red Mud. In addition, they observed an increase in the solid product in case of pyrolysis of real plastic waste due to the presence of many inorganic components.

Nearly similar yield values were reported in Sebestyén et al. (2017) study while using another zeolite-based catalyst HZSM-5 under same operating conditions. It should be pointed out that the ratio of catalyst to plastic in the feed was increased in the given study only to distinctively monitor the effect of catalyst on the pyrolysis process due to low reactor volume (Sebestyén et al. 2017). Zhang and Lei (2016) reported one of the highest liquid yields for LDPE using ZSM-5 catalyst at lower temperature of 375°C in a microwave-assisted reactor. The liquid yield was reported as 66.18 wt.% at a plastic to catalyst ratio of 10:1. Interestingly, the authors proved that the pyrolysis liquid could be upgraded under mild conditions (250°C and 2 hours) to jet fuel range hydrocarbons using another catalyst- RANEY® nickel. The resulted JP-5 navy drop-in fuel had a composition of 31.23% aliphatic alkanes, 53.06% cycloalkanes and 15% aromatics.

To protect the catalyst from contamination and consequently deactivation, Muhammad et al. (2015) used a two-stage fixed bed reactor to pyrolyze single plastics as well as a mixture of many plastic types. HZSM-5 was used in the second stage, utilizing PE, PP, PS and PET as a plastic feedstock from real municipal wastes. Other studies (Chen et al. 2020, Serrano et al. 2012) supported

Table 2. Catalytic pyrolysis of different plastic types.

Catalyst Category	Type of catalyst used	Reactor Type	Type of Plastic used as raw material	Plastic to Catalyst Ratio	Operating conditions	Yield of liquid product, % by weight Liquids/Gases/Solids	Reference
Zeolite-based Catalysis	Natural-Synthetic Zeolites	Batch	PS	10:1	450°C, 75 min	54/12.8/32.8 50/22.6/27.4	(Rehan et al. 2017)
	Y-zeolites	Batch	HDPE	10:1	430°C, 3 hrs	75/19/6	(Kunwar et al. 2016)
	Hβ and HY-zeolite	Conical spouted bed reactor (CSBR)	HDPE	30:1	500°C	73/24/3	(Elordi et al. 2009)
	HZSM-5	Batch	PE, PP, and PET	10:1	550°C at 20°C/min	70.6/18.8/6.5	(Sebestyen et al. 2017)
		Semi-Batch	PE, PP, PS, PET and PVC	10:1	440°C–30 min	56.9/40.4/3.2	(Lopez-Urionabarrenechea et al. 2012)
		Fluidized Bed	HDPE	0.93:1 (optimum)-9.2:1	350°C–550°C (Optimum was 450°C)	27.4/72.6/0	(Mastral et al. 2006)
		Batch	PP, PE	20:1	400°C–15 min and 450°C–30 min	92/5/3	(Santos et al. 2019)
		Conical spouted bed reactor (CSBR)	HDPE	30:1	500°C	58/7/1.9	(Elordi et al. 2009)
	ZSM-5	Semi-Batch	PE, PP, PS, PET and PVC	10:1	440°C and 500°C 30 min	56.9/40.4/3.2 39.8/58.4/1.8	(Lopez et al. 2011a)
		Semi-Batch	HDPE, LDPE, PP and PS	4:1	350°C–450°C for 30 min–60 min	PP: 92/8/2 PS: 75.6/9.4/15 LDPE: 60/15/25 HDPE: 50/25/25	(Mangesh et al. 2020)
		Microwave	LDPE	10:1	375°C	66.18/32.01/1.81	(Zhang and Lei 2016)

Table 2 contd. ...

...Table 2 contd.

Catalyst Category	Type of catalyst used	Reactor Type	Type of Plastic used as raw material	Plastic to Catalyst Ratio	Operating conditions	Yield of liquid product, % by weight Liquids/ Gases/Solids	Reference
	US-Y	Semi-Batch	LDPE	4:1	417°C and 15 min	90/9/1	(Akpanudoh et al. 2005)
	Zn-ZSM-11	Fixed-bed	LDPE	0.5:1–2.0:1	500°C at 20 and 60 min (optimum 20 min)	81.77/16.95/1.28	(Renzini et al. 2009)
Natural based catalysis	Red Mud	Semi-Batch	PE, PP, PS, PET and PVC	10:1	440°C and 500°C 30 min	76.2/21.6/2.2 57/41.3/1.7	(Lopez et al. 2011a)
		Semi-Batch	PE, PP, PS, PET and PVC	10:1	500°C, 20°C/min, 30 min	57/42.4/0.6	(Adrados et al. 2012)
	Activated Carbon	Fixed bed	LDPE, and mixed plastics (PS, PE, PP, PET)	1.7:1–0.4:1 (optimum was 1.7)	430°C–571°C (optimum was 500°C)–20 min	LDPE: 84/14/1 Mixed: 70.51/29.35/0.14	(Zhang et al. 2019)
MCM-41 Catalysis	Mesoporous MCM-41	Batch	PE	-	450°C–60 min	8.5/82.4/9.1	(Chandran et al. 2020)
FCC Catalyst	FCC-R1	Batch	Mixture of PE/ PP/PS/PVC	-	390°C	3.8/82.4/11.7	(Chandran et al. 2020)
Bimetallic Catalysts and Others	$CaCO_3$	Batch	HDPE	5:1	460°C at 90 min	94/5/1	(Chandran et al. 2020)
	Molten $AlCl_3$-NaCl eutectic salt	Batch	HDPE	2.3:1	300°C–500°C (optimum at 300°C)	91.7/5.1/3.2	(Su et al. 2019)
	Silica-Alumina	Batch/Continuous	HDPE, PP and PS and mixture of all three	10:1 and 5:1 (the optimum is 5:1)	450°C	HDPE: 48.28/51.72/<0.3 PP: 59.57/40.43/<0.3 PS: 50/50/<0.3 Mix: 45/55/<0.3	(Owusu et al. 2018)
	Iron Pillared Clay (Fe-PiLC-Fe)	Batch	HDPE	10:1	450°C	60/18/22	(Faillace et al. 2017)
	Sulphated zirconium hydroxide	Batch	PP, LDPE, HDPE and a mixture of three	10:1	500°C, 20°C/min	PP: 84/15/1 LDPE: 84/15/1 HDPE: 80/20/0 Mix: 82/18/0	(Panda et al. 2020)

the benefits of using a two-stage pyrolysis process for ease of catalyst recovery and better control of reaction conditions. A mixture of PP and PE of a mass ratio of 1:1 was pyrolyzed at 450°C for 30 min using different catalysts: ZSM-5, US-Y, NH_4ZSM-5 and their corresponding alkaline-treated and acid leached forms (Santos et al. 2018a). The results depicted that alkaline treatment of ZSM-5 zeolites improved two features of the catalyst: (i) generating mesopores, and (ii) increasing the density of the low strength acid sites, through promoting the removal of silicon from the zeolite, which led to higher yield of lighter fractions of products. With respect to the quality of liquid products, it was reported that all studied catalysts led to the increase in olefin content within the liquid fuel. This phenomenon was explained by the tendency of the intermediate bulky molecules to access the acid sites inside the pore structure of the catalyst with higher mesoporous volume which would also contribute to the increase of liquid yield.

When comparing acidic catalyst (Y-zeolite) to basic catalyst ($MgCO_3$) during batch pyrolysis of HDPE at a ratio of plastic to catalyst of 10:1, it was found that the use of Y-zeolite produced more liquid products in the gasoline range whereas $MgCO_3$ yielded higher liquid products in the diesel range (Kunwar et al. 2016). Similar results were reported by Santos et al. (2019) when studying the effect of HZSM-5 and Y-zeolites on the pyrolysis of PP and PE at 450°C (Santos et al. 2019). An optimization study by Wang et al. (2019) showed an increase in the production of benzene and ethylbenzene by about 33 folds in the presence of Y-zeolite catalyst for the pyrolysis of PS at an optimum PS to catalyst ratio of 0.7:1 and a temperature of 650°C.

Some other researchers looked for less expensive alternatives to zeolite-based catalysts such as pillared clays. Due to their mesoporous nature, clays play an important role as a catalyst in the petrochemical industry (Faillace et al. 2017). Pillared clays were synthesized by fractionation and treated to remove silt, salt as well as soluble and organic fractions. After that, they were intercalated with iron and calcined to form pillars at a temperature that reached 300°C. The experiments were conducted in a fixed bed tubular reactor to pyrolyze HDPE at 450°C and 500°C. The results of the study depicted that high yield of liquid products could be obtained (higher than 80 wt%) at 500°C using iron pillared clay (FE-PICL-FE) as a catalyst at a plastic to catalyst ratio of 10:1. Moreover, it was shown that the highest yield of linear hydrocarbons and aromatics were obtained at 300°C. This was attributed to different reasons: (i) smaller pore size and lower mesopore volume, (ii) lower contact time between polymer molecules and catalyst acid sites at lower temperatures, and (iii) increased secondary cracking, aromatization, at lower temperatures because they are exothermic.

FCC catalyst was reported in many studies with the observation that it generally increases the yield of liquid product. This could reach 80%–90%, as indicated by Kyong et al. (2002) in their study when they examined the effect of FCC on the yield of liquid products from pyrolysis of PS, LDPE, HDPE and PP. Olazar et al. (2009) obtained an aromatic content of about 54% of the liquid product in the diesel fraction while studying the catalytic pyrolysis of HDPE using FCC. An optimum ratio of catalyst to plastic waste in the feed was determined where maximum conversion to liquid product is reached; beyond this ratio, yield of gas and char products may increase (Abbas-Abadi et al. 2014).

Many alumina derived catalysts were used in different studies of waste plastic pyrolysis. Su et al. (2019) showed in their study on pyrolysis of HDPE that $AlCl_3$-NaCl eutectic salt catalyst improved the pyrolysis process in many ways: reduction in initial reaction temperature from 400°C to 200°C, increase in the reaction speed, enhancement of the quality of the produced liquid fuel by reducing the heavy oil components and reducing the olefin content, and lastly, the produced char has better pore structure which makes it very useful in adsorption applications. The optimum catalyst concentration in the catalyst-HDPE mixture is 30% by weight (Su et al. 2019). HDPE was pyrolyzed in a batch reactor at 350°C for 30 min using metal oxides impregnated on a waste brick kiln dust (WBKD) (Ahmad et al. 2017). The effect of different catalyst to plastic ratio was studied and based on the results, it was chosen to investigate further the composition of liquid products using 5 wt.% ZnO/WBKD due to highest liquid yield of about 93.6 wt.% among all tested ratio of catalysts to plastic. The composition of paraffins, olefins and naphthenes in the liquid products was measured as 74.09%,

22.21% and 3.7%, respectively. Moreover, the use of 5 wt% ZnO/WBKD increased the formation of middle range hydrocarbons C13-C16 by 4% and C17–C20 by 16% and decreased that of low range hydrocarbons C6-C12 by 19%. Kadapure et al. (2016) investigated the effect of other types of catalysts on the pyrolysis process. The authors compared the effect of chemical catalysis such as activated carbon and calcium hydroxide versus bio-based catalysis such as potato peel on pyrolysis of LDPE at a plastic to catalyst ratio of 6.7:1. The experiments were conducted in a batch reactor for a total of 120 min. The results concluded that potato peels led to better liquid yields of 74.52% by weight when compared to 65.16% and 55.16% for activated carbon and calcium hydroxide, respectively. Gonzalez et al. (2011) showed that activated carbon enhanced the aromatic content of the liquid product of pyrolysis process of PE. Sun et al. (2018) demonstrated in their studies that the activation technique of the catalysis has an effect on the physicochemical properties of the activated carbon. $ZnCl_2$-activated biochar catalyst boosted the aromatization processes (PP, PE and PS) leading to about 90.7% of the two-ring aromatics selectivity.

In summary, the use of catalysts during pyrolysis of waste plastics lowered the reaction temperature. Zeolite-based catalysis increase the yield of gas products versus that of liquid products which generally contains more aromatics. On the other hand, it was reported that the yield of liquid products may be increased by using base catalyst. The liquid products in this case tend to have hydrocarbons in the diesel range. FCC and other natural catalysts such as clays and brick kiln dust may also increase the yield of liquid products. They are favored due to their lower cost which increase their industrial application in large scale.

4. MECHANISM OF CATALYTIC PYROLYSIS

The mechanism of catalytic pyrolysis was described earlier by many scientists (Gopinath et al. 2020, Panti et al. 2013, Panda et al. 2010, Singh and Sharma 2008) in two possible ways: *ionic mechanism* or *free radical mechanism*. During the initial stages under the ionic mechanism, ions are produced through a reaction between the polymer molecule and the acidic sites of the catalyst. These sites are either *Lewis* where a hydride ion is removed or *Brønsted* where a hydrogen proton is added. Due to the strong positive charge densities of the *Lewis* acids, they can extract a hydride ion from a saturated hydrocarbon molecule to form a carbonium ion, while the *Brønsted sites* have active hydrogen ion that is added to the double bond of the polymer molecule to produce carbocations.

During the free radical mechanism, the long chain of the polymer molecule is subject to random successions under the effect of heat which involves breaking of the C-C bonds of the backbone of the polymer. This random chain succession results in the formation of primary alkyl free radicals (Fig. 2). This step is followed by hydrogen transfer among the primary radicals to form more stable secondary radicals as shown in Fig. 3 (Gornall 2011).

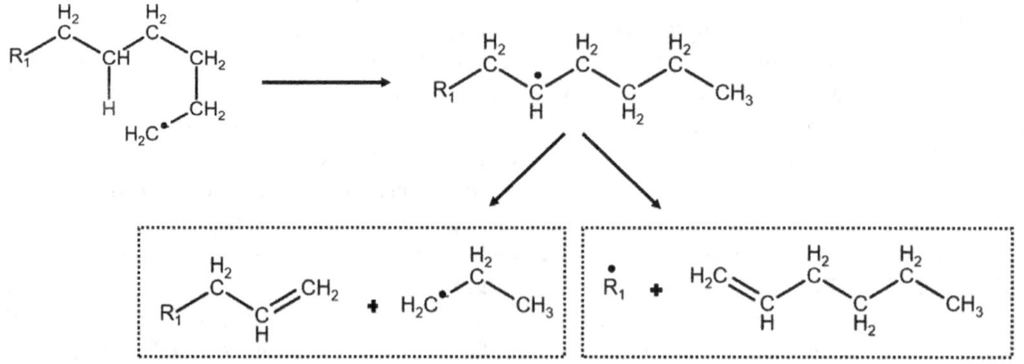

Figure 2. Free radical mechanism of PE.

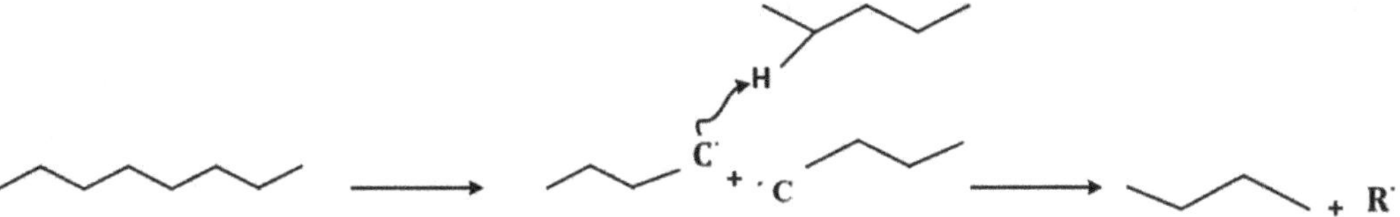

Figure 3. Formation of secondary radicals during free radical mechanism.

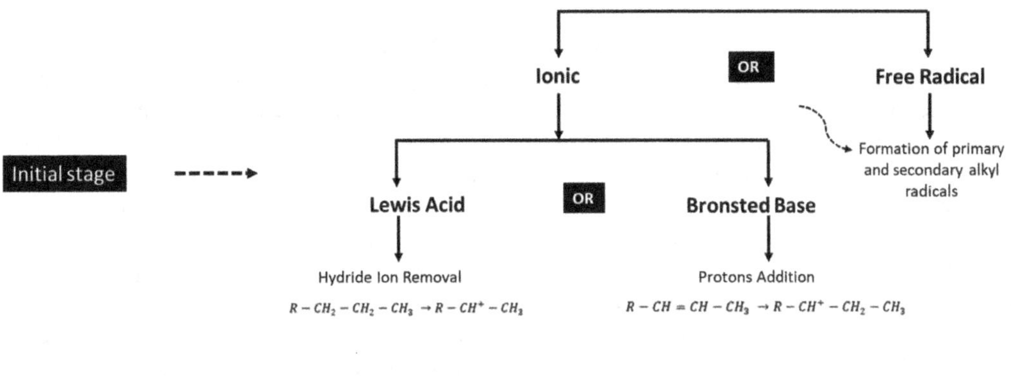

Figure 4. Schematic Diagram of the Catalytic Pyrolysis Mechanism.

The later stages in both ionic and free radical possible mechanisms include a combination of other steps like initiation, propagation, and termination. In addition, the ionic mechanism has additional steps such as isomerization and cyclization.

Figure 4 shows a schematic diagram of the steps involved in the mechanism of catalytic pyrolysis process. During initiation, hydrocarbon radicals (known as carbonium ions R`) are produced as a result of breaking the polymer main chain C-C bond. The hydrocarbon radicals formed in the previous step degrade to lower chain length molecules during propagation. During the last step, termination, recombination of two radicals occurs. Isomerization refers to different rearrangements that may occur by hydrogen atom, carbon atom, and methyl group shifts. Lastly, cyclization reactions might take place by some carbonium ions to form aromatics.

5. PYROLYSIS REACTOR TECHNOLOGIES

Several pyrolysis reactor configurations have been extensively studied in the literature and at different scales. The selection of the optimum configuration will impact not only the yield of products, their quality and their distribution but also the energy consumption and hence the cost of the process and the potential scale-up (Qureshi et al. 2020, Al-Salem et al. 2017). It is noteworthy that due to the sticky nature of plastics upon fusion, special attention should be given to the reactor design and its feeding system to avoid operational problems such as clogging (Lopez et al. 2017). In this section, the main pyrolysis configurations used in experimental studies will be discussed highlighting the pros and cons of each configuration and addressing the effect of using them on the resulting products.

5.1 Batch and semi-batch reactors

Batch and semi batch reactors may be considered the most common pyrolysis reactor configuration due to their simple design and ease of operation in addition to flexibility of particle size of the waste plastic that can be used. Batch and semi batch reactors are non-continuous configurations that involve frequent feed loading, restarting and discharging of products. Disadvantages include poor heat transfer rate and the difficulty associated with controlling the operating parameters, especially temperature (Lopez et al. 2017). It was recommended that batch and semi-batch reactors are not the best modes of operation to study the effect of temperature on the performance of catalytic pyrolysis of plastics. This was attributed to the fact that both plastic and catalyst have the same temperature profiles, which makes it difficult to determine the product distribution as a function of the reaction temperature independently from the catalyst bed temperature (Gulab et al. 2010). The way batch and semi-batch reactors are operated is that waste plastic is mixed together with the catalyst at specific calculated ratio and then the mixture is loaded to the reactor. The reactor is heated at specific heating rate in a closed atmosphere until the target reaction temperature of interest is reached. The reaction mixture is kept at this temperature for enough time to ensure the plastics are fully degraded. The pyrolysis products could be differently recovered. One approach is to allow the recovery of the products after reaction is completed where the reaction vessel is completely closed during the whole process (Wong et al. 2015). This approach allows for further secondary reactions to occur due to the presence of the volatile products in contact with the catalyst for extended periods of times. The second approach allows the volatiles to be extracted from the reaction vessel while formed (Lopez et al. 2017). It should be pointed out here that semi-batch reactors involve the presence of a continuous stream of inert gas (usually nitrogen) to remove the volatile products (Santos 2018b).

5.2 Fixed bed catalytic reactors (FBCRs)

Fixed bed catalytic reactors are one of the simplest reactor designs employed in pyrolysis of plastics. A column of solid catalyst is loaded to the reactor tube/vessel in the form of packed bed. In spite of the simple design, there are many drawbacks of the FBCR which include problems associated with plastic feeding related to its irregular shape, its high viscosities, and low heating rates. Some researchers chose to overcome these difficulties by using the FBCR after an initial thermal pyrolysis step where only liquid and gaseous products of the thermal degradation step is passed through the catalyst bed (Al-Salem 2019). Achilias et al. (2007) investigated the use of fixed bed catalytic reactor loaded with FCC in the pyrolysis of LDPE, HDPE, and PP at 450°C. The results showed that the main constituents of the liquid product were aliphatic compounds with LDPE leading to hydrocarbons in the gasoline range. Another study demonstrated fixed bed catalytic pyrolysis of PET in presence of CaO as a catalyst to improve the benzene yield. The authors succeeded in obtaining about 84% of benzene yield in the products at 600°C and minimized the deposition of solids (Kumagai et al. 2011).

5.3 Fluidized bed reactors (FBRs)

Fluidized bed reactors are continuous reactors that provide improved characteristics over batch and semi-batch reactors in spite of their complex design and operation. Due to their higher surface area of contact, rates of mixing and heat transfer are increased. In addition, fluidized bed reactors require lower maintenance and have lower rates of catalyst replacement (Gopinath et al. 2020). Additionally, catalyst may be replaced if deactivated without the need to stop the process which saves time, effort, and cost (Wong et al. 2015). There are two main types of fluidized bed reactors: bubbling fluidized bed and circulating fluidized bed, illustrated in Fig. 5. Both offer high heat transfer and shorter residence time and suitability for large scale operations. The circulating fluidized bed reactors features

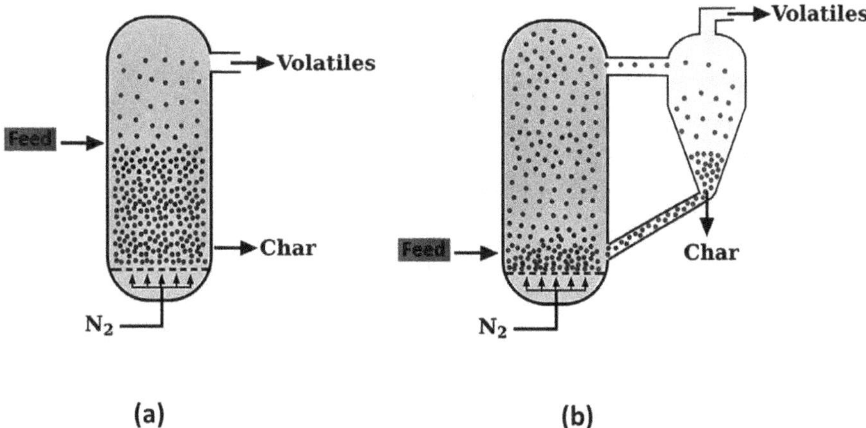

Figure 5. Fluidized Bed Reactors: (a) Bubbling Bed and (b) Circulating Bed, (Arabiourrutia et al. 2020) Reproduced or/and reprinted with permission from Elsevier; Copyright 2021.

additional capability in their design to separate the formed char and circulate it back to the reactor (Bhoi et al. 2020).

In most cases, nitrogen gas is used as the fluidizing agent; few studies reported the use of steam instead to minimize tar deposition due to its higher heat capacity (Pandey et al. 2020). Sand is mainly used as a fluidizing medium with caution to avoid defluidization of the bed. It is worth mentioning that the ratio of sand to plastic feed is required to be high enough to minimize the thickness of melted plastic coating the sand, hence decrease defluidization. This in turn increases the energy requirement and thus the operating cost (Orozco et al. 2021).

Laboratory scale fluidized bed reactors were employed in some catalytic pyrolysis studies. Those studies pointed out some disadvantages of the use of fluidized bed reactors in catalytic pyrolysis of plastics such as potential melting of the plastic material in the feeding system to the reactor, which was avoided by co-pyrolysis using biomass at a proper biomass to plastic ratio (Pandey et al. 2020). Authors raised concern about de-fluidization associated with the FBRs, which would be overcome by proper design and selection of the minimum fluidization velocity. It is worth mentioning that controlling the residence time in the reactor is a key to reach a uniform mass and energy transfer within the reactor and consequently uniform and narrow product distribution (Al-Salem et al. 2017).

5.4 Conical spouted bed reactors (CSBRs)

Spouted reactors are very similar to circulating and bubbling fluidized bed reactors with the difference of having the spout geometry as illustrated in Fig. 6. Unlike fluidized bed reactors, conical spouted bed reactors (CSBR) offer improved operation during pyrolysis of waste plastics that lead to avoiding particle agglomeration and de-fluidization. Due to their versatility in gas flow, they are considered suitable to handle solid feed particles with irregular shape as well as providing stable moving bed of particles over wide range of gas flows (Lopez et al. 2017, Wong et al. 2015). Moreover, they have low bed segregation, lower pressure drop, lower abrasion of catalyst particles, higher heat transfer and improved performance due to solid flow patterns (Elordi et al. 2009). The catalyst bed is loaded in certain particle size range while plastic is continuously fed to the reactor at target reaction temperature. Elordi et al. (2009) studied the performance of CSBR in catalytic pyrolysis of HDPE at 500°C. Three catalysts were used in their study: HZSM-5, HY and Hβ zeolite based with a mass flowrate to the reactor of 1 g/min. The product distribution was classified into 7 different groups under the three major classes: liquids, gases and solids. The results of the study showed high selectivity of HZSM-5 catalyst to light olefins, about 58% wt (C_2-C_4 in the gas stream), due to the catalyst's

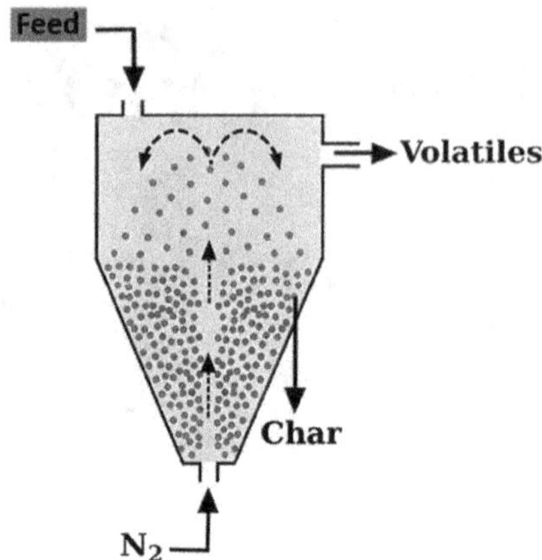

Figure 6. A schematic diagram of the conical spouted bed reactors (Arabiourrutia et al. 2020). Reproduced or/and reprinted with permission from Elsevier; Copywrite 2020.

small pore size. As the particle size of the catalyst gets bigger like those in HY and Hβ, the product selectivity is shifted towards heavier hydrocarbons such as waxes and leads to deactivation of the catalyst by the coke formed.

The catalytic cracking of HDPE was studied using FCC catalyst in a CSBR at 500°C (Elordi et al. 2012). They reported no bed de-fluidization experience during operation and the results showed high yield of gasoline fraction (50 wt.%) and olefins (28 wt.%) within the liquid products and no solid product formation. Ability of FCC to minimize secondary reactions yielding low paraffins or aromatics was illustrated. This was attributed to its moderate acidity as a result of the different reaction-regeneration cycles.

The suitability of the use of CSBR in the catalytic pyrolysis of HDPE was highlighted in the study with regard to the physical steps of coating the catalyst particles with fused plastic in addition to the improved mass and heat transfer rates. A study was conducted recently by Orozco et al. (2021) to determine the optimum operating conditions of catalytic pyrolysis of different waste plastics in a CSBR under a stable hydrodynamic operation. The group introduced a modified design to the normal CSBR, a fountain confined CSBR which led to outstanding performance using FCC as a catalyst. While the effect of different operating parameters on the minimum temperature to reach stable operation was investigated for PP, PS, HDPE, LDPE, PET and PMMA, under thermal degradation (without a catalyst), the use of catalyst was only employed in case of HDPE. The results showed about 13% reduction in the minimum temperature to reach stable operation at higher catalyst concentrations.

5.5 Vacuum reactors

Some studies have recently been reporting investigation of running catalytic pyrolysis of plastics under vacuum environment. It was reported that the negative pressure can facilitate the degradation of plastics through decreasing the activation energy of the pyrolysis reactions. Moreover, the vacuum environment helps in cooling and condensing the products and thus efficiently removing them from reactor (Mahari et al. 2018, Dong et al. 2017). An example of a vacuum pyrolysis reactor system is illustrated in Fig. 7.

The pyrolysis of LDPE in the presence of different catalysts was studied under constant vacuum pressure of 40 kPa and temperatures in the range of 370°C–430°C (Olivera et al. 2020), whereas Tekade et al. (2020) studied the effect of varying the vacuum pressure on the pyrolysis of PE, HDPE, and LDPE using zinc oxide as a catalyst. It was reported in the study that the yield of liquid product increased when the pressure was reduced to 50 kPa under atmospheric pressure. The highest liquid yield at optimum plastic to catalyst ratio was 78.65 % for HDPE and 72.68% for LDPE.

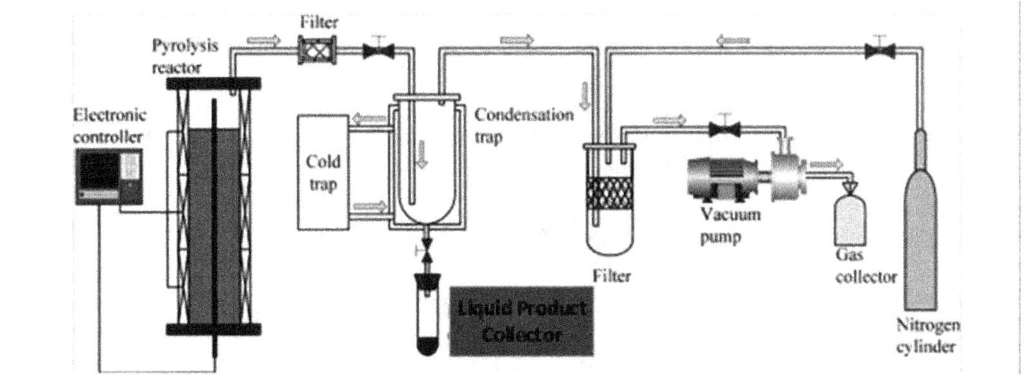

Figure 7. Schematic diagram of a vacuum pyrolysis system (Fan et al. 2014). Reproduced or/and reprinted with permission from Elsevier; Copywrite 2014.

5.6 Screw kiln reactors (SKR)

Screw Kiln reactors (SKRs) or sometimes known as Auger reactors are typically composed of a tubular reactor and a screw conveyor which run in continuous mode as illustrated in Fig. 8. Controlling the residence time within those types of reactors si usually done by either changing the length of the tubular reactor or varying the speed of rotation of the screw. The main advantages of the SKRs are the flexibility in operation and the ease of polymer handling. Moreover, they offer better temperature control through their good heat transfer as well as higher efficiencies (Qureshi et al. 2019, Santos 2018b).

The results of studying the catalytic degradation of LDPE in a SKR showed lower formation of gaseous products and a reduction in over cracking of heavy hydrocarbons using MCM-41 (alumina-silicate) (Serrano et al. 2000). The authors attributed the reason of the better performance to the fact that all the product fractions that exist within the reactor are subject to same residence time. Moreover, there is no continuous removal of the volatiles (olefinic gases) which are subjected to oligomerization reaction to yield a selective gasoline range fraction. Due to the continuous feeding

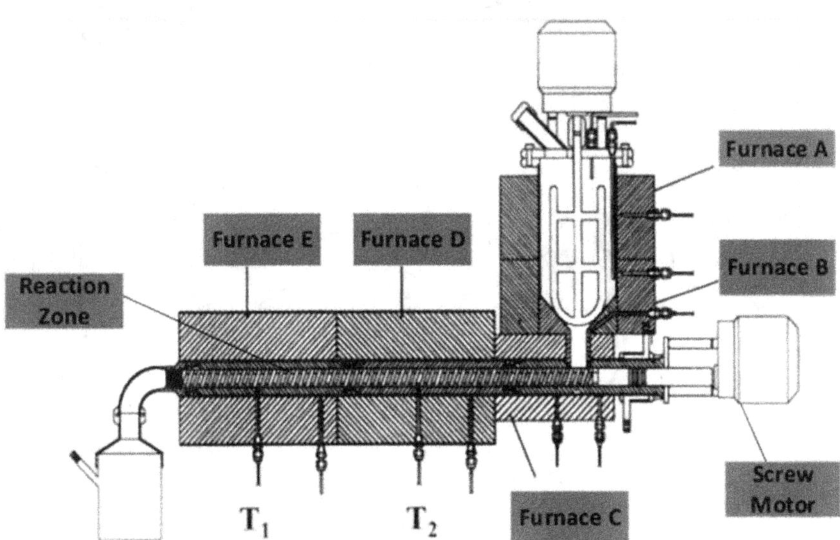

Figure 8. Schematic diagram for the SKR (Serrano et al. 2000) Reproduced or/and reprinted with permission from Elsevier; Copywrite 2000.

of the mixed catalyst and plastics, the liquid yield in the gasoline fraction was reported as 80 wt.% with minor gas production. Similar results were obtained by Aguado et al. (2002) in their catalytic degradation of LDPE using Al-MCM-41 catalyst. The authors investigated the effect of changing the speed of the screw on the yield of products. It was reported in the study that conversion decreased, with lower gasoline range product selectivity, by increasing the screw speed corresponding to shorter residence times along with a possible catalytic deactivation.

5.7 Microwave-assisted reactors (MAR)

Microwave-assisted pyrolysis has attracted much attention in recent years due to their proven ability of enhancing energy efficiency with better controlled heating, improving liquid product selectivity and decreasing reaction residence time when compared to conventional heated reactors (Ding et al. 2019). The advantages of using microwave-assisted technologies have been reported in many studies of catalytic pyrolysis of LDPE (Fan et al. 2017, Zhang et al. 2015), HDPE (Russell et al. 2012, Ludlow-Palafox and Chase 2001), PS (Undri et al. 2014), PP and HDPE (Undri et al. 2013). However, microwave-assisted reactors are still at early stages of scale-up to commercialization and show evidence of higher capital costs. It was reported in many studies that the use of MAR would help in dealing with problematic types of plastic such as PVC during pyrolysis (Kobayashi et al. 2019, Moriwaki et al. 2006). As PVC is known to be unfavored plastic due to its release of HCl gases during pyrolysis, MAR is used to selectively heat PVC first via irradiation to enable separating the dichlorination reactions from the degradation reactions. In a recent study, Zhou et al. (2021) used a large continuous scale (10 kg/hr) MAR as shown in Fig. 9 to investigate the effect of operating temperatures and plastic composition on the product yield of HDPE using ZSM-5 catalyst. The novel design of the MAR presented in their study showed a promising potential of scaling up to industrial applications.

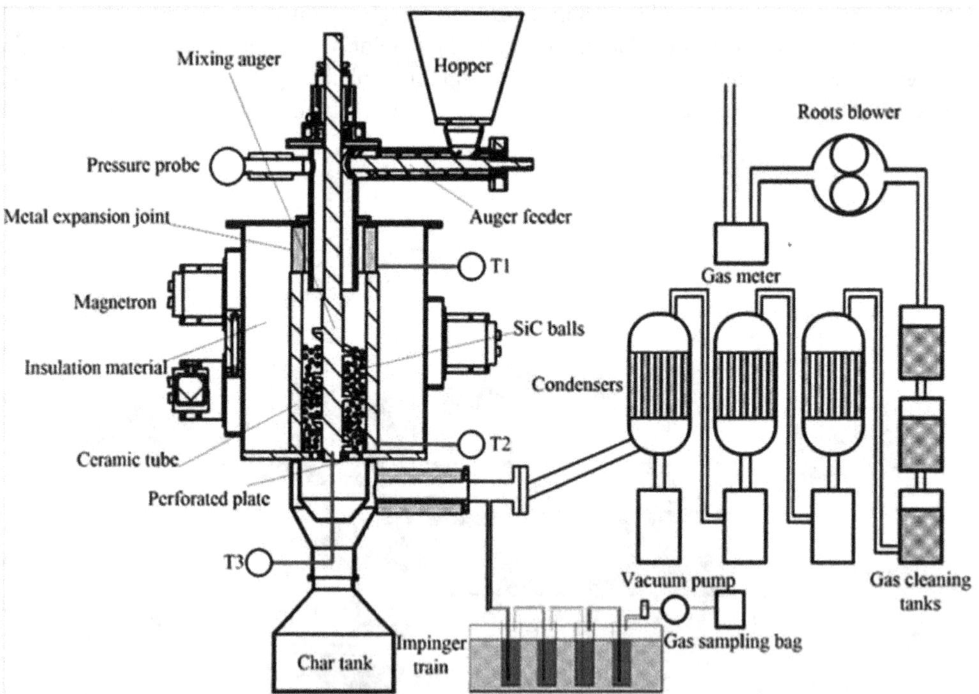

Figure 9. Schematic diagram of MAR experimental setup as used in Zhou et al. (2021) study (Zhou et al. 2021) Reproduced or/and reprinted with permission from Elsevier; Copywrite 2021.

5.8 Plasma-assisted reactors (PARs)

Plasma is the fourth state of matter consisting of electrons, ions and neutral particles. The degree of ionization of plasma is a function of the number of atoms that have gained or lost electrons (Khongkrapan et al. 2014). Plasma-assisted technology has been utilized essentially for hazardous and medical waste treatment. It started to gain some attention recently in research and laboratory studies around early 1990s (Nema et al. 2016). For classification, there are two major plasma-assisted approaches: (i) high temperature plasma, and (ii) low temperature plasma which is divided into thermal and non-thermal (cold) (Huang and Tang 2007). Thermal plasma comprises of a plasma torch/arc in providing the heat source to the reactants. Thermal plasma pyrolysis offers effective treatment due to the high temperature and fast heating rate, rapid start-up and shut-down as well as low potential of pollution emissions (Cai and Du 2021, Khongkrapan et al. 2014).

6. INFLUENCE OF OPERATING CONDITIONS

The catalytic pyrolysis of waste plastic is affected by many operating conditions such as temperature, residence time, heating rate, and feed to catalyst ratio. These operating conditions play an important role in the yield and quality of products. Optimizing the process condition is the key in better control of process and ease of targeting certain product distribution. It also helps in reducing catalyst deactivation and minimizing cost.

6.1 Temperature

Temperature is referred to as having the most significant impact on the pyrolysis reaction rate and conversion efficiency. It was reported in many studies (Sebestyén et al. 2017, Lopez et al. 2017, Miandad et al. 2016, Panda et al. 2010) that the increase in pyrolysis temperature will generally increase the yield of pyrolysis products. In the presence of a catalyst, the effect is less significant (Hernandez et al. 2005). Due to the complex structure of PE and PP, these plastics (PE and PP) require higher temperatures for the complete degradation when compared to PS plastics (Miskolczi et al. 2009). PS degrade at lower temperature and has lower activation energy than other plastics in addition to being very active during pyrolysis due to the presence of the styrene monomer in its building block (Al-Salem et al. 2017, Westerhout et al. 1997). Lopez et al. (2011a) investigated the pyrolysis temperature in the range of 400°C to 500°C and reported nearly 90 wt.% conversion at 425°C below which 30% solid was deposited (mixture of unconverted plastics plus char). They considered 440°C the lowest, thus the optimum temperature for complete conversion to gases and liquids using ZSM-5 catalyst.

In the study of the effect of temperature on the performance of the pyrolysis reaction of HDPE using US-Y catalyst (Gulab et al. 2010), activity of catalyst at its higher concentrations in the feed may be compensated by increasing the reaction temperatures. The effect of two temperatures, 395°C and 420°C, were studied under two reaction times, 15 min and 25 min, at a plastic to catalyst ratio of 8:1. It was observed that the temperature not only has an effect on the yield of the product but also on the product quality and selectivity. The liquid produced at lower temperature has lower fraction of middle boiling range hydrocarbons (C8–C9) and higher fractions of heavy hydrocarbons (C14-C18) than that produced at higher temperature due to the occurrence of more decomposition reactions at the higher temperature. This was also reported by Santos (2018b) during their study of catalytic pyrolysis of HDPE over H-ZSM-5 in an FBR where the yield of lighter products increased with the increase in temperature. When pyrolysis of HDPE was studied in a batch reactor using Y-zeolite and $MgCO_3$ as catalysts (Kunwar et al. 2016), results showed low liquid yields at 400°C and 410°C even after a total reaction time of 3 hours. The temperature was increased to 430°C in case of Y-Zeolite and to 450°C in case of $MgCO_3$ to recover the unconverted material and increase yields.

6.2 Heating rate

The amount of energy supplied to the pyrolysis process per unit time is known as the heating rate. As a general rule of thumb, the yield of liquid and char products increases as the heating rate increases. It is important to determine the optimum heating rate for each process (depending on other factors as well such as catalyst, feedstock type and type of reactor) as after a certain point, if the heating rate is too high, the gas yield will start to increase losing some of the liquid yields (Al-Salem et al. 2017). Marcilla et al. (2007) studied the effect of two different heating rates of the yield of the pyrolysis of HDPE in a FBR using HZSM-5 catalyst. The results showed that at low heating rates, alkanes and aromatics are dominant while at high heating rates, olefins and paraffins are dominating.

6.3 Residence time

Residence time has less significant effect on the yield and distribution of products in pyrolysis when compared to other operating conditions (Lopez et al. 2011b). The average amount of time the feed material spends within the reactor is known as residence time. In general, the longer the residence time, the more the conversion of the primary products, thus increasing the yield of shorter chain hydrocarbon and non-condensable gases (Al-Salem 2019). In batch and semi batch reactors where reactants and products co-exist, the secondary reactions are enhanced yielding more gases and char (Pandey et al. 2020, Hernandez et al. 2005).

6.4 Plastic to catalyst ratio

The plastic feed to catalyst ratio is another important parameter that influences pyrolysis process yield and product distribution. Zeolite-based catalyst, US-Y, was used in the pyrolysis of a single type plastic (LDPE) at different waste to catalyst ratios (Akpanudoh et al. 2005). An exceptionally high liquid yield value of close to 90% was reported at a plastic to catalyst ratio of 4:1 at a temperature of 417°C. It was suggested from the results of the study that beyond this percentage of catalyst in the feed, an adverse effect on liquid yield was observed due to further cracking to smaller molecules that goes into the vapor phase as gaseous products. The above plastic to catalyst ratio of 4:1 was assumed to be the optimum ratio. Similar conclusions were reported in Gulab et al. (2010) while studying the performance of the pyrolysis process of HDPE in a semi batch reactor using the same US-Y catalyst. As the plastic to catalyst ratio increases, there was less yield of liquid products due to over-cracking (secondary reactions). The liquid yield started to decrease beyond the plastic to catalyst ratio of 8:1 at the highest studied temperature of 420°C. It is worth mentioning that the higher concentration of catalyst not only increased the gaseous products at the expense of the liquid products, but also changed the distribution of the liquid products toward lighter hydrocarbons.

7. INDUSTRIAL-SCALE CATALYTIC PYROLYSIS PLANTS

Commercialization of catalytic pyrolysis of plastic wastes started as early as mid-70s despite the economic and technological challenges encountered. There are several large-scale facilities that have been in operation since 2015; some focused on one or few plastic types while others are accepting all types of plastic wastes or mixture of plastics. In cases where PVC is accepted as a feedstock, some de-chlorination steps must be employed before feeding the polymer to the reactor to avoid the corrosive effect of HCl on the downstream equipment (Pohjakallio and Vuorinen 2020, Butler et al. 2011). Table 3 summarizes the main industrial processes and their key features.

The availabilities of information on the large-scale plants are scarce in the literature as most of the technologies contain trade secret components. It should be noted that there are more commercially stable thermal pyrolysis plants than catalytic pyrolysis. Catalyst is being opposed considering the

Table 3. Commercial Plants for the Catalytic Pyrolysis of Waste Plastics.

Process	Country	Company	Operating Temperature,°C	Catalyst	Accepted Plastic Type	Capacity, tons per day	Yield of liquid products, %Wt.	Reference
Smuda	Poland	-	300–450	Nickel and Ferrous Silicates	PP, PE, PET, PVC	28–30	95	(Butler et al. 2011)
Reentech	Korea	-	350–370	Nickel	PP, PE, PS	-	75	(Butler et al. 2011, Solis 2018)
Fuji	Japan	Fuji Seal International, INC	390–400	HZSM5	PE, PP, PS		75–85	(Solis 2018)
Sapporo Plastic Recycling (SPR)	Japan	Sapporo Plastics Recycling Co., Ltd.			Waste plastic mixture	50	90	KleanIndustries, www. kleanindustries.com)
NanoFuel or KDV	Germany	Alphakat GmbH	270–370	HY	PE, PP, PVC		94	(Butler et al. 2011), (Solis 2018)
Zadgaonkar	India	-	350	-	PE, PP, PS, PET, PVC	5	75	(Butler et al. 2011), (Solis 2018)
Thermofuel	UK	Cynar Plc	350–425	Metal catalyst	PP, PE, PS	10–20	80	(Solis 2018)
Beston	China	Beston (Henan) Machinery Co., LTD.	-	unknown	Waste plastic mixture, rubber, tyre	6–24	-	Beston (Henan) Machinery Co., LTD (https:// bestonpyrolysisplant.com/)
Doing	China	Henan Doing Environmental Protection Technology Co., LTD	400–450	-	Waste plastic mixture, rubber, tyre		40–45	Doing (www.doinggroup. com)
OURSUN	China	Anhui Oursun Resources Technology Co., Ltd	-	-	PP, PE and PS	10–500	50–90	Waste plastic to oil machine (www. wasteplastictooilmachine. com)

extra cost and operational challenges associated with catalyst deactivation and need for regeneration. In this section, aspects of some of the industrial plants shown in Table 3 are discussed to emphasize key challenges for commercialization of catalytic pyrolysis of plastic wastes.

Sapporo Plastics Recycling Co., Ltd. (SPR) commercial operation of plastic waste mixture started in 2000 in Japan. The plant started with thermal pyrolysis operation at 400–50°C in addition to uniquely being able to install a de-chlorination process for pretreating PVC wastes. The plant experiences corrosion and clogging issues due to the formation of benzoic acid and terephthalic acid as a result of thermal degradation of PET. In order to overcome these issues, the process was converted to catalytic pyrolysis using hydrated lime [$Ca(OH)_2$]. Stable operation was achieved with reportedly high yields up to 93% (Fukushima et al. 2009). NanoFuel, on the other hand, uses a chlorine-binding catalyst for dichlorination of PVC wastes. Chlorine binds with salts in the plastic waste feed in the liquid form and is removed with the deactivated catalyst using an auger. This process is claiming the possibility of handling very high PVC concentrations without any problem. Zadgaonkar plant in India is using a patented additives and coal as a catalyst to perform dichlorination, which occurs during the pyrolysis process over affixed bed of this catalyst (Butler et al. 2011). One of the world's largest plastic waste pyrolysis plants operating in Poland is based on the Korean technology Smuda where catalyst is directly mixed with the waste plastic feed. The catalyst used in this process is nickel and ferrous silicate which is added to the melted plastic in the heated extruder feeding system. Although very high yield of liquid product is reported along with ability of handling all types of plastic wastes (including PET and PVC), other issues are noted such as the existence of some secondary reactions and heat gradients which may require post separation processes in addition to frequent maintenance required due to the nature of operating in a stirred tank reactor (Pohjakallio and Vuorinen 2020, Butler et al. 2011).

One of the major European technologies is Thermo Fuel, a process that converts plastic waste to low sulfur diesel fuel. The technology is characterized by its ability to handle unsorted raw plastic waste efficiently. The process compromised of multiple steps: pretreatment, thermal pyrolysis followed by catalytic conversion of the gaseous pyrolysis products to liquid hydrocarbons in the diesel range (Kiran and Naveen 2018). The plant has an energy efficiency of 95% using a power co-generation system. Similar vapor upgrading concept is followed by the Fuji technology in Japan which uses catalytic reforming of the pyrolysis vapors to upgrade to gasoline, diesel and kerosene (Butler et al. 2011). Recently, there has been several commercial scale catalytic pyrolysis companies in China (Table 3) that offer multiple sizes of catalytic pyrolysis modular systems to accommodate different market needs. It should be noted that the US department of energy (DOE) in collaboration with the bioenergy technologies office (BETO) have been developing multiple catalytic fast pyrolysis technologies on a pilot scale. The purpose of these projects is to establish a biorefinery based on waste plastic as a feedstock that is commercially viable (Schaidle et al. 2017).

In summary, catalytic pyrolysis of waste plastic is advancing on a commercial scale. The capacities of the current available technologies are yet below the levels that can help achieving circular economy and solve the problem of handling billions of toes of waste plastic generated every year. More efforts are needed from different governments to support the increase in plant capacities while maintaining economic viability.

8. CONCLUSIONS

Plastic consumption is increasing significantly worldwide leading to generating wastes in huge amount more than the environment can tolerate. The review of the work done on most recent catalytic pyrolysis of different plastic wastes has proven that pyrolysis is a very efficient method for completely degrading the plastics and replacing incineration and landfill applications. The proper selection of catalyst along with the optimization of the operating conditions can help targeting a certain liquid fuel product. Moreover, understanding the importance of improving process design and energy utilization would help reduce the process cost. From the different research studies that were reviewed in the present chapter, the following conclusions can be made:

- Zeolite-based catalysts, in particular ZSM-5, were utilized the most frequently in the pyrolysis of plastics among all investigated catalysts. Their small pore size, high acidity, and high BET surface boost secondary reactions, resulting in lower liquid yields, which are rich in aromatics and gasoline range hydrocarbons C6–C12, but higher gas yields, which are rich in light olefins.
- Alkaline treatment of zeolite-based catalysis led to the generation of mesopores that promoted the olefin content of the liquid products, while low mesopores present in clay-based catalysis increase the yield of linear hydrocarbons and aromatics.
- In some cases, extra refining techniques may be required to recover some value-added products such as styrene, benzene and toluene. The need for these extra steps may be avoided if PE and PP are utilized due to their composition which is free from oxygen (polymers containing C, H only).
- Research is still needed to solve the dilemma of finding cheap catalyst with longer lifetime and lower deactivation characteristics to aid the economic viability of commercialization. Several global efforts are increasing to reach the 2030 goals of recycling all plastics by moving towards pyrolysis-based biorefineries.
- Catalyst has significant effect on quality and distribution of liquid products. The more acidity of the catalyst, the more yield of light alkanes and single ring aromatics which play a vital role in drop-in fuels.
- The most common type of waste plastic utilized in experimental catalytic pyrolysis is LDPE while PVC was not preferred as it releases HCl which is corrosive to process equipment.
- Little studies were performed using real waste plastics or plastic mixture from actual MSW stream due to higher contamination levels and impurities.
- Batch and semi-batch reactors were favored over all other types due to their simple design and ease of operation, particularly in laboratory scale research. More research is needed to improve reactor design and investigate the recent advances in reactors with proven promising results. Those include CSBR which improved mass and heat transfer rates and avoided defluidization encountered in FBR, microwave assisted reactors with better heating control and improved product selectivity and conducting catalytic pyrolysis under vacuum environment to increase yield of liquid products at even lower temperatures.
- Cold plasma technology is still at early stages. More research is needed to help establish the technology application in catalytic pyrolysis of plastic wastes and facilitate the scale up to commercial.

LIST OF ABBREVIATIONS

Al-Zn	:	Aluminum-Zinc
$BaCO_3$	:	Barium carbonate
BET	:	Brunauer-Emmett-Teller (a measure of specific surface area of a catalyst)
CaO	:	Calcium Oxide
CSBR	:	Conical Spouted Bed Reactor
FBCR	:	Fixed Bed Catalytic Reactor
FBR	:	Fluidized Bed Reactor
FCC	:	Fluid Catalytic Cracking
Fe_2O_3	:	Ferric Oxide
FE-PICL-FE	:	Iron Pillared Clay

HCl	:	Hydrochloric Acid
HDPE	:	High-Density Polyethylene
HHV	:	High Heating Value
HTL	:	Hydrothermal Liquefaction
JP-5	:	Jet Propellant-5 (high flash point kerosene that meets the US military specifications)
LDPE	:	Low-Density Polyethylene
MCM-41	:	Mobil Composition of Matter No. 41 (mesoporous material with a hierarchical structure from a family of silicate and alumosilicate solids)
$MgCO_3$	:	Magnesium Carbonate
MSW	:	Municipal Solid Waste
NH_4ZSM-5	:	Ammonium exchanged ZSM-5
PAR	:	Plasma-Assisted Reactor
PET	:	Polyethylene Terephthalate
PP	:	Polypropylene
PS	:	Polystyrene
PVC	:	Polyvinyl Chloride
SKR	:	Screw Kiln Reactors
TiO_2	:	Titanium Oxide
US-Y	:	Ultra Stable Y
WBKD	:	Waste Brick Kiln Dust
$ZnCl_2$	:	Zinc Chloride
ZnO	:	Zinc Oxide
ZSM-5	:	Zeolite Socony Mobil–5

REFERENCES

Abbas-Abadi, M.S., Haghighi, M.N., Yeganeh, H. and McDonald, A.G. 2014. Evaluation of pyrolysis process parameters on polypropylene degradation products. Journal of analytical and applied pyrolysis 109(1): 272–277.

Achilias, D.S., Roupakias, C., Megalokonomos, P. and Antonakou, E.V. 2007. Chemical recycling of plastic wastes made from polyethylene (LDPE and HDPE) and polypropylene (PP). Journal of Hazardous Materials 19(3): 536–542.

Adrados, A., De Marco,I., Caballero, B.M., Lopez, A., Laresgoiti, M.F. and Torres, A. 2012. Pyrolysis of plastic packaging waste: A comparison of plastic residuals from material recovery facilities with simulated plastic waste. Waste Management 32(5): 826–832.

Aguado, J., Serrano, Escola, J.M. and Garagorri, E. 2002. Catalytic conversion of low-density polyethylene using a continuous screw kiln reactor. Catalysis Today 75(1–4): 257–262.

Ahmad, I., Khan, M.I., Khan, H., Ishaq, M., Khan, R., Gul, K. and Ahmad, W. 2017. Pyrolysis of HDPE into fuel like products: Evaluating catalytic performance of plain and metal oxides impregnated waste brick kiln dust. Journal of Analytical and Applied Pyrolysis, 124(3): 195–203.

Akpanudoh, N.S., Gobin, K. and Manos, G. 2005. Catalytic degradation of plastic waste to liquid fuel over commercial cracking catalysts Effect of polymer to catalyst ratio/acidity content. Journal of Molecular Catalysis A: Chemical 235 (1–2): 67–73.

Al-Salem, S. 2019. Plastics to Energy: Fuel, Chemicals, and Sustainability Implications. San Diego: Elsevier Science & Technology Books, ISBN -978-0-12-813140-4.

Al-Salem, S.M., Antelava, A., Constantinou, A., Manos, G. and Dutta, A. 2017. A review on thermal and catalytic pyrolysis of plastic solid waste (PSW*)*. Journal of Environmental Management 197(7): 177–198.

Amjad, M. and Fatemi, A. 2021. Creep behavior and modeling of high-density polyethylene (HDPE). Polymer Testing 94(1): 1–9.

Arabiourrutia, M., Lopez, G., Artetxe, M., Alvarez, J., Bilbao, J. and Olazar, M. 2020. Waste tyre valorization by catalytic pyrolysis – A review. Renewable and Sustainable Energy Reviews, 129: 1–24.

Artetxe, M., Lopez, M., Amutio, M., Elordi, G., Billbao, J. and Olazar, M. 2013. Cracking of high density polyethylene pyrolysis waxes on HZSM-5 catalysts of different acidity. Industrial and Engineering Chemistry Research 52: 10637–10645.

Beston (Henan) Machinery Co., LTD (https://bestonpyrolysisplant.com/).

Bhoi, P.R., Ouedraogo, A.S., Soloiu, V. and Quirino, R. 2020. Recent advances on catalysts for improving hydrocarbon compounds in bio-oil of biomass catalytic pyrolysis. Renewable and Sustainable Energy Reviews 121: 1–13.

Bonifazi, G., Capobianco, G. and Serranti, S. 2018. A hierarchical classification approach for recognition of low-density (LDPE) and high-density polyethylene (HDPE) in mixed plastic waste based on short-wave infrared (SWIR) hyperspectral imaging. Spectrochimica Acta Part A: Molecular and Biomolecular Spectroscopy 198: 115–122.

Butler, E., Devlin, G. and McDonnell, K. 2011. Waste Polyolefins to Liquid Fuels via Pyrolysis: Review of Commercial State-of-the-Art and Recent Laboratory Research. Waste Biomass Valorization 2(3): 227–255.

Cai, X. and Du, C. 2021. Thermal Plasma Treatment of Medical Waste. Plasma Chemistry and Plasma Processing 41(5): 1–46.

Chandran, M., Tamilkolundu, S. and Murugesan, C. 2020. Conversion of plastic waste to fuel. pp. 385–399. *In*: T.M. Letcher (ed.). Plastic Waste and Recycling: Environmental Impact, Societal Issues, Prevention, and Solutions. Elsevier Inc. Academic Press, London.

Chen, Z., Zhang, X., Che, L., Peng, H., Zhu, S., Yang, F. and Zhang, X. 2020. Effect of volatile reactions on oil production and composition in thermal and catalytic pyrolysis of polyethylene. Fuel 271(1): 117308–117316.

Ding, K., Liu, S., Huang, Y., Liu, Sh., Zhou, N., Peng, P., Wang, Y., Chen, P. and Ruan, R. 2019.Catalytic microwave-assisted pyrolysis of plastic waste over NiO and HY for gasoline-range hydrocarbons production. Energy Conversion and Management 196(4): 1316–1325.

Doing Henan Doing Environmental Protection Technology Co., Ltd (www.doinggroup.com)

Dong, L., Huang, Z., Ruan, J., Zhu, J., Huang, J, Huang, M. Kong, S. and Zhang, T. 2017 Pyrolysis routine of organics and parameter optimization of vacuum gasification for recovering hazardous waste toner. ACS Sustainable Chemical Engineering 11(5): 10038–10045.

Elordi, G., Olazar, M., Castano, P., Artetxe, M. and Bilbao, J. 2012. Polyethylene Cracking on a Spent FCC Catalyst in a Conical Spouted Bed. Industrial and Engineering Chemistry Research 51(43): 14008–14017.

Elordi, G., Olazar, M., Lopez, G., Amutio, M., Artetxe, M., Aguado, R. and Bilbao, J. 2009. Catalytic pyrolysis of HDPE in continuous mode over zeolite catalysts in a conical spouted bed reactor. Journal of Analytical and Applied Pyrolysis 85(1–2): 345–351.

Erdogan, S. 2020. Recycling of Waste Plastics into Pyrolytic Fuels and Their Use in IC Engines. *In*: B. Llamas, M.F. Ortega Romero and E. Sillero (eds.). Sustainable Mobility, Intech Open. DOI: 10.5772/intechopen.90639

Faillace, J.G., de Melo, C.F., de Souza, S.P.L. and Marques, M.R. 2017. Production of light hydrocarbons from pyrolysis of heavy gas oil and high density polyethylene using pillared clays as catalysts. Journal of Analytical and Applied Pyrolysis 126(5): 70–76.

Fan, L., Zhang, Y., Liu, S. and Ruan, R. 2017. Ex-situ catalytic upgrading of vapors from microwave-assisted pyrolysis of low-density polyethylene with MgO. Energy Conservation and Management 149: 432–441.

Fan, Y., Cai, Y., Li, X. and Yin, H. 2014. Rape straw as a source of bio-oil via vacuum pyrolysis: Optimization of bio-oil yield using orthogonal design method and characterization of bio-oil. Journal of Analytical and Applied Pyrolysis 106: 63–70.

Fukushima, M., Shioya, M., Wakai, K. and Ibe, H. 2009. Toward maximizing the recycling rate in a Sapporo waste plastics liquefaction plant. Journal of Material Cycles and Waste Management 11: 11–18.

Geyer, R., Jambeck, J.R. and Law, K.L. 2017. Production, use, and fate of all plastics ever made. Science Advances 3(7): 1–5.

Gonzalez, Y.S., Costa, C., Marques, M.C. and Ramos, P. 2011. Thermal and catalytic degradation of polyethylene wastes in the presence of silica gel, 5A molecular sieve and activated carbon. Journal of Hazardous Material 187(1–3): 101–112.

Gopinath, K.P., Nagarajan, V.M., Krishnan, A. and Malodan, R. 2020. A critical review on the influence of energy, environmental and economic factors on various processes used to handle and recycle plastic wastes: Development of a comprehensive index. Journal of Cleaner Production 274(7): 1–16.

Gornall, T. 2011. Catalytic Degradation of Waste Polymers: Subtitle [Doctoral dissertation, University of Central Lancashire]. http://clok.uclan.ac.uk/4886/2/Gornall%20Tina%20Final%20e- Thesis%20%28Master%20Copy%29.pdf

Gulab, H., Jan, M.R. Shah, J. and Manos, G. 2010. Plastic catalytic pyrolysis to fuels as tertiary polymer recycling method: Effect of process conditions. Journal of Environmental Science and Health Part A, Tox. Hazard Subst. Environ. Eng. 45(7): 908–915.

Hazrat, M.A., Rasul, M.G. and Khan, M.M.K. 2015. A Study on Thermo-Catalytic Degradation for Production of Clean Transport Fuel and Reducing Plastic Wastes. Procedia Engineering, 105: 865–876.

Hernández, M.R., García, Á.N. and Marcilla, A. 2005. Study of the gases obtained in thermal and catalytic flash pyrolysis of HDPE in a fluidized bed reactor. Journal of Analytical and Applied Pyrolysis 73(2): 314–322.

Huang, H. and Tang, L. 2007. Treatment of organic waste using thermal plasma pyrolysis technology. Energy Conversion and Management 48(4): 1331–1337.

Kadapure, S.A., Arush, K., Sagar, C., Shreshtha, S., Sangeeta, M. and Devatt, R.T. 2016. Tertiary recycling of waste plastic using catalytic cracking into crude oil. Energy Sources, Part A: Recovery, Utilization, and Environmental Effects 38(19): 2942–2948.

Kasar, P., Sharma, D.K. and Ahmaruzzaman, M. 2020. Review: Thermal and catalytic decomposition of waste plastics and its co-processing with petroleum residue through pyrolysis process. Journal of Cleaner Production 265: 1–25.

Khongkrapan, P., Thanompongchart, P., Tippayawong, N. and Kiatsirriroat, T. 2014. Microwave plasma assisted pyrolysis of refuse derived fuels. Central European Journal of Engineering, 4(1): 72–79.

Kiran, T.P. and Naveen, A. 2018. Wastage From Plastic to Diesel Fuel-Thermofuel. International Journal of Advance Research in Science and Engineering 7(5): 228–236.

KleanIndustries, (www.kleanindustries.com)

Kobayashi, J., Hori, M., Kobayashi, N., Itaya, Y., Hatano, S. and Mori, S. 2019. Selective dechlorination of polyvinyl chloride by microwave irradiation. Journal of Chemical Engineering Japan 52: 656–661.

Koch, E. 1977. Non-Isothermal Reaction Analysis. Materials Science, Academic Press, London.

Kumagai, S., Grause, G., Kameda, T., Takano, T., Horiuchi, H. and Yoshioka, T. 2011. Improvement of the Benzene Yield During Pyrolysis of Terephthalic Acid Using a CaO Fixed-Bed Reactor. Industrial and Engineering Chemistry Research 50(11): 6594–6600.

Kunwar, B., Kwon, B., Moser, B.R., Chandrasekaran, S.R., Rajagopalan, N. and Sharma, B.K., 2016. Catalytic and thermal depolymerization of low value post-consumer high density polyethylene plastic. Energy 111: 884–892.

Kwon B.G., Saido, K., Koizumi, K., Sato, H., Ogawa, N., Chung, S.Y., Kusui, T., Kodera, Y. and Kogure, K. 2014. Regional distribution of styrene analogues generated from polystyrene degradation along the coastlines of the North-East Pacific Ocean and Hawaii. Environmental Pollution 188(2): 45–49.

Kyong, H.L., Noh, N.-S., Shin, D.-H. and Sao, Y. 2002. Comparison of plastic types for catalytic degradation of waste plastics into liquid product with spent FCC catalyst. Polymer Degradation and Stability 78(3): 539–544.

Lopez, G., Artetxe, M., Amutio, M., Bilbao and Olazar, M. 2017. Thermochemical routes for the valorization of waste polyolefinic plastics to produce fuels and chemicals. A review. Renewable and Sustainable Energy Reviews 73: 346–368.

Lopez-Urionabarrenechea, A., de Marco, I., Caballero, B.M., Laresgoiti, M.F. and Adrados, A. 2012. Catalytic stepwise pyrolysis of packaging plastic waste. Journal of Analytical and Applied Pyrolysis 96: 54–62.

Lopez, A., de Marco, I. Caballero, B.M., Laresgoiti, M.F., Adrados, A. and Aranzabal, A. 2011a. Catalytic pyrolysis of plastic wastes with two different types of catalytic: ZSM-5 zeolite and Red Mud. Applied Catalysis B: Environmental 104(3-4): 211–219.

Lopez, A., de Marco, I., Caballero, B.M., Adrados, A. and Laresgoiti, M.F. 2011b. Deactivation and regeneration of ZSM-5 zeolite in catalytic pyrolysis of plastic wastes. Waste Management 31(8): 1852–1858.

Ludlow-Palafox, C. and Chase, H.A. 2001. Microwave-Induced Pyrolysis of Plastic Wastes. Industrial and Engineering chemistry Research 40(22): 4749–4756.

Mahari, W.A., Chong, C.H., Lam, W.H., Anuar, T.N., Ma, N.L., Ibrahim, M.D. and Lam, S.S. 2018. Microwave co-pyrolysis of waste polyolefins and waste cooking oil: Influence of N_2 atmosphere versus vacuum environment. Energy Conversion and Management 171: 1292–1301.

Mangesh, V.L., Padmanabhan, S., Tamizhdurai, P. and Ramesh, A. 2020. Experimental investigation to identify the type of waste plastic pyrolysis oil suitable for conversion to diesel engine fuel. Journal of Cleaner Production 246(10): 1–12.

Marcilla, A., Ferez, M. D., Cortes, A. N. G. (2007) Study of the polymer–catalyst contact effectivity and the heating rate influence on the HDPE pyrolysis. Journal of Analytical and Applied Pyrolysis, 79 (1): 424-432.

Mastral, J.F., Berrueco, C., Gea, M. and Ceamanos, J. 2006. Catalytic degradation of high density polyethylene over nanocrystalline HZSM-5 zeolite. Polymer Degradation and Stability, 91(12): 3330–3338.

Miandad, R., Barakat, M.A., Aburiazaiza, A.S., Rehan, M. and Nizami, A.S. 2016. Catalytic pyrolysis of plastic waste: A review. Process Safety and Environmental Protection 102: 822–838.

Miskolczi, N., Angyal, A., Bartha, L. and Valkai, I. 2009. Fuel by pyrolysis of waste plastics from agricultural and packaging sectors in a pilot scale reactor. Technology 90(7–8): 1032–1040.

Moriwaki, S., Machida, M., Tatsumoto, H., Otsubo, Y., Aikawa, M. and Ogura, T. 2006. Dehydrochlorination of poly vinyl chloride by microwave irradiation. Applied Thermal Engineering 26(7): 745–750.

Muhammad, C., Onwudili, J. and Williams, P.T. 2015. Thermal Degradation of Real-World Waste Plastics and Simulated Mixed Plastics in a Two-Stage Pyrolysis−Catalysis Reactor for Fuel Production. Energy and Fuel 92(5): 2601–2609.

Murthy, K., Shetty, R.J. and Shiva, K. 2020. Plastic waste conversion to fuel: a review on pyrolysis process and influence of operating parameters. Energy Sources, Part A: Recovery, Utilization and Environmental Effects (Accepted/In press) https://doi.org/10.1080/15567036.2020.1818892

Nanda, S. and Berruti, F. 2021. Thermochemical conversion of plastic waste to fuels: a review. Environmental Chemistry Letters 19: 123–148.

Nema, S.K., Jain, V.N., Ganeshprasad, K.S., Sanghariyat, A., Soni, S., Patil, C., Chauhan, V. and John, P.I. 2016. Plasma Pyrolysis Technology and its Evolution at FCIPT. Gandhinagar, India: Facilitation Centre for Industrial Plasma Technologies (FCIPT): Institute for Plasma Research. Technical Report, 1-27.

Olazar, M., Lopez, G. Amutio, M., Elordi, G., Aguado, R. and Bilbao, J. 2009. Influence of FCC catalyst steaming on HDPE pyrolysis product distribution. Journal of Analytical and Applied Pyrolysis 85(1–2): 359–365.

Olivera, M., Musso, M., De León, A., Volonterio, E., Amaya, A., Tancredi, N. and Bussi, J. 2020. Catalytic assessment of solid materials for the pyrolytic conversion of low-density polyethylene into fuels. Heliyon 6(9): 1–8.

Orozco, S., Alvarez, J., Lopez, G., Artetxe, M., Bilbao, J. and Olazar, M. 2021. Pyrolysis of plastic wastes in a fountain confined conical spouted bed reactor: Determination of stable operating conditions. Energy Conversion and Management 229: 1–11.

Owusu, P.A., Banadda, N., Zziwa, A., Seay, J. and Kiggundu, N. 2018. Reverse engineering of plastic waste into useful fuel products. Journal of Analytical Applied Pyrolysis 130: 285–293.

Panda, A.K., Alotaibi, A., Kozhevnikov, I.V. and Shiju, N.R. 2020. Pyrolysis of Plastics to Liquid Fuel Using Sulphated Zirconium Hydroxide Catalyst. Waste and Biomass Valorization 11: 6337–6345.

Panda, A.K., Singh, R.K. and Mishra, D.K. 2010. Thermolysis of waste plastics to liquid fuel. A suitable method for plastic waste management and manufacture of value-added products—a world prospective. Renewable and Sustainable Energy Reviews 14(1): 233–248.

Pandey, U., Stormyi, J., Hassani, A., Jaiswal, R., Haugen, H.H. and Moldestad, B.M.E. 2020. Pyrolysis of plastic waste to environmentally friendly products. Energy Production and Management in the 21st Century IV 246: 61–74.

Panti, N., Shah, P., Agarwal, S. and Singhal, P. 2013. Alternate Strategies for Conversion of *Waste* Plastic to Fuels. International Scholarly Research Notices 2013: 1–8.

PETRA. 2015. Retrieved from PET Resin Association: http://www.petresin.org/news_introtoPET.asp (Last Accessed on: September 2021).

Pohjakallio, M. and Vuorinen, T. 2020. Plastic Waste and Recycling: Environmental Impact, Societal Issues, Prevention, and Solutions, (pp. 359–384). Elsevier Inc. Academic Press, London.

Qureshi, M., Oasmaa, A., Pihkola, H., Deviatkin, I., Tenhunen, A., Mannila, J., Minkkinen, H., Pohjakallio, M. and Laine-Ylijoki, J. 2020. Pyrolysis of plastic waste: Opportunities and challenges. Journal of Analytical and Applied Pyrolysis 152: 104804–104815.

Qureshi, M.S., Oasmaa, A. and Lindfors, C. 2019. Thermolysis of plastic waste: Reactor comparison, Pyroliq 2019: Pyrolysis and Liquefaction of Biomass and Wastes. Engineering Conferences International ECI Digital Archives 1–27.

Rahimi, A. and Garcia, J. 2017. Chemical recycling of waste plastics for new materials production. Nature Reviews—Chemistry 1: 1–12.

Rajendran, K.M., Chintala, V., Sharma, A., Pal, S., Pandey, J.K. and Ghodke, P. 2020. Review of catalyst materials in achieving the liquid hydrocarbon fuels from municipal mixed plastic waste (MMPW). Materials Today Communications 24: 1–14.

Rehan, M., Miandad, R., Barakat, M.A., Ismail, I.M.I., Almeelbi, T., Gardy, J., Hassanpour, A., Khan, M.Z., Demirbas, A. and Nizami, A.S. 2017. Effect of zeolite catalysts on pyrolysis liquid oil. International Biodeterioration & Biodegradation 119: 162–175.

Renzini, M.S, Sedran, U. and Pierella, L.B. 2009. H-ZSM-11 and Zn-ZSM-11 zeolites and their applications in the catalytic transformation of LDPE. Journal of Analytical and Applied Pyrolysis 86(1): 215–220.

Roy, P.K., Titus, S., Surekha, P., Tulsi, E., Deshmukh, C. and Rajagopal C. 2008. Degradation of abiotically aged LDPE films containing pro-oxidant by bacterial consortium. Polymer Degradation and Stability 93(10): 1917–1922.

Russell, A., Antreou E.I., Lam S.S., Ludlow-Palafox, C. and Chase, H.A. 2012. Microwave-assisted pyrolysis of HDPE using an activated carbon bed. Royal Society of Chemistry, Communications, RSC Advances 2: 6756–6760.

Sadat-Shojai, M. and Bakhshandeh, G.-R. 2011. Recycling of PVC wastes. Polymer Degradation and Stability 96(4): 404–415.

Santos, B.P.S., Almeida, D.D., Marques, M.F.V. and Henriques, C. 2019. Degradation of Polypropylene and Polyethylene Wastes Over HZSM-5 and USY Zeolites. Catalysis Letters, 149: 798–812.

Santos, B.P.S., Almeida, D., Marques, M.V. and Henriques, C.A. 2018a. Petrochemical feedstock from pyrolysis of waste polyethylene and polypropylene using different catalysts. Fuel 215: 515–521.

Santos, E. 2018b. Nanostructure Materials as Catalysts for the Degradation of Polyolefins, [Doctoral dissertation., Technical University of Lisbon].

Schaidle, J., Magrini, K. and Wang, H. 2017. Catalytic Fast Pyrolysis. https://www.chemcatbio.org/pdfs/catalytic-fast-pyrolysis-2017.pdf: U.S. Department of Energy (DOE) and Bioenergy Technologies Office (BETO).

Sebestyén, Z. Barta-Rajnai, E., Bozi, J., Blazsó, M., Jakab, E., Miskolczi, N., Sója, J. and Czégény, Zs. 2017. Thermo-catalytic pyrolysis of biomass and plastic mixtures using HZSM-5. Applied Energy 207: 114–122.

Sen, S.K. and Raut, S. 2015. Microbial degradation of low density polyethylene (LDPE): A review. Journal of Environmental Chemical Engineering 3(1): 462–473.

Serrano, D., Aguado, J. and Escola, J. 2012. Developing Advanced Catalysts for the Conversion of Polyolefinic Waste Plastics into Fuels and Chemicals. Applied Catalysis B: Environmental, 285: 1924–1941.

Serrano, D.P., Aguado, J., Escola, J.M. and Garagorri, E. 2000. Conversion of low density polyethylene into petrochemical feedstocks using a continuous screw kiln reactor. Journal of Analytical and Applied Pyrolysis 58–59: 789–801.

Sharuddin, S.D.A., Abnisa, F., Daud, W.M.A. and Aroua, M.K. 2016. A review on pyrolysis of plastic wastes. Energy Conversion and Management 115: 308–326.

Singh, B. and Sharma, N. 2008. Mechanistic implications of plastic degradation. Polymer Degradation and Stability 93(3): 561–584.

Solis, M. 2018. Potential of chemical recycling to improve the recycling of plastic waste. [Master's thesis, KTH School of Industrial Engineering and Management]. https://kth.diva-portal.org/smash/get/diva2:1233729/FULLTEXT01.pdf

Su, J., Fang, C., Yang, M., You, C., Lin, Q., Zhou, X. and Li, H. 2019. Catalytic pyrolysis of waste packaging polyethylene using $AlCl_3$-NaCl eutectic salt as catalyst. Journal of Analytical and Applied Pyrolysis 139: 274–281.

Sun, K., Huang, Q., Chi, Y. and Yan, J. 2018. Effect of $ZnCl_2$-activated biochar on catalytic pyrolysis of mixed waste plastics for producing aromatic-enriched oil. Waste Management 81: 128–137.

Tekade, S.P., Gugale, P.P., Gohil, M.L., Gharat, S.H., Patil, T., Chaudhari, P.K., Patle, D.S. and Sawarkar, A.N. 2020. Pyrolysis of waste polyethylene under vacuum using zinc oxide. Energy Sources, Part A: Recovery, Utilization, and Environmental Effects 1–16.

Undri, A., Luca, R., Frediani, M. and Frediani, P. 2013. Efficient disposal of waste polyolefins through microwave assisted pyrolysis. Fuel 116: 662–671.

Undri, A., Frediani, M., Luca, R. and Frediani, P. 2014. Reverse polymerization of waste polystyrene through microwave assisted pyrolysis. Journal of Analytical and Applied Pyrolysis 105: 35–42.

Wang, J., Jiang, J., Sun, Y., Zhong, Z., Wang, X., Xia, H., Liu, G., Pang, S., Wang, K., Li, M., Xu, J., Ruan, R. and Ragauskas, A.J. 2019. Recycling benzene and ethylbenzene from *in-situ* catalytic fast pyrolysis of plastic wastes. Energy Conversion and Management 200: 112088–112097.

Waste plastic to oil machine (www. wasteplastictooilmachine.com)

Westerhout, R.W.J., Waanders, J., Kuipers, J.A.M. and van Swaaij, W.P.M. 1997. Kinetics of the Low-Temperature Pyrolysis of Polyethene, Polypropene, and Polystyrene Modeling, Experimental Determination, and Comparison with Literature Models and Data. Industrial Engineering Chemistry Research 36: 1955–1964.

Wong, S.L., Ngadi, N., Abdullah, T.A.T. and Inuwa, I.M. 2015. Current state and future prospects of plastic waste as source of fuel: A review. Renewable and Sustainable Energy Reviews 50: 1167–1180.

Zhang, Y., Duan, D., Lei, H., Villota, E. and Ruan, R. 2019. Jet fuel production from waste plastics via catalytic pyrolysis with activated carbon. Applied Energy 251: 113337-113353.

Zhang, X. and Lei, H. 2016. Synthesis of high-density jet fuel from plastics via catalytically integral processes. RSC Advances 6(8): 6154–6163.

Zhang, X., Lei, H., Yadavalli, G., Zhu, L., Wei, Y. and Liu, Y. 2015. Gasoline-range hydrocarbons produced from microwave-induced pyrolysis of low-density polyethylene over ZSM-5. Fuel 144: 33–42.

Zhou, N., Dai, L., Lv, Y., Li, H., Deng, W., Guo, F., Chen, P., Lei, H. and Ruan, R. 2021. Catalytic pyrolysis of plastic wastes in a continuous microwave assisted pyrolysis system for fuel production. Chemical Engineering Journal 418: 129412–129423.

11

AGRO-RESIDUE WASTE MANAGEMENT

SOURCES, RECENT PRACTICES, AND APPLICATIONS

Madhu Agrawal, Karishma Maheshwari, Neha Pal* and *Pushpendra Kushwaha*

1. INTRODUCTION

The continuous rise in solid waste is generating significant worldwide concern about how to handle the trash sustainably, and researchers have explored several municipal solid waste (MSW) management options to date (Jara-Samaniego et al. 2017). MSW is a complex substrate of various elements such as agro-waste, paper, plastic waste, electronic waste, unused or expired household pharmaceuticals, etc. (Priti and Mandal 2019, Tang et al. 2020). MSW production in the world is currently over 1.3 billion tonnes per year, with a projected growth to around 2.2 billion tonnes per year by 2025, as shown in Fig. 1, portraying regional-wise solid waste generated per year (Kumar and Agrawal 2020, Lima et al. 2018). According to India's central pollution control board (CPCB), per capita, waste creation has grown exponentially (0.26 kg/day to 0.85 kg/day) (Kumar and Agrawal 2020). About 147613 MT waste has been generated in India as per the reports till January 2020 (Singh et al. 2019). It is believed that 80 to 90% in volume of MSW is dumped in landfills without suitable management procedures or open burning, resulting in contamination of air, groundwater, and soil (Kumar and Agrawal 2020). So, the requirement of a new technique to manage the solid waste is gaining attention.

MSW composition is influenced by culture, economic development, climate, and energy sources, affecting the typology of trash and their relative percentages of occurrence. In terms of economics, MSW in developing countries consists mainly of organic waste, but in developed countries, paper, plastic, and other inorganic materials account for most MSW (Lima et al. 2018). The majority of MSW (70–75% in volume) is organic, and the proportion of inorganic waste is progressively altering

Department of Chemical Engineering, Malaviya National Institute of Technology Jaipur, Jaipur 302017, India.
* Corresponding author: madhunaresh@gmail.com; magrawal.chem@mnit.ac.in

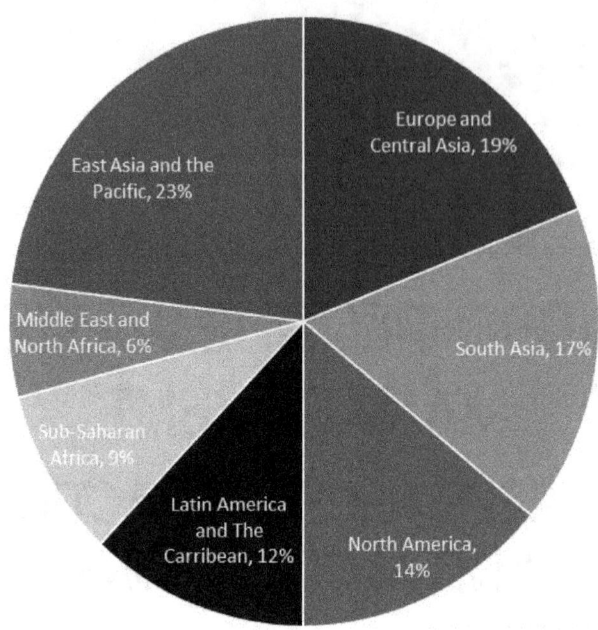

Figure 1. Major contributors of solid waste region wise statistics till January 2020 (data taken from Singh et al. 2019).

and is anticipated to change further in the future (Ramachandra et al. 2018, Sharholy et al. 2007). Nevertheless, MSW has loopholes due to a lack of waste separation at the source, treatment, re-use, recycling, and proper disposal. Improper MSW management, such as open burning, open dumping, and unclean landfilling, contributes to various environmental issues, including global warming, ozone layer depletion, impacts on human health, ecosystem harm, and biogenic resource depletion, among others (Yousefloo and Babazadeh 2020, Laurent et al. 2014). Uncontrolled waste processing, such as burning, is linked to air pollution, crucial pollutants like particle matter, and the creation of heavy metals and volatile organic compounds, all of which contribute to severe health consequences in terms of mortality and morbidity.

A study by Kanhai et al. (2021) discussed the case study of Accra, situated in Ghana, which suffers many environmental and health issues due to improper generated urban municipal solid waste and imported electronic waste (E-waste) management (Kanhai et al. 2021). According to the article by Boateng et al. (2019) in Oti landfill site, Kumasi heavy metal (Pb, Fe, Cd, and Cr) concentration was found above the acceptable limit of the World Health Organization for drinking water. Another study by Vongdala et al. (2019) discusses the municipal solid waste landfills in Vientiane, Laos; the effect of soil quality, groundwater quality, and number of plants in MSW landfill was studied. Little imprints of surface water contamination were seen, whereas Cr and Pb levels in the groundwater in both seasons were significantly above the World Health Organization(WHO) standards. In the dump soils and their surroundings, contamination produced by Cd and Cu reached the environmental risk level. The study noticed that Ipomoea aquatic vegetable is highly affected by environmental pollution. The high concentration of heavy metals such as Cu, Cr, Zn and Pd are reported in the Ipomoea Aquatica vegetable. Also, the grass Pennisetum purpureum (elephant grass) has exceptionally high levels of heavy metals in all plant sections (Vongdala et al. 2019). The above study shows direct disposal of municipal solid waste in the landfill contaminates the groundwater by leaching heavy metals and leads to air pollution (emitting greenhouse gases). Therefore, it has become necessary to manage solid waste from the perspective of the environment and human health protection.

In this chapter, the emergence of agro-residue waste and its impact on the environment and human health have been discussed. Moreover, several types of waste from significant sources have been highlighted. The article endows the worldwide statistics and Indian waste scenario of agricultural, industrial, and residential waste. The authors compile the literature on prominent agro-waste to be converted into valuable chemicals. The trash from several sectors used by investigators to develop coagulants and adsorbents have been briefed in the article. The chapter also covers the pathways which convert solid waste into valuable products and their applications for wastewater treatments.

2. SOURCES OF SOLID WASTE

Solid waste is mainly produced from households (apartment complexes, residences, compounds, and multi-story complex), business districts (marketplace, commercial complexes, guest houses, and recreational gardens), institutions such as colleges, jailhouses, playgrounds, healthcare facilities, and governmental institutions or organizations), factories, and urban facilities such as road sweepings, litter pickup, and drainage debris clearing (Ojowuro et al. 2019). Out of all the types of waste, domestic or municipal solid waste is extensively produced from various human activities. Most municipal solid waste created in growing nations comes from households (55–80%), second by marketplaces or commercial sectors (10–30%), while the rest are from streets, industries, and academics (Nabegu 2010). The distribution of MSW may inculcate food, paper, wood, plastic, etc., as reported by the World Bank shown in Fig. 2. The composition of MSW at sources of generation and collection points on a weight basis consists of a higher organic fraction (40–60%), ash and fine earth (30–40%), paper (3–6%), and plastic, glass, and metals (each less than 1%) (Sharholy et al. 2008). In the below section, a brief elaboration about massively generated electronic waste, agricultural waste and plastic waste has been discussed with the relevant literature references.

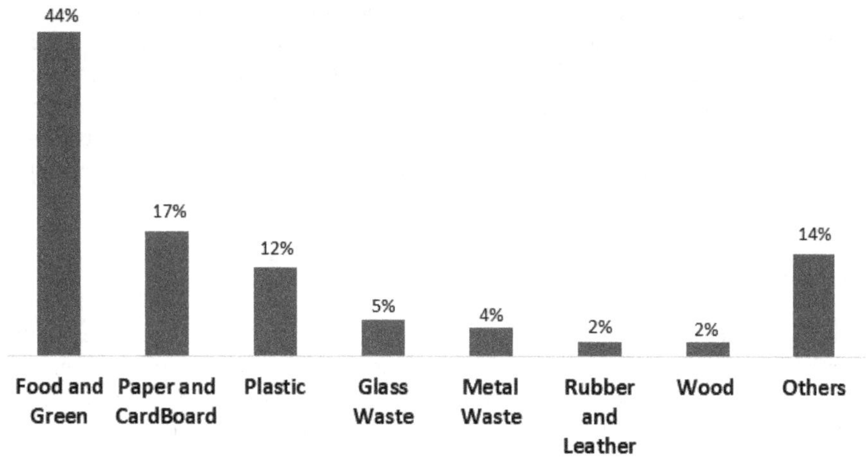

Figure 2. General composition of municipal solid waste (Source: The World Bank 2021).

2.1 Electronic waste

In both emerging and developed countries, rising demand for electrical and electronic goods on the domestic and worldwide market has resulted in significant amounts of electronic garbage (e-waste) (Cole et al. 2019). E-waste, also known as waste electrical and electronic equipment (WEEE), is a combination of plastics, metals, and composites that refers to end-of-life (EoL) electronic equipment (Rene et al. 2021). Today, there is indeed greater awareness about environmental pollution increasing as a consequence of a growth in electrical and electronic product waste (de Albuquerque et al. 2020). Computers and their peripherals (monitors, central processing units, printers, and keyboards), typewriters, cell phones and chargers, audio equipment, remote controls, batteries, compact discs, local compact disc/plasma television sets, air conditioners, refrigerators, and fluorescent and other mercury-containing lamps are all examples of e-waste (Awasthi et al. 2018). However, increased sales of computers and other information technology accessories throughout COVID-19, along with the limited life span of many e-devices, may increase e-waste production. The unorganized sectors in impoverished countries are responsible for most e-waste (Rene et al. 2021). The necessity of reuse has long been accepted in national and international environmental policies. Direct reuse, such as transferring on or selling to a second owner, is termed as waste minimization approach (Cole et al. 2019). In developing nations, the fast growth of e-waste has become a challenge in environmental sustainability (Awasthi et al. 2018). Another adverse effect of the electronics sector on the environment is the generation of e-waste, which is growing much faster than any other type of waste worldwide, specifically in industrialized countries like the United States, South Korea, and Europe (de Albuquerque et al. 2020).

2.2 Agricultural waste

Agricultural waste is generated by agriculture farming, industry, egg production, and forests. Those wastes are abundant in biomass and compostable elements which might have the possibilities to be reused. Waste from agricultural areas is a source of harmful substances such as chemicals and pesticides used in farming activities that pollute groundwater and surface water (Banerjee et al. 2018). Furthermore, agricultural operations involve chemical pesticides produced in empty containers categorized as hazardous waste due to the chemical they contain and classified as agricultural solid waste due to the source of their generation (Buenrostro et al. 2001). Such wastage was produced by the misuse of fertilizers on fields where nutrient enrichment of surface water has occurred in the deposition of nitrogen (N) and phosphorus (P). The effects of nutrient enrichment have included a population explosion and destruction to the water body's natural system. Poultry waste causes pollution of groundwater and surface water by discharging toxic metals, pesticides, and infections into the land. Most crucially, poultry excrement includes nitric nitrogen (NO_3–N), which promotes the blue baby disease in newly born (Banerjee et al. 2018). Agriculture accounts for 60% of all exports, either directly or indirectly. It employs 67% of the working population. It is vital to economic development and planning and provides several services to manufacturing and services (Suryanarayana and Rao 2013).

2.3 Plastic waste

Plastic waste is a kind of waste and recyclable material that solely consists of materials and goods produced entirely or mainly made of plastic. Plastic waste pollutes all of the world's main ecosystems, expressing concern about its potential effects on biodiversity and human health as more minor and more extensive plastic products are formed in natural and artificial environments (Law et al. 2020). For more than 50 years, plastics have played a vital role in a wide range of industrial applications, apparel, toys, domestic applications, engineering applications, and construction materials, including packaging, automobile, architecture, pharmaceutical, and telecoms (Faraca and Astrup 2019). Plastics used in packaging have a lot of room for expansion, which means that demand for plastics will rise. According to the ministry of housing and urban affairs, approximately 70% of plastic packaging products are converted into plastic waste. Furthermore, the use of plastics has resulted in declining environmental implications due to their manufacture and disposal. In 2016, the planet produced 242 Mt of plastic waste (Liang et al. 2021). In 2010, an expected 5 to 13 million metric tonnes (Mt) of plastic waste was dumped into the ocean by developing nations with inadequate solid waste infrastructure and developed countries with significant waste creation. In 2016, the U.S. generated the most plastic waste in the world, which was 42.0 Mt. The following top plastic waste producers were India and China. Both have large populations; however, the EU-28 countries collectively created more plastic waste than India or China while having just 40% of the population (Law et al. 2020). Asia produces more than 50% of total plastic made worldwide, and it has taken up nearly all of the world's exported plastic waste in recent years. However, in Asia, the condition of management of plastic waste is not optimal, and several countries lack the requisite recycling technology and infrastructure (Liang et al. 2021). To estimate plastic pollution in 2030 for 173 countries, researchers examined the impact of three primary management techniques: plastic waste minimization, management of waste, and ecosystem recovery, at various levels of effort. Plastic waste has sparked widespread interest worldwide, with research mainly focusing on recycling methods and the pollution it produces (Liang et al. 2021).

2.4 Other wastes

MSW stream includes a substantial amount of household waste. MSW comprises household, commercial waste, biodegradable, non-biodegradable, and partially biodegradable materials (Lalremruati and Devi 2021). The household waste composition also indicates the waste reuse/recycling activities performed as an informal activity in many developing nations. In this practice, households (primarily and middle-class society) separate a few waste substances of economic/reusable/recyclable value for further sale to interim buyers, street vendors, or fast-food shops. As a result, this technique impacts the original composition of the household waste at society's solid waste treatment and collection stations (Suthar and Singh 2015). During the last few decades, the amount of MSW generated in developing country cities has increased many times. High populace growth in urban areas due to rural immigrants, changing lifestyles, economic development, social improvements among society groupings in metropolitan areas, and other factors contribute to this massive increase in MSW amount (Suthar and Singh 2015). Waste is generated from public and private offices, educational institutions, museums, libraries, archaeological sites, and entertainment centres such as movie theatres and stadiums (Buenrostro et al. 2001). As a result of the municipal authorities' inconsistent waste collection, there is extensive and uncontrolled waste dumping in institutions. Infectious or hazardous waste is produced in the laboratories of medical institutions, metal shops, and repair shops (Tang et al. 2020a).

Industrialization is the essential tool for developing countries to boost their economies; nevertheless, it also increases the volume of generated industrial solid waste (ISW). Many sources of industrial waste, such as power generation units, produce coal ash iron; steel mills produce both blast furnace slag and steel melting slag (Tang et al. 2020a); and the zinc industry has a large amount of jarosite (Linsong et al. 2020) as a waste by-product. A large amount of industrial solid waste produced from various sectors, such as agro-processing, meat processing, poultry production, the sugar factories, milk processing units, paper industries, and many more, have a significant amount of organic by-products that contribute to the industrial waste (Abdel-Shafy and Mansour 2018, Daâssi et al. 2016). Industrial solid waste is classified as either common industrial solid waste (CISW) or hazardous industrial solid waste (HISW) based on the harmful properties for public health and environmental issues. HISW can be corrosive, combustible, explosive, oxidizable, poisonous, reactive, or spread infectious diseases (Tang et al. 2020, Daâssi et al. 2016). Since leather sector utilizes meat processing industry waste as raw products, it is an example of the circular economy. Unfortunately, the tanning process generates a significant amount of organic matter. Zuazua-ros et al. (2021) reported the situation of a tannery in Navarra, Spain, that is dedicated to a sustainable industrial and environmental standard and produces approximately 2 tonnes of meat shavings and buffing dust per day, and also 10 tonnes of discarded bullock hair per day.

Healthcare facilities, laboratories, hospitals, medical centres, and research facilities produce medical or hospital waste. Because the nature of medical waste differs from municipal or other types of waste, it should not be disposed of with municipal solid waste. According to biological waste laws, medical waste must be handled appropriately because they pose a higher risk of infection and significant injury to the environment and human health (Karimi et al. 2020a, Swati et al. 2017). Exposure to medical waste causes high levels of skin and blood infections and respiratory problems,

Hepatitis B, HIV, and tumors; therefore, safe disposal of these type of wastes is a necessity (Bokhoree et al. 2014).

These wastes lead to effects on human and environmental resources which may result into severe hazardous health effects and increased level of toxicity. Therefore, next topic of concern is how these solid wastes affect the environment and what are its future consequences. Details on these topics of concerns have been highlighted in preceding sections.

3. IMPACT OF SOLID WASTE ON HEALTH AND ENVIRONMENT

3.1 Effect of solid waste disposal on human health

Improper solid waste management has a detrimental impact on the environment and human health. Employees in this industry are exposed to the most significant direct health risks, and they should be kept as far away from hazardous material as possible (Di Maria et al. 2021). Handling hospital and clinical waste comes with its own set of threats, such as disease vector breeding, and poses health concerns for the general population (Hossain et al. 2015, 2011). Hazardous industrial contaminants are uncontrollably mixing up with municipal waste that endangers human health; heavy metal concentrations in the food chain represent a particular risk, as shown by a problem involving MSW and rich metal-containing liquid industrial effluents (Ihedioha et al. 2017). There are effluents directly disposed to a sewerage/drainage or an open municipal solid waste disposal site which leads to colossal issues for ecosystem which has been reported by several investigators. In the same string, researchers examined various dumping sites in different states of the world where these kinds of waste have been disposed of unmonitored and few of the latest reports have been briefed in the present section. Another investigation by Yay and Suna (2015) examined the life cycle assessment of discarded solid waste in Sarakya, Turkey, for about one year. The study reveals that one-tonne waste was generated from Sarakya city this waste contained a significant portion of kitchen waste (42.3%), plastic (13.3%), combustibles (12.2%), and other wastes from the yard, non-combustibles, metals, cartons, and paper. The data signifies that solid waste has a noticeable impact on the heath of living organisms due to heavy metal leaching.

Another investigation by Singh et al. (2021) examined the health risk due to open Deonar dumping site situated in Mumbai by propensity score matching (PSM). The outcome shows that the air-quality index revealed high levels of PM10, PM2.5, NO_2 and CO exposure due to this greater risk of respiratory sickness (12%), ocular irritation (8%), and stomach troubles (7%). Solid waste, directly or indirectly, affects human health by leaching metals in groundwater and emitting hazardous gases due to improper handling of solid waste; therefore, solid waste management is needed.

3.2 Effect of solid waste disposal on the environment

Industrial and commercial growth often overlooks environmental problems caused by solid waste production. Because of the adverse effects on many aspects of terrestrial ecosystems, the open-dumping of solid waste has widespread in low and middle income countries (Liu et al. 2015, Ali et al. 2014). Solid waste is dumped or piled in open places with no barriers to keep wildlife out of the waste. Open dumps are often utilized in emerging and developing nations since they are easy to operate and require little financial investment. Numerous open dumping sites can be found all around the globe, but their precise count is impossible to update. The International Solid Waste Organization assessed some larger open dumps as shown in Table 1 (Das et al. 2019).

In the dumpsites, soil and surrounding area are polluted due to metals leaching out in the soil and toxic gas emission. In this content, Ali et al. (2014) examined the impact of open dumping on soil pollution and plant variety in the Capital Development Authority (CDA) and sub-sectors of the H-belt of Islamabad situated in Pakistan. The observation shows that the presence of a higher amount

Table 1. List of open dumping landfill.

Open dump landfill	Country	Area	Reference
Apex regional	Las Vegas, US	2200 acre	(Toet al.2019)
Bordo poniente	Mexico City, Mexico	927 acre	(Bravo et al. 2019)
Leogang	Shanghai, chain	830 acre	(Feng et al. 2017)
Malagrotta	Rome, Italy	680 acre	(Barbieri et al. 2014)
Sudokwon	Incheon, South Korea	570 acre	(Chun and Bae 2012)
Xinfeng	Guangzhou, China	227 acre	(Cheng et al. 2015)
La Chureca	Managua, Nicaragua	1729 acre	(Hartmann 2018)
West New Territories	Hong Kong	272 acre	(Zhang et al. 2011)
Deonar	Mumbai, India	326 acre	(Singh et al. 2021)
Delhi landfill	Delhi, India	500 acre	(Swati et al. 2017)
Sudokwon	South Korea	570 acre	(Chun 2017)
Malagrotta	Italy	680 acre	(Barbieri et al. 2014)
Hyper-arid coast	Peru	-	(Ziegler-Rodriguez et al. 2019)
Andean highlands	Peru	-	(Ziegler-Rodriguez et al. 2019)
Amazon Rainforest	Peru	-	(Ziegler-Rodriguez et al. 2019)

of heavy metals (Zn, Pd, Cu, Ni) at the dumping site and the open-dumping solid waste destroyed 44 plant species(Ali et al. 2014). Solid waste dumping destroys the soil quality as well as groundwater quality. Solid waste creates hazardous pollutants into the soil system, delaying the physiological root system of plants. Additional tests showed that pollutants leached from solid waste contaminate surface and groundwater through metal discharge and percolation (Das et al. 2019, Preziosi et al. 2019). The investigation by Mavimbela et al. (2019) shows the soil quality of Bloemfontein's southern landfill in South Africa; the result shows that the pH value enhances as depth increases and the concentration of Cr reaches beyond the limit in groundwater (Mavimbela et al. 2019). Another investigation by Abd El-Salam and Abu-Zuid (2015) noticed the high concentrations of Mn (0.749–1.39 mg L^{-1}), Cd (0.03–0.26 mg L^{-1}), Cr (0.06–0.09 mg L^{-1}), Ni (0.173–0.359 mg L^{-1}), Fe (0.426–11.4 mg L^{-1}), and Mn (0.749–1.39 mg L^{-1}) in the groundwater. Metal species such as Ni Cr, Cd, and Pb are inherent pollutants in solid wastes transported to plants through the roots and block the vascular system, depriving them of vitality in the long run (Dhar et al. 2016). Dumpsites damage soil and groundwater quality because metals leach out into the soil. Leached metals enhance the concentration of the metals in groundwater which creates a serious threat for humans and the environment. Solid waste generates severe environmental threats; therefore, advanced technologies are needed to treat solid waste.

4. WAYS ADOPTED TO UTILIZE WASTE FOR SYNTHESIZING VALUE-ADDED PRODUCTS

4.1 Incineration

The incineration process is a waste treatment process in which solid waste is directly burned in the presence of oxygen at high temperatures (above 800°C) to convert waste into heat, ash, and flue gas (da Silva et al. 2020, Zhanget al. 2020, Li et al. 2019). This process needs a small installation area because this plant can be installed within the city limit, which reduces the transportation cost and reduces the waste up to 90% in volume and 75% in weight (Li et al. 2019). Moreover, the complete combustion of municipal solid waste (MSW) produces heat in the form of steam, which is then used to generate electricity via steam turbines (da Silva et al. 2020). In energy and economic terms, Tan et al. (2015) obtained better results (higher energy production and lower cost per tonne of waste) for incineration than for gasification when analyzing waste management in a Malaysian city. However, incineration is not an ideal solution for managing MSW because the resulting fly ashes consider hazardous waste due to the presence of several toxic elements such as As, Cd, Cr, Cu, Ni, Pd, Zn, Hg and dioxin. The presence of various harmful elements in fly ash threatens the environment as well as human health (Wang et al. 2022, Zhang et al. 2020). Many researchers are working to convert fly ash into a valuable product (construction materials) since fly ash is rich in Ca, Si, and Al (Marietaet al. 2021, Wang et al. 2022, 2021, Zhang et al. 2020). However, chlorides in fly ash are very harmful to construction materials, which can reduce the strength of concrete due to electrochemistry corrosion.

Moreover, valuable metals (Zn and Pb) can be recovered from fly ash to alleviate the global shortage of metal resources using (thermal separation, hydrometallurgical methods, and electrochemical) processes (Zhang et al. 2021). To extract valuable metals (Zn, Pd, and Cu), Lane et al. (2020) used a selective valorization/condensation process at 900–1100°C and obtained a valorization of Zn (55.9 ± 12.5% w/w) and Pb (4.7 ± 1.1% w/w); still, the valorization of Cu is significantly less. All the metals extraction from the fly ash sill have technical barriers. Moreover, in the case of MSW incineration, limiting polychlorinated dibenzo-p-dioxins and polychlorinated dibenzofurans (PCDD/Fs) emissions is a significant environmental pollutant (Cai et al. 2021, Cudjoe and Acquah 2021); therefore, a new alternative is required to convert waste into a valuable product.

4.2 Gasification

Gasification is a process that turns carbon-containing raw materials such as coal, biomass, and some waste streams into syngas (H_2, CO, CO_2, and CH_4), char, and condensable hydrocarbons (tar) (Cai et al. 2021, Hameed et al. 2021, Jeong et al. 2020). These products can be created with complete ranges of energy or energy carriers such as biomethane, heat, hydrogen, energy, bio-oils, and chemical compounds (Hameed et al. 2021). Compared to combustion, oxygen usage in a gasification reactor is substantially lower than the theoretical oxygen required for fuel combustion. The fundamental elements of biomass gasification are the highest possible and reasonable process productivity (Hameed et al. 2021). MSW gasification minimizes and avoids corrosion and emissions by preserving alkali and heavy metals (excluding mercury and cadmium), sulphur, and chlorine within the process residues, eliminating substantially PCDD/F production decreasing thermal NOx generation owing to lower temperatures and lower conditions (He et al. 2009).

In recent years, gasification has received extensive attention for the sustainable production of H_2 from MSW and biomass. In this context, Zheng et al. (2018) produced hydrogen-rich syngas via a gasification process using a drop tube reactor (DTR) at temperatures (1000–1100°C), CO_2/steam (0.5–3). The highest amount of hydrogen (18.82 mol/Kg-MSW^{-1}) was obtained at CO_2/steam ratio = 2.5 and 1000°C (Zheng et al. 2018). In another study by Chen et al. (2020a), the gasification process was used to produce hydrogen-rich syngas (15.96 mol kg-MSW^{-1}) at 757.6°C temperature, 22.26 min residence time, respectively. However, tar content in syngas significantly impacts product gas application, which is a major technical impediment to commercializing gasification technology. For reducing tar concentration, the researcher focused on using catalyst gasification. Irfan et al. (2019) used waste marble powder as a catalyst in a fixed bed reactor at 700–900°C for producing hydrogen-rich syngas and reducing the tar concentration (Irfan et al. 2019). Another investigation by Irfan et al. (2021) used Ni-CaO catalyst to improve hydrogen concentration from 212 mL/g to 442 mL/g and reduce tar concentration. Removal of tar is still a technical barrier for industrial application of gasification technology; therefore, environment-friendly approach for MSW treatment is still needed.

4.3 Composting

In the composting technique, the complete cycle of organic matter biodegradation generated compost, used in agriculture and soil remediation (Tang et al. 2020b, Pergola et al. 2018). Consequently, compost is an essential method for dealing with organic solid waste that is environmentally beneficial and effective (Tang et al. 2020, Jara-Samaniego et al. 2017, Wu et al. 2017).Compost organic solid fraction separation from MSW is critical for encouraging solid waste recycling and reduction. In this context, Jara-Samaniego et al. (2017) used therapeutic approach for the long-term treatment and recycling of municipal solid waste, generated in the Chimborazo region of Ecuador. A fraction of the compost was collected as seedling growth media and six piles of municipal solid wastes were prepared: Piles 1, 2, and 3 were prepared from Riobamba landfill municipal solid wastes, whereas Piles 4, 5, and 6 from source-separated market waste and urban pruning wastes. Composts C3, C4, and C5 (from Piles 3, 4, and 5, respectively) were mixed with peat in various ratios (0, 25, 50, and 75%, volume/volume) for seedling development of tomato, courgette, and pepper after composting. The composts produced had appropriate physical and chemical properties and a suitable degree of stability and maturity and were phytotoxic-free. In terms of their suitability for use as growing media, substrates containing 25% (volume/volume) of compost C3, C4, or C5 had the best qualities, with only C3 lowering seed germination in the courgette crop. Furthermore, the substitution of peat with compost at a rate of 25% can result in a cost reduction of 23% in the substrate. This reduction may be as much as 21% in the most cautious scenario, with a 2.9% improvement in the company contribution margin (Jara-Samaniego et al. 2017). The MSW compost used in agriculture has grown even more due to limited availability and high prices for small landholdings, and increasing attention in organic farming. Composite of MSW is helpful for increasing soil organic matter, enhancing soil structure and microbial activity, and supplying critical components like nitrogen (N) for plant nutrition/growth (Shah et al. 2019). Nevertheless, human health consequences may occur due to hazardous element transport and absorption in the soil and their entrance into the food chain.

For reducing the leaching of hazardous elements in the soil, Shah et al. (2019) examined the different composting methods (aerobically (AC),anaerobically (ANC), and aerobic-anaerobically (AANC)) on the cultivation of carrot and spinach for 85 and 90 days, respectively. The results show that yield and N uptake of vegetables(carrot and spinach) were in order of ANC > AANC> AC and for metal absorption (cadmium (Cd) and lead (Pb)), was lower in ANS as compared to AANC and AC. Another investigation by Lima et al. (2018) also used different processes for converting MSW into composite, which are windrow composting (WC), wire mesh composting bin (WMCB), and passively aerated static pile composting (PASP) for analysing the removal capacity of Pd, Zn, and Cd as compared to MSW. These findings revealed similar inherent characteristics in different composts, indicating that the various composting procedures influence the physiochemical properties of the end product. The removal of Pd, Cd, and Zn was 94.0–99.6%, 55.4–89.8%, and 22.1–64.0%, respectively. The researcher used different approaches to remove toxic metals from compost, but the total removal of metals from compost is still under investigation. Therefore, an urgent need is felt to carry out more research on traditional composting techniques for producing high-quality humidified organic matter in composts, which can enhance their adsorption stability, desorption potential, long-term applicability, and minimal environmental risk.

4.4 Pyrolysis

The pyrolysis process implies the thermal decomposition of complex organic compounds into valuable products in an oxygen-limited atmosphere at relatively high temperatures. The method produces a combination of highly flammable gases (such as CH_4, H_2, and CO), char, and pyrolysis oil (Nanda and Berruti 2021, Lu et al. 2020). There are three kinds of pyrolysis based on the amount of time spent burning (i.e., slow, flash, and fast) (Nanda and Berruti 2021, Lu et al. 2020). Fast and flash pyrolysis processes result in high yield of bio-oil owing to high heating rate and short residence time during the reaction (Nanda and Berruti 2021). On the other hand, slow pyrolysis is distinguished by its moderate temperatures, low heating rates, and prolonged residence durations, all of which boost product quality (Nanda and Berruti 2021). Syngas and pyrolysis oil products from slow pyrolysis of MSW have been used as alternative energy sources and feedstocks for the synthesis of fine chemicals and biochar (solid product of pyrolysis) has been used as carbon sinks, adsorbents, chemical catalysts, and electron conductors in enhanced anaerobic digestion processes (Chen et al. 2020b, Lu et al. 2020, Hu and Gholizadeh 2019, Czajczyńska et al. 2017). For converting waste to energy, Song et al. (2020) used the pyrolysis process to convert solid waste into a valuable product (H_2, CO, and CO_2 gas) using iron ore as a catalyst in a fixed bed reactor. Another study by Mlonka-Mędrala et al. (2021) pyrolyzed agriculture waste and obtained 19% gas fuel and 56.9% high carbon content char at 600°C under nitrogen environment (Mlonka-Mędrala et al. 2021). To improve the product yield and quality, researchers used different catalysts such as zeolite, bimetallic, dolomite, and metal oxides (Li et al. 2020). In the same string, Saeaung et al. (2021) optimized the effect of plastic waste conversion into valuable chemicals using a different catalyst (spend FCC, MgO, Zeolite) and temperature (400°C–600°C); results show that the 79% Lactide was obtained at 400°C using zeolite catalyst. For improving yield and product quality, Veses et al. (2020) investigated the combined effect of pyrolysis and catalytic cracking of municipal solid waste and produced more than 80 vol.% CO and H_2 at 900°C. During the pyrolysis process, various products are obtained using the pyrolysis process, such as biofuel, chemicals, and adsorbents. It reduces the waste before landfill; therefore, this process promises approaches to enhance to produce a valuable product from waste. Product yield is dependent on operating circumstances (temperature, heating rate, vapour residence time, reaction time, reactor type, and MSW characteristics). MSW requires drying to increase the energy content of bio-oil. The drying stage might improve the overall cost of the procedure(Nanda and Berruti 2021).

4.5 Hydrothermal carbonization

Hydrothermal carbonization is an exothermally thermochemical technique that can be utilized as an energy-efficient approach to convert solid waste into high-value-added products (hydro-char and biofuel) under high temperatures (180–50°C) and pressure (2–6 MPa) (Wang et al. 2019). The exothermic nature of the waste materials and the low energy requirements for setting up the unit makes it simple to run (Lucian et al. 2020). For instance, Shen et al. (2016) used hydrothermal carbonization process for dechlorination of poly(vinyl chloride) (PVC) wastes with lignin and then washed with water to enhance the hydrochar concentration from 26.15 to 31.80 MJ/kg (solid fuel) at 210°C. Another study by Mendoza et al. (2021) converted agro forest waste (giant bamboo, coffee wood, eucalyptus, and coffee parchment) into hydro-char using hydrothermal carbonization at different temperatures (180–240°C), and results show that 24.6–29.2 MJ kg hydrochar was obtained at ≥ 220°C. Moreover, study by Lucian et al. (2018) obtained bioethanol from the organic fraction of municipal solid waste (OFSW) using hydrothermal carbonization liquid and slurry into an anaerobic digester, which enhanced the bioethanol production 37% and 363% as compared to (OFSW) at 180°C, 1 h reaction time. Hydrothermal carbonization of waste generates energy nevertheless, additional modifications are needed to make it a commercial technology due to the complicated reaction mechanisms and operational obstacles (a considerable amount of toxic water generated and gas, mainly CO_2) (Lucian et al. 2018).

4.6 Bioconversion

Anaerobic digestion and fermentation processes follow a bioconversion approach to convert organic waste (kitchen waste, agro-waste) into valuable products such as energy resources and chemicals (Fang et al. 2020, Yaashikaa et al. 2020). These conversion methods benefit organic solid wastes with a high moisture content since they stimulate microbial activity; in this process, microbial agents are utilized in the hydrolysis of solid waste (Soomro et al. 2020). Many researchers work on bioconversion of solid waste because it is an environmentally friendly approach and provides valuable products (biofuel, biofertilizer, and valuable chemicals). The research by Meinita et al. (2019) produced ethanol (13.9g L–1) from solid carrageenan waste (extract from red seaweed) using the fermentation process by Saccharomyces cerevisiae ATCC 200062. Another investigation by Fang et al. (2020) converted agriculture waste into poly-c-glutamic acid by B. amylo liquefaciens JX-6 from 10.00 kg to 37.50 kg using a solid-state fermentation process. Also, the research by Akhtar et al. (2020) converted agriculture waste into butyric acid (yielded 5.63, 5.5, 4.26 mg/100 g) using a solid-state fermentation process by Clostridium tyrobutyricum. Moreover, Bibi et al. (2021) converted agro waste into bio-fertilizer using Burkholderia cenocepacia strain. Another study by Agrawal et al. (2019) obtained ethanol (9.68%) using the fermentation process by Saccharomyces cerevisiae MTCC 4780 at 30°C, 48 h residence time and pH 6. In a similar manner, Farmanbordar et al. (2018) obtained ABE (acetone, butanol, and ethanol) 8 g/L concentration using fermentation process by Clostridium acetobuty licumused from municipal solid waste (starchy and lignocellulosic materials). The bioconversion process minimizes the weight and volume of waste disposed of and recovering energy. However, the bioconversion process is limited to using only organic components.

5. APPLICATION OF PRODUCTS FOR VARIOUS ENGINEERING PROCESSES

To protect the environment, the application of hazardous industrial waste is a difficult task. Advances in waste management, reprocessing, and long-term attention are considered critical, and a conceptual framework is now needed to maximize environmental, economic, and economic benefits. Vast amounts of toxic waste are generated all over the world as a result of various industrial activities. Currently, India produces around 960 million tonnes of solid waste per year due to municipal, mining, industrial, agricultural, and other industries. 4.5 million tonnes of this waste isrisky (Piskunov et al. 1988). The highly generated waste is agricultural waste, used to synthesize chemicals significantly for wastewater treatment. The most commonly synthesized products are coagulants and adsorbents. They are brief as follows:

5.1 Synthesizing adsorbents

Nowadays, many researchers focus on the synthesis and performance of adsorbents. For instance, Liu et al. (2014) reported that the bottom ash of incinerators is now reused in several developed countries. Likewise in European countries, the average recycling rate of incinerator bottom ash is around 60%. The bottom ash of incinerators contains a lot of silica, so it can be used as a source of silica for synthesizing mesoporous silica structures. According to the findings, the surface area and pore size distribution of the produced silica materials were 992 m^2 /g and 2–3.8 nm, respectively. There were several investigations in support of the possibility of municipal waste to be utilized as raw material for generating valuable product.

Qiu et al. (2018) investigated that the municipal solid waste incinerator (MSWI) fly ash was modified using a thermally hydrothermal technique to produce zeolitic. The temperature of the hydrothermal process was 200°C for 30 minutes. The modified fly ash used to have an absorption capacity of 0.498 meq/g, which was 22 times greater than the raw MSWI fly ash. Adsorption isotherm and kinetics studies were conducted to determine the adsorption behaviour of the modified fly ash. Although the modified fly ash's adsorption capability improved, MSWI fly ash might be used as the basic adsorbents. A recent study by Karimi et al. (2020b) observed that the CO_2 adsorption of a variety of activated carbon (AC) adsorbents synthesis from compost generated from the mechanical/biological treatment of municipal solid waste was studied. Sulfuric acid and N_2 calcination at

400 and 800°C were used to characterize the AC samples. Dynamic breakthrough tests in a range of temperature of 40–100°C and pressures up to 3 bar were used to assess CO_2 adsorption capacities. The AC sample, which were treated with sulfuric acid and activated at 800°C, provided maximum CO_2 absorption capacity with an amount adsorbed approximately 2.6 mol/kg at 40°C. Tang et al. (2021) investigated that study using leather processing solid waste produced MgO-doped biochar (MgO/BC) adsorbent. Acid orange II adsorption capability by synthesized sorbent MgO/BC was potentially high as 448.4 mg/g. The high specific surface area and enhanced electrostatic interaction of MgO/BC resulted in a significant adsorption capability for acid orange II. Moreover, MgO/BC provided excellent anionic dye removal from real leather processing effluent. Therefore, with the stated facts and findings it can be proposed that municipal waste has strong potential to be used for synthesis of valuable chemicals which further can be implemented to various engineering processes.

5.2 Synthesizing coagulants

Recently, the researchers focused on the synthesis of coagulants from the waste materials. Zhang et al. (2015) examined how to produce polyaluminum ferric chloride (PAFC) coagulant with blast furnace dust. In turbidity removal, charge neutralization was a key coagulation mechanism of PAFC. Other coagulation mechanisms, such as bridge connection and sweep flocculation, played essential roles in the decolourization process in addition to charge neutralization (Zhang et al. 2015). Yang et al. (2019) studied a waste pickle liquid from a steel plant that was used as a significant raw ingredient to create a compound coagulant: polymeric ferric aluminium sulphate chloride (PFASC), which is rich in iron and acid. Also, PFASC was utilized in the biological treatment of paper manufacturing effluent, along with polyacrylamide (PAM), to reduce COD and chromium. When PFASC and PAM were joined and treated to a wastewater sample from a paper mill, COD and chromium were decreased by 65.3% and 71.2%, respectively. The method has been tested in a paper mill, and the findings are in line with those achieved in laboratory.

Furthermore, Monney et al. (2019) evaluated the contamination removal potential of synthesized coagulant from bauxite slime waste compared to industrial-grade coagulant [$Al_2(SO_4)_3.18H_2O$] used for the treatment of car wash wastewater with standard jar test. The synthesized coagulant used in the bench-scale flocculation–flotation system completely removed turbidity (100%) and COD by up to 99%, respectively. Li and Yanmei (2021) synthesized a polymeric aluminium ferric sulfate (PAFS) using high-sulfur bauxite flotation tailings and red mud through roasting, acid leaching and polymerization. leaching conditions were followed: sulfuric acid concentration 4.5 mol/L, leaching temperature 100ºC, leaching time 90 min and liquid-solid ratio 6 ml/g. The wastewater disposal test of prepared PAFS coagulant showed that COD, turbidity, and chromaticity removal rates are 45.61%, 75%, and 94.18%, respectively.

6. EXPERIMENTAL STUDIES FOR SYNTHESIZING USEFUL PRODUCTS FROM BIO-WASTE

6.1 Carbonization of bio-waste from different methods

Agricultural products, including dry fruits, vegetables, fruits, etc., are generally consumed in massive quantities due to their health and immunity enhancing attributes. They have high lignocellulosic content, lignin, and other extractive components. There are versatile approaches to carbonize bio-waste to develop valuable products. The central focus has been intended towards synthesizing low-cost activated carbon (Yeow et al. 2021). Approaches adopted by investigators include muffle furnace activation, hydrothermal carbonization, and microwave irradiation (Huang et al. 2019). The varying raw materials and the characteristic attributes of developed material are tabulated in Table 2.

Table 2. Several ways to carbonize agro-waste, properties and applications.

Method	Raw Material	Process parameters	Specific surface area	Application of the synthesized product and the outcomes	Reference
Muffle furnace carbonization	Palm kernel shells	Sample à Grounded à Sieved à carbonized (500–550°C) à *Activated carbon*	1058 m^2/g	Methylene Blue removal with uptake capacity of 225 mg/g via adsorption	(García et al. 2018)
	Babassu coconut	Sample à Grounded à Sieved (28 mesh particle size) à carbonized à*Activated carbon*	625 m^2/g	Treatment of tartrazine with outcome of sorption capacity of 92 mg/g via adsorption	(Reck et al. 2018)
	Date palm leaflets	Sample à Grounded à Sieved (28 mesh particle size) à carbonized (500°C for 2 h in tube furnace) à*Activated carbon*	604 m^2/g	Desalination of concentrated stream reporting 5.38 mg/g capacity via electro-adsorption	(Kyaw et al. 2021)
Microwave irradiation	Camellia oleifera residue	Sample à dried à pre- carbonized (400°C for 2 h in argon atmosphere) à microwaved (2450 MHz for 2 min) à carbonized (800°C for 2 h) *Activated carbon*	1726 m^2/g	As pseudo-capacitor with 275 F/g capacitance resulting into active electro-chemical stabilities	(Bo et al. 2019)
	Spent coffee grounds	Sample à freeze (25°C overnight) à microwaved (5 h) à*Activated carbon*	-	For methylene blue removal yield 18.7 mg/g capacity	(Francaet al.2009)
	Walnut shells	Sample à microwaved (at 600 W for 20 min and 5 sequential cycle) à*Activated carbon*	436 m^2/g	For congo red dye treatment revealing 75.75 mg/g	(Maheshwari et al. 2020)
Hydrothermal approach	Coffee husk	Stainless steel reactors at heating rate of 20°C/min	33.30 m^2/g	Removal of methylene blue with capacity of 31.3 mg/g	(Ronix et al. 2017)
	Bamboo Sawdust	Autoclave was used to carbonize 200°C for 2 h	57.74 m^2/g	Treatment of congo red dye reporting 90.51 mg/g	(Li et al. 2018)
	Bamboo	Teflon autoclave was heated at 473 K for 24 h	5.03 m^2/g	Removal of methylene blue reporting 717 mg/g sorption capacity	(Lv et al. 2022)

As shown in Table 2, conventional muffle furnace methods have been the most adapted method by investigators especially for developing activated carbon for wastewater treatment. Moreover, hydrothermal approach is a recent innovation which needs improvisations to yield higher surface characteristics. Sample containing high surface area resulted in excellent removal capacities. Therefore, it is more in trend due to good results for treatment process.

6.2 Characterization of developed product

The developed valuable product is usually characterized for its structural and functional assessment. The commonly used methods include X-ray photoelectron spectra (XPS) for analysing the oxidation state, scanning electron microscope (SEM) assessment for morphology, energy dispersion spectrometer (EDS) for relative quantity of elements, Fourier transform infrared instrument (FTIR) for detecting the functional groups, Brunauer-Emmett-Teller (BET) for analysing surface attributes, proximate analysis for the compositional analysis, etc. The biomass derived activated carbon generally has a porous structure with significant presence of carbon and traces of other elements. For instance, a study by Li et al. (2017) proposed a facile method for preparing activated carbon from gulfweed and then activating it with the help of a chemical method using Potassium hydroxide (KOH). The chemically activated carbon demonstrated large amount of pores on the surface. Moreover, a strong peak of carbon was found in EDS analysis with sharp inclinations in XRD patterns with diffractions at 23° and 44° attributing to (002) and (100) planes indicating crystalline morphology. Our previous investigations reported effortless fabrication of activated carbon from wheat bran revealing porous SEM images with massive amount of C (72%) as reported in EDS pattern. In addition to this, XPS revealed that developed carbon has amorphous characteristics (Agarwal and Singh 2017). Furthermore, IR spectrum revealed strong presence of C=O, C-O and alcohol vibrations (Li et al. 2017). Another investigation revealed that denim fabric waste based AC resulted in an excellent surface area of 1582 m^2/g with porous SEM morphology having longitudinal wire shaped strands ranging from 5 to 15 μm (Silva et al. 2018). A recent study revealed that date palm leaflets could result in microporous carbon with surface area of 604 m^2/g with strong troughs at binding energy of 285 eV corresponding to C-O. Also, the IR spectrum revealed sharp elevations around 1100 cm^{-1} for C-O stretching vibrations (Kyaw et al. 2021).

From the several investigations highlighted, it can be inferred that structural morphology of the developed valuable product out of agricultural waste may be microporous (< 2 nm), mesoporous (2–50 nm), and macro porous (> 50 nm) with varying surface area distributions revealing specific surface area in the range of 500–3000 m^2/g (Maheshwari and Agrawal 2020). The literature data has been assessed as tabulated in Table 3 for the structural analysis of the synthesised product and other adhering characteristic from agricultural waste.

6.3 Application of developed product into various areas of water treatment

For water and wastewater treatment, there are several pollutant removal techniques implemented by many investigators in literature (Renu et al. 2018), namely, adsorption (Gupta and Khatri 2019), coagulation (Bazrafshan et al. 2016), pressure driven methods (Saad et al. 2020), electro-coagulation (Thakur and Chauhan 2018), advanced oxidation process (Neves et al. 2020), etc. For instance, adsorption has been examined utilizing waste walnut shells in our previous work (Maheshwari et al. 2020) revealing the energy required for adsorption process to be around 3.4 kWh for treating congo red dye with initial concentration of 10 mg/L providing the contact time of 2 h having a small batch volume of 50 mL revealing 97.7% efficacy. In same string, biomass was used for deionizing the RO reject outlet which resulted into energy efficient system with maximum sorption performance of 35 mg/g (Maheshwari et al. 2021). Moreover, enormous instances have been reported by various

Table 3. Structural statistics of the developed chemical from agro-waste.

Agro-waste selected	Surface characteristic of developed chemical	Nature of the pores	Other characterizations	References
Beetroot-based activated carbons	BET surface area of 2281 m^2/g	Microporous surface	Pore volume of 0.31 m^2/g	(Veerakumar et al. 2016)
Left over coconut shell	BET surface area of 2022 m^2/g	Microporosity in the major sections of material	2.54 nm was the avg. pore diameter	(Gupta and Khatri 2019)
Used tea leaves based activated carbon	Specific surface area of 3.23 m^2/g	Microporous	Rough surface with broken flat surface	(Wong et al. 2018)
Jute stick derived chemical product	Irregular distribution of pores	Micro and mesoporous	Jute stick derived AC contained 73% carbon	(Ghosh et al. 2019)
Citrus limetta peel based activated carbon	Unevenly distributed cavities	Contains pores of various sizes	The infrared spectroscopy reveals that there was functional presence of carboxyl, methylene, hydroxyl and methyl groups	(Shakoor and Nasar 2016)
Almond shell waste	452.2 m^2/g	Microporous ordered structure having diameter less than 1 nm	177 cm^3/g cumulative surface area for volume of gas adsorbed	(Maniscalco et al. 2020)
Xylose	408.1 m^2/g	Microporous structure	Pore volume of 0.278 cm^3/g	(Lu et al. 2021)

investigators utilizing agro-waste for developing adsorbent which further has been reported for treating effluent stream majorly comprising dye molecules and heavy metals. The critical review of fabricated product has been provided in preceding discussions.

García et al. (2018) utilized palm kernel shells to synthesize activated carbon with a specific surface area of 1058 m^2/g and investigated its application in removal of methylene blue. The synthesized product exhibited a good sorption performance with a capacity of 225.3 mg/g. Another study by Liu et al. (2020b) explored the application of waste straw based sorbent in the field of dye and heavy metal removal. The study reported a strong adsorption capacity of 3053 mg/g for methyl orange dye. In the same string, a study focused on agro residue corn cob for synthesizing AC reporting mesoporous structural appearance and examined successful removal of methylene blue with the sorption ability of 183.3 mg/g (Jawad et al. 2020). Furthermore, an investigation deals with direct, pre, post and modified lotus seedpod bio-based material for the treatment of organic contaminant 17 β-estradiol aqueous solution reporting surface area of 363.3 m^2/g for modified material revealing a capacity of 100.6 mg/g (Liu et al. 2020a). Therefore, it can be suggested that biomass has tremendous potential to be used as raw material for producing valuable chemical to treat water and waste water.

7. CONCLUSIONS

The chapter briefed the impact of solid waste, disposed from various sources, on the environment and health. According to the world bank (2021), food and green waste contribute 44% compared to other waste in municipal solids, which includes the waste from agricultural sectors due to huge consumption, whereas industrial waste is the most toxic nature and highly contaminates the ecological system imparting metals, dyes, etc., leading to hazardous diseases like cancer. The authors consolidate the attempts made by investigators to convert the massively produced bio-waste to be used further for different applications like wastewater treatment. Recently, majority of the agro-waste has found its application in developing porous adsorbent for treating effluent streams comprising dyes and heavy metals. According to the author's perspective, agro-waste is highly suitable for developing biochar due to its inherent chemical composition having massive lignocellulosic content which leads to porous material synthesis on activation. Moreover, it can be seen that 70% of its application has been explored for pollutants' removal from waste streams as the developed material has active sites attracting contaminants with strong efficacies. With aforementioned facts, India, being a hub for agricultural activities, has significant potential for developing valuable products from solid waste.

REFERENCES

Abd El-Salam, M.M. and Abu-Zuid, G.I. 2015. Impact of Landfill Leachate on the Groundwater Quality: A Case Study in Egypt. Journal of Advanced Research 6(4): 579–586.

Abdel-Shafy, H.I. and Mansour, M.S.M. 2018. Solid Waste Issue: Sources, Composition, Disposal, Recycling, and Valorization. Egyptian Journal of Petroleum 27(4): 1275–1290.

Agarwal, M. and Singh, K. 2017. Removal of Copper, Cadmium, and Chromium from Wastewater by Modified Wheat Bran Using Box – Behnken Design : Kinetics and Isotherm. Separation Science and Technology 53: 1476–1489.

Agrawal, T., Jadhav, S.K. and Quraishi, A. 2019. Bioethanol Production from an Agrowaste, Deoiled Rice Bran by Saccharomyces Cerevisiae MTCC 4780 via Optimization of Fermentation Parameters. EnvironmentAsia 12(1): 20–24.

Akhtar, T., Hashmi, A.S., Tayyab, M., Anjum, A.A., Saeed, S. and Ali, S. 2020. Bioconversion of Agricultural Waste to Butyric Acid Through Solid State Fermentation by Clostridium Tyrobutyricum. Waste and Biomass Valorization 11(5): 2067–2073.

Ali, S.M., Pervaiz, A., Afzal, B., Hamid, N. and Yasmin, A. 2014. Open Dumping of Municipal Solid Waste and Its Hazardous Impacts on Soil and Vegetation Diversity at Waste Dumping Sites of Islamabad City. Journal of King Saud University - Science 26(1): 59–65.

Awasthi, A.K., Wang, M., Wang, Z., Awasthi, M.K. and Li, J. 2018. E-Waste Management in India : A Mini-Review. Waste Management Research, 36(5): 408–414.

Banerjee, P., Hazra, A., Ghosh, P., Ganguly, A., Murmu, N.C. and Chatterjee, P.K. 2018. Solid Waste Management in India: A Brief Review. Waste Management and Resource Efficiency, 1027–1049 (Conference Paper).

Barbieri, M., Sappa, G., Vitale, S., Parisse, B. and Battistel, M. 2014. Soil Control of Trace Metals Concentrations in Landfills: A Case Study of the Largest Landfill in Europe, Malagrotta, Rome. Journal of Geochemical Exploration 143: 146–154.

Bazrafshan, E., Alipour, M.R. and Mahvi, A.H. 2016. Textile Wastewater Treatment by Application of Combined Chemical Coagulation, Electrocoagulation, and Adsorption Processes. Desalination and Water Treatment 57: 9203–9215.

Bibi, F., Ilyas, N., Arshad, M., Khalid, A., Saeed, M., Ansar, S. and Batley, J. 2021. Formulation and Efficacy Testing of Bio-Organic Fertilizer Produced through Solid-State Fermentation of Agro-Waste by Burkholderia Cenocepacia. Chemosphere 291: 132762.

Bo, X., Xiang, K., Zhang, Y., Shen, Y., Chen, S., Wang, Y., Xie, M. and Guo, X. 2019. Microwave-Assisted Conversion of Biomass Wastes to Pseudocapacitive Mesoporous Carbon for High-Performance Supercapacitor. Journal of Energy Chemistry 39: 1–7.

Boateng, T.K., Opoku, F. and Akoto, O. 2019. Heavy Metal Contamination Assessment of Groundwater Quality: A Case Study of Oti Landfill Site, Kumasi. Applied Water Science 9(2): 1–15.

Bokhoree, C., Beeharry, Y., Makoondlall-Chadee, T., Doobah, T. and Soomary, N. 2014. Assessment of Environmental and Health Risks Associated with the Management of Medical Waste in Mauritius. APCBEE Procedia 9: 36–41.

Bravo, A.K.G., Serrano, D.A.S, Jiménez, G.L., Nirmalkar, K., Murugesan, S., García-Mena, J., Castillo, M.E.G. and Gálvez, L.R.T. 2019. An inoculum to digest organic waste. Energies 12(12): 2343.

Buenrostro, O., Bocco, G. and Cram, S. 2001. Classification of Sources of Municipal Solid Wastes in Developing Countries. Resources, Conservation and Recycling 32(1): 29–41.

Cai, J., Zeng, R., Zheng, W., Wang, S., Han, J., Li, K., Luo, M. and Tang, X. 2021. Synergistic Effects of Co-Gasification of Municipal Solid Waste and Biomass in Fixed-Bed Gasifier. Process Safety and Environmental Protection 148: 1–12.

Chen, D., Cen, K., Cao, X., Zhang, J., Chen, F. and Zhou, J. 2020. Upgrading of Bio-Oil via Solar Pyrolysis of the Biomass Pretreated with Aqueous Phase Bio-Oil Washing, Solar Drying, and Solar Torrefaction. Bioresource Technology 305: 123130.

Chen, G., Jamro, I.A., Samo, S.R., Wenga, T., Baloch, H.A., Yan, B. and Ma, W. 2020. Hydrogen-Rich Syngas Production from Municipal Solid Waste Gasification through the Application of Central Composite Design: An Optimization Study. International Journal of Hydrogen Energy 45(58): 33260–33273.

Cheng, J., Ding, C., Li, X., Zhang, T. and Wang, X. 2015. Heavy Metals in Navel Orange Orchards of Xinfeng County and Their Transfer from Soils to Navel Oranges. Ecotoxicology and Environmental Safety 122: 153–158.

Chun, S.K. 2017. Mechanism of Hydrogen Sulfide Generation from a Composite Waste Landfill Site: A Case Study of the 'Sudokwon Landfill Site', Korea. Journal of Material Cycles and Waste Management 19(1): 443–452.

Chun, S.K. and Bae, Y.S. 2012. An Impact Analysis of Landfill for Waste Disposal on Climate Change: Case Study of 'Sudokwon Landfill Site 2nd Landfill' in Korea. Korean Journal of Chemical Engineering 29(11): 1549–1555.

Cole, C., Gnanapragasam, A., Cooper, T. and Singh, J. 2019. Assessing Barriers to Reuse of Electrical and Electronic Equipment, a UK Perspective. Resources, Conservation & Recycling: X, 1, 100004.

Cudjoe, D. and Acquah, P.M. 2021. Environmental Impact Analysis of Municipal Solid Waste Incineration in African Countries. Chemosphere 265: 129186.

Czajczyńska, D., Anguilano, L., Ghazal, H., Krzyżyńska, R., Reynolds, A.J., Spencer, N. and Jouhara, H. 2017. Potential of Pyrolysis Processes in the Waste Management Sector. Thermal Science and Engineering Progress 3: 171–197.

Daâssi, D., Zouari-Mechichi, H., Frikha, F., Rodríguez-Couto, S., Nasri, M. and Mechichi, T. 2016. Sawdust Waste as a Low-Cost Support-Substrate for Laccases Production and Adsorbent for Azo Dyes Decolorization. Journal of Environmental Health Science and Engineering 14(1): 1–12.

da Silva, L.J. de V.B., dos Santos, I.F.S., Mensah, J.H.R., Gonçalves, A.T.T. and Silva, R.M. B. 2020. Incineration of Municipal Solid Waste in Brazil: An Analysis of the Economically Viable Energy Potential. Renewable Energy 149: 1386–1394.

Das, S., Lee, S.H., Kumar, P., Kim, K.H., Lee, S.S. and Bhattacharya, S.S. 2019. Solid Waste Management: Scope and the Challenge of Sustainability. Journal of Cleaner Production 228: 658–678.

de Albuquerque, C.A., Mello, C.H.M., de Freitas Gomes, J.H., dos Santos, V.C. and Zara, J. V. 2020. E-Waste in the World Today : An Overview of Problems and a Proposal for Impovement in Brazil. Environmental Quality Management 29: 63–72.

Dhar, H., Kumar, P., Kumar, S., Mukherjee, S. and Vaidya, A.N. 2016. Effect of Organic Loading Rate during Anaerobic Digestion of Municipal Solid Waste. Bioresource Technology 217: 56–61.

Di Maria, F., Mastrantonio, M. and Uccelli, R. 2021. The Life Cycle Approach for Assessing the Impact of Municipal Solid Waste Incineration on the Environment and on Human Health. Science of the Total Environment 776: 145785.

Fang, J., Huan, C.C., Liu, Y., Xu, L. and Yan, Z. 2020. Bioconversion of Agricultural Waste into Poly-γ-Glutamic Acid in Solid-State Bioreactors at Different Scales. Waste Management 102: 939–948.

Faraca, G. and Astrup, T. 2019. Plastic Waste from Recycling Centres: Characterisation and Evaluation of Plastic Recyclability. Waste Management 95: 388–398.

Farmanbordar, S., Karimi, K. and Amiri, H. 2018. Municipal Solid Waste as a Suitable Substrate for Butanol Production as an Advanced Biofuel. Energy Conversion and Management 157: 396–408.

Feng, S.J., Gao, K.W., Chen, Y.X., Li, Y., Zhang, L.M. and Chen, H.X. 2017. Geotechnical Properties of Municipal Solid Waste at Laogang Landfill, China. Waste Management 63: 354–365.

Franca, A.S., Oliveira, L.S. and Ferreira, M.E. 2009. Kinetics and Equilibrium Studies of Methylene Blue Adsorption by Spent Coffee Grounds. Desalination 249(1): 267–272.

García, J.R., Sedran, U., Zaini, M.A.A. and Zakaria, Z.A. 2018. Preparation, Characterization, and Dye Removal Study of Activated Carbon Prepared from Palm Kernel Shell. Environmental Science and Pollution Research 25(6): 5076–5085.

Ghosh, R.K., Ray, D.P., Debnath, S., Tewari, A. and Das, I. 2019. Optimization of Process Parameters for Methylene Blue Removal by Jute Stick Using Response Surface Methodology. Environmental Progress and Sustainable Energy 38: 13146.

Gupta, K. and Khatri, O.P. 2019. Fast and Efficient Adsorptive Removal of Organic Dyes and Active Pharmaceutical Ingredient by Microporous Carbon: Effect of Molecular Size and Charge. Chemical Engineering Journal 378: 122218.

Hameed, Z., Aslam, M., Khan, Z., Maqsood, K., Atabani, A.E., Ghauri, M., Khurram, M.S., Rehan, M. and Nizami, A.S. 2021. Gasification of Municipal Solid Waste Blends with Biomass for Energy Production and Resources Recovery: Current Status, Hybrid Technologies and Innovative Prospects. Renewable and Sustainable Energy Reviews 136: 110375.

Hartmann, C. 2018. Waste Picker Livelihoods and Inclusive Neoliberal Municipal Solid Waste Management Policies: The Case of the La Chureca Garbage Dump Site in Managua, Nicaragua. Waste Management 71: 565–577.

He, M., Hu, Z., Xiao, B., Li, J., Guo, X., Luo, S., Yang, F., Feng, Y., Yang, G. and Liu, S. 2009. Hydrogen-Rich Gas from Catalytic Steam Gasification of Municipal Solid Waste (MSW): Influence of Catalyst and Temperature on Yield and Product Composition. International Journal of Hydrogen Energy 34(1): 195–203.

Hossain, M.S., Rahman, N.N.N.A., Balakrishnan, V., Alkarkhi, A.F.M., Rajion, Z.A. and Kadir, M.O.A. 2015. Optimizing Supercritical Carbon Dioxide in the Inactivation of Bacteria in Clinical Solid Waste by Using Response Surface Methodology. Waste Management 38(1): 462–473.

Hossain, M.S., Santhanam, A., Nik Norulaini, N.A. and Omar. A.K.M. 2011. Clinical Solid Waste Management Practices and Its Impact on Human Health and Environment—A Review. Waste Management 31(4): 754–766.

Hu, X. and Gholizadeh, M. 2019. Biomass Pyrolysis: A Review of the Process Development and Challenges from Initial Researches up to the Commercialisation Stage. Journal of Energy Chemistry 39: 109–143.

Huang, Q., Song, S., Chen, Z., Hu, B., Chen, J. and Wang, X. 2019. Biochar-Based Materials and Their Applications in Removal of Organic Contaminants from Wastewater: State-of-the-Art Review. Biochar 1(1): 45–73.

Ihedioha, J.N., Ukoha, P.O. and Ekere, N.R. 2017. Ecological and Human Health Risk Assessment of Heavy Metal Contamination in Soil of a Municipal Solid Waste Dump in Uyo, Nigeria. Environmental Geochemistry and Health 39(3): 497–515.

Irfan, M., Li, A., Zhang, L., Ji, G. and Gao, Y. 2021. Catalytic Gasification of Wet Municipal Solid Waste with HfO2 Promoted Ni-CaO Catalyst for H2-Rich Syngas Production. Fuel 286: 119408.

Irfan, M., Li, A., Zhang, L., Wang, M., Chen, C. and Khushk, S. 2019. Production of Hydrogen Enriched Syngas from Municipal Solid Waste Gasification with Waste Marble Powder as a Catalyst. International Journal of Hydrogen Energy 44(16): 8051–8061.

Jara-Samaniego, J., Pérez-Murcia, M.D., Bustamante, M.A., Pérez-Espinosa, A., Paredes, C., López, M., López-Lluch, D.B., Gavilanes-Terán, I. and Moral, R. 2017. Composting as Sustainable Strategy for Municipal Solid Waste Management in the Chimborazo Region, Ecuador: Suitability of the Obtained Composts for Seedling Production. Journal of Cleaner Production 141: 1349–1358.

Jawad, A.H., Bardhan, M., Islam, M.A., Islam, M.A., Syed-Hassand, S.S.A., Surip, S.N., ALOthman, Z.A. and Khan, M.R. 2020. Insights into the Modeling, Characterization and Adsorption Performance of Mesoporous Activated Carbon from Corn Cob Residue via Microwave-Assisted H_3PO_4 Activation. Surfaces and Interfaces 21: 100688.

Jeong, Y.S., Choi, Y.K., Kang, B.S., Ryu, J.H., Kim, H.S., Kang, M.S., Ryu, L.H. and Kim, J.S. 2020. Lab-Scale and Pilot-Scale Two-Stage Gasification of Biomass Using Active Carbon for Production of Hydrogen-Rich and Low-Tar Producer Gas. Fuel Processing Technology, 198: 106240.

Kanhai, G., Fobil, J.N., Nartey, B.A., Spadaro, J.V. and Mudu, P. 2021. Urban Municipal Solid Waste Management: Modeling Air Pollution Scenarios and Health Impacts in the Case of Accra, Ghana. Waste Management 123: 15–22.

Karimi, H., Herki, B.M.A., Gardi, S.Q., Galali, S., Hossini, H., Mirzaei, K. and Pirsaheb, M. 2020. Site Selection and Environmental Risks Assessment of Medical Solid Waste Landfill for the City of Kermanshah-Iran. International Journal of Environmental Health Research 32: 155–167.

Karimi, M., Zafanelli, L.F.A.S., Almeida, J.P.P., Ströher, G.R., Rodrigues, A.E. and Silva, J.A.C. 2020. Novel Insights into Activated Carbon Derived from Municipal Solid Waste for CO 2 Uptake: Synthesis, Adsorption Isotherms and Scale-Up. Journal of Environmental Chemical Engineering 8(5): 104069.

Kumar, A. and Agrawal, A. 2020. Recent Trends in Solid Waste Management Status, Challenges, and Potential for the Future Indian Cities—A Review. Current Research in Environmental Sustainability 2: 100011.

Kyaw, H.H., Al-mashaikhi, S.M., Tay, M., Myint, Z., Al-harthi, S. and Al-abri, M. 2021. Activated Carbon Derived from the Date Palm Leaflets as Multifunctional Electrodes in Capacitive Deionization System. Chemical Engineering & Processing: Process Intensification 161: 108311.

Lalremruati, M. and Devi, A.S. 2021. Changes in Physico-Chemical Properties during Composting of Three Common Household Organic Solid Wastes Amended with Garden Soil. Bioresource Technology Reports 15: 100727.

Lane, D.J., Hartikainen, A., Sippula, O., Lähde, A., Mesceriakovas, A., Peräniemi, S. and Jokiniemi, J. 2020. Thermal Separation of Zinc and Other Valuable Elements from Municipal Solid Waste Incineration Fly Ash. Journal of Cleaner Production 253: 120014.

Laurent, A., Bakas, I., Clavreul, J., Bernstad, A., Niero, M., Gentil, E., Hauschild, M.Z. and Christensen, T.H. 2014. Review of LCA Studies of Solid Waste Management Systems - Part I: Lessons Learned and Perspectives. Waste Management 34(3): 573–588.

Law, K.L., Starr, N., Siegler, T.R., Jambeck, J.R., Mallos, N.J. and Leonard, G.H. 2020. The United States' Contribution of Plastic Waste to Land and Ocean. Science Advance 6: 1–8.

Li, J. and Yanmei, J. 2021. Preparation of Polymeric Aluminum Ferric Sulphate from Waste Residue of Aluminum Industry. E3S Web of Conferences 271: 04005.

Li, Q., Faramarzi, A., Zhang, S., Wang, Y., Hu, X. and Gholizadeh, M. 2020. Progress in Catalytic Pyrolysis of Municipal Solid Waste. Energy Conversion and Management 226: 113525.

Li, R., Zhang, B., Wang, Y., Zhao, Y. and Li, F. 2019. Leaching Potential of Stabilized Fly Ash from the Incineration of Municipal Solid Waste with a New Polymer. Journal of Environmental Management 232: 286–294.

Li, S., Han, K., Li, J., Li, M. and Lu, C. 2017. Preparation and Characterization of Super Activated Carbon Produced from Gulfweed by KOH Activation. Microporous and Mesoporous Materials 243: 291–300.

Li, Y., Meas, A., Shan, S., Yang, R., Gai, X., Wang, H. and Tsend, N. 2018. Hydrochars from Bamboo Sawdust through Acid Assisted and Two-Stage Hydrothermal Carbonization for Removal of Two Organics from Aqueous Solution. Bioresource Technology 261: 257–264.

Liang, Y., Tan, Q., Song, Q. and Li, J. 2021. An Analysis of the Plastic Waste Trade and Management in Asia. Waste Management 119: 242–253.

Lima, J.Z., Raimondi, I.M., Schalch, V. and Rodrigues, V.G.S. 2018. Assessment of the Use of Organic Composts Derived from Municipal Solid Waste for the Adsorption of Pb, Zn and Cd. Journal of Environmental Management 226: 386–399.

Linsong, W., Peng, Z., Yu, F., Sujun, L., Yue, Y., Li, W. and Wei., S. 2020. Recovery of Metals from Jarosite of Hydrometallurgical Nickel Production by Thermal Treatment and Leaching. Hydrometallurgy 198: 105493.

Liu, A., Ren, F., Lin, W.Y. and Wang, J.Y. 2015. A Review of Municipal Solid Waste Environmental Standards with a Focus on Incinerator Residues. International Journal of Sustainable Built Environment 4(2): 165–188.

Liu, N., Liu, Y., Zeng, G., Gong, J., Tan, X., Wen, J., Liu, S., Jiang, L., Li, M. and Yin, Z. 2020. Adsorption of 17β-Estradiol from Aqueous Solution by Raw and Direct/Pre/Post-KOH Treated Lotus Seedpod Biochar. Journal of Environmental Sciences (China) 87: 10–23.

Liu, Q., Li, Y., Chen, H., Lu, J., Yu, G. and Möslang, M. 2020. Superior Adsorption Capacity of Functionalised Straw Adsorbent for Dyes and Heavy-Metal Ions. Journal of Hazardous Materials 382(130): 121040.

Liu, Z.-s., Li, W-k. and Huang, C-y. 2014. Synthesis of Mesoporous Silica Materials from Municipal Solid Waste Incinerator Bottom Ash. Waste Management 34(5): 893–900.

Lu, J.S., Chang, Y., Poon, C.S. and Lee, D.J. 2020. Slow Pyrolysis of Municipal Solid Waste (MSW): A Review. Bioresource Technology 312: 123615.

Lu, T., Liu, Y., Xu, X., Pan, L., Alothman, A.A., Shapter, J., Wang, Y. and Yamauchi, Y. 2021. Highly Efficient Water Desalination by Capacitive Deionization on Biomass-Derived Porous Carbon Nanoflakes. Separation and Purification Technology 256: 117771.

Lucian, M., Volpe, M., Gao, L., Piro, G., Goldfarb, J.L. and Fiori, L. 2018. Impact of Hydrothermal Carbonization Conditions on the Formation of Hydrochars and Secondary Chars from the Organic Fraction of Municipal Solid Waste. Fuel, 233, 257–268.

Lucian, M., Volpe, M., Merzari, F., Wüst, D., Kruse, A., Andreottola, G. and Fiori, L. 2020. Hydrothermal Carbonization Coupled with Anaerobic Digestion for the Valorizatio of the Organic Fraction of Municipal Solid Waste. Bioresource Technology 314: 123734.

Lv, B.W., Xu, H., Guo, J.Z., Bai, L.Q. and Li, B. 2022. Efficient Adsorption of Methylene Blue on Carboxylate-Rich Hydrochar Prepared by One-Step Hydrothermal Carbonization of Bamboo and Acrylic Acid with Ammonium Persulphate. Journal of Hazardous Materials 421: 126741.

Maheshwari, K., Agrawal, M. and Gupta, A.B. 2021. Experimental Investigation for Treating the RO Reject Stream through Capacitive Deionization. Separation and Purification Technology 276: 119261.

Maheshwari, K., Agrawal, M. 2020. Advances in Capacitive Deionization as an Effective Technique for Reverse Osmosis Reject Stream Treatment. Journal of Environmental Chemical Engineering 8(6): 104413.

Maheshwari, K., Solanki, Y.S., Ridoy, M.S.H., Agarwal, M., Dohare, R. and Gupta, R. 2020. Ultrasonic Treatment of Textile Dye Effluent Utilizing Microwave-Assisted Activated Carbon. Environmental Progress and Sustainable Energy 39: 13410.

Maniscalco, M.P., Corrado, C., Volpe, R. and Messineo, A. 2020. Evaluation of the Optimal Activation Parameters for Almond Shell Bio-Char Production for Capacitive Deionization. Bioresource Technology Reports 11: 100435.

Marieta, C., Guerrero, A. and Leon, I. 2021. Municipal Solid Waste Incineration Fly Ash to Produce Eco-Friendly Binders for Sustainable Building Construction. Waste Management 120: 114–124.

Mavimbela, S.S.W., Ololade, O.O., van Tol, J.J. and Aghoghovwia, M.P. 2019. Characterizing Landfill Leachate Migration Potential of a Semi-Arid Duplex Soil. Heliyon 5(10): e02603.

Meinita, M.D.N., Marhaeni, B., Jeong, G.T. and Hong, Y.K. 2019. Sequential Acid and Enzymatic Hydrolysis of Carrageenan Solid Waste for Bioethanol Production: A Biorefinery Approach. Journal of Applied Phycology 31(4): 2507–2515.

Mendoza, M., Lisseth, C., Sermyagina, E., Saari, J., de Jesus, M.S., Cardoso, M., de Almeida, G. M. and Vakkilainen, E. 2021. Hydrothermal Carbonization of Lignocellulosic Agro-Forest Based Biomass Residues. Biomass and Bioenergy 147: 106004.

Mlonka-Mędrala, A., Evangelopoulos, P., Sieradzka, M., Zajemska, M. and Magdziarz, A. 2021. Pyrolysis of Agricultural Waste Biomass towards Production of Gas Fuel and High-Quality Char: Experimental and Numerical Investigations. Fuel 296: 120611.

Monney, I., Buamah, R., Donkor, E.A., Etuaful, R., Nota, H.K. and Ijzer, H. 2019. Treating Waste with Waste : The Potential of Synthesized Alum from Bauxite Waste for Treating Car Wash Wastewater for Reuse. Environmetnal Science Pollution Research International 26: 12755–12764.

Nabegu, A.B. 2010. An Analysis of Municipal Solid Waste in Kano Metropolis , Nigeria. Journal of Human Ecology 31: 111–119

Nanda, S. and Berruti, F. 2021. A Technical Review of Bioenergy and Resource Recovery from Municipal Solid Waste. Journal of Hazardous Materials 403: 123970.

Neves, N.S.C.S., Barbosa, A.A., Santana, I.L.S., Pereira, P.M.N., Pacheco, J.G.A., Benachour, M. and Rocha, O.R.S. 2020. Treatment of Bicomponent Textile Dyes Using Combined Photocatalysis and Adsorption Process Made from Residue-Based Reactor and Adsorbent Material. Chemical Engineering Communications 208: 1523–1542.

Ojowuro, O.M., Olowe, B. and Aremu, A.S. 2019. Waste Management and Resource Efficiency. Springer Singapore.

Pergola, M., Persiani, A., Palese, A.M., Meo, V.D., Pastore, V., D'Adamo, C. and Celano, G. 2018. Composting: The Way for a Sustainable Agriculture. Applied Soil Ecology 123: 744–750.

Piskunov, V.M., Matveev, A.F. and Yaroslavtsev, A.S. 1988. Utilizing Iron Residues from Zinc Production in the U.S.S.R. JOM 40(8): 36–39.

Preziosi, E., Frollini, E., Zoppini, A., Ghergo, S., Melita, M., Parrone, D., Rossi, D. and Amalfitano, S. 2019. Disentangling Natural and Anthropogenic Impacts on Groundwater by Hydrogeochemical, Isotopic and Microbiological Data: Hints from a Municipal Solid Waste Landfill. Waste Management 84: 245–255.

Priti, Mandal, K. 2019. Review on Evolution of Municipal Solid Waste Management in India: Practices, Challenges and Policy Implications. Journal of Material Cycles and Waste Management 21(6): 1263–1279.

Qiu, Q., Jiang, X., Lv, G., Chen, Z., Lu, S. and Ni, M. 2018. Adsorption of Heavy Metal Ions Using Zeolite Materials of Municipal Solid Waste Incineration Fl y Ash Modi Fi Ed by Microwave-Assisted Hydrothermal Treatment. Powder Technology 335: 156–163.

Ramachandra, T.V., Bharath, H.A., Kulkarni, G. and Han, S.S. 2018. Municipal Solid Waste: Generation, Composition and GHG Emissions in Bangalore, India. Renewable and Sustainable Energy Reviews 82: 1122–1136.

Reck, I.M., Paixão, R.M., Bergamasco, R., Vieira, M.F. and Vieira, A.M.S. 2018. Removal of Tartrazine from Aqueous Solutions Using Adsorbents Based on Activated Carbon and Moringa Oleifera Seeds. Journal of Cleaner Production 171: 85–97.

Rene, E., Sethurajan, R.M. and Kumar, V. 2021. Electronic Waste Generation , Recycling and Resource Recovery: Technological Perspectives and Trends. Journal of Hazardous Materials 416: 125664.

Renu, Agarwal, M. and Singh, K. 2018. Removal of Copper, Cadmium, and Chromium from Wastewater by Modified Wheat Bran Using Box–Behnken Design: Kinetics and Isotherm. Separation Science and Technology 53(10): 1476–1489.

Ronix, A., Pezoti, O., Souza, L.S., Souza, I.P.A.F., Bedin, K.C., Souza, P.S.C., Silva, T.L., Melo, S. A.R., Cazetta, A.L. and Almeida, V.C. 2017. Hydrothermal Carbonization of Coffee Husk: Optimization of Experimental Parameters and Adsorption of Methylene Blue Dye. Journal of Environmental Chemical Engineering, 5(5): 4841–4849.

Saad, M.S., Balasubramaniam, L., Wirzal, M.D.H., Halim, N.S.A., Bilad, M.R., Nordin, N.A. H. M., Putra, Z.A. and Ramli, F.N. 2020. Integrated Membrane–Electrocoagulation System for Removal of Celestine Blue Dyes in Wastewater. Membranes 10(8): 184–195.

Saeaung, K., Phusunti, N., Phetwarotai, W., Assabumrungrat, S. and Cheirsilp, B. 2021. Catalytic Pyrolysis of Petroleum-Based and Biodegradable Plastic Waste to Obtain High-Value Chemicals. Waste Management 127: 101–111.

Shah, G.M., Tufail, N., Bakhat, H.F., Ahmad, I., Shahid, M., Hammad, H.M., Nasim, W., Waqar, A., Rizwan, M. and Dong, R. 2019. Composting of Municipal Solid Waste by Different Methods Improved the Growth of Vegetables and Reduced the Health Risks of Cadmium and Lead. Environmental Science and Pollution Research 26(6): 5463–5474.

Shakoor, S. and Nasar, A. 2016. Removal of Methylene Blue Dye from Artificially Contaminated Water Using Citrus Limetta Peel Waste as a Very Low Cost Adsorbent. Journal of the Taiwan Institute of Chemical Engineers 66: 154–163.

Sharholy, M., Ahmad, K., Mahmood, G. and Trivedi, R.C. 2008. Municipal Solid Waste Management in Indian Cities—A Review. Waste Management 28(2): 459–467.

Sharholy, M., Ahmad, K., Vaishya, R.C. and Gupta, R.D. 2007. Municipal Solid Waste Characteristics and Management in Allahabad, India. Waste Management 27(4): 490–496.

Shen, Y. 2016. Dechlorination of Poly(Vinyl Chloride. Wastes via Hydrothermal Carbonization with Lignin for Clean Solid Fuel Production. Industrial and Engineering Chemistry Research 55(44): 11638–11644.

Silva, T.L., Cazetta, A.L., Souza, P.S.C., Zhang, T., Asefa, T. and Almeida, V.C. 2018. Mesoporous Activated Carbon Fibers Synthesized from Denim Fabric Waste: Efficient Adsorbents for Removal of Textile Dye from Aqueous Solutions. Journal of Cleaner Production 171: 482–490.

Silva, R.V., De Brito, J., Lynn, C.J. and Dhir, R.K. 2017. Use of Municipal Solid Waste Incineration Bottom Ashes in Alkali-Activated Materials, Ceramics and Granular Applications: A Review. Waste Management 68: 207–220.

Singh, M. 2019. Solid Waste Management in Urban India: Imperatives for Improvement. Journal of Contemporary Issues in Business and Government 25(1): 87–92.

Singh, S.K., Chokhandre, P., Salve, P.S. and Rajak, R. 2021. Open Dumping Site and Health Risks to Proximate Communities in Mumbai, India: A Cross-Sectional Case-Comparison Study. Clinical Epidemiology and Global Health 9: 34–40.

Song, Q., Zhao, H., Jia, J., Yang, L., Lv, W., Bao, J., Shu, X., Gu, Q. and Zhang, P. 2020. Pyrolysis of Municipal Solid Waste with Iron-Based Additives: A Study on the Kinetic, Product Distribution and Catalytic Mechanisms. Journal of Cleaner Production 258: 120682.

Soomro, A.F., Abbasi, I.A., Ni, Z., Ying, L. and Liu, J. 2020. Influence of Temperature on Enhancement of Volatile Fatty Acids Fermentation from Organic Fraction of Municipal Solid Waste: Synergism between Food and Paper Components. Bioresource Technology 304: 122980.

Suryanarayana, B. and Rao, V.S. 2013. Credit To Agriculture Sector In India. Indian Streams Research Journal 3(4): 72–80.

Suthar, S. and Singh, P. 2015. Household Solid Waste Generation and Composition in Different Family Size and Socio-Economic Groups: A Case Study. Sustainable Cities and Society, 14(1): 56–63.

Swati, Ghosh, P. and Thakur, I.S. 2017. An Integrated Approach to Study the Risk from Landfill Soil of Delhi: Chemical Analyses, *in Vitro* Assays and Human Risk Assessment. Ecotoxicology and Environmental Safety 143: 120–128.

Tan, S.T., Ho, W.S., Hashim, H., Lee, C.T., Taib, M.R. and Ho., C.S. 2015. Energy, Economic and Environmental (3E) Analysis of Waste-to-Energy (WTE) Strategies for Municipal Solid Waste (MSW) Management in Malaysia. Energy Conversion and Management 102: 111–120.

Tang, J., Qunwei, W. and Gyunghyun, C. 2020. Science of the Total Environment Ef Fi Ciency Assessment of Industrial Solid Waste Generation and Treatment Processes with Carry-over in China. Science of the Total Environment 726: 138274.

Tang, Y., Zhao, J., Zhang, Y., Zhou, J. and Shi, B. 2021. Chemosphere Conversion of Tannery Solid Waste to an Adsorbent for High-Ef Fi Ciency Dye Removal from Tannery Wastewater: A Road to Circular Utilization. Chemosphere 263: 127987.

Tang, Z., Xi, B., Huang, C., Tan, W., Li, W., Zhao, X., Liu, K. and Xia, X. 2020. Mobile Genetic Elements in Potential Host Microorganisms Are the Key Hindrance for the Removal of Antibiotic Resistance Genes in Industrial-Scale Composting with Municipal Solid Waste. Bioresource Technology 301: 122723.

Thakur, S. and Chauhan, M.S. 2018. Treatment of Dye Wastewater from Textile Industry by Electrocoagulation and Fenton Oxidation: A Review. Water Quality Management 79: 117–129.

To, S., Coughenour, C. and Pharr, J. 2019. The Environmental Impact and Formation of Meals from the Pilot Year of a Las Vegas Convention Food Rescue Program. International Journal of Environmental Research and Public Health 16(10): 1718.

Veerakumar, P., Muthuselvam, I.P., Hung, C.T., Lin, K.C., Chou, F.-C. and Liu, S.B. 2016. Biomass-Derived Activated Carbon Supported Fe_3O_4 Nanoparticles as Recyclable Catalysts for Reduction of Nitroarenes.ACS Sustainable Chemical Engineering 4: 6772–6782

Veses, A., Sanahuja-Parejo, O., Callén, M.S., Murillo, R. and García, T. 2020. A Combined Two-Stage Process of Pyrolysis and Catalytic Cracking of Municipal Solid Waste for the Production of Syngas and Solid Refuse-Derived Fuels. Waste Management 101: 171–179.

Vongdala, N., Tran, H.D., Xuan, T.D., Teschke, R. and Khanh, T.D. 2019. Heavy Metal Accumulation in Water, Soil, and Plants of Municipal Solid Waste Landfill in Vientiane, Laos. International Journal of Environmental Research and Public Health 16(1): 1–13.

Wang, L., Zhang, Y., Chen, L., Guo, B., Tan, Y., Sasaki, K. and Tsang, D.C.W. 2022. Designing Novel Magnesium Oxysulfate Cement for Stabilization/Solidification of Municipal Solid Waste Incineration Fly Ash. Journal of Hazardous Materials 423: 127025.

Wang, L., Chang, Y. and Li, A. 2019. Hydrothermal Carbonization for Energy-Efficient Processing of Sewage Sludge: A Review. Renewable and Sustainable Energy Reviews 108: 423–440.

Wang, X., Gao, M., Wang, M., Wu, C., Wang, Q. and Wang, Y. 2021. Chloride Removal from Municipal Solid Waste Incineration Fly Ash Using Lactic Acid Fermentation Broth. Waste Management 130: 23–29.

Wong, S., Tumari, H.H., Ngadi, N., Mohamed, N.B., Hassan, O., Mat, R. and Amin, N.A.S. 2018. Adsorption of Anionic Dyes on Spent Tea Leaves Modified with Polyethyleneimine (PEI-STL). Journal of Cleaner Production 206: 394–406.

Wu, J., Zhao, Y., Zhao, W., Yang, T., Zhang, X., Xie, X., Cui, H. and Wei, Z. 2017. Effect of Precursors Combined with Bacteria Communities on the Formation of Humic Substances during Different Materials Composting. Bioresource Technology 226: 191–199.

Yaashikaa, P.R., Kumar, P.S., Saravanan, A., Varjani, S. and Ramamurthy, R. 2020. Bioconversion of Municipal Solid Waste into Bio-Based Products: A Review on Valorisation and Sustainable Approach for Circular Bioeconomy. Science of the Total Environment 748: 141312.

Yang, S., Li, W., Zhang, H., Wen, Y. and Ni, Y. 2019. Separation and Puri Fi Cation Technology Treatment of Paper Mill Wastewater Using a Composite Inorganic Coagulant Prepared from Steel Mill Waste Pickling Liquor. Separation and Purification Technology 209: 238–245.

Yay, E. and Suna, A. 2015. Application of Life Cycle Assessment (LCA) for Municipal Solid Waste Management: A Case Study of Sakarya. Journal of Cleaner Production 94: 284–293.

Yeow, P.K., Wong, S.W. and Hadibarata, T. 2021. Removal of Azo and Anthraquinone Dye by Plant Biomass as Adsorbent—A Review. Biointerface Research in Applied Chemistry 11(1): 8218–8232.

Yousefloo, A. and Babazadeh, R. 2020. Designing an Integrated Municipal Solid Waste Management Network: A Case Study. Journal of Cleaner Production 244: 118824.

Zhang, J., Zhang, S. and Liu, B. 2020. Degradation Technologies and Mechanisms of Dioxins in Municipal Solid Waste Incineration Fly Ash: A Review. Journal of Cleaner Production 250: 119507.

Zhang, X., Wang, Q., Liu, Y., Wu, J. and Yu, M. 2011. Application of Multivariate Statistical Techniques in the Assessment of Water Quality in the Southwest New Territories and Kowloon, Hong Kong. Environmental Monitoring and Assessment 173(1–4): 17–27.

Zhang, Y., Li, S., Wang, X. and Li, X. 2015. Coagulation Performance and Mechanism of Polyaluminum Ferric Chloride (PAFC) Coagulant Synthesized Using Blast Furnace Dust. Separation and Purification Technology 154: 345–350.

Zhang, Y., Wang, L., Chen, L., Ma, B., Zhang, Y., Ni, W., Tsang, D. C. W. 2021. Treatment of Municipal Solid Waste Incineration Fly Ash: State-of-the-Art Technologies and Future Perspectives. Journal of Hazardous Materials, 411, 125132.

Zheng, X., Ying, Z., Wang, B. and Chen, C. 2018. Hydrogen and Syngas Production from Municipal Solid Waste (MSW) Gasification via Reusing CO2. Applied Thermal Engineering, 144: 242–247.

Ziegler-Rodriguez, K., Margallo, M., Aldaco, R., Vázquez-Rowe, I. and Kahhat, R. 2019. Transitioning from Open Dumpsters to Landfilling in Peru: Environmental Benefits and Challenges from a Life-Cycle Perspective. Journal of Cleaner Production 229: 989–1003.

Zuazua-ros, A., Vidaurre-arbizu, M. and Silvia, P. 2021. From the Leather Industry to Building Sector : Exploration of Potential Applications of Discarded Solid Wastes. Journal of Cleaner Production 291: 125960.

12

PRODUCTION OF BIOFUELS FROM WASTE BIOMASS VIA FISCHER-TROPSCH SYNTHESIS

AN OVERVIEW

Cibele Halmenschlager, Felix Link* and *Natalia Montoya Sanchez*

1. INTRODUCTION

The production of biomass waste is significant. In 2018, 3.35 kg of agricultural waste were produced per day and per person, amounting to around 9 billion tons per year globally. This is about four times higher than the rate of municipal solid waste generation and makes agriculture the second largest waste producer after industry (Kaza et al. 2018). Some agricultural waste is useful as fertilizer, soil amendment, or animal feed. However, dumping, burning, or landfilling are often used to dispose of agricultural waste. Because of its biogenic nature, the CO_2 released during combustion and decomposition is not accounted for in global warming potential (GWP) calculations. Nevertheless, the associated N_2O and CH_4 emissions contribute to global warming. In 2018, the decomposition and burning of crop residues, and the decomposition of animal manures led to N_2O and CH_4 emissions that are estimated to be equivalent to 1.04 Gt CO_2 emissions worldwide (Faostat 2021).

Controlled decomposition of agricultural waste is favorable to reduce these emissions. Anaerobic digestion is a commonly used technology for agricultural waste management, which not only reduces emissions associated with decomposition but also produces methane. Methane as a product can only be used locally for power and heat generation or requires a developed gas grid for transportation. In addition, lignin-rich feedstocks, like straw or wood, are not easily processed via anaerobic digestion (Fernandes et al. 2009). An alternative thermochemical approach for the management of

Department of Chemical and Materials Engineering, University of Alberta, 9211-116th Street, Edmonton, Alberta, T6G 1H9, Canada.

* Corresponding author: melohalm@ualberta.ca

agricultural waste is the combination of gasification and Fischer-Tropsch (F-T) synthesis to produce liquid hydrocarbons that can be used as "drop in" transportation fuels with a low carbon footprint (Ail and Dasappa 2016). This pathway is attractive because it converts the readily available biomass waste feedstocks into valuable products. Furthermore, it is aligned with current climate and political considerations, which increasingly support the production of transportation fuels from renewable sources.

The production of transportation fuels from biomass and biomass waste feedstocks via F-T synthesis involves three process steps: (i) conversion of the feed material to synthesis gas (syngas), (ii) conversion of syngas to synthetic crude oil (syncrude), and (iii) refining of syncrude to transportation fuels. This chapter presents a technical overview of these process steps. In this way, the requirements, reaction chemistry, and available technologies for each conversion are discussed. Furthermore, the status of the technology and the challenges for its implementation in terms of economic, political, and environmental considerations are also presented.

2. FEEDSTOCK AND PRETREATMENT

2.1 Waste biomass for syngas production

Possible feedstocks for biomass-to-liquid (BTL) processes are diverse. Biomass refers to any organic material of plant or animal origin and includes agricultural crop, forestry, and wood processing residues, algae, industrial and municipal solid waste (MSW). Biomass is the third largest source of energy behind coal and petroleum-derived fuels (Tumuluru et al. 2011). The amount of biomass available in the world was estimated to be about 550 Gt of carbon in 2018, where 80% come from plants (Bar-On et al. 2018).

Agricultural waste comprises field and process residue. Field residues include corn stover, wheat straw, oat straw, barley straw, sorghum stubble, rice straw, animal waste, and grass (Sadh and Duhan 2018, Widjaya et al. 2018). Process residue refers to the unused part after the primary biomass was transformed into more valuable products. These residues include molasses, husks, bagasse, seeds, stem, straw, stalk, shell, pulp, peel, among others (Sadh and Duhan 2018). On the other hand, forestry residues correspond to the material that is left from logging timber such as limbs and tops or the whole-tree harvest, e.g., unmerchantable tree like dead or poorly formed tree (*Source: US Department of Energy*). Wood processing residue includes unused sawdust, bark, branches, leaves, and needles.

Municipal waste (MSW) includes commercial and residential garbage such as yard trimming, paper and paperboard, plastic, textiles and food waste. It can also include wet waste like food, sewage sludge, manure, and organic waste from industrial operations. Contamination of these materials must be considered during processing. For example, while agricultural waste tends to have high levels of nitrates, the waste produced by the construction industry might be contaminated with protective agents, bonding primers, paints, among others (Moskalik et al. 2012). Sewage waste very often has a high concentration of nitrogen, phosphorous, and heavy metals (Adhikari and Chakraborty 2018).

Although carbon, hydrogen, and oxygen typically make up for most of the biomass composition, other elements such as nitrogen, sulfur, and chlorine are also found in smaller amounts. Trace components such as phosphorous, chlorine, and heavy metals vary among different biomass feedstocks. The diversity between biomass feedstocks leads to variation in the elemental composition, making the processing of this type of material more challenging than traditional feedstocks such as coal or petroleum. Table 1 shows the proximate and elemental analysis of different biomass feedstocks. The heterogeneity of these materials often creates the need for a pretreatment before conversion to syngas.

Table 1. Average chemical composition of different biomass feedstocks. Data from Vassilev et al. 2010.

Feedstock	Moisture content	Volatile matter [a]	Fixed carbon [a]	Ash content [a]	Elemental analysis [a]				
					C	O	H	N	S
	(wt%)	(wt%)	(wt%)	(wt%)	(wt%)	(wt%)	(wt%)	(wt%)	(wt%)
Wood biomass	33.8	77.9	19.3	8.3	52.9	38.7	7.8	0.4	0.2
Agricultural waste	26.2	72.4	25.2	9.7	50.3	41.6	6.2	1.75	0.3
Animal waste	5.9	55.5	13.6	30.9	58.9	23.1	7.4	9.2	1.5
Sewage sludge	6.4	48	5.7	46.3	50.9	33.4	7.3	6.1	2.3

[a] - on a dry basis.

2.2 Pretreatment of waste biomass for syngas production

Biomass is a low energy density, bulky material with a low heating value. In contrast to fossil fuels, it is unlikely that biomass will be collected from a single concentrated location. Therefore, the logistical network from feedstock site to the processing plant should be well planned and economically accounted for. Size, moisture content, and density are important parameters of biomass feedstocks that affect syngas production. Thus, depending on the type of reactor and its feedstock specifications, biomass needs to be screened, crushed or milled, and dried (Basu 2018). Furthermore, conversion to other more energy dense feedstocks such as bio-oils might be necessary. In this way, biomass pretreatment for syngas production includes physical processes such as densification, size reduction, and drying, as well as thermochemical processes such as pyrolysis, hydrothermal liquefaction (HTL), and torrefaction.

2.3 Physical processes

The low density of biomass poses a challenge to transportation, handling, storage, and processing. This problem can be addressed by densification. Mechanical densification consists in applying pressure to physically densify the material. The most popular mechanical densification processes are bailing, pelletizing, and briquetting. Bales are commonly used to densify the biomass in the field. They can be square, rectangular or round depending on the equipment used (Darr and Shah 2012). Round bales range from 1.2 m x 1.5 m to 1.5 m x 1.5 m while rectangular bales measure 0.9 m x 0.9 m x 1.8 m (*Source: Ontario- Ministry of Agriculture, Food and Rural Affairs*). Pellets are formed using a piston press. The standard shape of a biomass pellet is cylindrical, having a length smaller than 38 mm and diameter around 7 mm (*Source: Ontario- Ministry of Agriculture, Food and Rural Affairs*). Briquettes are similar to pellets but they differ in size. Briquettes are usually cylindrical with a diameter and length ranging from 25 to 100 mm and 10 to 400 mm, respectively (Kpalo et al. 2020). As the size of mechanically densified biomass is typically too large for gasification, a size reduction step is required at the gasification site.

Size reduction involves grinding, milling, or chipping and it is an energy intensive process. For instance, a hammer mill can use from 5 to 60 kWh t^{-1} (Tumuluru et al. 2014). Depending on the syngas production reactor technology (Section 3), biomass needs to be crushed to a specific particle size. A fixed-bed gasifier can accept few tens of millimeter-sized particles, while an entrained flow gasifier can accept only micron-sized biomass (Basu 2018). Grinding of biomass usually takes place in two steps. The first grinding breaks the wood or agricultural bales, pellets or briquettes into smaller size material to better flow into the dryers. The second step grinding further breaks the biomass further into smaller pieces (Tumuluru et al. 2019). Figure 1 shows the pretreatment steps required before the biomass is suitable for syngas production (gasification).

Moisture content is a parameter that needs to meet the reactor specifications for syngas production and can vary greatly depending on the biomass source (Table 1). A high biomass moisture content will increase the consumption of energy and oxygen in a gasifier in order to maintain reaction temperature (Luk et al. 2013). Reducing the moisture content is a very energy intensive process as each kilogram of moisture requires around 2300 kJ for vaporization (Basu 2018, Widjaya et al. 2018). Biomass can be dried using direct or indirect heat. In direct heat dryers, the biomass gets direct contact with hot fluid, either hot air or hot steam, while in indirect heat driers, the feedstock is kept separated by a heat exchanger surface from the fluid providing heat (Roos 2008).

2.4 Thermochemical processes

The aim of thermochemical processes is to transform biomass feedstocks to a homogeneous, more energy dense product with a lower water content. In the case of pyrolysis and HTL, this product is a viscous bio-oil that is easier and more economic to transport. In the case of torrefaction, the product is a solid similar to coal.

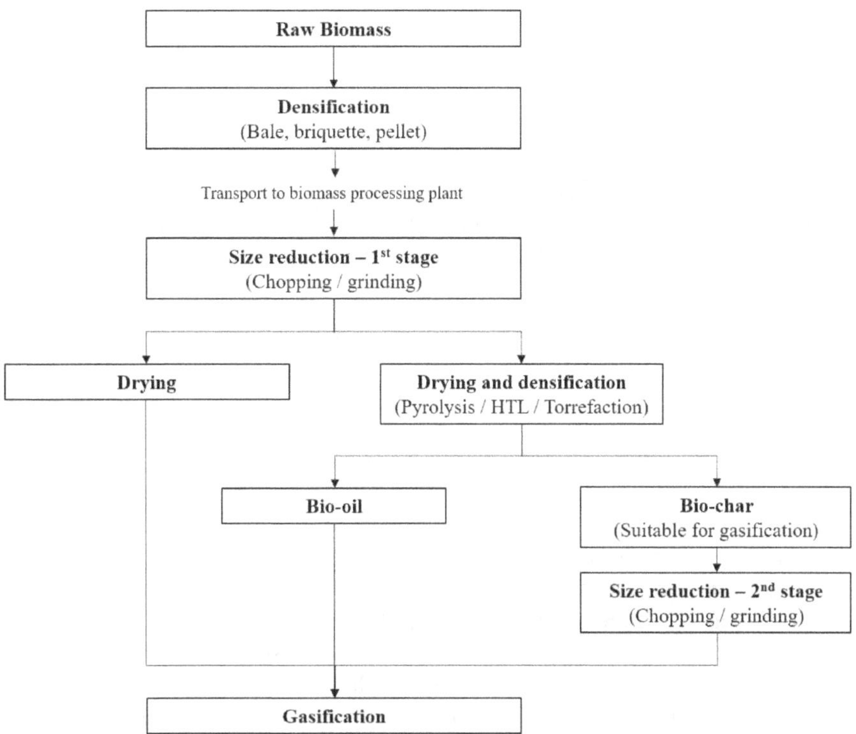

Figure 1. Simplified flow diagram of biomass pretreatment for syngas production.

Biomass pyrolysis is mostly used for conversion of lignocellulosic biomass (wood and agricultural waste) (Wang et al. 2020) and refers to thermal decomposition in the absence of oxygen. Thermal decomposition of biomass starts between 350°C and 550°C and continues up to 800°C. The long chain molecules in the biomass break down to smaller molecules to form gases, condensable vapors, and solid (biochar). A simplified diagram of the process is shown in Fig. 2. The products of biomass pyrolysis include pyrolytic gas (CH_4, H_2, CO, CO_2, C_2H_2, C_2H_4, C_2H_6), solid (char or carbon), and liquid (tars, heavier hydrocarbons, and water) (Basu 2018). The rate and the extent of solid, liquid, and gas formation depend on biomass heating rate, pressure, reactor configuration, feedstock composition, etc. Depending on the heating rate and temperature, pyrolysis can be classified into slow pyrolysis and fast pyrolysis. Slow pyrolysis favors the production of biochar and gas. Fast pyrolysis favors production of bio-oil (yield on a dry basis is up to 75 wt%) (Bridgewater 2004). Comprehensive reviews on biomass pyrolysis can be found in (Wang et al. 2020, Hu and Gholizadeh 2019, Kan et al. 2016).

Hydrothermal liquefaction enables the use of wet biomass and uses the moisture from the feed as a conversion medium. The feedstock is added to a pressurized reactor (5 to 28 MPa) (Sahu et al. 2020) along with inert gases (N_2) or reducing gases (H_2 or CO). The conversion at intermediate temperatures (250–375°C) produces bio-oil. The reaction pathway of HTL consists of three major steps: (i) depolymerization; (ii) decomposition; and (iii) recombination (Gollakota et al. 2018). Depolymerization reduces the chain length of the lignocellulose molecules in the feedstock. It mimics the natural geological process of producing fossil fuels. Decomposition involves decarboxylation (loss of water and CO_2) and deamination (removal of amino acids). The recombination process is the reverse of the first step and can lead to char formation. The economics of the HTL process are still questionable due to its high-pressure requirements (Gollakota et al. 2018). Detailed studies on HTL of biomass can be found in (Gollakota et al. 2018, Dimitriadis and Bezergianni 2017).

Torrefaction is a process that can homogenize the biomass by producing a stable, energy-dense, solid fuel (Yan et al. 2009). Torrefaction increases the energy density, water resistance, and

Gas cleaning and treatment

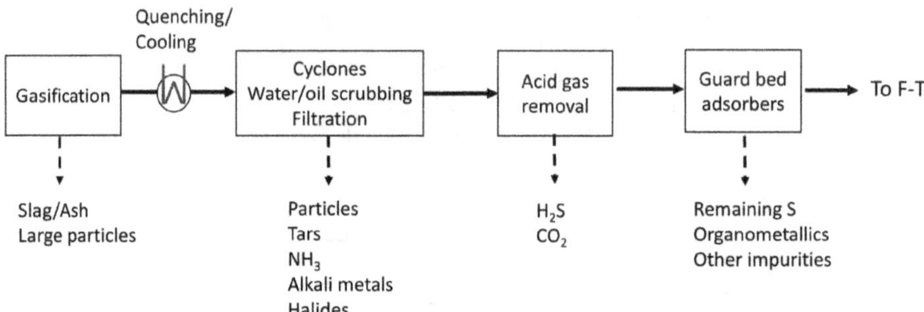

Figure 2. Generic scheme of the removal of contaminants from syngas. The conditioning of the syngas via water-gas shift reaction is omitted here and is discussed in Section 4.2.

grindability, which ultimately makes the transport and handling easier and more economical (Acharya et al. 2012). It also reduces the moisture content, increases the heating value and lowers the O/C ratio of biomass, which makes it more efficient for applications, including gasification and combustion. Torrefaction consists of slowly heating biomass in an inert atmosphere, under atmospheric pressure to a maximum temperature of 300°C. In torrefaction, the initial heating of biomass removes free water and further heating results in removal of bound water through chemical reactions (Couhert et al. 2009). A torrefied solid typically contains up to 90% of the initial energy content and only 70% of its initial weight (Zhiwen et al. 2018).

Dry torrefaction (or low temperature pyrolysis) involves temperatures between 200–300°C. This process results in two products: gas and char. The char has 70–90% of the energy content of the original biomass, and about 60–80% of its mass. The gases make up for the rest of the mass balance (Stelt et al. 2011). In wet torrefaction (or hydrothermal treatment), the biomass is treated in hot compressed water, resulting in three products including gases, aqueous chemicals, and a solid fuel. The temperature is in the range of 200–260°C, and pressures are up to 4.8 MPa. The solid product contains 80–95% of the energy content of the initial biomass and about 55–90% of its mass. The gas product makes up about 10% of the feedstock mass and the water, which contains mainly sugars, makes up for the rest of the mass balance. This process is advantageous when the biomass has a high moisture content (Bach and Tran 2015).

3. SYNGAS PRODUCTION

Conversion of biomass-derived feedstocks to syngas, which is a mixture of carbon monoxide (CO) and hydrogen (H_2), is not only an energy-intensive operation but also the most expensive step in BTL processes. The main technologies for syngas production are gasification and reforming (Maitlis and de Klerk 2013).

3.1 Gasification

Gasification refers to the high-temperature conversion of solid or liquid carbon-based feedstocks into a gaseous product, with a usable heating value, in an oxygen-deficient environment (Maitlis and de Klerk 2013). Gasification is an intermediate process between pyrolysis and combustion. Consequently, both pyrolysis and combustion reactions are part of the overall gasification chemistry (Table 2). In a typical process, feedstock that has been previously conditioned is

Table 2. Typical gasification reactions for biomass feedstocks.

Eq.	Reaction types	Enthalpy of reaction[a] (kJ·mol^{-1})
	Heterogeneous reactions	
	Pyrolysis	
(1)	$C_xH_yO_z \rightarrow aCO_2 + bH_2O + cCH_4 + dCO + eH_2 + char + tar$	b
	Gasification	
(2)	$C + CO_2 \leftrightarrow 2CO$ (Boudouard)	+172
(3)	$C + H_2O \leftrightarrow CO + H_2$ (Water-gas)	+131
(4)	$C + 2H_2 \leftrightarrow CH_4$ (Hydrogasification)	–74.8
	Combustion	
(5)	$C + O_2 \rightarrow CO_2$	–394
(6)	$C + \frac{1}{2} O_2 \rightarrow CO$	–111
	Homogeneous reactions	
(7)	$CO + H_2O \leftrightarrow CO_2 + H_2$ (Water-gas shift)	–41.2
(8)	$2CO + 2H_2 \rightarrow CH_4 + CO_2$ (Methanation)	–247
(9)	$CH_4 + H_2O \leftrightarrow CO + 3H_2$ (Steam reforming of methane)	+172.6
(10)	$CH_4 + 0.5O_2 \rightarrow CO + 2H_2$ (Partial oxidation of methane)	–36
(11)	$CH_4 + CO_2 \leftrightarrow 2CO + 2H_2$ (Dry reforming of methane)	–74.9
(12)	$C_nH_m + 2_{nH}2O \rightarrow \left(2n + \frac{m}{2}\right) H_2 + nCO_2$ (Steam reforming of tar)	

[a] ΔHr0 at 298 K
[b] Exothermic reaction

fed to a gasifier, where the remaining moisture is evaporated. As the temperature increases (200–500°C), thermal degradation or pyrolysis takes place, and gaseous products, char, and ash are produced (Eq. 1). Gaseous products include carbon monoxide, carbon dioxide, methane, hydrogen, water vapor, and tars. Tars are major primary products during pyrolysis and contain oxygenates and condensable hydrocarbons. Further increase in temperature (500–900°C) leads to different homogeneous reactions among pyrolysis products, as well as with the gasifying agent (Eq. 7–12). These homogeneous reactions are much more rapid than heterogeneous char gasification (Eq. 2–4), which takes place at even higher temperatures (> 900°C) and is the rate-controlling step of the overall gasification process. Both types of reactions contribute to the final gasification products, which include CO, H_2 (desired products), CO_2, H_2O (combustions products), and H_2S, NH_3, ash/slag, unburnt char (undesired by-products) (Speight 2020). Due to the endothermic nature of drying, pyrolysis, and most gasification reactions, most commercial gasifiers allow a certain amount of feed combustion to supply the heat required for reactions to occur. In this way, combustion and partial oxidation (Eq. 5–6) take place along with char gasification. A comprehensive discussion on gasification chemistry and reactor technology can be found in (Zhiwen et al. 2018, Higman and Burgt 2003).

Gasification of biomass and the corresponding gas product distribution are largely dependent on the gasifying agent. The use of oxygen, air, or steam is possible. Most biomass gasification processes use air instead of oxygen for economic reasons. The use of oxygen requires an air separation unit (ASU), which has a significant impact on the cost of the process (Maitlis and de Klerk 2013). Yet, depending on the final application, the use of oxygen is justified. Production of syngas for F-T synthesis benefits from the use of oxygen instead of air, as the syngas is not diluted with nitrogen. Having nitrogen would increase the size of the equipment downstream (increased volumetric flow), decrease the concentration of reactive species and would have repercussions in the gas recycling during syngas conversion.

Compared to coal, the additional complexity of biomass-derived feedstocks makes gasification more challenging. In this case, the compositional variation, lower energy density, and heterogeneous nature among possible feedstocks, as well as their impact on the existing gasification technology need to be addressed (Speight 2015). Irrespective of the feedstock, the selection of the gasification technology has to consider its compatibility with the feed as well as the syngas requirements for downstream operation.

According to the contacting mode and the temperature of reaction, gasifiers are classified in three main groups: fixed-bed or moving-bed, fluidized-bed, and entrained flow reactors. Table 3 presents some characteristics of these gasification technologies. Fixed-bed reactors operate at lower temperatures compared to other gasifiers, which leads to co-production of pyrolysis liquids. Fixed-bed gasification is typically used in power production at small-medium scales. Fluidized-bed gasifiers are known for their excellent mixing and uniform temperature distribution. They operate at medium temperatures, which significantly decreases production of pyrolysis liquids as part of the products. Fluidized-bed gasification falls in the middle range capacity and is popular for combined production of heat and power (CPH), as well as synthesis applications. Entrained flow gasifiers operate at high temperatures and produce a clean gas free of pyrolysis products, which simplifies the downstream gas cleaning steps. Entrained flow gasifiers require operation at large capacity and are intended for integrated combined cycle plants (IGCC) and synthesis applications (IEA 2019, Basu 2018, de Klerk 2011b).

Depending on the particular feedstock, gasifier type, gasifying agent, and operating conditions, gasification of biomass-derived feedstocks can produce a wide range of syngas compositions, with H_2/CO ratios varying between 0.5–1.2. Furthermore, any biomass-derived syngas contains contaminants like hydrogen sulfide, ammonia, dust, and alkalis. Therefore, cleaning and conditioning of syngas is required.

Table 3. Characteristics of different gasification technologies.

	Gasification Technology		
Parameters	**Fixed-bed**	**Fluidized-bed**	**Entrained flow**
Gas exit temperature (°C)	425–650	800–1000	1200–1600
Reaction zone temperature (°C)	-	800–1000	~ 1990
Feed particle size (mm)	6–50	< 10	< 0.1
Oxygen requirement	Low	Moderate	High
Pyrolysis products in gas	Yes	Possibly	Low
Nature of ash produced	Dry	Dry	Slag
Cold gas efficiency (%)	80	80	89
Application	Small scale	Medium Scale	Large Scale

3.2 Reforming

Conventionally, reforming applies to the conversion of natural gas into syngas (Maitlis and de Klerk 2013). In the context biomass-derived feedstocks, bio-gas would be the equivalent of natural gas. Bio-gas is produced during anaerobic digestion of biomass-derived feedstocks and consists of high concentrations of methane and carbon dioxide, and trace amounts of hydrogen sulfide. Over the past decades, reforming of biomass has drawn significant attention for production of hydrogen, rather than production of syngas for synthesis applications, e.g., F-T synthesis. This is mainly due to the non-centralized availability of bio-gas feedstocks that is required for economical operation of F-T processes. Nevertheless, bio-gas to liquid routes via F-T synthesis have been suggested and facilities that produce fuels from (fossil) natural gas via F-T are in operation (Arutynov et al. 2020,

Okeke and Mani 2017, de Klerk 2011b). The chemistry involved in reforming is similar to that of gasification. There are two major routes of reforming, namely steam or dry reforming and partial oxidation. Steam or dry reforming are endothermic catalytic processes that convert methane to syngas primarily through reactions given in Eq. 7, 9, and 11. Partial oxidation is an exothermic catalytic or non-catalytic process that mainly follows reaction given by Eq. 10. A comprehensive discussion can be found in (Rostrup-Nielsen and Christiansen 2011).

4. SYNGAS CLEANING AND CONDITIONING

4.1 Syngas cleaning

As discussed, the product gas from gasification of biomass consists mainly of CO, CO_2, H_2 and H_2O. However, particles and trace compounds like tars, acid gases, organic species and volatile organometallic species remain in the syngas. These compounds have the potential to deactivate the downstream F-T catalysts and therefore, need to be removed to match the specifications shown in Table 4. While some of the impurities are partly removed along with the condensed water during syngas cooling, the low concentrations of impurities required for F-T synthesis need to be achieved via a series of dedicated gas cleaning steps (Woolcock and Brown 2013). An overview of these steps is shown in Fig. 2.

Since coal gasification yields similar impurities as biomass gasification, commercially proven technologies are available for syngas cleaning. However, the higher content of alkali metal species and the possible presence of other components like phosphorous species in agricultural waste need to be considered in the design of the syngas cleaning process. An overview of the main syngas cleaning and conditioning steps is given in the following. More detailed descriptions of the processes are available in the literature (Acharya 2018, Abdoulmoumine et al. 2015, Woolcock and Brown 2013, Higman and Burgt 2003).

Table 4. Specifications for syngas to be used in F-T synthesis. Data from Goransson et al. 2011.

Trace compound group	Required concentration for F-T synthesis
Sulfur compounds (H_2S + COS + CS_2 + others)	< 1 ppmv
Nitrogen compounds (NH_3 + HCN + others)	< 1 ppmv
HF + HCl + HBr	< 10 ppbv
Alkali metals	< 10 ppbv
Tar	Below dewpoint
Particles	Essentially removed

4.1.1 Removal of particulates and tars

Particles such as ash and tars from incomplete gasification are entrained in the syngas stream exiting the gasifier. Their concentration and composition largely depend on the feedstock composition and gasification parameters. Typically, more than 99% of these compounds need to be removed for syngas applications such as F-T synthesis (Woolcock and Brown 2013). The available technologies can be grouped into hot ($T > 300°C$) cold ($T < 100°C$) and warm gas cleaning processes depending on the temperature level.

Hot gas cleaning processes for particle removal include cyclone separators as well as various filtration systems. Cyclones can only remove between 95 and 99% of the particles but have the advantage of low maintenance compared to filters. Candle filters can achieve higher separation efficiencies, e.g., 99.5% (Seville 1997). Tars can lead to clogging of filters and therefore, need to be removed upstream via catalytic or thermal tar reforming. Catalytically coated filter elements can combine tar and particle removal (Nacken et al. 2015, 2012).

Particulates can also be removed via wet scrubbing, which is classified as a cold gas cleaning process and has the disadvantage that the used scrubbing medium needs to be treated (Woolcock and Brown 2013). Traditionally, water has been used as scrubbing medium, but alkaline water, acidic water and oil scrubbing have also been applied (Chornet 2012, Zwart et al. 2009). NH_3, alkali metal and halide compounds that have not condensed out during syngas cooling are effectively removed by aqueous scrubbing along with particles (Woolcock and Brown 2013).

4.1.2 Bulk sulfur removal

Sulfur species in the syngas consist mainly of hydrogen sulfide (H_2S) with lower levels of carbonyl sulfide (COS) and carbon disulfide (CS_2). The removal of sulfur species from gas streams has been an issue in syngas applications, natural gas production and integrated gasification combined cycle power plants (Qayyum et al. 2020, Modal et al. 2011). Hence, a variety of commercially proven processes for this task is available. Cold gas cleaning processes based on physical or chemical absorption of sulfur species have been commercially applied to obtain the low limit for sulfur species in syngas for F-T synthesis (Table 4) (Dry et al. 1987).

Chemical absorption processes make use of reversible chemical reactions between acid gases and an absorber molecule while in physical absorption processes, the acid gas is dissolved in a solvent. For both processes, the absorption typically occurs in an absorber column at elevated pressure and relatively low temperature. The rich absorber solution containing the acid gas species is then fed to a second column where the acid gas is removed from the solvent at the opposite conditions, i.e., elevated temperature and a lower pressure. The lean absorber solution is recirculated to the absorber column. Commercial processes with different solvents and conditions differ in their maximum H_2S removal efficiency, their ability to effectively remove COS and CS_2, and their selectivity for H_2S removal over CO_2 removal. Typically, CO_2 removal is beneficial for downstream F-T synthesis as it reduces the total volume of gas flow. An overview of some commercial absorption processes is given in Table 5 (Edwards 1979).

Hot or warm gas cleaning methods for sulfur removal have been developed as well. These processes typically rely on the reversible adsorption of H_2S and acidic species on solid adsorbent materials (Cheah et al. 2009). While warm syngas desulfurization using metal oxide based materials has been demonstrated successfully in an IGCC power plant (RTI International 2018), it has not been used commercially in F-T syngas pretreatment. Independent of the removal method, the removed acid

Table 5. Properties of selected bulk acid gas absorption technologies. Data from Edwards 1979, Cheah 2009.

Absorber solution	Commercial name	Principle of absorption	Removal of H_2S to	Removal of organic sulfur to	Absorption of CO_2
Methanol	Rectisol	physical	0.02 ppm	COS < 0.1 ppm	< 1 ppmv
Propylene carbonate	Fluor solvent	physical	4 ppm	[a]	< 1%
N-methyl-2-pyrrolidone	Purisol	physical	4 ppm	[a]	< 1%
Dimethylether polyethylene glycol	Selexsol	physical	1 ppm	1 ppm	< 1%
Potassium carbonate	Benfield	chemical	4 ppm	[a]	< 1.5% (up to 0.8 in two stage)
Monoethanolamine	MEA	chemical	1 ppm	[a]	ppm level
Diethanolamine	DEA	chemical	4 ppm	[a]	ppm level

[a] Partial removal or complete removal of some organic sulfur species

gas stream needs to be treated in a sulfur recovery unit. Most commonly, sulfur species are converted to elemental sulfur either via gaseous (Claus process) or via liquid redox processes (Cover et al. 1985).

4.1.3 Removal of trace compounds

Depending on the origin of the biomass feedstock, trace compounds other than the ones listed in Table 4 can be present in the syngas and can be harmful to downstream processes. These include metal compounds, e.g., Hg, As, Se, Cd and Zn. In addition, Fe and Ni carbonyl species can also be present. For the removal of these compounds along with traces of remaining sulfur, nitrogen or halide species, the syngas is passed through adsorber beds. These contain adsorbent materials like activated carbon, metal oxides and zeolite materials, among others (Layne et al. 2007). The trace compounds remain on the adsorbent material and are removed from the syngas with high efficiency. If the maximum capacity of an adsorber bed is reached, the material in the bed needs to be regenerated or replaced.

4.2 Adjustment of H_2/CO ratio

While syngas cleaning aims at removing compounds different from CO, CO_2, H_2 and H_2O present in the syngas, the adjustment of the H_2/CO ratio is often required or is beneficial for high conversions during F-T synthesis. The syngas obtained from biomass gasification typically exhibits a H_2/CO ratio of 0.5–1.2 (Section 3). However, from the stoichiometry of the F-T reaction, it can be seen that a H_2/CO ratio of 2 is optimal for complete conversion (Eq. 12). Thus, the H_2/CO ratio in syngas can be adjusted via the water-gas shift (WGS) reaction, in which H_2 is formed at the expense of CO according to Eq. 7.

WGS is carried out at temperatures between 310–450°C over Fe_2O_3-Cr_2O_3 catalysts (high temperature WGS) or 200–250°C over CuO-ZnO/Al_2O_3 catalysts (low temperature WGS). The WGS reaction is an exothermic equilibrium reaction and H_2 formation is favored at low temperatures (Smirniotis and Gunugunuri 2015). Steam is added to the reactor to drive the equilibrium towards H_2 generation. While maximum hydrogen formation is obtained by in-series operation of high temperature and low temperature WGS reactors, the incorporation of the low temperature reactors is often not required to obtain a H_2/CO ratio of 2. Alternatively, WGS can be carried out over sulfide Co-Mo/Al_2O_3 catalysts, which tolerate percentage levels of sulfur species in the syngas while remaining active (sour WGS) (Ratnasamy and Wagner 2009). Sour WGS can be operated before the bulk acid gas removal. This has the additional advantage that the CO_2 formed during WGS is removed from the syngas in the bulk acid gas removal step.

5. FISCHER-TROPSCH SYNTHESIS

5.1 Fischer-tropsch product composition and distribution

Once syngas from biomass has been cleaned and conditioned, the next step in the BTL process is the conversion of syngas to syncrude. F-T synthesis, which is a well-known indirect liquefaction process, makes this conversion possible. It proceeds via a combined carbon monoxide polymerization and hydrogenation reaction of highly exothermic nature (heat of reaction is of the order of –165 $kJ.mol^{-1}$ of CO converted). Control of the reaction temperature requires efficient heat removal, which is typically achieved by steam generation (de Klerk 2011b).

The product from F-T synthesis consists of a mixture of hydrocarbons, oxygenates, and water. Alkanes (particularly, *n*-alkanes), alkenes (particularly, *n*-1-alkenes), and alcohols are the most abundant compounds classes in syncrude, and they are formed according to Eq. 12–14 (de Klerk 2020). Carboxylic acids and aldehydes are also obtained as primary products, although to a lesser extent.

Both the reaction network and mechanism of F-T synthesis are complex and have been extensively investigated. A comprehensive discussion on the topic can be found in (Davis 2001, Dry 1981).

$$\text{Alkanes: } n\,CO + (2n+1)\,H_2 \rightarrow C_nH_{2n+2} + n\,H_2O \qquad \text{Eq. (12)}$$

$$\text{Alkenes: } n\,CO + 2nH_2 \rightarrow C_nH_{2n} + n\,H_2\,O \qquad \text{Eq. (13)}$$

$$\text{Alcohols: } n\,CO + 2nH_2 \rightarrow C_nH_{2n} + 1\,OH + (n-1)\,H_2\,O \qquad \text{Eq. (14)}$$

For F-T applications targeting liquid fuels, the product distribution, particularly the carbon number distribution or the chain length of the hydrocarbon products, becomes an important part of the discussion. The probability of chain growth, also known as the α-value of the catalyst, determines the carbon number distribution of syncrude. The Anderson-Schulz-Flory (ASF) distribution (Eq. 15), which relates the carbon number (n) and the probability of chain growth (α) to the molar fraction of each carbon number in the product (x_n), is widely used to estimate the F-T product distribution (de Klerk 2011b). At constant conditions, the α-value of different catalysts indicates which one is more likely to promote chain propagation. For instance, an α-value close to the unity indicates favored production of long-chain hydrocarbons such as waxes (Speight 2015).

$$x_n = (1-\alpha)\cdot\alpha^{(n-1)} \qquad \text{Eq. (15)}$$

The composition of syncrude in terms of carbon number distribution and oxygenate content depends on the F-T synthesis conditions, e.g., temperature, pressure, and syngas composition (H_2/CO ratio), as well as on the catalyst formulation and reactor technology. Therefore, the development of an F-T synthesis process needs to consider the relationship between these variables to produce a syncrude of suitable composition that can be refined to desired products. Based on the operating conditions, there are two types of industrial F-T processes: high-temperature Fischer-Tropsch (HTFT) synthesis, which employs Fe-based catalysts and operates in the temperature range of 300–350°C, and low-temperature Fischer-Tropsch (LTFT) synthesis, which employs Fe- or Co-based catalysts and operates in the temperature range of 200–240°C (de Klerk 2016). HTFT synthesis favors production of lighter hydrocarbons and alkenes, while LTFT synthesis gives a product richer in waxy hydrocarbons (Table 5) (Choudhury 2015). In the case of HTFT synthesis, syncrude is composed of three different product phases, namely, gaseous, organic liquid, and aqueous. A fourth phase (organic solids or waxes) is observed in syncrude from LTFT synthesis (de Klerk 2011b).

The different nature of Fe- and Co-based catalysts has an effect on the syncrude composition and process operation (Table 6). Fe-based catalysts are less active towards hydrogenation, producing a syncrude with a higher content of alkenes and oxygenates. In addition, alkali compounds in Fe-based catalysts are used to promote CO adsorption, which favors chain growth probability. Fe-based catalysts are active towards the WGS reaction, while Co-based catalysts have no significant activity in this respect. This difference allows Fe-based catalysts to convert a wider range of syngas compositions, with H_2/CO ratios < 2, into syncrude (Zhang et al. 2010). In other words, while the H_2/CO ratio needs

Table 6. Typical sync rude composition for low and high temperature F-T synthesis based on compound classes. Data from de Klerk 2016.

	Syncrude composition by compound classes (wt%)	
Carbon number distribution	**LTFT** C_1–C_{120}	**HTFT** C_1–C_{30}
Alkanes	> 70	20–30
Cycloalkanes	< 1	< 1
Alkenes	15–20	> 50
Aromatics	< 1	1–5
Oxygenates	~ 5	10–15

to be adjusted strictly to a value of 2 via WGS for Co-based catalysts, Fe-based catalysts offer some flexibility in terms of the H_2/CO ratio of the feed due to their "intrinsic" WGS activity. For HTFT, Fe-based catalysts produce a syncrude with a more significant aromatic content. F-T reaction rates using Fe-based catalysts decrease strongly in the presence of water vapor (de Klerk 2011b). Fe-based catalysts can operate over a wider temperature range than the Co-based ones (Rahimpour and Elekaei 2009), which enables better opportunities for electricity co-generation from reaction heat.

5.2 Fischer-tropsch reactor types

Since F-T synthesis is highly exothermic and the product selectivity is sensitive to the reaction temperature, reactors technology is strongly determined by heat management requirements (Matilis and de Klerk 2013). Furthermore, selection of a reactor type needs to consider other aspects such as syngas composition, number of phases present at reaction conditions, strategy for catalysts deactivation and replacement, process robustness, and turndown ratio. A detailed discussion on fundamental aspects as well as a critical comparison of reactor types for F-T synthesis is found in (Matilis and de Klerk 2013, Sie and Krishna 1999, De Swart et al. 1997).

The three main types of reactors employed for F-T synthesis are fixed-bed, fluidized-bed, and slurry-bed. Fixed-bed reactors approximate plug flow reactors (PFRs), while fluidized-bed and slurry-bed are better described as continuous stirred tank reactors (CSRTs). The selection of a reactor type has consequences for both the F-T synthesis and the downstream refining operations (Table 7) (de Klerk 2011b). For instance, the reaction phase does not limit the α-value of the catalyst in fixed-bed reactors but it does in the case of fluidized-bed and slurry-bed reactors. At typical operating conditions (2.5 MPa and > 320°C), fluidized-bed reactors involve reaction in the gas phase; thus, the α-value of the catalyst needs to be small (0.7 or lower) to avoid liquid formation (de Klerk 2011b). Slurry-bed operation, on the contrary, involves formation of liquid product and requires a higher α-value of the catalyst.

Other aspects such as catalyst-product separation, catalyst mechanical strength requirements, and catalyst replacement strategy also vary according to the reactor type (Table 7) (Matilis and de Klerk 2013). In fixed-bed reactors, the catalyst remains in the bed facilitating the product separation. For this type of reactor, the crushing strength of the catalyst particles is particularly important during catalyst loading. Furthermore, because loading and unloading of the catalyst is a time-consuming task, fixed-bed reactors that employ catalysts that deactivate slowly or can be regenerated *in-situ* are preferred. Multi-tubular fixed-bed technology is the preferred one for industrial LTFT synthesis (Jess and Kern 2009). Product separation in fluidized-bed and slurry-bed reactors increases in complexity as gas-solid and liquid-solid separations, respectively, are necessary. In fluidized-bed reactors, separation of the catalyst from the gaseous stream is achieved using cyclones. In slurry-bed reactors, separation of catalyst particles from the heavy waxes that remain in the slurry requires an

Table 7. Typical sync rude composition for low and high temperature F-T synthesis using Fe- and Co-based catalysts. Data from de Klerk 2020.

		Syncrude composition (wt%)		
Product fraction	**Carbon number range**	**Fe-LTFT**	**Co-LTFT**	**Fe-HTFT**
Tail gas	C_1-C_2	6	7	23
Liquid petroleum gas	C_3-C_4	8	5	24
Naphtha	C_5-C_{10}	12	20	33
Distillate	C_1-C_{22}	20	22	7
Paraffin wax	C_{22+}	50	44	-
Aromatics	-	-	-	3
Oxygenates	C_1-C_5	4	2	10

in-situ filtration (Matilis and de Klerk 2013). The design of this liquid-solid separation is one of the main issues of the slurry-bed technology. Catalysts resistant to attrition that can handle the constant movement and impact during fluidized-bed and slurry-bed operation are required. In both cases, on-line catalyst replacement is possible (de Klerk 2011b). Fluidized-bed reactors are only used for HTFT synthesis (Matilis and de Klerk 2013).

In F-T synthesis, the CO conversion affects both the reactor and gas loop design. CO conversion is never complete, and so reactor configuration usually entails some recycle. Furthermore, F-T reactors can be configured to operate in series and/or parallel. Reactors in series can use intermediate product separation and conditioning to manipulate the product selectivity. For fluidized- and slurry-bed reactors, which approximate CSTR behavior, in series configuration can improve productivity (de Klerk 2011b). Operation in series can also reduce catalyst deactivation by too high water partial pressure and can increase single-pass conversion.

6. REFINING OF FISCHER-TROPSCH SYNCRUDE

6.1 Property requirements for synthetic fuel products

Liquid fuel products are intended for combustion in an engine or a turbine. Their properties affect the combustion characteristics and have been regulated to ensure reliable and stable operation of the engine or turbine and auxiliary systems in the vehicle, as well as the transport and storage stability of the fuel. The requirements for fuel products are summarized in fuel specifications for motor gasoline, jet fuel and diesel. These standards are specific to certain markets and regions and not all of the specifications listed in the standards are critical for the use of F-T products as liquid fuels. In the subsections 6.1.1–6.1.3 selected fuel specifications are discussed with a focus on the specifications that impose the need for refining of F-T syncrude.

6.1.1 Motor gasoline

Motor gasoline is a complex mixture of hydrocarbons in the C_4–C_{10} range with a maximum final boiling point of 210°C. The most prominent regulated specification for motor gasoline is the octane number. It quantifies the resistance of a fuel against auto-ignition and is determined by the molecular structure. It decreases with increasing molecular weight (hydrocarbon chain length) and generally increases in the order of *n*-alkanes <*n*-alkenes < iso-alkanes and iso-alkenes < aromatics (Lovell 1958). Cycloalkanes exhibit similar octane numbers as the corresponding alkenes obtained by ring-opening. Since auto-ignition is detrimental to the engine stability, fuel specifications set minimum limits for the octane number, e.g., RON > 95 in the European gasoline standard EN 228 (*Source: European Committee for Standardization (CEN), 2013*). Products from F-T synthesis consist mainly of *n*-alkanes and *n*-1-alkenes, both of which have low octane numbers. For blending of F-T products to motor gasoline, the content of these needs to be reduced significantly via refining processes (de Klerk 2011b). Other specifications relevant for the refining of F-T products in motor gasoline are the alkene and aromatic (benzene) content. The maximum limits for these compounds must also be considered during the refining process.

6.1.2 Jet-Fuel

Jet-fuel consists of a mixture of hydrocarbons in the approximate range C_9–C_{16}, with a maximum final boiling point of 300°C. The specifications for civilian jet-fuel (Jet A-1) are internationally harmonized and there are specific standards for jet-fuel containing synthetic hydrocarbons (ASTM D7566) (ASTM 2020). As of today, the use of synthetic hydrocarbons is mostly limited to its use as blend (up to 50%) with petroleum-derived kerosene.

One of the most critical specifications for jet-fuel from F-T synthesis is the density, which is limited to 775–840 kg· m^{-3}. The density of alkanes and alkenes, which are the major constituents of F-T syncrude, is lower than the specification limit in the kerosene range (Jurgens et al. 2019). An increase of the density can be obtained by increasing the mass fraction of cycloalkanes and aromatics via refining processes. While this would be a requirement for the use of F-T products as 100% synthetic jet fuels, it can also be achieved by blending with high density petroleum-derived kerosene.

Another specification that leads to the need for refining of F-T products for the use of jet-fuel is the freezing point. In-flight, freezing of the fuel could lead to severe consequences and therefore the maximum freezing point for Jet-A1 fuel is limited to –47°C. Kerosene range material from F-T synthesis has a high amount of linear hydrocarbons, which exhibit relatively high freezing points. Hence, the amount of isomerized material needs to be increased during refining to satisfy the freezing point specification (de Klerk 2011a).

Due to a change of the swelling behavior of seals in the fuel system in contact with low aromatic fuels, there is a specification for a minimum 8 vol.-% aromatics in the synthetic jet-fuel (ASTM 2020). Particularly, LTFT syncrude contains no aromatics and therefore the aromatics content in the final fuel needs to be adjusted either by blending with petroleum-derived kerosene or refining techniques if the syncrude is intended to be used as 100% synthetic jet-fuel.

6.1.3 Diesel fuel

Diesel fuels, which are hydrocarbon mixtures in the C_{11}–C_{22} range, exhibit the highest maximum boiling point of the three main transportation fuel types (up to ca. 370°C). For diesel fuels, the combustion behavior is determined by the compression ignition delay, which is quantified by the cetane number. The quality of the diesel increases with its cetane number and minimum specifications for cetane numbers are defined in the specifications for diesel fuels (*Source: European Committee for Standardization (CEN), 2013*). Components that have a high octane number typically have a low cetane number and vice versa. Therefore, F-T syncrude in the diesel boiling range exhibits high cetane numbers and the limit set in the specifications is not critical for F-T diesel. While the large amount of linear hydrocarbons in F-T syncrude is beneficial in terms of cetane number, these compounds lead to a high freezing point. To meet the specifications for diesel fuels, diesel range F-T syncrude needs to be refined to increase the amount of isomerized species at the expense of linear species. Isomerized species have a lower freezing point. However, isomerized compounds negatively affect the cetane number. Therefore, the degree of isomerization needs to be limited, which needs to be considered in the refinery design (de Klerk 2011a). Similar to jet-fuel, diesel has a minimum density requirement that is not easily met by refining of F-T syncrude. The density can be increased by refining, however, at the expense of distillate yield or cetane number. The better option to achieve the density specification is blending with petroleum-derived diesel (de Klerk 2020, 2011a).

6.2 Refining of fischer-tropsch products

F-T products yield gaseous and liquid hydrocarbons that can most often not be used directly as fuels or blending components for fuels. Therefore, refining of the products is required. The main technical objectives of the refining process for fuels from F-T syncrude are:

(i) *To increase the yield of liquid fuel products.* If the gaseous F-T products cannot be used in other nearby chemical processes, e.g., polyethylene production from the ethene produced during F-T, their use is limited to fuel gas. Transformation of some of these gaseous products to liquid fuel products can therefore be beneficial. Similarly, if there is no market for the heavy wax fraction of the liquid F-T product, refining steps to form liquid fuel from these waxes can be incorporated.

(ii) *To improve the fuel properties.* As discussed in Section 6.1, refining is required to adjust the properties of the liquid fuel products to satisfy the specifications that are in place for motor gasoline, jet-fuel, and diesel.

The technology for refining of petroleum products is commercially proven and available. However, using a refining approach based on petroleum refining is difficult because the composition of F-T syncrude is different from that of petroleum crude. Compared to petroleum crude, F-T syncrude has (i) percentage levels of oxygenates; (ii) high amounts of alkenes; (iii) high amounts of linear hydrocarbons; (iv) almost no sulfur contamination (de Klerk 2020).

6.2.1 Motor gasoline

The production of on-specification motor gasoline from F-T products has been carried out commercially in gas-to-liquid (GTL) and coal-to-liquid (CTL) processes (de Klerk 2011b). Some properties of the motor gasoline products from F-T refining are shown in Table 8. It is worth mentioning that all commercial F-T based motor gasolines are products from co-refining coal tar or natural gas condensates along with F-T products. This is mainly due to economic reasons; however, concepts for motor gasoline refining purely from F-T products have been established (de Klerk 2011b). These rely on the increase of naphtha range products from oligomerization, the increase of the octane number of naphtha range products via isomerization or hydroisomerization as well as the production of high octane number aromatics from aromatization and naphtha reforming (Halmenschlager et al. 2016, Guo et al. 2011, Dancuart et al. 2004, Tamm et al. 1988). HTFT produces a higher mass fraction of naphtha range molecules and very little waxes (see Section 5.1). Therefore, it has been the primary

Table 8. Characteristics of main reactor types employed in F-T synthesis. Data from de Klerk 2011b.

	Fixed-bed	Fluidized-bed	Slurry-bed
Nature of the reactor	PFR	CSTR	CSTR
Reaction phase	Gas or gas + liquid	Gas	Gas + liquid
Catalyst particle size (mm)	> 2	< 0.1	< 0.1
Mass transfer limitation	High	Medium-low	Medium
Heat transfer limitation	High	Medium-low	Low
Catalyst-product separation	Easy	Fairly easy	Challenging
Catalyst strength requirement	Low [a]	High [b]	Medium [b]
On-line catalyst replacement	Not possible	Possible	Possible
Scale-up risk (lab to plant)	Low	Medium	Medium

[a] Catalyst needs to avoid crushing.
[b] Catalyst needs to be resistant towards attrition.

Table 9. Selected properties of refined Fe-HTFT motor gasoline compared to current EN 228 specifications. Data from CEN 2012, de Klerk 2011b, 2020.

Fuel Property	Specification EN 228	Unrefined straight-run Fe-HTFT naphtha	Refined Fe-HTFT CTL gasoline (Sasol Synfuels)	Refined Fe-HTFT GTL gasoline (PetroSA)
Density/kg m^{-3}	720–775	-	729	748
Boiling range/°C	< 210	35–204[a)]	35–200	35–195
RON	95 (min)	68	93	95
MON	85 (min)	62	83	85

C_5-400 °F

technology used commercially for motor gasoline production via F-T (Dancuart et al. 2004). If LTFT is chosen as F-T pathway, fluid catalytic cracking or hydrocracking are employed to produce lower boiling point products from waxes and heavy distillates (de Klerk 2011b). These are then employed as feedstocks for the above-mentioned refining pathways.

6.2.2 Jet fuel

ASTM approved fully synthetic jet fuel has been produced from refined F-T product blended with coal tar in the commercial scale. The refining product used for jet fuel bending is isoparaffinic kerosene which is obtained from hydroisomerization and hydrocracking of distillate and wax products from F-T. The refining process and fuel properties are explained in (de Klerk 2016). The aromatics and cycloalkanes in coal tar are required to meet density and aromatics specifications (see Section 6.1.2). If a biomass-based upgraded pyrolysis product or a different biomass-based high density cycloalkane and aromatics-rich stream could be used instead of coal-tar, a fully renewable jet-fuel production would be possible from biomass derived feedstocks. Such an aromatics-rich kerosene range blending feedstock could also be obtained from F-T gas and naphtha, e.g., by aromatization and aromatic alkylation (de Klerk 2016, Sakuneka et al. 2008).

6.2.3 Diesel fuel

Diesel fuels that fulfill specifications have only been produced commercially in co-refining scenarios with coal tar and natural gas liquids (see Table 10) (Leckel 2009). The reason is that on-specification diesel fuel strictly from F-T products can only be obtained with significant yield loss (see Section 6.1.3) (de Klerk 2009). Therefore, other large-scale commercial efforts are aiming for a jet-fuel and diesel blendable feedstock from natural gas via F-T synthesis rather than a final fuel (de Klerk 2011b). LTFT synthesis is chosen due to its high liquid and waxy product yield. Essentially all liquid and waxy F-T product is hydrocracked and hydroisomerized in one reaction step (Bouchy et al. 2009, Sie et al. 1991,Eilers et al. 1990). An oxygenate free distillate product is formed during hydrocracking with a selectivity of ca. 80%. It consists of branched alkanes and therefore has good freezing and cold flow properties while retaining a high cetane number (see Table 10). Mainly due to its low density, the blending feedstock needs to be blended with high density material from other sources to fulfill final diesel specifications. This concept has been suggested in use with biomass feedstocks such as agricultural waste for gasification and F-T (Viguie et al. 2013).

Table 10. Selected properties of commercial Fe-HTFT-based diesel and Cu-LTFT-based diesel blending feedstock compared to current EN 590 specifications. Data from CEN 2013, Leckel 2009, de Klerk 2011b.

Fuel Property	Specification EN 590	Refined Fe-HTFT diesel (CTL, Sasol)[a]	Cu-LTFT based hydrocracked diesel blending feedstock (GTL, Shell SMDS)
Density ($kg \cdot m^{-3}$)	820–845	829	776
Boiling range (°C)	-[b)]	178–374	184–357
Sulfur content ($mg\ kg^{-1}$)	< 10	< 1	< 2
Cetane number	51	51	81

[a] Contains 25% coal-tar derived distillate along with refined HTFT distillate

[b] Indirectly limited by flash point and 95% distillation recovery

6.3 Refinery technology selection

The refinery technology selection for F-T products to fuels is not trivial and depends on several factors. Key criteria affecting the refinery design are listed in the following. Refinery designs for final fuels from F-T products based on these principles have been proposed (de Klerk 2011b).

(i) The design of the refinery needs to consider the nature of the F-T syncrude. Refinery technology has to be selected according to the F-T technology and the corresponding syncrude properties. As an example, refinery conversions that require alkenes in the feedstock are less suitable for Co-LTFT syncrude, which contains little alkenes.
(ii) The refinery design is also determined by the intended product and location. For instance, depending on the local market situation, some refineries might only have one target fuel, e.g., motor gasoline, merely producing jet fuel and diesel as by-products.
(iii) All refinery steps have a certain selectivity for specific classes of fuel products, but hydrocarbons in boiling ranges of other fuel products will be co-produced. Hence, a typical refinery will have multiple fuel classes as output.
(iv) Chemicals co-production along with fuels can be beneficial from an economic point of view.
(v) In some scenarios, it can be more efficient to produce fuel blending feedstocks that only satisfy fuel specifications in part instead of final fuels (see Section 6.2). This has the advantage of a lower refining effort for the F-T products and is particularly interesting for small scale operations or remote locations without direct market access.
(vi) Aqueous product refining to fuel additives (ethanol, fuel ethers) as well as wastewater treatment need to be considered.

7. TECHNICAL, ECONOMICAL, AND ENVIRONMENTAL PERSPECTIVES OF LIQUID FUEL PRODUCTION VIA FISCHER-TROPSCH SYNTHESIS

Conversion of biomass and biomass waste into liquid fuels via F-T synthesis has gained increased attention in recent years. However, the current and future scenario for commercial application of this technology depends on a number of interrelated parameters based on technical, economic, political, and environmental considerations. Some comments on these are presented in the following.

Discussion on the technical feasibility of BTL processes is significant. The knowledge gained from coal-to-liquids (CTL) and gas-to-liquids (GTL) processes (de Klerk 2011b) is the foundation of this discussion. These processes have shown that fuel production via F-T synthesis is technically feasible; however, its commercial application is limited. Reasons for that lie in the complexity of the process, which requires a relatively large scale to be able to operate economically. As agricultural waste is not produced in one central place, spoke-and-hub concepts that rely on feeding a central, larger-scale F-T synthesis facility with densified bio-oil have been proposed as a more economical alternative (Lan et al. 2014). On the technical side, it is worth mentioning that the long-term operability of syngas production and syngas cleaning with biomass as a feedstock, rather than coal, pose a challenge to commercialization of the technology (see Section 3 and 4). In other words, modifying the existing technology as well as understanding the different challenges associated to biomass conversion is to some extent a slow and complex process.

As mentioned before, the economics of liquid fuel production via F-T synthesis have an effect on current and future scenarios for commercialization. The investment cost of a facility producing 178 million liters per year of fuel product was estimated to be around 500 million US dollars. The production cost of the fuel in such a refinery would be around 1.1–1.2 US dollars per liter (Swanson et al. 2010). Larger scales would enable a lower production cost, but would require a higher initial investment (Li et al. 2015). Given that the price of, for instance, fossil jet fuel is in the range of 0.5 dollars per liter, it is clear that market implementation of F-T based bio-fuels is and will continue to be strongly reliant on subsidies and thereby on the political situation. The current drive to shift from fossil fuels to fuels from renewable and waste sources is strong and suggest that policies to aid process development will be part of the future.

This transition in the preferred source for fuel production is closely linked to environmental and climate change considerations. Thus, greenhouse gas (GHG) emissions are one of the key parameters to demonstrate the benefits of using biomass and waste biomass as a feedstock for fuel production. It has been estimated that the overall fuel chain GHG emissions for biomass-based F-T fuels is low (64 $g_{CO2\text{-}eq} \cdot km^{-1}$) (Marano and Ciferno 2001) compared to that of gasoline (230 $g_{CO2\text{-}eq}$ km^{-1}) or diesel (270 $g_{CO2\text{-}eq}$ km^{-1}) *(Source: Natural Resources Canada 2014)* from fossil fuels. Yet, it should not be considered CO_2 emission free (Kazulis et al. 2018). Similarly, when compared with production of fuels from conventional sources, e.g., coal, the integration of biomass into the existing F-T infrastructure could have benefits. For instance, it has been estimated that production of F-T diesel from coal has a GHG footprint of 583.47 $g_{CO2\text{-}eq}$ km^{-1} (Marano and Ciferno 2001). However, when coal is co-processed with 20% of biomass in the same process, the GHG footprint decreases to 249.79 g_{CO2eq} km^{-1} (57% reduction) (Marano and Ciferno 2001).

Furthermore, F-T synthesis is not the only pathway to obtain liquid transportation fuels from biomass or waste biomass. Other technologies based on direct liquefaction, e.g., hydrothermal liquefaction or pyrolysis combined with hydrogenation, or indirect liquefaction, e.g., methanol synthesis, are also used. These pathways have been reviewed in (Fatih Demirbas et al. 2011, Serrano-Ruiz and Dumesic 2011, Naik et al. 2010, Bridgewater 2003). Comparison between processes has shown that conversion of biomass via F-T synthesis is advantageous. Table 11 presents a summary of the estimated GHG emissions of different pathways to produce renewable jet-fuel (de Jong et al. 2017). Results from this study showed that production of jet-fuel via F-T synthesis has the highest GHG emissions savings (86–104%), followed by HTL (77–80%), pyrolysis (54–75%), direct sugar-to-hydrocarbons (71–75%) and corn stover-based alcohol-to-jet (60–75%) processes. It has also been shown that the GHG savings potential of F-T derived jet fuels is higher than that of traditional vegetable oil hydrotreating (O'Connel et al. 2019).

The technical feasibility as well as the estimated environmental benefits from conversion of biomass and waste biomass to fuels, via F-T synthesis, has led to several ongoing and planned demonstration projects ***(Chapter 13: Waste to Fuels via the Fischer-Tropsch Process: A Modularized Approach).*** Fulcrum Bioenergy is a developing technology to produce fuel using MSW gasification followed by F-T synthesis. The process is expected to reduce the GHG emissions by up to 80% compared to conventional crude oil production (Shahabuddin et al. 2020). Their first plant was built for an annual feedstock capacity of over 175,000 tons of household garbage, which will produce up to 39.75 million liters of fuel per year starting from 2021 (*Source: Fulcrum Bioenergy*). Similarly,

Table 11. GHG emissions of different pathways to produce renewable jet fuel from biomass. Data from De Jong et al. 2017.

Feedstock	Technology pathway to Renewable Jet Fuel	CO_2 emission ($g_{CO2eq} \cdot MJ^{-1}$)
Willow, Poplar, Corn stover, Forestry residues	Fischer-Tropsch	(–3)–4[a]
Forestry residues	Hydrothermal Liquefaction	18–20
Forestry residues	Pyrolysis	22–37
Corn, Corn stover, Sugar cane	Alcohol-to-Jet	22–71
Sugar cane	Direct Sugars to Hydrocarbons	45–75
Fossil fuel Diesel (baseline)	N.A.	91.8
Fossil Fuel Gasoline (baseline)	N.A.	93.3

[a]- negative CO_2 emission indicates that the process removed more CO_2 from the environment than it emitted.

Altalto Immingham plans to convert over 500,000 tons of commercial waste per year into jet-fuel (Shahabuddin et al. 2020).

Aided by government incentives towards renewable energy, the use of syngas from biomass and waste biomass gasification for production of fuels (particularly jet-fuel) via F-T synthesis will likely reach the technology maturity required to create new industry that contributes to GHG emissions reduction in the future.

REFERENCES

Abdoulmoumine, N., Adhikari, S., Kulkarni, A. and Chattanathan, S. 2015. A Review on Biomass Gasification Syngas Cleanup. Applied Energy 155: 294–307.

Acharya, B., Sule, I. and Dutta, A. 2012. A review on advances of torrefaction technologies for biomass processing. Biomass conversion & Biorefinery 2: 349–369.

Acharya, B. 2018. Cleaning of Product Gas of Gasification. pp. 373–391. *In*: Basu, P., Biomass Gasification, Pyrolysis and Torrefaction: Practical Design and Theory. Academic Press. United Kingdom. https://doi.org/10.1016/B978-0-12-812992-0.00010-8

Adhikari, S. and Chakraborty, J.P. 2018. Conversion of Solid Wastes to Fuels and Chemicals Through Pyrolysis. pp. 239–263. *In*: Maddela, N.R., Cruzatty, L.C. and Chakraborty, S. (eds.). Advances in the Domain of Environmental Biotechnology: Microbiological Developments in Industries, Wastewater Treatment and Agriculture. Springer Nature. Amsterdam, Netherlands. https://doi.org/10.1016/b978-0-444-63992-9.00008-2

Ail, S.S. and Dasappa, S. 2016. Biomass to liquid transportation fuel via Fischer-Tropsch synthesis-Technology review and current scenario. Renewable and Sustainable Energy Reviews 58: 267–286.

Arutyunov, V., Nikitin, A., Strekova, L., Savchenko, V. and Sedov, I. 2021. Utilization of Renewable Sources of Biogas for small-Scale Production of Liquid Fuels. Catalysis Today 379: 23–27.

ASTM D7566 - Standard Specification for Aviation Turbine Fuel Containing Synthesized Hydrocarbons; ASTM International, West Conshohocken, PA 2020. https://doi.org/10.1520/D7566-20C

Bach, Q.-V., Tran, K.-Q. (2015) Dry and wet torrefaction of woody biomass- A comparative study on combustion kinetics. Energy Procedia, 75, 150-155.

Bar-On, Y.M., Philips, R. and Milo, R. 2018. The biomass distribution on Earth. Proceedings of the National Academy of Sciences 115: 6506–6511.

Basu, P. 2018. Biomass Gasification, Pyrolysis and Torrefaction: Practical Design and Theory (third edition). Academic Press, United Kingdom. https://doi.org/10.1016/B978-0-12-812992-0.00014-5

Bouchy, C., Hastoy, G., Guillon, E. and Martens, J.A. 2009. Fischer-Tropsch Waxes Upgrading via Hydrocracking and Selective Hydroisomerization. Oil Gas Science and Technology 23(4): 91–112.

Bridgewater, A.V. 2003. Renewable fuels and chemicals by thermal processing of biomass. Chemical Engineering Journal 91: 87–102.

Bridgwater, A.V. 2004. Biomass Fast Pyrolysis. Thermal Science 8: 21–49.

Cheah, S., Carpenter, D. and Magrini-Bair, K. 2009. Review of Mid- to High Temperature Sulfur Sorbents for Desulfurization of Biomass- and Coal-Derived Syngas. Energy & Fuels, 23(11), 5291–5307.

Chornet, E., Valsecchi, B., Drolet, G., Gagnon, M. and Nguyen, B. 2012. United States Patent No. US8137655B2. Retrieved May 18: 2021.

Choudhury, H., Sankar, C. and Vijayanand, S. 2015. Biomass gasification integrated Fischer-Tropsch synthesis: perspectives, opportunities and challenges. pp. 383–435. *In*: Pandey, A., Stocker, M., Bhaskar, T. and Sukumaran, R.K. (eds.). Recent Advances in Thermochemical Conversion of Biomass. Amsterdam, Netherlands: Elsevier.

Couhert, C., Salvador, S. and Commandre, J.-M. 2009. Impact of torrefaction on syngas production from wood. Fuel 88(11): 2286–2290.

Cover, A., Hubbard, D., Jain, S., Shah, K., Koneru, P. and Wong, E. 1985. Review of selected sulfur recovery processes. Houston: Gas Research Institute. Retrieved in June 7th, 2021. https://citeseerx.ist.psu.edu/viewdoc/download?doi=10.1.1.629.2990&rep=rep1&type=pdf

Dancuart, L., De Haan, R. and de Klerk, A. 2004. Processing of Primary Fischer-Tropsch Products. In Studies in Surface Science and Catalysts 152: 482–532. Amsterdam: Elsevier. https://doi.org/10.1016/s0167-2991(04)80463-4

Darr, M.J. and Shah, A. 2012. Biomass storage: an update on industrial solutions for baled biomass feedstocks. Biofuels 3(3): 321–332.

Davis, B. 2001. Fischer-Tropsch synthesis: current mechanism and futuristic needs. Fuel Processing and Technology 71: 157–166.

De Jong, S., Antonissen, K., Hoefnagels, R., Lonza, L., Wang, M., Faaij, A. and Junginger, M. 2017. Life-Cycle analysis of greenhouse gas emissions from renewable jet fuel production. Biotechnology for Biofuels 10(64): 1–18.

de Klerk, A. 2009. Can Fischer-Tropsch Syncrude Be Refined to On-Specification Diesel Fuel? Energy and Fuels 23: 4593–4604.

de Klerk, A. 2011a. Fischer-Tropsch Fuels Refinery Design. Energy & Environmental Science 4: 1177–1205.

de Klerk, A. 2011b. Fischer-Tropsch Refining. Wiley-VCH Verlag & Co. KGaA. Germany de Klerk, A. 2016. Aviation Turbine Fuels Through the Fischer-Tropsch Process. *In*: Christopher Chuck (ed.). Biofuels for Aviation: Feedstocks, Technology, and Implementation (first ed., pp. 241-259). Academic Press. San Diego, USA.

de Klerk, A. 2020. Biomass-, Coal-, Gas- and Waste-to-Liquids Processes. *In*: Letcher, T.M. (ed.). Future Energy: Improved, Sustainable, and Clean Options for our Planet. (third ed., pp. 199-226). Elsevier Oxford, United Kingdom.

De Swart, J.W.A., Krishna, R. and Sie, S.T. 1997. Selection, design and scale up of the Fischer-Tropsch reactor. Studies in Surface Science and Catalysis 107: 213–218.

Dimitriadis, A. and Bezergianni, S. 2017. Hydrothermal liquefaction of various biomass and waste feedstocks for biocrude production: A state of the art review. Renewable and Sustainable Energy Reviews 68: 113–125.

Dry, M. 1981. pp. 159–255. *In*: Catalysis Science and Technology. Springer-Verlag, Berlin, Germany.

Dry, M., De, H. and Erasmus, W. 1987. Update of the Sasol Synfuel Process. Annual Review of Energy 12: 1–21.

Edwards, M. 1979. H2S removal processes for low BTU coal gas. US Department of Energy, Springfield, USA.

Eilers, J., Posthuma, S. and Sie, S. 1990. The Shell Middle Distillate Synthesis Process (SMDS). Catalysis Letters 7(1–4): 253–269.

European Committee for Standardization (CEN), 2013. EN 228—Automotive Fuels—Unleaded Petrol—Requirements and Test Methods.

Fatih Demirbas, M., Balat, M. and Balat, H. 2011. Biowastes-to-biofuels. Energy Conversion and Management 52: 1815–1828.

FAOSTAT statistical database. Food and Agriculture Organization of the United Nations. (2021). Retrieved June 4, 2021, from http://www.fao.org/faostat/en/

Fernandes, T., Klaasse Bos, G., Zeeman, G., Sanders, J. and van Lier, J. 2009. Effects of thermo-chemical pre-treatment on anaerobic biodegradability and hydrolysis of lignocellulosic biomass. Bioresource Technology 100(9): 2575–2579.

Fulcrum Bioenergy. Retrieved May 27, 2021, from https://fulcrum-bioenergy.com/company/projects/

Gollakota, A.R., Kishore, N. and Gu, S. 2018. A review on hydrothermal liquefaction process. Renewable and Sustainable Energy Reviews 81: 1378–1392.

Goransson, K., Soderlind, U., He, J. and Zhang, W. 2011. Review of Syngas Production via Biomass DFBGs. Renewable Sustainable Energy Reviews 15(1): 482–492.

Guo, X., Liu, G. and Larson, E. 2011. High-Octane Gasoline Production by Upgrading Low-Temperature Fischer-Tropsch Syncrude. Industrial & Engineering Chemistry Research 50(16): 9743–9747.

Halmenschlager, C.M., Brar, M., Apan, I. and de Klerk, A. 2016. Oligomerization of Fischer-Tropsch Tail Gas over H-ZSM-5. Industrial & Engineering Chemistry Research 55(51): 13020–13031.

Higman, C. and Burgt, M. 2003. Auxiliary Technologies. pp. 323–367. *In*: Gasification. Gulf Professional Publishing. https://doi.org/10.1016/b978-075067707-3/50008-5.

Hu, X. and Gholizadeh, M. 2019. Biomass pyrolysis: A review of the process development and challenges from initial researches up to the commercialization stage. Journal of Energy Chemistry 39: 109–143.

IEA report. 2019. Global Energy Review 2019. Retrieved June 7th, 2020. https://iea.blob.core.windows.net/assets/dc48c054-9c96-4783-9ef7-462368d24397/Global_Energy_Review_2019.pdf

Jess, A., Kern, C. (2009) Modeling of Multi-Tubular Reactors for Fischer-Tropsch Synthesis. Chemical Engineering & Technology 32(8): 1164–1175.

Jurgens, S., Obwald, P., Selinsek, M., Piermartini, P., Schwab, J., Pfeifer, P. and Kohler, M. 2019. Assessment of Combustion Properties of Non-Hydroprocessed Fischer-Tropsch Fuels for Aviation. Fuel Processing Technology 193: 232–243.

Kan, T., Strezov, V. and Evans, T.J. 2016. Lignocellulosic biomass pyrolysis: A review of product properties and effects of pyrolysis parameters. Renewable and Sustainable Energy Reviews, 57: 1126–1140.

Kaza, S., Yao, L., Bhada-Tata, P. and Van Woerden, F. 2018. What a waste 2.0: A Global Snapshot of Solid Waste Management to 2050. World Bank. Washington, DC, USA. https://doi.org/10.1596/978-1-4648

Kazulis, V., Vigants, H., Veidenbergs, I. and Blumberga, D. 2018. Biomass and natural gas co-firing- evaluation of GHG emissions. Energy Procedia 147: 558–565.

Kpalo, S.Y., Zainuddin, M.F., Manaf, L.A. and Roslan, A.M. 2020. A Review of Technical and Economic Aspects of Biomass Briquetting. Sustainability 12(11): 4609–4639.

Lan, P., Lan, L., Xie, T. and Liao, A. 2014. The preparation of Syngas by the Reforming of Bio-Oil in a Fluidized-bed Reactor. Energy Sources, Part A,: Recovery, Utilization, and Enviromental Effects 36: 242–249.

Layne, A., Alvin, M., Granite, E., Pennline, H., Keairns, D. and Newby, R. 2007. Overview of Contaminant Removal from Coal-Derived Syngas. IMECE2007 (pp. 397–407). Seattle. https://doi.org/10.1115/IMECE2007-42165

Leckel, D. 2009. Diesel Production from Fischer-Tropsch: The Past, the Present, and the New Concepts. Energy and Fuels 23(5): 2342–2358.

Li, Q., Zhang, Y. and Hu, G. 2015. Techno-economic analysis of advanced biofuel production based on bio-oil gasification. Bioresource Technology 191: 88–96.

Lovell, W. (1958) Knocking Characteristics of Hydrocarbons. ASTM, West Conshohocken, USA.

Luk, H., Lam, T., Oyedun, A., Gebreegziabher, T. and Hui, C. 2013. Drying of biomass for power generation: A case study on power generation from empty fruit bunch. Energy 63: 205–215.

Maitlis, P.M. and de Klerk, A. 2013. Greener Fischer-Tropsch Processes for Fuels and Feedstocks. Wiley-VCH Verlag CmbH & Co. Weinheim, Germany.

Marano, J.J. and Ciferno, J.P. 2001. Life-Cycle Greenhouse-Gas Emissions Inventory For Fischer-Tropsch Fuels. U.S. Department of Energy-National Energy Technology Laboratory. Retrieved May 11, 2021, from https://www.eesi.org/files/netl_emissions_060001.pdf

Modal, P., Dang, G. and Garg, M.O. 2011. Syngas Production through Gasification and Cleanup for Downstream Applications-Recent Developments. Fuel Processing Technology 92(8): 1395–1410.

Moskalik, J., Skvaril, J., Stelcl, O., Balas, M. and Lisy, M. 2012. Enery Recovery from Contaminated Biomass. Acta Polytechnica 52(3): 77–82.

Nacken, M., Ma, L., Heidenreich, S., Verpoort, F. and Baron, G. 2012. Development of a Catalytic Ceramic Foam for Efficient Tar Reforming of a Catalytic Filter for Hot Gas Cleaning of Biomass-Derived Syngas. Applied Catalysis B 125: 111–119.

Nacken, M., Baron, G., Heidenreich, S., Rapagna, S., D'orazio, A., Galluci, K. and Foscolo, P. 2015. New DeTar Catalytic Filter with Integrated Catalytic Ceramic Foam: Catalytic Activity under Model and Real Bio Syngas Conditions. Fuel Processing Technology 134: 98–106.

Naik, S., Goud, V., Rout, P. and Dalai, A. 2010. Production of first and second generation biofuels: A comprehensive review. Renewable and Sustainable Energy Reviews 14(2): 578–597.

Natural Resources Canada. 2014. Retrieved May 13th, 2021, https://www.nrcan.gc.ca/sites/www.nrcan.gc.ca/files/oee/pdf/transportation/fuel-efficient-technologies/autosmart_factsheet_9_e.pdf

O'Connell, A., Kousoulidou, M., Lonza, L. and Weindorf, W. 2019. Considerations on GHG emissions and energy balances of promising aviation biofuel pathways. Renewable and Sustainable Energy Reviews 101: 504–515.

Okeke, I. and Mani, S. 2017. Techno-Economic Assessment of Biogas to Liquid Fuels Conversion Technology via Fischer-Tropsch Synthesis. Biofuels, Bioprod. Biorefining 11(3): 472–487.

Ontario- Ministry of Agriculture, Food and Rural Affairs. (n.d.). Retrieved April 27, 2021, from http://www.omafra.gov.on.ca/english/engineer/facts/11-035.htm#:~:text=The%20low%20density%20of%20biomass,properties%20than%20the%20raw%20biomass

Qayyum, A., Ali, U. and Ramzan, N. 2020. Acid Gas Removal Techniques for Syngas, Natural Gas, and Biogas Clean up—A Review. Energy Sources, Part A,: Recovery, Utilization, and Environmental Effects 1–24.

Rahimpour, M. and Elekaei, H. (2009) A comparative study of combination of Fischer-Tropsch synthesis reactors with hydrogen perm selective membrane in GTL technology. Fuel Processing Technology, 90, 747-761.

Ratnasamy, C. and Wagner, J. 2009. Water Gas Shift Catalysis. Catalysis Review 51(3): 325–440.

Roos, C.J. 2008. Biomass Drying and Dewatering for Clean Heat & Power. Olympia: Northwesr CHP Application Center. Retrieved March 28, 2021, from https://d1wqtxts1xzle7.cloudfront.net/31037090/biomassdryinganddewateringforcleanheatandpower.pdf?1364305118=&response-content-disposition=inline%3B+filename%3DBiomass_Drying_and_Dewatering_for_Clean.pdf&Expires=1616970614&Signature=JUyx0VWJak0cv5KW1nIj~A

RTI International. 2018. RTI Warm Syngas Cleanup Operational Testing at Tampa Electric Company's Polk I IGCC Site, Final Report. Tampa Bay: No. DE-FE0026622. Retrieved June 7th, 2021. https://www.osti.gov/servlets/purl/1419426

Rostrup-Nielsen, J., Christiansen, L. J. (2011) Concepts in Syngas Manufacture (6th edition). Imperial College Press, United Kingdom.

Sadh, P.K. and Duhan, J.S. 2018. Agro-industrial wastes and their utilization using solid state fermentation: a review. Bioresources and Bioprocessing 5: 1–15.

Sahu, S.N., Sahoo, N.K., Naik, S.N. and Mahapatra, D.M. 2020. Advancements in hydrothermal liquefaction reactors: overview and prospects. pp. 195–213. *In*: Singh, L., Yousuf, A. and Mahapatra, D.M. (eds.). Bioreactors. Elsevier. Amsterdam, Netherlands. https://doi.org/10.1016/b978-0-12-821264-6.00012-7

Sakuneka, T., de Klerk, A., Nel, R. and Pienaar, A. 2008. Synthetic Jet Fuel Production by Combined Propene Oligomerization and Aromatic Alkylation over Solid Phosphoric Acid. Industrial & Engineering Chemistry Research 47(6): 1828–1834.

Serrano-Ruiz, J. and Dumesic, J. 2011. Catalytic routes for the conversion of biomass into liquid hydrocarbon transportation fuels. Energy & Environmental Science 4: 83–99.

Seville, J.P.K. 1997. Gas Cleaning in Demanding Applications (First edition). Springer. United Kingdom.

Shahabuddin, M., Alam, M., Krishna, B.B., Bhaskar, T. and Perkins, G. 2020. A review on the production of renewable aviation fuels from the gasification of biomass and residual wastes. Bioresources Technology 312: 123596–123611.

Sie, S. and Krishna, R. 1999. Fundamentals and Selection of Advanced Fischer-Tropsch Reactors. Applied Catalysis 186(1–2): 55–70.

Sie, S., Senden, M. and Van Wechem, H. 1991. Conversion of Natural Gas to Transportation Fuels via the Shell Middle Distillate Synthesis Process (SMDS). Catalysis Today 8(3): 371–394.

Smirniotis, P. and Gunugunuri, K. 2015. Water Gas Shift Reaction: Research Developments and Applications. Elsevier. Waltham, USA. https://doi.org/10.1016/C2013-0-09821-0

Speight, J. 2015. Synthetic liquid fuel production from gasification. pp. 147–174. *In*: Luque, R. and Speight, J.G. (eds.). Gasification for synthetic fuel production: Fundamentals, Processes, and Applications. Woodhead Publishing. Cambridge, United Kingdom.

Speight, J. 2020. Handbook of Gasification Technology: Science, Processes, and Applications, Scrivener Publishing, USA.

Stelt, M.v., Gerhauser, H., Kiel, J.H. and Ptasinski, K.J. 2011. Biomass upgrading by torrefaction for the production of biofuels: A review. Biomass and Bioenergy 35: 3748–3762.

Swanson, R., Satrio, J. and Brown, R. 2010. Techno-Economic Analysis of Biofuels Production Based on Gasification. NREL. Retrieved June 3, 2021, from https://www.nrel.gov/docs/fy11osti/46587.pdf

Tamm, P., Mohr, D. and Wilson, C. 1988. Octane Enhancement by Selective Reforming of Light Paraffins. Studies in Surface Science and Catalysts, 38, 335-353.

Tumuluru, J.S., Wright, C.T., Hess, J.R. and Kenney, K.L. 2011. A review of biomass densification systems to develop uniform feedstock commodities for bioenergy application. Biofuels, Bioproducts and Biorefining 5: 683–707.

Tumuluru, J.S., Tabil, L.G., Song, Y., Iroba, K.I. and Meda, V. 2014. Grinding energy and physical properties of chopped and hammer-milled barley. Biomass Energy 60: 58–67.

Tumuluru, J.S. and Heikkila, D. 2019. Biomass Grinding Process Optimization Using Response Surface Methodology and Hybrid Genetic Algorithm. Bioengineering 6: 12–32.

US Department of Energy. Energy Efficiency & Renewable Energy. Retrieved April 27, 2021, from https://www.energy.gov/eere/bioenergy/biomass-resources

Vassilev, S.V., Baxter, D., Andersen, L.K. and Vassileva, C.G. 2010. An Overview of the Chemical Composition of Biomass. Fuel 89: 913–933.

Viguie, J.-C., Ullrich, N., Porot, P., Bournay, L., Hecquet, M. and Rousseau, J. 2013. BioTfuel Project: Targeting the Development of Second-Generation Biodiesel and Biojet Fuels. Oil Gas Science & Technology 68(5): 789–946.

Wang, G., Dai, Y., Yang, H., Xiong, Q., Wang, K., Zhou, J. and Li, Y. 2020. A Review of Recent Advances in Biomass Pyrolysis. Energy & Fuels 34: 15557–15578.

Widjaya, E.R., Chen, G., Bowtell, L. and Hills, C. 2018. Gasification of non-woody biomass: A literature review. Renewable and Sustainable Energy Reviews 89: 184–193.

Woolcock, P. and Brown, R. 2013. A Review of Cleaning Technologies for Biomass-Derived Syngas. Biomass and Bioenergy 52: 54–84.

Yan, W., Acharjee, T.C., Coronella, C.J. and Vasquez, V.R. 2009. Thermal Pretreatment of Lignocellulosic Biomass. Environmental Progress & Sustainable Energy 28: 435–440.

Zhang, Q., Kang, J. and Wang, Y. 2010. Development of novel catalysts for Fischer-Tropsch synthesis: tuning the product selectivity. ChemCatChem 2: 1030–1058.

Zhiwen, C., Mingfeng, W., Yongzhi, R., Enchen, J., Yang, J. and Weizhen, L. 2018. Biomass torrefaction: A promising pretreatment technology for biomass utilization. IOP conference series: Earth and Environmental Science 113: 012201–012207.

Zwart, R., Van de Drift, A., Bos, A., Visser, H., Cieplik, M. and Konemann, H.W. 2009. Oil-Based Gas Washing-Flexible Tar Removal for High-Efficient Production of Clean Heat and Power as Well as Sustainable Fuels and Chemicals. Environmental Progress & Sustainable Energy 28(3): 324–335.

13

WASTE TO FUELS VIA THE FISCHER-TROPSCH PROCESS

A MODULARIZED APPROACH

Chelsea Tucker[1,2] and *Eric van Steen*[2,*]

1. INTRODUCTION

The past 50 years have been characterised by rapid industrialisation and urbanisation. Whilst this has increased the economic potential and living standards of millions across the globe, it has also increased the volume and complexity of the waste we generate.

Worldwide, up to 46% of all municipals solid waste (MSW) is organic. Organic waste (i.e. food, agricultural, biomass or animal waste) provides a unique challenge for waste management specialists, as it produces significant amounts of carbon dioxide (CO_2) and methane (CH_4) on decomposition in landfills. Whilst carbon dioxide is well-known as a greenhouse gas, methane is even more potent, with 27 times the global warming potential (Boucher et al. 2009). For this reason, diverting organic waste from landfill is a key step towards reducing spurious methane emissions and global decarbonization.

An economically and environmentally desirable option for the management of organic waste is to divert organic waste from landfill and convert it to energy. Waste-to-energy (WTE) is said to decrease greenhouse gas emissions by one tonne of CO_2 equivalent per tonne of waste (Arena 2015, Psomopoulos et al. 2009) when compared to landfill. Converting waste to energy is a particularly attractive option for lower income regions where widespread access to energy is limited, and proportions of organic waste within the municipal waste stream is high.

1.1 Technology options for waste-to-energy

There are many technological options available for converting organic waste to energy (see Fig. 1). The most common technology is direct combustion, i.e., burning organic waste under oxygen to

[1] Green Chemical Reaction Engineering, Engineering and Technology Institute Groningen, University of Groningen, Nijenborgh 4, 9747 AG Groningen, The Netherlands.

[2] Catalysis Institute, Department of Chemical Engineering, University of Cape Town, South Africa.

* Corresponding author: eric.vansteen@uct.ac.za

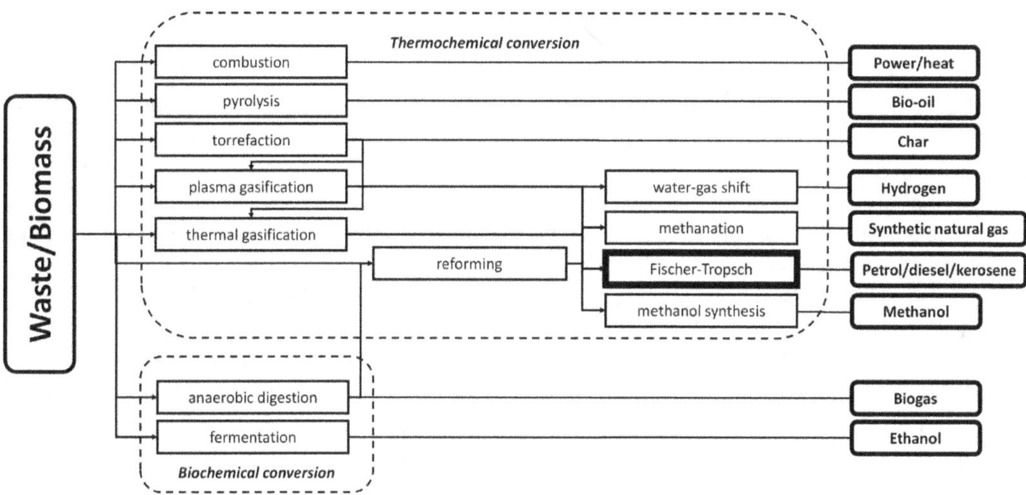

Figure 1. Conversion routes for waste or biomass to fuel, with Fischer-Tropsch synthesis highlighted.

form carbon dioxide, water and heat. This heat can be used for steam generation that can be used to generate electricity and contribute to heating.

The disadvantage of direct combustion is its low economic potential: the high capital costs of the process compared to low prices associated with its only product (electricity). These projects are typically only economical for situations where electricity and/or heat can be used by a coupled process, or where its primary purpose is waste management rather than energy production.

For this reason, there is economic incentive to convert waste into high-value material, rather than just electric power. There are a variety of processes that can be used (see Fig. 1) including, but not limited to, torrefaction (to char), pyrolysis (to bio-oil), thermal or plasma gasification (to syngas), anaerobic digestion (to biogas) and fermentation (to alcohols).

Torrefaction and pyrolysis (see Fig. 1) involve the decomposition of waste or biomass under oxygen-free conditions. Torrefaction, which occurs at temperatures between 200–350°C, allows for partial decomposition to char (a solid, partially carbonized fuel with a relatively low moisture content). Pyrolysis on the other hand occurs at higher temperatures between 400–600°C and produces both char and a complex mixture of oxygenated organic compounds called bio-oil (Zhang and Brown 2011). Bio-oil in its raw state can be used as a fuel for boilers and gas-turbines (Zhang and Brown 2011).

Thermal gasification involves the conversion of a carbonaceous species (organic waste in this case) under low-oxygen conditions at temperatures as high as 1200°C. This results in the formation of a mixture of syngas (CO and H_2) as well as CH_4, CO_2, N_2 and small quantities of hydrocarbons (Bridgwater 2003). Plasma gasification uses a plasma arc (6000–10000°C), rather than feedstock, as a heat source.

Anaerobic digestion refers to the decomposition of organic waste in oxygen-scarce environments. The product of this process is a mixture of CH_4 and CO_2 called bio gas with a composition of between 55%–75% and 30%–45%, respectively (Hilkiah Igoni et al. 2008), which could be reformed to syngas.

In some of the technological routes presented in Fig. 1, including gasification and coupled anaerobic digestion/reforming, a fuel is not produced directly but rather an intermediate mixture of useful gases called synthesis gas or syngas (a mixture of mainly CO and H_2) (Bridgwater 2003). The transformation of waste/biomass into synthesis gas opens up a number of different downstream process using different technologies. Syngas can be further processed to hydrogen, synthetic natural gas, methanol, petrol, diesel, or kerosene (jet fuel).

1.2 Fischer-tropsch process

The Fischer-Tropsch synthesis is the only method highlighted in Fig. 1 that is industrially used to directly convert syngas into fuels such as petrol, diesel, or jet fuel. These fuels are attractive products for the global energy market, especially in lower income regions, as they require no alterations to current generators, vehicle engines or fuel distribution networks (in contrast to the direct use of methanol, hydrogen, LNG or bio-oil as a fuel).

Processes utilizing the Fischer-Tropsch synthesis have been classified based on the feedstock for syngas generation: coal-to-liquid (CTL), gas-to-liquid (GTL), biomass-to-liquid (BTL) or waste-to-liquid (WTL). This terminology can be extended to any feedstock, and the versatility of the process is sometimes indicated with the abbreviation XTL, which stands for any carbon-containing material to liquid. Fischer-Tropsch technology has been developed around CTL and GTL-processes. Biomass-to-liquid (BTL) and waste-to-liquid (WTL) have not been implemented widely due to the high production costs (Liu et al. 2011). However, due to increasing environmental regulations in the past 20 years, widespread interest in the development of BTL and WTL plants has been seen worldwide (Ailand Dasappa 2016).

2. COMMERCIALIZATION OF FISCHER-TROPSCH BIOFUELS

Research into Fischer-Tropsch waste- and biomass-based fuels developed rapidly in the 21st century. Despite this, the progress for commercialization has, unfortunately, been slower than anticipated (Mawhood et al. 2016). Table 1 illustrates the Fischer-Tropsch biomass-to-liquid (BTL) and waste-to-liquid (WTL) plants that have either been planned, commissioned, operational or, in some cases, mothballed between 2008 and 2021.

2.1 Early projects (2008–2015)

In 2008, CHOREN set up the first commercial BTL plant with a capacity of 5000 bbl/day in Freiberg, Sachsen, Germany. The plant used a three-stage gasification process in combination with Fischer-Tropsch synthesis using the Shell Middle Distillate Synthesis (SMDS) technology. Despite initial success, the company filed for insolvency in July 2011 (ETIP Bioenergy 2017) due to financial difficulties associated with the start-up of the demonstration plant and operations at the plant were ceased by early 2012 (Dimitriou et al. 2018).

NSE Biofuels, a joint venture between Neste Oil and Stora Enso, set up a demonstration BTL facility in Varkaus, Finland in 2010 with a 14 bbl/day capacity with plans to upscale to a ca. 2000 bbl/day facility (ETIP Bioenergy 2017). The upscaled project, however, did not gain enough traction, or funding, to become a reality (ETIP Bioenergy 2017) and the demonstration plant eventually stopped operation (Mawhood et al. 2016).

Further plans to build BTL plants, such as Ajos BTL—with a capacity of ca. 2500 bbl/day—and Solena Fuels Green Sky project—with a capacity for bio jet and biodiesel of ca. 1157 bbl/day—were frozen due in part to lack of governmental and legislative support (ETIP Bioenergy 2017).

2.2 Recent developments (2015 – present)

Despite the initial lack of success of these plants, a new generation of demonstration and commercial Fischer-Tropsch BTL and WTL plants have been commissioned or planned between 2015 and 2021. Total's BioTfuel project plans to start the demonstration of their new biomass-to-fuel plant in Dunkirk (France) in 2021. The plant, which will be the largest of its kind in operation, will convert a mixture of straw, forest waste, energy crops and coal into diesel and jet fuel.

Table 1. Planned, operational and mothballedFischer-Tropsch waste and biomass to fuel plants.

Companies involved	Project Name	Location	Feedstock	Target Product	Syngas generation	ASU	Conversion Technology	Approx. capacity[1] (bbl/day)	Status	Ref.
Solena Fuels	GreenSky	Thurrock, Essex	MSW	Jet Fuel	Plasma gasification	No	Fischer-Tropsch	1157	Mothballed in 2016	(Ail and Dasappa 2016)
CEA/Air Liquide	SYNDIÈSE-BtS	Bure-Saudron, France	Forest and agricultural waste	Naptha, kerosene diesel	Thermal O_2 blown, high pressure gasifier	Yes	Fischer-Tropsch	530		(ETIP Bioenergy 2017)
CHOREN	Sigma Plant	Freiberg, Germany	Dry biomass	Diesel	Carbo-V gasification	Yes	Fischer-Tropsch	5000	Operation ceased in 2012	(Ail and Dasappa 2016)
Red Rock Biofuels, Velocys	Red Rock Biofuels	Oregon, USA	Woody Biomass	Jet Fuel	TCG gasifier-externally heated, air blown	No	Fischer-Tropsch	1100	Expected spring 2022	(Mawhood et al. 2016, Velocys 2018)
Fulcrum Bioenergy, BP, Johnson Matthey	Sierra Biofuels	Nevada, USA	MSW	Jet fuel	TRI externally heated gasifier/ steam reformer	No	Fischer-Tropsch	684	Operation expected to begin 2021	(Ail and Dasappa 2016, Fulcrum Bioenergy 2018)
Total	BioTFuel Project	Dunkirk, France	Biomass + Coal	Diesel, Jet fuel	Thermal gasifier	Yes	Fischer-Tropsch	5000	Demonstration expected 2021	(Mawhood et al. 2016)
Fulcrum BioEnergy	CenterPoint BioFuels Plant	Gary, Indiana	MSW	Jet fuel	TRI externally heated gasifier/ steam reformer	No	Fischer-Tropsch	2152	Currently finalizing site selection	(Fulcrum Bioenergy 2018)
Velocys, British Airways	UK Waste-to-Jet	North East England	MSW	Jet fuel	Unknown	N/A	Fischer-Tropsch	Unknown	Shell pulled support from project in 2021	(Hargreaves 2018)

[1] Converted from t/year, gal/year, m3/year where appropriate.

Fulcrum Bioenergy is also active in the waste-to-liquid market having signed a licencing deal with BP and Johnson Matthey in September 2018 to commission a ca. 2100 bbl/day WTL plant (Fulcrum Bioenergy 2018).

Velocys, a US-based company and a subsidiary of Oxford Catalysts Group, has been actively involved in developing a new highly active catalyst as well as microchannel reactors (Deshmukh et al. 2009). The BTL plant by Red Rock Biofuels, in Lakeview (Oregon, USA) will reportedly be ready for operation in late 2021(Kennedy 2018). It has been designed with Velocys technology and will supply Southwest Airlines and FedEx Express.

Velocys also announced a collaboration with Shell in 2018 to develop a WTL plant in the UK that will produce jet-fuel for British Airways (Hargreaves 2018, Velocys 2018). Unfortunately, in early 2021, Shell pulled out of this project.

The growth of Fischer-Tropsch BTL and WTL over the past 10 years should not be understated. However, the Fischer-Tropsch pathway to biofuels still has quite a way to go to establish itself as a viable, sustainable investment. Whilst the technology is undoubtedly mature, and the fuel quality is high, the capital costs involved as well as the physical construction of an operational small-to-medium scale Fischer-Tropsch plant remains a challenge (Mawhood et al. 2016).

3. ECONOMIC BARRIERS FOR FISCHER-TROPSCH BIOFUELS

Considering the history of false starts in commercializing BTL and WTL, it is important to consider the multifaceted economics of the Fischer-Tropsch synthesis. The viability of any major fuel venture is dependent on three factors: capital cost of the project, operational costs and selling price of the product. In the case of Fischer-Tropsch waste-to-liquid plants, these factors are dependent on the oil price, cost of construction, cost of the raw materials (including transportation to processing site), and scale (Zennaro 2013).

3.1 Oil price dependency

Fischer-Tropsch fuels are not only in direct competition with traditional crude oil fuels, but also beholden to the price of these fuels. This provides a unique challenge for the FT biofuels, as any changes in the cost of making FT fuels cannot be made back by changing their price. This also means that Fischer-Tropsch fuels become particularly enticing when the oil price is high, with the opposite effect when it is low. It has been noted that based on typical investment costs, current designs for CTL, GTL and BTL-processes are thought to be only competitive at crude oil prices of $74, $44 and $94 USD per barrel, respectively (van Steen et al. 2018).

Figure 2 illustrates the West Texas oil price from 1955 (van Steen et al. 2018). Large scale Fischer-Tropsch plants commissioned in this period are also shown. During times of oil price spikes in the 1980s and 2000s, there was significant investment into the Fischer-Tropsch process, with multiple plants being commissioned after these periods. At low oil prices, producing Fischer-Tropsch products as a part of the liquid fuel market is economically unsustainable.

Whilst the burden of the oil price on the overall profitability may be lessened by sound economic policies for waste or biomass-based fuels (Sorda et al. 2010), sustainable development of BTL and WTL require such processes to remain competitive within an economic climate in which oil prices remain low. This necessitates future Fischer-Tropsch BTL and WTL plants to be less capital and operation intensive than their predecessors.

3.2 Capital cost

Fischer-Tropsch technology has always been associated with very high capital investment. Table 2 illustrates the capital costs and capacities of the largest GTL and CTL plants worldwide. Large scale FT

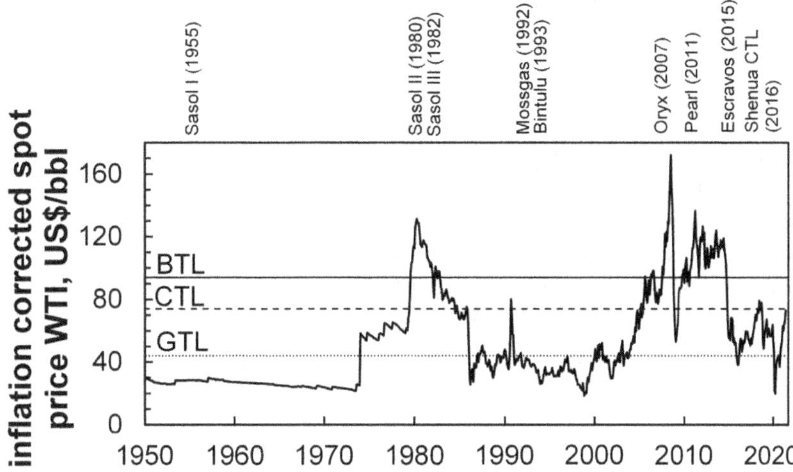

Figure 2. West Texas oil price (inflation corrected) as a function of year indicating FT plants commissioned between 1955 and 2021 and the price level beyond which current designs for GTL, CTL and BTL are competitive (van Steen et al. 2018).

Table 2. Capacity and capital costs of selected large-scale Fischer-Tropsch plants.

Operation	Feedstock	Capacity, bbl/day	Initial capital cost[1], 10^9 US$	Capital cost $10^3 \cdot$ US$/(bbl/day)
Sasol (Secunda)	CTL	154 000[3]	17[3]	108
Mossgas (Mossel Bay)	GTL	22 500[3,4,8]	4.7[4]	208
Bintulu (Malaysia)	GTL	12 500[3]	1.6[3]	128
Oryx (Qatar)	GTL	34 000[3]	1.5[3,5]	44
Pearl (Qatar)	GTL	140 000[3,9]	27[3]	193
Escavros (Nigeria)	GTL	33 000[4]	11[5,6]	333
Shenhua CTL (China)	CTL	100 000	11[7]	110

[1]Capital costs corrected for inflation to 2021, [2]Total liquid production, [3]Data from Glebova (2013), [4]Data from De Klerk (2011), [5]Data from van Steen (2018), [6]Data from Enerdata (2014), [7]Data from Reuters (2011), [8]Total output from Mossgas is 36 000 bbl/day equivalent including gas condensate, [9]Total liquid hydrocarbon product is 260 000 bbl/day including 120 000 bbl/d gas condensate, dry gas and 140 000 bbl/day of GTL product

plants have, in the past, cost between $1 and $30 billion USD (inflation adjusted to 2021) to build and commission—depending on capacity. The capital cost for the OryxGTL process was particularly low, which may be attributed to the state of the world economy at the time of construction. The high capital costs of GTL-plant in Escavros in Nigeria arepartially attributed to the long delay in construction.

Coal-to-liquid plants, such as Sasol's Secunda plant in South Africa and Shenhua CTL plant in Ningdong (China), typically cost significantly more than (Boerrigter 2006) than GTL plants due to the capital associated with gasification of solid feedstock as opposed to reforming of natural gas.

The capital cost requirement for waste-to-liquid (WTL) and biomass-to-liquid (BTL) plants are more comparable to coal-to-liquid (CTL) than gas-to-liquid (GTL) plants since they utilise a solid feedstock. Due to the variability of feedstock, and greater need for gasification pre-treatment (sorting, heating and drying), the capital cost of BTL is estimated to be 20% more than a hybrid CTL/BTL plant and as much as 70% more expensive than CTL (Liu et al. 2011). Organic waste has a slightly lower overall moisture content than biomass, but more variability in terms of its content, and thus the capital costs for a WTL-process are likely to be comparable to that of a BTL-process.

3.3 Cost of raw material and transportation

One of the major challenges involved in biomass-to-liquid is sourcing sustainably produced biomass, which has historically been far more expensive than more typical feedstocks such as coal or natural gas. It has been reported that the cost of biomass adds $ 12 USD/GJ (inflation adjusted to 2021) to diesel fuel costs (Boerrigter 2006). Waste is substantially less expensive, thus making WTL an economically more attractive option.

Raw material transportation costs, on the other hand, affect both BTL and WTL plants. Unlike coal which is mined centrally (close to the Fischer-Tropsch plant) and natural gas which can be transported via pipelines, waste and biomass feedstock is generated in a wide distribution, and is not easy to transport (Liu et al. 2011). This means that large-scale centralized Fischer-Tropsch BTL and WTL plants face a significant logistics challenge for feed procurement.

3.4 Scale

The delocalized nature of biomass and waste means that BTL and WTL plants will very likely operate at a much smaller scale than traditional plants. Designing a small-scale Fischer-Tropsch plant results in a significant loss of economies of scale (De Klerk 2016, Zennaro 2013), due to many inherent costs that are independent or not linearly-dependent on scale.

Figure 3 shows the relative total cost indicator (TCI) for various BTL plants as a function of capacity (bbl/day), based on calculations by Boerrigter (2006) using a 34,000 bbl/day GTL plant as a reference and scaling up for the cost of additional biomass gasification and logistics. Most large-scale industrial GTL and CTL plants have a capacity between 10,000 and 160,000 bbl/day (see Table 2). These large-scale capacities have a relative TCI of between 0.8 and 1.4. An increase or decrease in capacity within this range has only a moderate effect on the TCI. However, for plants with a capacity of < 5000 bbl/day, the relative TCI increases dramatically with even small decrease in capacity. The scale of all commercial BTL and WTL plants mentioned in Table 1 lie within the highlighted region of Fig. 3, with relative TCIs between 1.6 and 3.

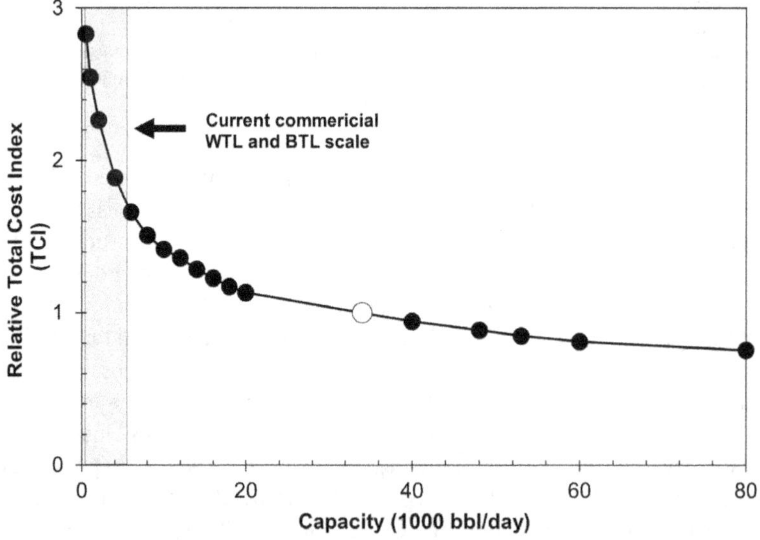

Figure 3. Relative total cost index (TCI) for a BTL plant calculated by Boerrigter (2006) using an industrial 34 000 bbl/day GTL plant as a reference (indicated as open circle). TCI considers scale-up for the additional cost of biomass gasification and pre-treatment.

4. A CASE FOR MODULARIZATION

There is an unavoidable loss of economies of scale for smaller-scale BTL and WTL plants that are designed to mimic large-scale CTL facilities. Despite this, almost all commercial plants using the Fischer-Tropsch synthesis to convert biomass or waste solids to fuel (see Table 1) have used this approach.

One way to offset the loss of economies of scale is to develop a modular plant design that can be shop fabricated and field assembled.

Modularized process technology came to prominence in the world of nuclear engineering, where the massive capital costs of large-scale nuclear reactors became a hindrance to the success of funding projects. The concept of modularisation is to design systems that are smaller in scale, simpler and repeatable that can be manufactured centrally and brought to a site for installation. This allows for less on-site construction, more relevance for remote contexts (no necessity to build up infrastructure in the surrounding region) and the bypassing of many initial financial barriers. Modularization combats the loss of economies of scale by the production of multiple smaller plants (ifthe market for the overall process is large enough).

A modularised WTL plant cannot simply be designed as a scaled down version of a large-scale CTL plant as the objectives of the two are different. In the classical CTL process, the objective is to get as much carbon as possible into desired products, whilst at the same time minimizing energy losses. This means that, for CTL, the product spectrum can be quite diverse with significant recycling and refining requirements. In the case of WTL, the primary objective is to convert organic waste to minimize methane emissions and to enable the creation of energy from localized feedstock in a decentralized manner.

The different objectives should result in a different design approach. Modular WTL requires a design that prioritizes simplicity, capital cost efficiency and utility self-sufficiency over carbon efficiency. This design requires a higher flexibility for the process to be built up over time in modules, thus reducing the strain on start-up finances.

5. DESIGN STRATEGIES FOR MODULAR FISCHER-TROPSCH BIOFUELS

5.1 Typical Fischer-Tropsch BTL and WTL plants

Whilst there is no 'standard' design when it comes to WTL and BTL plants, most of the current commercial designs (see Table 1) are similar to large-scale CTL plants. One such design of a Fischer-Tropsch BTL plant is shown in Fig. 4. This plant design includes the following:

- **Air separation unit** to provide pure oxygen to the syngas plant.
- **Gasifier** (oxygen-blown) to convert waste or biomass to syngas (H_2 and CO).
- **Water-gas shift reactor** to correct H_2/CO ratio for the Fischer-Tropsch process.
- **Fischer-Tropsch reactor** to covert H_2 and CO to a mixture of hydrocarbons.
- **Product recovery** to separate out the product phases.
- **Refining section** to convert mixture of hydrocarbons to on-spec fuels.
- **Acid gas removal** to avoid build-up of CO_2 in the recycle.

Boerrigter (2006) reported the capital cost breakdown for equipment of a BTL plant, with a similar configuration to that seen in Fig. 4, with the exception that acid gas removal was done during the gas clean-up stage. The estimated cost breakdown is shown in Table 3.

Of the total equipment costs, syngas manufacture is estimated to be responsible for the largest proportion (up to 47%) with air separation and gasification contributing 28% and 19%, respectively. It is noted by Boerrigter (2006) that the cost of both units could be halved if methane gas was used as

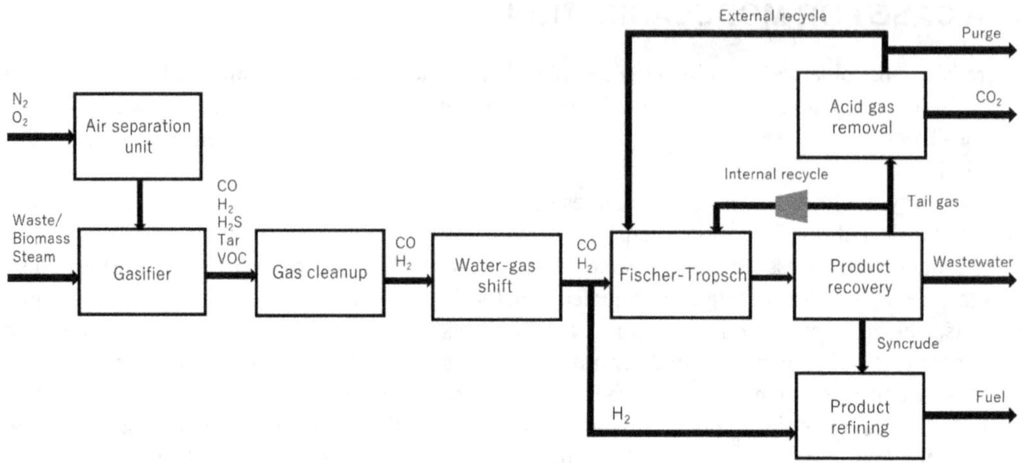

Figure 4. Typical large scale CTL derived configuration for a Fischer-Tropsch BTL/WTL plant.

Table 3. Estimated breakdown of main equipment costs for biomass to liquid plant (Boerrigter 2006).

Cost items	Percentage of cost
Air separation unit	28
Gasifier	19
H_2 manufacturing + water gas shift	6
Acid gas removal	22
Fischer-Tropsch synthesis	16
Product upgrading	9

a feedstock (as for gas-to-liquid). Acid gas removal via a Rectisol unit is estimated to cost up to 22% of the total equipment cost whilst the Fischer-Tropsch unit itself accounts for only 16%. Equipment required for product upgrading, H_2 manufacturing and a water-gas shift unit contribute a total of 15%.

5.2 Syngas generation from biomass and waste

5.2.1 Gasification: The 'golden standard?

Thermal gasification is typically regarded as the golden standard for the conversion of coal, biomass, waste, and other solid carbonaceous species to syngas (Bridgwater 2003). Gasification is a stepwise process that occurs in a reactor with a temperature gradient between 100°C up to 1200°C (Brown 2011, Bridgwater 2003). The first step, occurring at 100–300°C, is the heating and drying of waste material. The second step occurs from 225°C–500°C and involves dried waste being pyrolyzed to char, tar and permanent gases such as CO, CO_2, H_2 and CH_4.

The char produced during pyrolysis then undergoes a series of solid-gas reaction, resulting in the production of more CO and CH_4.

Carbon-oxygen reaction	$C + CO_2 \leftrightarrow 2CO$
Boudouard reaction	$C + H_2O \leftrightarrow H_2 + CO$
Carbon-water reaction	$C + 2H_2 \leftrightarrow CH_4$
Hydrogenation reaction	$C + ½\, O_2 \leftrightarrow CO$

The final step, and most important for determining the final composition of the exit gas, is a series of gaseous reactions including the water-gas shift reaction and methanation.

Water-gas shift reaction $CO + H_2O \leftrightarrow H_2 + CO_2$

Methanation $CO + 3H_2 \leftrightarrow CH_4 + H_2O$

Fixed fluidized bed gasification technology is well developed and commercially established, making it an attractive option for engineers and investors alike (Basu 2010). To date, all industrial implementations of WTL and BTL have used gasification as their syngas generation technique (see Table 1), and most authors who propose new designs use the same strategy (Bernical et al. 2013, Trippe et al. 2013, Ng and Sadhukhan 2011, Tijmensen et al. 2002).

However, there are significant challenges when it comes to gasification, the simplest of which is the cost. Boerrigter (2006) estimates that gasification costs up to 19% of equipment costs for biomass-to-liquid (Table 3) whilst Shah et al. (2014) estimates this can cost up to 70% when all the supporting units are included in the cost.

The energy requirements for gasification pre-processing are enormous. Waste fed to a gasifier cannot have more than 10–15% moisture (Tijmensen et al.2002). Drying of waste, with a typical moisture content between 30 and 60 wt.-% (Fagernäs et al. 2010), requires between 0.4–1 GJ/tonne of waste (albeit a part of this heat can be obtained by passing hot gas from the gasifier over the waste). Considering the typical energy value of municipal solid waste is around 10 GJ/tonne (Levaggi et al. 2020), the energy required is around 4–10% of the feed energy content.

Furthermore, for the purpose of converting municipal solid waste, gasifiers may not be ideal as they are not flexible when it comes to the type of feedstock. Changes to ash content, ash fusion temperature, feed reactivity, agglomeration and particle size in the feed may require an entirely different gasifier design, rather than simply process variable changes (De Klerk 2011).

Gasifiers generate syngas with major impurities. Depending on the type of biomass or waste used, and the type of gasifier employed, syngas can contain particulates, sulphur compounds such as hydrogen sulphide, carbonyl sulphide and organic sulphur, nitrogen compounds such as ammonia and HCN (especially for air-blown gasification), volatile metals, halogens, tars, hydrocarbons and aromatics (Brodeur et al. 2017). The presence of these impurities increases the complexity of required syngas clean-up steps.

5.2.2 An alternative solution: anaerobic digestion and biogas reforming

Unfortunately, as waste and biomass are both solid feedstocks, reforming (which is a less expensive, more flexible technology) cannot be used directly as it is only possible for gases. However, waste and biomass can be converted into a gas using a biochemical method, anaerobic digestion, which has been used extensively worldwide.

Anaerobic digestion involves the conversion of carbon material to biogas (CO_2 and CH_4) and proceeds via four key stages which are facilitated through microorganisms. Each stage has its own process chemistry and relevant conditions. The stages, hydrolysis, acidogenesis, acetogenesis and methanogenesis, are described as follows:

- **Hydrolysis:** First step in anaerobic digestion involving breaking down of complex long-chain molecules such as carbohydrates (via water addition) into amino acids, sugars and fatty acids that can undergo successive reactions (Anukam et al. 2019). This is a slow reaction, and often termed rate-limiting (Kannah and Banu 2012). The hydrolysis of cellulose to glucose is shown below:

$$(C_6H_{10}O_5)\,n + nH_2O \text{ -> } C_6H_{12}O_6$$

- **Acidogenesis:** Intermediate products are further degraded by acidogenic microorganisms to form volatile fatty acids (such as butyric, propionic and acetic acid) as well as carbon dioxide (Meegoda et al. 2018, Anukam et al. 2019).

$$C_6H_{12}O_6 \leftrightarrow 2CH_3CH_2OH + 2CO_2$$

$$C_6H_{12}O_6 + 2H_2 \leftrightarrow 2CH_3CH_2COOH + 2H_2O$$

$$C_6H_{12}O_6 \leftrightarrow 3CH_3COOH$$

- **Acetogenesis:** Long chain fatty acids, volatile fatty acids and alcohols are broken down into acetic acid (*CH_3COOH*), hydrogen and $CO_2$by acetogenic microorganisms (Kannah and Banu 2012).

$$CH_3CH_2COO^- + 3H_2O \leftrightarrow CH_3\ COO^- + H^+ + HCO_3^- + 3H_2$$

$$C_6H_{12}O_6 + 2H_2O \leftrightarrow 2CH_3\ COOH + 2CO_2 + 4H_2$$

$$CH_3CH_2OH + 2H_2O \leftrightarrow CH_3\ COO^- + 2H_2 + H^+$$

- **Methanogenesis:** During this phase, methane is produced, either by converting acetic acid or hydrogen and carbon dioxide into methane, respectively.

$$CH_3COOH \rightarrow CH_4 + CO_2$$

$$CO_2 + 4H_2 \rightarrow CH_4 + 2H_2O$$

$$2CH_3CH_2OH + CO_2 \rightarrow CH_4 + 2CH_3\ COOH$$

The anaerobic digestion of municipal-solid waste, wet biomass, sludge, human and farm waste produces a gas called biogas—a mixture of primarily CH_4 and CO_2 with small contaminants biogas (see Table 4). After a relatively simple cleaning step to remove hydrogen sulphide, biogas can be fed into a reformer (Navas-Anguita et al. 2019, Zhao et al. 2019, Herz et al. 2017, Izquierdo et al. 2013) and used to generate syngas.

Reforming of biogas to syngas can be achieved either via dry reforming with no additional feeds, bi-reforming with the addition of steam or tri-reforming with the addition of steam and oxygen.

Dry reforming

$$CH_4 + CO_2 \leftrightarrow 2CO + 2H_2$$

$$CO_2 + H_2 \leftrightarrow CO + H_2O$$

Bi-reforming

$$3CH_4 + CO_2 + H_2O \leftrightarrow 4CO + 8H_2$$

Tri-reforming

$$CH_4 + CO_2 \leftrightarrow 2CO + 2H_2$$

$$CH_4 + H_2O \leftrightarrow CO + 3H_2$$

$$CH_4 + \tfrac{1}{2}O_2 \leftrightarrow CO + 2H_2$$

$$CO + H_2O \leftrightarrow CO_2 + H_2$$

Table 4. Concentration of components of biogas in vol.% (Deublein and Steinhauser 2008).

Component	Typical Composition (vol.%)
Methane (CH_4)	55–70
Carbon Dioxide (CO_2)	30–45
Hydrogen Sulphide (H_2S)	0–0.5
Ammonia (NH_3)	0–0.05
Water Vapour (H_2O)	1–5
N_2	0–5

The concept of using biogas as a feed for syngas generation within the Fischer-Tropsch process is relatively novel (Zhao et al. 2019, Naqi 2018), but the implications for remote fuel production from waste are significant. Firstly, anaerobic digestion is a well-established technology that is modular and scalable. In addition, gas reforming typically costs up to 60% less than gasification (Boerrigter 2006) and unlike gasification, biogas generation allows for feed flexibility. Whilst the concentration of the various components of biogas do change with different feedstocks, the CH_4:CO_2 ratio rarely drops below 1, irrespective of the type of raw material used (Izquierdo et al. 2013). This means that changes to the feedstock composition can be managed online simply by adjusting the feed rate of steam and/or air. Furthermore, whilst biogas reforming requires a sizable amount of heat, waste does not have to be dried (digestion typically requires wet material sludge) thus decreasing energy requirements significantly.

Economic analyses for biogas reforming in the context of fuel production are scarce. However, an investigation into the techno-economic comparability between gasification and biogas reforming for hydrogen production, conducted by Yao et al. (2017), showed that the capital cost of the entire process (incl. cleaning, anaerobic digestion) for a dual fluidized gasifier system was 22% higher than for biogas/steam reforming system, whilst attaining the same output.

Figure 5 illustrates a schematic of the once-through BTL/WTL plant if biogas reforming were used. A notable change to the block flow is the location of the gas clean up section. Biogas may contain a relatively high concentration of H_2S, which has to be reduced, since H_2S is a known poison of standard reforming catalysts (Chattanathan et al. 2014). Hence, the gas clean-up stage needs to occur upstream from the reformer.

Another difference in Fig. 5 is the removal of the water-gas shift reactor. Biogas reforming using oxygen and water (called tri-reforming) has the benefit of tuning the H_2/CO ratio with relative ease with low steam/carbon (S/C) ratios (Song and Pan 2004). Whilst gasification of biomass typically produces an H_2/CO ratio of 1, tri-reforming can ensure a H_2/CO ratio of 2 (i.e., adequate for the downstream Fischer-Tropsch synthesis) at a temperature as low as 750°C (Tucker 2020). This makes the water-gas shift unit somewhat redundant (albeit it may still be incorporated to enhance the robustness of the operation).

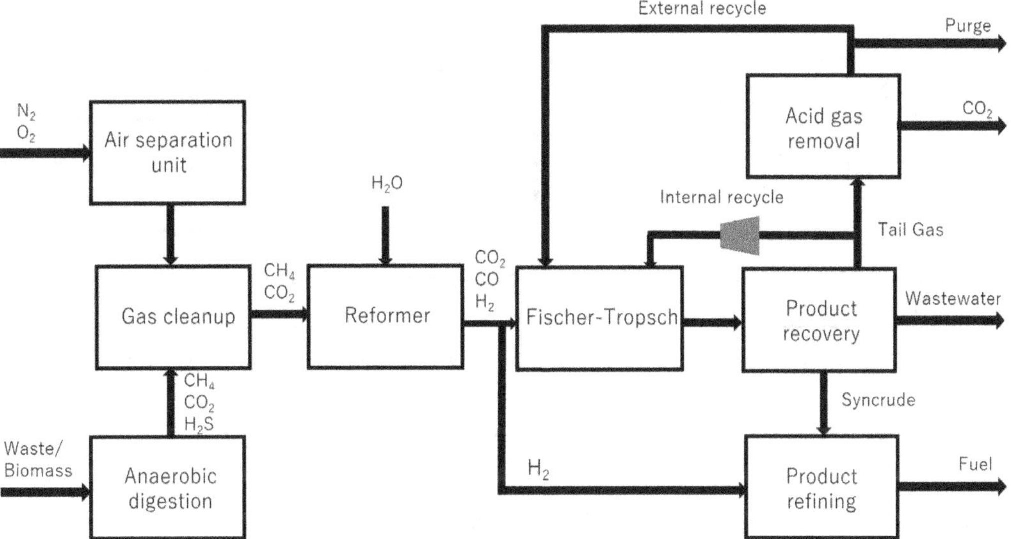

Figure 5. Schematic of the small-scale BTL/WTL plant using anaerobic digestion and a biogas reformer as the syngas generation technique.

5.3 Removing air separation unit and recycle

One of the most significant costs towards a Fischer-Tropsch CTL, BTL or WTL plant is the air separation unit (ASU), which is used to separate oxygen from air to be used in the gasifier/ reformer. It has been reported that the ASU contributes anywhere between 8–28% to the overall capital cost of Fischer-Tropsch plant (Zennaro 2013, Boerrigter 2006, Wilhelm et al. 2001). Thus, an important factor to consider for modular BTL and WTL is the removal of the expensive and compression-intensive ASU.

Without the air separation unit, syngas generation must occur using air rather than pure oxygen. The use of air is only possible using a once-through (single pass) synthesis scheme, without internal and external recycle streams. This is done to avoid a build-up of nitrogen as the syngas stream would now contain approximately 51 vol.% N_2 (Jess et al. 1997). The scheme for a single-pass Fischer-Tropsch process without an ASU or recycle loops is shown in Fig. 6.

Operating the Fischer-Tropsch process in single pass mode has some direct benefits. Firstly, single pass operation removes the necessity for costly and energy-intensive recompression, as well as acid-gas removal, in the recycle loop. Excess nitrogen can also play an important role in heat removal from the exothermic Fischer-Tropsch reaction (Jess et al. 1997), especially if a fixed bed reactor system is used. In addition, as biogas contains a small amount of N_2, the removal of the recycle stream allows for a simplification of the biogas-cleaning step and/or purge stream losses.

Single pass operation also adds additional challenges. Higher volumetric flow rates (due to higher volumes of diluent nitrogen) inevitably result in the need for larger equipment, and higher power requirements for the Fischer-Tropsch feed compressor. High nitrogen partial pressures result in decreased thermal efficiency and reaction rates. The largest challenge with single pass mode, however, is the loss of fuel yield at low-moderate Fischer-Tropsch conversions. Thus, to maintain higher fuel yields for single pass systems, the Fischer-Tropsch synthesis should be operated at higher conversions than typically seen in industry.

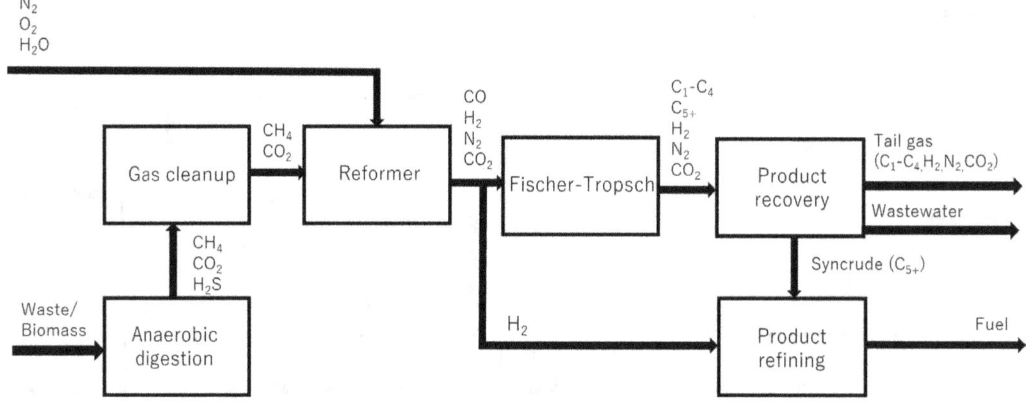

Figure 6. Schematic of the small-scale WTL/BTL plant with neither an ASU nor internal/external recycles.

5.4 Fischer-tropsch synthesis at high conversion

The Fischer-Tropsch synthesis is heart of the waste-to-fuel process, converting syngas into long chain hydrocarbons in a fixed or slurry bed reactor, using an iron or cobalt based catalyst. The FT reaction to paraffins and olefins can formally be represented as follows.

Alkane formation $\quad nCO + (2n{+}1)H_2 \rightarrow C_nH_{(2n+2)} + nH_2O$

Alkene formation $\quad nCO + 2nH_2 \rightarrow C_nH_{2n} + nH_2O$

A critical factor to consider for the single pass design is the choice of conversion associated with the Fischer-Tropsch reaction. Whilst a recycle system maintains overall conversions of carbon monoxide of above 90%, a typical per pass conversion for the Fischer-Tropsch synthesis is only ca. 55–65% (Rytter and Holmen 2015). Thus, for this system to be economically viable, the Fischer-Tropsch synthesis would also have to be operated at a higher conversion than typically seen in industry. This means that the reaction system will operate under a high partial pressure of hydrocarbons and water (products), and low partial pressures of carbon monoxide and hydrogen (reactants). These harsh conditions have been reported (Tucker and van Steen 2020, van Steen et al. 2018, Bukur et al. 2012, Ma et al. 2011) to significantly affect the activity, selectivity, and stability of Fischer-Tropsch catalysts. The choice of catalyst, reactor type and operating conditions must thus be chosen with this constraint in mind.

5.4.1 Catalyst choice

The only metals that have the required CO hydrogenation activity necessary for the Fischer-Tropsch process are ruthenium, nickel, cobalt and iron (Dry 2002). However, ruthenium has limited commercial availability and is costly (Dry 2002) whilst nickel has a high selectivity towards methane at elevated temperatures and has been reported to produce volatile carbonyls under high pressures and low temperature (Dry 2002). Iron and cobalt are therefore considered to be the only practical catalysts available for the Fischer-Tropsch synthesis.

Khodakov et al. (2007) compared iron and cobalt based Fischer-Tropsch catalysts in terms of their advantages and disadvantages as well as suitability to different Fischer-Tropsch conditions (see Table 5). Iron is significantly less expensive than cobalt, despite comparable activity. However, iron-based catalysts are also less resistant to deactivation via coking, carbon deposition and attrition. This means that cobalt requires less downtime for catalyst replacement. Both catalysts can operate under conditions required for the formation of long-chain hydrocarbons (typically termed low-temperature Fischer-Tropsch or LTFT), with slightly different ideal H_2: CO ratios.

Whilst the effect of high conversion (in the Fischer-Tropsch synthesis) on the activity, selectivity and deactivation of cobalt has recently been studied in detail (Tucker 2020, Tucker and van Steen 2020, Tucker et al. 2020, Bukur et al. 2012, Ma et al. 2011), literature on the effect of these conditions on iron is sparse (van Steen et al. 2018). The critical factors to consider when choosing a catalyst to operate at high conversion in a modular N_2 and CO_2 rich system arewater-gas shift activity, reverse water-gas shift activity and methanation.

Table 5. Comparison of cobalt and iron as Fischer-Tropsch synthesis catalysts adapted from Khodakov et al. (2007).

Parameter	Cobalt	Iron
Relative Cost (2016)	220	1
Deactivation	Resistant to deactivation	Less resistant to deactivation (coking, carbon deposition)
Activity at low conversion	Comparable	
Temperature of FTS	Only LTFT	HTFT and LTFT
High conversion consequences	Lower effect of water on WGS reaction	Water increases CO_2 formation and magnetite formation which deactivates the catalyst
Extent of WGS reaction	Insignificant except at very high conversions	Significant throughout conversion range
Flexibility to temperature and pressure	Not flexible. Significant effect of temperature and pressure on hydrocarbon selectivity	Flexible. Methane selectivity low even at high temperatures
H_2/CO Ratio	~ 2	0.5–2.5
Attrition resistance	Good	Not very resistant

- **Water-gas shift activity (WGS)**

  Iron has strong activity for the water-gas shift reaction, represented as follows.

  $$CO + H_2O \leftrightarrow H_2 + CO_2$$

  This means that under high partial pressures of water (as seen at high conversion) there is a strong propensity towards the formation of carbon dioxide, thus lowering the selectivity for the formation of long chain hydrocarbons.

  A carbon dioxide selectivity of between 25–50% was reported for CO conversion levels of 80% using various iron based catalysts ($Fe/Ru/K_2O/Al_2O_3$, $Fe/Cu/K_2O/SiO_2$ and $Fe/K_2O/Al_2O_3$) (van Steen et al. 2018). Cobalt is known to have a much lower water-gas shift activity, thus is typically regarded as more suitable for operating at high conversion.However, recent studies (Tucker and van Steen 2020, Ma et al. 2011) show that at CO conversions higher than 80%, in a slurry bed reactor, the selectivity towards CO_2 significantly increases for cobalt-based catalysts despite their low WGS activity. Levels of CO_2 selectivity between 2–20% were reported at CO conversions of 80–95% for $Pt-Co/Al_2O_3$(Tucker and van Steen 2020).

- **Reverse water-gas shift (RWGS)**

  The reverse water-gas shift (RWGS) reaction is, as named, the reverse reaction for the water-gas shift and is represented as follows.

  $$H_2 + CO_2 \leftrightarrow CO + H_2O$$

  Like for the water-gas shift reaction, the activity of iron-based catalysts for the reverse RWGS is higher than the activity of cobalt-based catalysts. Due to the large equilibrium constant for the WGS reaction at Fischer-Tropsch temperatures applicable to long-chain hydrocarbons (LTFT), the RWGS is not typically considered for this reaction. However, at high conversion conditions, the partial pressures within the reactor change so significantly that there may be a possibility for conversion of CO_2 to CO via the RWGS reaction. This would be especially true in situations where there is a high CO_2 partial pressure in the feed (as would be the case for a single pass WTL system without CO_2 removal). In these cases, iron's strong RWGS activity may be beneficial for internal CO_2 hydrogenation and increasing overall yield for the formation of hydrocarbons. This concept could be facilitated with the aid of *in-situ* water removal to shift the equilibrium conversion towards the RWGS.

- **Methanation**

  Methanation, represented as follows, also becomes a critically important reaction under high CO conversions within the Fischer-Tropsch reaction.

  $$CO + 3H_2 \leftrightarrow CH_4 + H_2O$$

  The selectivity towards methane is reported to increase significantly when operating the Fischer-Tropsch synthesis under high conversion conditions. This appears to be true for both cobalt (Tucker and van Steen 2020, Ma et al. 2011)and iron based catalysts (van Steen et al. 2018) due to enhanced H_2/CO ratios under these conditions (Tucker and van Steen 2020). However, as cobalt has a lower chain growth probability, the increase in the methane selectivity at high conversion is far more significant than seen for iron. At CO conversion levels between 80–95%, the methane selectivity of a $Pt-Co/Al_2O_3$ catalyst was found to be between 10–55% (Tucker and van Steen 2020), as opposed to iron that had a methane selectivity of between 3–25% (van Steen et al.2018) in the same conversion range.

5.4.2 Reactor choice

The choice of Fischer-Tropsch reactor influences the catalyst shape, mass transfer conditions, pressure drop, temperature control, and (most importantly for single pass operation) the partial pressure distribution of products and reactants. Two reactor types are common in modern commercial Fischer-Tropsch operations—fixed bed and slurry bed reactors. Table 6 shows a summary of advantages and disadvantages of each (De Klerk et al. 2013).

Slurry bed technology is based on a continuous stirred tank reactor design, with a liquid medium (wax for Fischer-Tropsch) to carry the catalyst throughout the reactor. The slurry bed design was first successfully operated at the Rheinpreusseb-Koppers demonstration plant in 1952 (Davis 2002) and has since been used for large scale applications by Sasol Ltd. (1993 onwards).

Slurry reactors offer good temperature control—with almost perfect isothermal behaviour due to mixing. The reactor also experiences low pressure drop with no dependency on catalyst size. This means that catalyst particles can be small and powder-like, reducing both mass and heat transfer limitations. For this reason, the catalyst consumption per tonne product is said to be 25–100% less than for fixed bed reactors (De Klerk et al. 2013). A significant benefit for modular operation is that slurry reactors are less costly than multitubular reactors, for the same capacity (De Klerk et al. 2013).

On the other hand, there has been significant issues in the scale-up of slurry bed reactors caused by fines. Mixing causes abrasion and attrition of the catalyst which often carries the catalyst out the reactor resulting in a loss of activity. The well-mixed nature of slurry bed reactors also means that internal concentrations are homogeneous. This means that high concentrations of water at high conversion affect the entire catalyst bed, rather than just a portion.

Fixed bed reactors were used as a primary Fischer-Tropsch reactor during the 1930s and 1940s. At this time, these reactors consisted essentially of a rectangular box filled with catalyst and cooled by tubes and cooling plates interlaced throughout the bed (De Klerk et al. 2013, Krishna and Sie 2000). Not long after did the design of a multitubular fixed bed reactor come to fruition. This reactor was essentially designed as a set of double concentric tubes (Krishna and Sie 2000) where the catalyst was loaded into the annular space. This provided good temperature control, with a catalyst bed thickness of only 9 mm (De Klerk et al. 2013); however, it was generally considered an overly-complicated design. In addition to this, loading and unloading catalyst with this design was very difficult.

The next generation of **multi-tubular reactors** consisted of multiple single tubes of ca. 50 mm diameter in which the catalyst was loaded. A shell surrounding the tubes was filled with flowing cooling water/steam to keep the system isothermal (De Klerk et al. 2013) (similar to a shell and

Table 6. Advantages and disadvantages of different reactor types for Fischer-Tropsch synthesis adapted from De Klerk et al. (2013).

	Fixed Bed	**Slurry Bed**
Catalyst shape required	Large pellets	Fine robust particles
Extent of mass transfer limitations	High	Low
Pressure drop	Large pressure drop if catalyst size not controlled	Small pressure drop despite catalyst particle size
Temperature control	Difficult to control temperature	Good heat control
Response to poisons	Poison absorbed in first layer of catalyst bed	Poison spreads throughout reactor
Catalyst replacement	Offline, hindered	Online
Selectivity to methane	Higher	Lower
Deactivation due to water	Constrained to bottom part of bed	Deactivation throughout catalyst

tube heat exchanger). This design allowed for simpler catalyst loading and unloading as well as far higher throughputs.

Modern fixed bed multi-tubular technology is well developed and robust. These reactors have been a commercial success in Shell's Pearl GTL in Qatar with a total capacity of 260 000 bbl/day. This design also has the added benefit of a concentration gradient which limits poisoning and/or oxidation to certain layers of the catalyst bed (De Klerk et al. 2013). The concentration gradient is also beneficial when targeting operation at high conversion, as the effect of high water-partial pressures on the stability and selectivity is limited to only catalyst particle near the end of the reactor.

However, there are some downsides to this technology. Due to the geometry of the reactor, large catalyst pellets are required to reduce pressure drop. This creates internal mass transport limitations which limits the effectiveness factor of the catalyst and decrease the productivity per gram of catalyst. Large pellets also create heat transfer limitations which makes controlling temperature even more tricky. This also makes control of selectivity problematic as temperature increases significantly increase selectivity towards methane, especially when using cobalt-based catalysts.

To limit these shortfalls, companies such as Velocys have developed novel microchannel reactor technology. Microchannel reactors are essentially multi-tubular fixed beds with much smaller tube diameters, smaller catalyst particles and shorter bed lengths (to avoid high pressure drop) (Deshmukh et al. 2009). This design limits heat and mass transfer issues associated with typical multi-tubular fixed bed technology. These reactors are also reportedly modular, allowing the plants to easily scale up to suit demand,making them ideal for small-to-medium scale operation.

5.5 Refining vs upgrading

The Fischer-Tropsch synthesis produces a mixture of linear hydrocarbons, waxes and oxygenates called syncrude (Calemma and De Klerk 2013). Depending on the required product (gasoline, diesel, jet fuel, etc.), syncrude can undergo multiple refining steps such as oligomerization, isomerization, hydrotreating, or hydrocracking to reach its final form. This makes refining a difficult and expensive process.

The choice regarding whether Fischer-Tropsch syncrude from modular BTL and WTL should be refined to final product on site, or simply upgraded and sold to central refineries is non-trivial, and locality will be an important factor in this consideration. Calemma et al. (2013) describes the four business cases for syncrude post-processing as follows:

- Syncrude product
- Upgrading
- Partial refining
- Refining

5.5.1 Syncrude product

To sell the raw syncrude product, one must have a centralized refinery that is willing to purchase and refine to final products. For the purposes of decentralized,modular fuel production, this may be impractical as it is unlikely that there will be refineries near enough that have the technical capabilities to process syncrude (multiple phases and different carbon distribution to crude oil).

5.5.2 Upgrading products

Low temperature Fischer-Tropsch products typically contain a large proportion of waxes that are solid at ambient temperature. Upgrading involves converting this wax into liquid hydrocarbons. Mild hydrocracking is the most common technique used for this purpose (De Klerk 2011). Hydrocracking

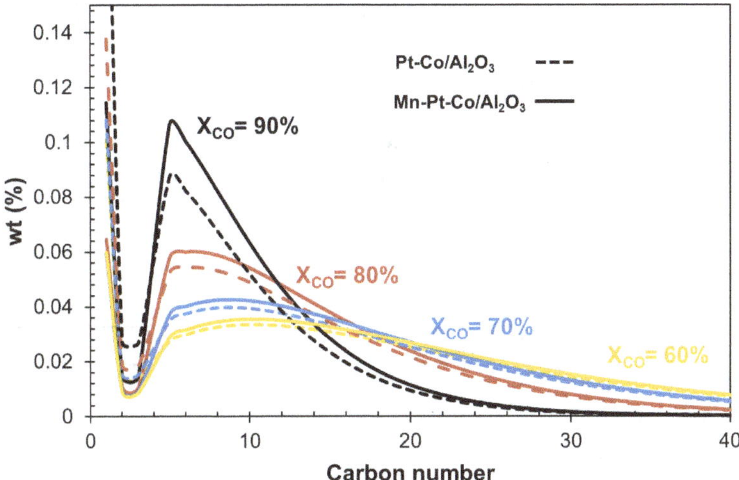

Figure 7. Carbon number distribution for Pt-Co/Al_2O_3 and Mn-Pt-Co/Al_2O_3 in a slurry bed reactor at various conversions based on data from (Tucker and van Steen 2019, 2020).

is a combination of catalytic cracking and hydrogenation. The reaction pathway includes multiple chemical reaction steps such as dehydrogenation, protonation, isomerisation, cracking, deprotonation and hydrogenation. Despite this complexity, Fischer-Tropsch products typically only require a single hydrocracking step to achieve high diesel-fraction yields (Möller et al. 2009). The product distribution emerging from a hydrocracker is, however, dependent on the extent of reaction in the hydro-cracking unit. Recent studies (Tucker 2020, Tucker and van Steen 2020) have shown operating at high Fischer-Tropsch conversions result in less wax production (C_{22+}).

Figure 7 shows the distributions of organic product compounds for Pt-Co/Al_2O_3 and Mn-Pt-Co/Al_2O_3 (Mn:Co = 0.14) between X_{CO} =60 and X_{CO} = 90% (Tucker and van Steen 2020). At a conversion of X_{CO} = 60%, the product distribution is wide for both catalysts and evenly distributed over the C_5–C_{15} range, with a significant fraction of this stream present as waxes (C_{22+}). Increasing the conversion to X_{CO} = 70% and X_{CO} = 80%, decreases the width of the product distribution, as well as shifting the distribution towards lighter hydrocarbons, as a consequence of a lowering of the chain growth probability upon increasing the conversion. Thus, at higher conversions wax hydrocracking is not strictly necessary to enhance diesel yields, whilst hydrocracking may be critical for systems operating at low conversion.

5.5.3 Complete refining

Complete refining would entail trying to separate out all possible products (distillate, kerosene, naphtha, etc.), and process via multiple reactions including oligomerization, hydroisomerisation, aromatisation and hydrotreating to produce multiple drop-in fuels such as diesel, petrol, jet fuel as well as chemicals (Calemma and De Klerk 2013). This approach is typically not used in modern Fischer-Tropsch plants due to the complexity and cost of each fuel refining step.

5.5.4 Partial refining

Partial refining is the most practical route for modular operation, focusing on refining a subset of wanted products and some blending materials (Calemma and De Klerk 2013). Many Fischer-Tropsch facilities (including larger scale facilities like Oryx in Qatar and Shell SMDS in Bintulu) make use of partial refining on-site. It is important to note that whilst this process is simpler than total refining,

Table 7. Regional diesel specifications for the US, EU, Africa and South Africa.

	Diesel specifications				
	ASTM D975: 15b (US)[1]	**EN 590:2014 (EU)[1]**	**EN 15940:2016 (EU)[2]**	**AFRI-5 2016 (AU)[3]**	**SANS 342:2016 (ZA)[4]**
Diesel density at 15°C	-[a]	820	765–800	820–880	805–850
Sulphur, ppm	< 15	< 10.0	< 5.9	50	10[b]
Polycyclic aromatic hydrocarbons (max mass%)	-[a]	-[a]	8	-[a]	8
Water, ppm	500	200	< 200	-[a]	250
Cetane number min	49	51	70	49	51
Viscosity at 40°C mm^2/s	1.9–4.1	2.0–4.5	2.0–4.5	-[a]	2.2–5.3
Flash Point,°C	55	55	55	-[a]	
Cloud Point,°C					
Winter	–5 to –34	–10 to –34	–10 to –34	-[a]	+6
Summer					-[a]
Distillation					
T_{95}	370	360	360	-[a]	362
T_{90}	338				360

[a] No limit specified
[b] Based on low sulphur diesel. Standard diesel is rated at 500 mg/kg.
[1](Neste Corporation 2015), [2](De Klerk 2009), [3](The African Refiners Association 2020),
[4](South African National Standards 2016)

the resulting liquid is not diesel fuel, but rather a distillate. Distillate is a mixture of roughly C_{10}–C_{22} straight-chain hydrocarbons, whilst diesel is a distillate that meets region-legislated specifications of cetane number, density, sulphur content, boiling point and distillation range, etc. (Calemma and De Klerk 2013, De Klerk 2011, 2009).

Table 7 shows the regional fuel specifications for the United States (ASTM D975:15b), European Union (EN 590:2014 and EN 15940:2016), African Union (AFRI-5 2016) and South Africa (SANS 342:2016). These specifications change with time and are updated regularly.

A potential issue for partially refined diesel obtained from a Fischer-Tropsch process is the cloud point, which represents the temperature at which waxy hydrocarbons begin to solidify and a cloud is visible in the diesel. This specification is set to avoid blockages in fuel filters and engine injectors. It is inherently based on the minimum temperatures within a region. This specification can vary significantly between 6°C (South Africa) and –34°C (lowest in EU and US). The rest of Africa has no cloud point specification. As Fischer-Tropsch LTFT fuels have a typically high cloud point (at 0°C (de Klerk 2009)), fuels may need to undergo isomerization in order to be sold as diesel on the EU and US markets.

The specification that partially refined distillate often cannot meet is density. The distillate product from Fischer-Tropschsynthesis typically has a density below 780 kg/m³ which does not fall within the EU (EN 590), African (AFRI-5) or South African (SANS 342:2016) specifications (De Klerk 2009). However, the United States of America (ASTM D975) has no such specification, and new EU specifications from 2016 were designed with a lower density limit in order to take Fischer-Tropsch based fuels into account (de Klerk 2009, de Klerk 2011).

5.6 Co-generation electricity and fuels

A consequence of the single pass design, and operation at high but not complete Fischer-Tropsch conversions, is the inevitable surplus of tail-gas containing CO, H_2, methane and other light hydrocarbons.

To recover the economic losses associated with lower carbon yields, the product spectrum may need to be diversified from only fuel to a combination of fuel and electricity (see Fig. 8). This concept has been reported to save operating costs for small-scale BTL systems (Liu et al. 2011). A study by Liu et al. (2011) used a detailed process simulation, life cycle analysis and a cost analysis to study 16 designs for BTL. This study showed that the levelized liquid fuel production costs for once-through CTL, CBTL hybrid and BTL systems can be reduced by as much 47%, 37% and 10%, respectively, by introducing electricity as a coproduct and providing decarbonized electricity to the grid as an independent energy producer (Liu et al. 2011).

In addition to the economic benefits of co-generation, this also allows the system to be energy self-sufficient, as the co-produced power can be used for compressors and pumps which would otherwise add to operating costs.

Producing electricity from Fischer-Tropsch gaseous products and unconverted reactants (containing C_1–C_4, CO_2, CO and H_2) can be achieved using a steam turbine or gas turbine. Alternatively, a hydrogen fuel cell can be used though this would require an extra reforming step.

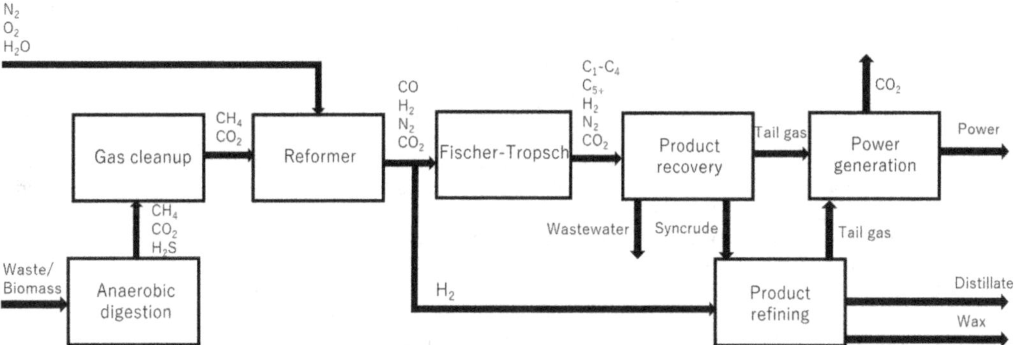

Figure 8. Schematic of the small-scale, once-through decentralized diesel system.

6. CONCLUDING REMARKS

Designing a modular plant for Fischer-Tropsch waste-to-liquid or biomass-to-liquid is by no means trivial. This chapter has focused on a few design choices that could enhance the simplicity or cost effectiveness of a waste-to-fuel plant. There is, however, no cookie-cutter design when it comes to biomass and waste-based technology. These decisions depend entirely on the type of feedstock available, local fuel market, capacity, plant location and local infrastructure.

It is critical to understand that every unit removed or simplified could result in further complications downstream. For instance, the removal of a Fischer-Tropsch recycle results in the requirement for electricity generation downstream. These options must therefore be weighed up with comprehensive economic analyses based on the site design in question.

The Fischer-Tropsch catalyst is also fundamentally important. Current catalyst designs cannot withstand the harsh conditions of high conversion which results in further design complexity. Future modifications to support and promoter materials may improve hydrothermal stability and selectivity to extents that make these additions redundant.

ACKNOWLEDGEMENTS

The scientific guidance from Prof. Arno De Klerk, Dr. Braam van Dyk and Dr. Ankur Bordeloi towards this chapter is gratefully acknowledged. This research was in part supported by the South African Research Chair Initiative (SARChI), UID 114606, and the BRICS-program (BRIC160624174073; UID 110487).

LIST OF ABBREVIATIONS

BTL	:	Biomass-to-liquid
CTL	:	Coal-to-liquid
FT	:	Fischer-Tropsch
GTL	:	Gas-to-liquid
MSW	:	Municipal solid waste
WTL	:	Waste to liquid

REFERENCES

Ail, S.S. and Dasappa, S. 2016. Biomass to liquid transportation fuel via Fischer Tropsch synthesis—technology review and current scenario. Renewable and Sustainable Energy Reviews 58: 267–286.

Anukam, A., Mohammadi, A., Naqvi, M. and Granström, K. 2019. A Review of the Chemistry of Anaerobic Digestion: Methods of Accelerating and Optimizing Process Efficiency. Processes 7(8): 1–19.

Arena, U. 2015. Process and technological aspects of municipal solid waste gasification: a review, Waste Management 32(4): 625–639.

Basu, P. 2010. Biomass gasification and pyrolysis: Practical design and theory. Academic Press. Massachusetts, United States.

Bernical, Q., Joulia, X., Borgne, I.N., Floquet, P., Baurens, P. and Boissonnet, G. 2013. Sustainability assessment of an integrated high temperature steam electrolysis-enhanced biomass to liquid fuel process. Industrial and Engineering Chemistry Research: 52(22): 7189–7195.

Boerrigter, H. 2006. Economy of Biomass-to-Liquids (BTL) plants: An engineering assessment. Energy Research Centre of the Netherlands, Report ECN-C-06-019, 2006.

Boucher, O., Friedlingstein, P., Collins, B. and Shine, K.P. 2009. The indirect global warming potential and global temperature change potential due to methane oxidation. Environmental Research Letters 4(4): 1–5.

Bridgwater, A.V. 2003. Renewable fuels and chemicals by thermal processing of biomass. Chemical Engineering Journal 91(2–3): 87–102.

Brodeur, G., Ramakrishnan, S. and Hsu, C.S. 2017. Biomass to liquid (BTL) fuels. pp. 1117–1132. *In*: C. Hsu and P. R. Robinson (eds.). Springer Handbook of Petroleum Technology. Springer. New York, United States.

Bukur, D.B., Pan, Z., Ma, W. and Jacobs, G. 2012. Effect of CO conversion on the product distribution of a Co/Al_2O_3 Fischer–Tropsch synthesis catalyst using a fixed bed reactor. Catalysis Letters 142(11): 1382–1387.

Calemma, V. and De Klerk, A. 2013. Fischer-Tropsch syncrude: to refine or to upgrade? *In*: Maitlis, P.M. and De Klerk, A. (eds.). Greener Fischer-Tropsch Processes for Fuels and Feedstocks. Wiley-VCH Verlag Gmb& Co. Weinheim, Germany.

Chattanathan, S.A., Adhikaria, S., McVey, M. and Fasinaa, O. 2014. Hydrogen production from biogas reforming and the effect of H_2S on CH_4 conversion. International Journal of Hydrogen Energy 39(35): 19905–19911.

Davis, B.H. 2002. Overview of reactors for liquid phase Fischer-Tropsch synthesis. Catalysis Today 71(3–4): 249–300.

De Klerk, A. 2016. Small-Scale Fischer-Tropsch Gas-to-Liquid Facilities. pp. 379–399. *In*: Davis, B.H. and Occelli, M.L. (eds.). Fischer-Tropsch Synthesis, Catalysts and Catalysis. Advances and Applications. CRC Press, Taylor & Francis Group. Florida, United States.

De Klerk, A. 2011. Fischer-Tropsch refining, Wiley-VCH Verlag Gmb& Co. Weinheim, Germany.

De Klerk, A. 2009. Can Fischer-Tropsch syncrude be refined to on-specification diesel fuel? Energy and Fuels 23(9): 4593–4604.

De Klerk, A., Li, Y.-W. and Zennaro, R. 2013. Fischer-Tropsch Technology. *In*: Maitlis, P.M. and De Klerk, A. (eds.). Greener Fischer-Tropsch Processes for Fuels and Feedstocks. Wiley-VCH Verlag Gmb& Co. Weinheim, Germany.

Deshmukh, S.R., LeViness, S.C., Robota, H.J. and Davis, M. 2009. Commercializing an Advanced Fischer-Tropsch Synthesis Technology. pp. 361–377. *In*: Davis, B.H. and Occelli, M.L. (eds.). Fischer-Tropsch Synthesis, Catalysts and Catalysis Advances and Applications. CRC Press, Taylor & Francis Group. Florida, United States.

Deublein, D. and Steinhauser, A. 2008. Biogas from Waste and Renewable Resources. WILEY-VCH Verlag GmbH & Co. KGaA. Weinheim, Germany.

Dimitriou, I., Goldingay, H. and Bridgwater, A.V. 2018. Techno-economic and uncertainty analysis of Biomass to Liquid (BTL) systems for transport fuel production. Renewable and Sustainable Energy Reviews 88: 160–175.

Dry, M.E. 2002. High quality diesel via the Fischer-Tropsch process—A review. Journal of Chemical Technology and Biotechnology 77(1): 43–50.

Dry, M.E. 2002. The Fischer-Tropsch process: 1950–2000. Catalysis Today 71: 227–241.

Enerdata 2014. The future of Gas-To-Liquid (GTL) industry. Available at: https://www.enerdata.net/publications/executive-briefing/future-gas-liquid-gtl-industry.html. (Accessed 20 July 2021).

ETIP Bioenergy. 2017. Discontinued BtL projects. Available at: https://www.etipbioenergy.eu/value-chains/conversion-technologies/advanced-technologies/biomass-to-liquids/discontinued-btl-projects. (Accessed 16 June 2021).

Fagernäs, L., Brammer, J., Wiléna, C., Lauer, M. and Verhoeff, F. 2010. Drying of biomass for second generation synfuel production. Biomass and Bioenergy 34(9): 1267–1277.

Fulcrum Bioenergy. 2018. Fulcrum targets northwest Indiana for the location of its next Waste-to-Fuel plant. Available at: https://fulcrum-bioenergy.com/wp-content/uploads/2018/12/2018-12-13-Fulcrum-Centerpoint-Announcement-FINAL.pdf. (Accessed 3 June 2021).

Glebova, O. 2013. Gas to Liquids: Historical development and future prospects. The Oxford Institute for Energy Studies. Oxford, United Kingdom.

Herz, G., Reichelt, E. and Jahn, M. 2017. Design and evaluation of a Fischer-Tropsch process for the production of waxes from biogas. Energy 132: 370–381.

Hilkiah Igoni, A. Ayotamuno, M.J., Eze, C.L., Ogaji, S.O.T. and Probert, S.D. 2008. Designs of anaerobic digesters for producing biogas from municipal solid-waste. Applied Energy, 85(6): 430–438.

Izquierdo, U., Barrio, V.L., Requires, J., Cambra, J.F., Güemez, M.B. and Arias, P.L. 2012. Tri-reforming: A new biogas process for synthesis gas and hydrogen production. International Journal of Hydrogen Energy 38(18): 7623–7631.

Jess, A., Popp, R. and Hedden, K. 1997. From natural gas to liquid hydrocarbons. Part 4: Production of diesel oil and wax by Fischer-Tropsch-synthesis using a nitrogen-rich synthesis gas—Investigations on a semi-technical scale. Erdoel Erdgas Kohle 113: 531–540.

Kannah, R.Y. and Banu, J.R. 2012. Introductory Chapter: An Overview of Biogas' *In*: Banu, J.R. (ed.). Anaerobic Digestion, Intech Open. https://doi.org/10.5772/intechopen.82198.

Kennedy, H. 2018. Red Rock Biofuels facility breaks ground in Lakeview, Biofuels digest. Available at: http://www.biofuelsdigest.com/bdigest/2018/07/29/red-rock-biofuels-facility-breaks-ground-in-lakeview/. (Accessed December 2018).

Khodakov, A.Y., Chu, W. and Fongarland, P. 2007. Advances in the development of novel cobalt Fischer-Tropsch catalysts for synthesis of long-chain hydrocarbons and clean fuels. Chemical Reviews 107(5): 1692–1744.

Krishna, R. and Sie, S.T. 2000. Design and scale-up of the Fischer-Tropsch bubble column slurry reactor. Fuel Processing Technology 64(1): 73–105.

Levaggi, L., Levaggi, R., Marchiori, C. and Trecroci, C. 2020. Waste-to-Energy in the EU: The effects of plant ownership, waste mobility, and decentralization on environmental outcomes and welfare. Sustainability 12(14): 5743.

Liu, G., Larson, E.D., Williams, R.H., Kreutz, T.G. and Guo, X. 2011. Making Fischer−Tropsch fuels and electricity from coal and biomass: Performance and cost analysis. Energy and Fuels 25(1): 415–437.

Ma, W., Jacobs, G., Ji, Y., Bhatelia, T., Bukur, D.B., Khalid, Syed. and Davis, B.H. 2011. Fischer–Tropsch synthesis: Influence of co conversion on selectivities, H_2/CO usage ratios, and catalyst stability for a Ru promoted Co/Al_2O_3 catalyst using a slurry phase reactor. Topics in Catalysis 54 (13–15): 757–767.

Mawhood, R., Gazis, E., de Jong, S. and Hoefnagels, R. 2016. Production pathways for renewable jet fuel: a review of commercialization. Biofuels, Bioproducts and Biorefinery 10: 462–484.

Meegoda, J.N., Li, B., Patel, K. and Wang, L.B. 2018. A review of the processes, parameters, and optimization of anaerobic digestion. International Journal of Environmental Research and Public Health 15(10): 2224.

Möller, K., le Grange, P. and Accolla, C. 2009. A two-phase reactor model for the hydrocracking of Fischer-Tropsch-derived wax. Industrial and Engineering Chemistry Research 48(8): 3791–3801.

Naqi, A. 2018. Conversion of biomass to liquid hydrocarbon fuels via anaerobic digestion: A feasibility study. [MSc dissertation, University of South Florida]. Scholar Commons. https://scho larcommons.usf.edu/ cgi/viewcontent.cgi?article=8836&context=etd

Navas-Anguita, Z., Cruz, P.L., Martín-Gamboa, M., Iribarren, D. and Dufour, J. 2019. Simulation and life cycle assessment of synthetic fuels produced via biogas dry reforming and Fischer-Tropsch synthesis. Fuel 235: 1492–1500.

Neste Corporation 2015. Neste Renewable Diesel Handbook. Neste, pp. 1–33. https://doi.org/10.1016/S0997-7538(02)01247-0.

Neville Hargreaves. 2018. Altalto waste-to-jet fuel project. Available at: https://worldwastetoenergy.com/wp-content/uploads/2019/05/Neville-Hargreaves-Velocys.pdf. (Accessed July 2021).

Ng, K.S. and Sadhukhan, J. 2011. Techno-economic performance analysis of bio-oil based Fischer-Tropsch and CHP synthesis platform. Biomass and Bioenergy 35(7): 3218–3234.

Petro, S.A. 2020. Operations and Refinery. Available at: http://www.petrosa.co.za/innovation_in_action/pages/operations-and-refinery.aspx. Accessed 19 July 2021.

Psomopoulos, C.S., Bourka, A. and Themelis, N.J. 2009. Waste-to-energy: A review of the status and benefits in USA. Waste Management 29(5): 1718–1724.

Reuters. 2011 . China Shenhua coal-to-liquids project profitable-exec. Available at: https://www.reuters.com/article/shenhua-oil-coal-idUKL3E7K732020110908. (Accessed June 2021).

Rytter, E. and Holmen, A. 2015. Deactivation and regeneration of commercial type Fischer-Tropsch Co-catalysts—A mini-review. Catalysts 5(2): 478–499.

Shah, V., Kuehn, N.J. and Turner. M.J. 2014. Cost and performance baseline for fossil energy plants, Volume 4: Coal-to-liquids via Fischer-Tropsch synthesis. Available at: https://www.netl.doe.gov/energy-analysis/details?id=736. (Accessed June 2019).

Song, C. and Pan, W. 2004 Tri-reforming of methane: A novel concept for catalytic production of industrially useful synthesis gas with desired H2/CO ratios. Catalysis Today 98(4): 463–484.

Sorda, G., Banse, M. and Kemfert, C. 2010. An overview of biofuel policies across the world. Energy Policy 38(11): 6977–6988.

South African National Standards (2016) SANS 342 : 2016.

The African Refiners Association (ARA) (2020). ARA policy on African gasoline and diesel specifications. 2006–2007.

Tijmensen, M.J.A., Faaij, A.P.C., Hamelinck, C.N. and van Hardeveld, M.R.M. 2002. Exploration of the possibilities for production of Fischer Tropsch liquids and power via biomass gasification. Biomass and Bioenergy 23(2): 129–152.

Trippe, F., Fröhling, M., Schultmann, F., Stahl, R., Henrich, E. and Dalai, A. 2013. Comprehensive techno-economic assessment of dimethyl ether (DME) synthesis and Fischer-Tropsch synthesis as alternative process steps within biomass-to-liquid production. Fuel Processing Technology 106: 577–586.

Tucker, C.L. 2020. Waste to Fuel: Designing a cobalt-based catalyst and process for once-through Fischer-Tropsch synthesis operated at high conversion [Doctoral dissertation, University of Cape Town]. Open UCT. http://hdl.handle.net/11427/33059.

Tucker, C.L. and van Steen, E. 2019. Manganese Promotion of Pt-Co/Al2O3 Catalyst for the Single Pass Fischer-Tropsch Biomass-to-Liquid Operation, in North American Catalysis Society Meeting.

Tucker, C.L., van Steen, E. 2020. Activity and selectivity of a cobalt-based Fischer-Tropsch catalyst operating at high conversion for once-through biomass-to-liquid operation. Catalysis Today, 342, 115–123.

Tucker, C.L., van Steen, E. and Claeys, M. 2020. Decoupling the deactivation mechanisms of a Pt-Co/Al_2O_3 Fischer-Tropsch catalyst operated at high conversion and "simulated" high conversion. Catalysis Science and Technology 10: 7056–7066.

Tucker, C.L., Ragoo, Y., Mathe, S., Macheli, L., Bordoloi, A., Rocha, T.C.R., Govender, S., Kooyman, P.J. and van Steen, E. 2022. Manganese promotion of a cobalt Fischer-Tropsch catalyst to improve operation at high conversion. Journal of Catalysis 411: 97–108.

van Steen, E., Claeys, M., Möller, K.P. and Nabaho, D. 2018. Comparing a cobalt-based catalyst with iron-based catalysts for the Fischer-Tropsch XTL-process operating at high conversion. Applied Catalysis A: General 549: 51–59.

Velocys 2018. Biorefineries. Available at: https://www.velocys.com/our-biorefineries/. (Accessed November 2019).

Wilhelm, D.J., Simbeck, D.R., Karp, A.D. and Dickenson, R.L. 2001. Syngas production for gas-to-liquid applications: technologies, issues and outlook. Fuel Processing Technology 71: 139–148.

Yao, J., Kraussler, M., Benedikt, F. and Hofbauer, H. 2017. Techno-economic assessment of hydrogen production based on dual fluidized bed biomass steam gasification, biogas steam reforming, and alkaline water electrolysis processes. Energy Conversion and Management 145: 278–292.

Zennaro, R. 2013. Fischer—Tropsch process economics. *In*: Maitlis, P.M. and De Klerk, A. (eds.), Greener Fischer-Tropsch Processes for Fuels and Feedstocks. Wiley-VCH Verlag Gmb& Co. Weinheim, Germany.

Zhang, X. and Brown, R.C. 2011. Introduction to thermochemical processing of biomass into fuels, chemicals and power. *In*: Brown, R.C. (ed.). Thermochemical Processing of Biomass: Conversion into Fuels, Chemicals and Power. John Wiley & Sons. New Jersey, United States.

Zhao, X., Naqi, A., Walker, D.M., Roberge, T., Kastelic, M., Joseph, B. and Kuhn, J.N. 2019. Conversion of landfill gas to liquid fuels through a TriFTS (tri-reforming and Fischer-Tropsch synthesis) process: A feasibility study. Sustainable Energy and Fuels 3(2): 539–549.

14

WASTE TO CARBON

A SUSTAINABLE APPROACH FOR CONVERTING AGRICULTURAL WASTES INTO BIO-BASED CARBON ADSORBENTS FOR WASTEWATER TREATMENT

George S. Morcos,[1] *Abdelkadr S. Ahmed,*[1]
Marwa M.H. El-Sayed[2] and *Mayyada M.H. El-Sayed*[1,*]

1. INTRODUCTION

Activated carbon (AC) or activated charcoal is a porous amorphous carbonaceous material, which is physically characterized by a turbostratic crystalline structure that exhibits a high degree of porosity, in addition to a high specific surface area. Chemically, AC shows a high degree of surface reactivity due to its surface chemistry that is influenced by its method of preparation. AC may thus possess various organic functional groups on its surface, such as carboxyl, carbonyl, carboxylic anhydride, lactone, quinone, phenol, and sulfonic acid groups, and these groups provide it with unique chemical and physical properties. With regard to its physical appearance, AC can be prepared in the form of powder, granules, pellets, fibers, or cloth. The chemical nature of AC renders it a versatile material that is utilized in diverse applications such as water and wastewater treatment, desalination, air purification, and catalysis. Various precursor materials have been utilized for the synthesis of AC, including natural/bio-based and synthetic materials. The former category comprises living and dead biomass, natural polymers, and agricultural wastes (AWs). Of particular interest are AWs which have been popular in the synthesis of AC due to their low cost, availability, and sustainability (Chen et al. 2020, Gao et al. 2020, Heidarinejad et al. 2020, Yahya et al. 2015).

[1] Department of Chemistry, School of Science and Engineering, The American University in Cairo, AUC Avenue, P.O. Box 74, New Cairo, 11835, Cairo, Egypt
[2] Department of Civil Engineering, Embry-Riddle Aeronautical University, Daytona Beach, FL 32114, USA
* Corresponding author: Mayyada@aucegypt.edu

AC is derived from various carbon-based sources that are classified into bio-based and coal-based sources (bituminous/subbituminous coal, lignite, anthracite). Bio-based sources include raw materials made up of cellulose, hemicellulose, or lignin, such as wood (Khezami et al. 2005), peat soil (Khadiran et al. 2015) or AWs. Due to the high costs and environmental concerns associated with AC production, AWs have been recently exploited as viable resources for AC production. Thus, AC is being produced from a myriad of AWs which can be grouped into pits, pith, seeds, stones, shells, peels, pods, stems, leaves, stalks, pulp, straws, tassels, husks, hulls, cobs, and bagasse (González-García 2018, Ioannidou and Zabaniotou 2007).

2. CONVERSION OF AWs TO AC

The process of converting AWs into AC involves, in essence, a series of steps that can be classified into three main stages: I) pretreatment, II) carbonization, and III) activation. The initial pre-treatment is required prior to carbonization to furnish the waste material by washing, drying and size reduction. The carbonization can be performed either individually or in combination with activation. A schematic flowchart of the conversion process is depicted in Fig. 1, and a detailed discussion of its stages is provided in the following sections. The flowchart shows the detailed process scheme from AWs raw material to AC production.

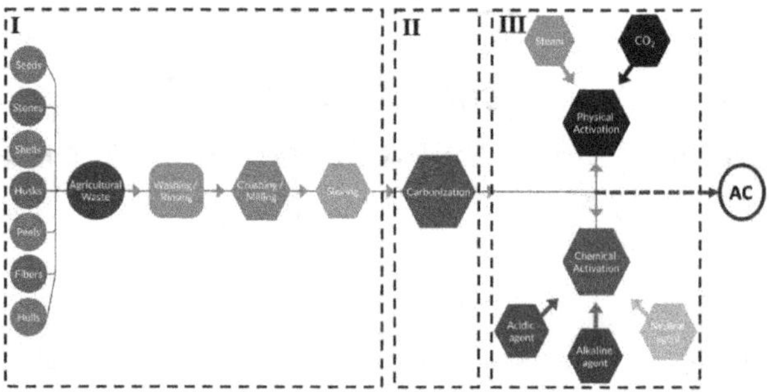

Figure 1. Flowchart showing the steps of preparing AC from AWs.

2.1 Pre-treatment

AWs are initially pre-treated prior to carbonization. Pre-treatment involves thorough washing, rinsing, and drying of AWs, followed by crushing and/or milling to convert them into powder. Particles are then fractioned by analytical sieves to obtain the particle size range desired for the subsequent carbonization stage (Castro et al. 2000). Particle size influences the carbonization process in which larger particles heat up slower than smaller ones and thus less volatile yields are expected, while the smaller particles would allow more uniform heat distribution among the particles bulk volume (Beis et al. 2002). Furthermore, the particle size of the raw precursor material influences the resulting physical properties of the AC. As reported in literature, the particle size of the precursor material can range widely between 177 μm to 4 mm (Bagheri and Abedi 2009, Şentorun-Shalaby et al. 2006). The smaller size range, in general, imposes limitations due to handling difficulties (Şentorun-Shalaby et al. 2006), or the clogging of gas pipes in the reactor (Ozdemir et al. 2014). In a couple of studies on producing AC by steam activation and by chemical activation with potassium hydroxide, it has been shown that reducing the precursor particle size decreases the yield of AC and increases its surface area; however, it had no significant effect on the micro pore volume. In another study involving zinc chloride activation, the effect of increasing the zinc loading on the AC porosity was more pronounced with larger sized particles (Caturla et al. 1991).

2.2 Carbonization

Carbonization is the heat-treatment process required for converting organic precursors into carbon material. In other words, it refers to the reduction of the volatile content associated with the raw material in order to produce char or charcoal, a high content of fixed carbon. This process involves heat treatment of precursor materials at high temperatures which typically range between 300 and 900°C. Such thermal treatment can take place under atmospheric or high pressures, in inert or oxidative environments, and in the presence or absence of catalysts. However, this treatment is commonly conducted in an inert atmosphere, in a process referred to as 'pyrolysis'. The selected carbonization method and its operating parameters dictate the physical properties of the produced activated carbon (Chen et al. 2020, Gao et al. 2020). It was suggested in previous literature that temperature and heating rates are among the most influencing carbonization factors (Pallarés et al. 2018).

Temperature has a prominent effect on the surface area of char. Elevated heating temperatures enhance the development of pores due to the destruction of more carbonaceous compounds as cellulose and lignin, and the removal of volatile matter. This, in turn, creates more pores in the material (Bouchelta et al. 2012). Moreover, increasing the pyrolysis temperature results in a less char yield and a higher surface area and pore volume, since more labile material is removed (Angın 2013, Brunner and Roberts 1980, Zhao et al. 2018).

In addition to temperature, the heating rate is another factor that influences the yield and physical properties of the produced AC. The associated heat and mass transfer during the heating process determines the physical properties of the final AC product. A slow heating rate limits the diffusion of the inert gas into the bulk of the carbonaceous material (CM). As the heating rate increases, the temperature elevates faster in the CM's bulk leading to enhanced extraction of volatile compounds and dissociation of cellulose, hemicellulose, or lignin. This, in turn, facilitates the access of the inert gas to the CM's bulk inducing more micropores (Safdari et al. 2019). However, increasing the heating rate, on the other hand, also increases the yield of tar and evolving gases, and subsequently decreases the char yield, the specific surface area, and the micropore volume. The decline in surface area and pore volume with fast heating rates has been attributed to the shorter time provided for the discharge of volatiles from the pores and their consequent entrapment within the char particles increasing the chance of pore blockage by carbon deposits (Angın 2013). High heating rates might also induce partial graphitization of the CM, which is not favorable for micropore formation (Bouchelta et al. 2012). In terms of the produced gases, high yield of volatiles associated with fast heating rates has been ascribed to the cracking of char as well as the decomposition of tar compounds (Brunner and Roberts 1980, Safdari et al. 2019). Therefore, it is instrumental to optimize the heating rate in accordance with the pyrolysis, carbonization or activation conditions. It has been reported that the optimum heating rate prior to which the char surface area increases proportionally until reaching a maximum, commonly ranges from 10 to 20°C/min (Bouchelta et al. 2012, Tay et al. 2001). It was also demonstrated that the effect of heating rate on the physical properties of the produced char is less pronounced with lower activation temperatures (Brunner and Roberts 1980).

2.3 Activation

Carbonization of organic material into char is followed by an activation stage wherein the physical and chemical attributes of the CM are manipulated. Activation affects the specific surface area and pore volume of the CM by introducing new pores and developing existing ones, and hence the process involves formation of pores, along with their expansion, combination, and collapse. The process also modifies the surface of the CM to target specific characteristics. Carbon can be activated via three main techniques, namely physical, chemical, and physio-chemical activation (Gao et al. 2020), each of which is detailed in the following sections.

2.3.1 Physical activation

Physical activation, also known as dry oxidation, utilizes an oxidizing gas such as air, carbon dioxide (CO_2), steam, or a mixture thereof. Initially, carbonization takes place through pyrolysis in the presence of an inert gas such as nitrogen or argon. This is followed by activation with an oxidant at high temperatures (400–900°C) to produce AC. Physical activation is considered a green production method since it does not involve the usage of any chemicals that may generate secondary waste. On the downside, physical activation requires higher thermal input and longer processing time as compared to other activation methods and, nonetheless, produces a low yield of AC that has a poor specific surface area (Ioannidou and Zabaniotou 2007, Yahya et al. 2015). Yield is defined as the ratio of the resulting AC to that of the original AWs dry weight (Diao et al. 2002). Examples of different AWs that have been activated by physical means are provided in Table 1. As shown, activation was generally performed at temperatures ranging from 400–900°C, time intervals of 0.5–2 h and using steam, carbon dioxide or nitrogen as activating agents.

As is clear from the studies given in the table, coconut shells possessed the highest surface areas among other physically activated AWs recording 1926 m^2/g (Li et al. 2008) when activated for 2 h using steam at 1000°C, and 1700 m^2/g (Guo et al. 2009) when activated with CO_2 at 600°C for the same time duration. Treatment of AWs with steam and CO_2 under the same carbonization and activation conditions showed that steam activation developed a microporous structure with a higher surface area than that obtained with CO_2 activation (Pallarés et al. 2018, Toles et al. 2000, Pastor-Villegas and Durán-Valle 2002). However, at lower gas flow rates (2500 cm^3/min), CO_2 activation improved the surface area and porosity in comparison to steam activation (Pallarés et al. 2018).

2.3.2 Chemical activation

Chemical activation is a single-step process in which the CM is carbonized and activated simultaneously. Chemical activation is also referred to as wet oxidation, as it involves the use of liquid activating agents throughout the entire carbonization process. This type of activation involves the dehydration, degradation, and oxidation of the CM through the action of activating agents, which accordingly leads to the *in-situ* formation of pores into the CM. The chemical nature of these activating agents dictates the specific physico-chemical properties of the resulting AC. Activating agents are classified into acidic, alkaline, neutral, and self-activating (Gao et al. 2020). Chemical activation is favored due to its low thermal requirements and short processing time. It produces a high yield of carbon characterized by a high specific surface area, and controlled porosity. The produced carbon, however, requires exhaustive washing to remove the highly corrosive chemical agents (Heidarinejad et al. 2020).

2.3.2.1 Alkaline activating agents

The most commonly used strong alkaline activating agents are potassium hydroxide (KOH), sodium hydroxide (NaOH), and potassium carbonate (K_2CO_3). KOH yields ACs with remarkably high surface areas, at a relatively low activation temperature since it increases the rate of pyrolysis and suppresses tar generation. During its activation process, KOH may dehydrate the CM by forming potassium oxide (K_2O) and water. The produced water may then react with carbon to produce hydrogen and carbon monoxide. Water may react, as well, with the produced carbon monoxide to produce hydrogen and carbon dioxide. K_2O, on the other hand, may react with CO_2 to produce potassium carbonate (K_2CO_3), and with carbon to produce metallic potassium, that diffuses into the vapor phase, intercalating into the carbonaceous matrix and inducing pores within (Cao et al. 2006, Gao et al. 2020). In spite of NaOH being cheaper and more environmentally friendly (Heidarinejad et al. 2020), KOH yields AC

Table 1. Previous studies reporting the synthesis of physically activated AWs showing their activation conditions of temperature and time, activating agents, and their surface areas.

AW	Activation conditions		Activating agent	Surface area (m^2/g)	Reference
	Temperature (°C)	Time (h)			
Coconut shell	1000	2	Steam	1926	(Li et al. 2008)
Olive waste cake	850	1	Steam	1271	(Baçaoui et al. 2001)
Apricot stones	800	1	Steam	1190	(Savova et al. 2001)
Almond tree pruning	850	0.5	Steam	1080	(González et al. 2009)
Almond shells	800	1	Steam	998	(Savova et al. 2001)
Cherry stones	800	1	Steam	875	(Savova et al. 2001)
Olive stone	850	0.5	Steam	813	(González et al. 2009)
Walnut shell	850	0.5	Steam	792	(González et al. 2009)
Date pits	800	1	Steam	702	(Awwad et al. 2013)
Corncob	700	2	Steam	786	(El-Hendawy et al. 2001)
Nut shells	800	1	Steam	743	(Savova et al. 2001)
Oat hull	800	2	Steam	625	(Fan et al. 2004)
Almond shell	850	0.5	Steam	601	(González et al. 2009)
Barley straw	700	1	Steam	552	(Pallarés et al. 2018)
Ceiba Pentandra hulls	-	-	Steam	521	(Rao et al. 2008)
Sawdust	800	1	Steam	516.3	(Malik 2003)
Grape seeds	800	1	Steam	497	(Savova et al. 2001)
Coconut buttons	400	1	Steam	479	(Anirudhan and Sreekumari 2011)
Corn stover	800	1	Steam	442	(Fan et al. 2004)
Phaseolus aureus hulls	400	1	Steam	325	(Raoet al. 2009)
Sugarcane bagasse	900	2	Steam	320	(Devnarain et al. 2002)
Rice husk	600	1	Steam	272.5	(Malik 2003)
Peanut hull	600	2	Steam	253	(Girgis et al. 2002)
Rice straw	650	1	Steam	122.9	(Daifullah et al. 2007)
Water hyacinth (*Eichhornia crassipes*)	800	1.5	Steam	97.68	(Halder et al. 2015)
Coconut shell	600	2	CO_2	1700	(Guo et al. 2009)
Almond shells	755	5	CO_2	1217.7	(Marcilla et al. 2000)
Pistachio-nutshell	800	2.5	CO_2	1064.2	(Yang and Lua 2003)
Pistachio-nutshell	900	0.5	CO_2/vacuum	896	(Lua and Yang 2004)
Barley straw	800	1	CO_2	789	(Pallarés et al. 2018)
Pistachio-nutshell	900	0.5	CO_2	778.1	(Lua et al. 2004)
Coconut coir	700	0.5	N_2	826	(Chaudhuri and Azizan 2012)

of higher micropore volume, porosity, pore size, and surface area than its counterpart produced by NaOH activation at the same conditions (Ahmed and Theydan 2014b, Olivares-Marín et al. 2009). Similar to KOH, K_2CO_3 produces AC with a higher proportion of micropores than mesopores (Foo and Hameed 2012, Márquez-Montesino et al. 2020). In contrast to KOH, NaOH produces AC which possesses a higher percentage of mesopore volume relative to micropore volume (Islam et al. 2017).

Various AWs have been activated chemically by KOH, NaOH, and K_2CO_3, as listed in Table 2. Activation was conducted at temperatures ranging between 250–1000°C using conventional heating, except for a single study (Ahmed and Theydan 2012) that deployed microwave (MW) radiation. Activation time ranged from 8 min up to 4 h and different impregnation ratios (i.e., weight ratios of the activating agent to the precursor (Hayashi et al. 2000)) were used depending on the nature of the AW, to obtain AC yields of 5–85%.

Table 2. Previous studies reporting AWs activated using alkaline activating agents. The activation parameters, surface areas and yields obtained are given in the table.

AW	Activating agent	Activation conditions			Surface area (m^2/g)	Yield (%)	Reference
		Temperature (°C)	Time (h)	Impregnation ratio			
Tremella	KOH	800	3	5:1	3760	-	(Guo et al. 2017)
Enteromorpha prolifera	KOH	500	1.5	-	3332	-	(Gao et al. 2014)
Rice husk (2-stage)	KOH	700	1	4:1	3263	-	(Liu et al. 2016)
Corncob	KOH	850	1	4:1	3227	-	(Liu et al. 2014a)
Corncob	KOH	850	3	-	3054	27.5	(Wang et al. 2015a)
Lignite	KOH	800	1	4:1	3036	-	(Xing et al. 2015)
Lignin of papermaking black liquor	KOH	750	1	3:1	2943	41.6	(Gao et al. 2013)
Kraft lignin	KOH	700	1	3:1	2920	-	(Vicinisvarri et al. 2014)
Spartina alterniflora	KOH	450	1.5	3:1	2825	32	(Liu et al. 2013)
Carrageenan (2-stage)	KOH	700	4	4:1	2800	-	(Nogueira et al. 2018)
Silkworm cocoon waste	KOH	800	3	1:1	2797	21.7	(Li et al. 2015)
Date press cake	KOH	750	1.5	3:1	2632.5	44.5	(Heidarinejad et al. 2018)
Cane pith	KOH	780	1	5:1	2207	8.2	(Tseng and Tseng 2006)
Fir wood	KOH	780	1	4:1	2179	17.8	(Wu et al. 2005)
Chickpea	KOH	850	-	2:1	2082	-	(Özsin et al. 2019)
Hazelnut shells (2-stage)	KOH	900	1	2:1	2031	-	(Kwiatkowski and Broniek 2017)
Bamboo	KOH	850	2	1:1	1896	-	(Hameed et al. 2007)
Pomelo peel	KOH	800	2.5	2:1	1892	-	(Li et al. 2016)
Albizia lebbeck	KOH	620W MW radiation	0.13	1:1	1824.88	22.48	(Ahmed and Theydan 2012)
Coffee grounds	KOH	800	1	-	1778	19	(Laksaci et al. 2017)
Vine shoots	KOH	700	1	2:1	1671	-	(Manyà et al. 2018)
Cherry stones	KOH	900	2	3:1	1624	5	(Olivares-Marín et al. 2009)
Walnut shell	KOH	250	2	1:1	1604	-	(Ansari et al. 2018)
Rice straw (2-stage)	KOH	800	1	4:1	1554	13.5	(Basta et al. 2009)
Paulownia flower	KOH	600	2	-	1471	-	(Chang et al. 2015)
Coconut husk	KOH	700	3	4:1	1448	-	(Talat et al. 2018)
Maize stalk	KOH	700	1	0.75:1	1414	13.1	(El-Hendawy 2009)
Olive stones	KOH	600	1	0.5:1	1281	85.15	(Alslaibi et al. 2013)

Wood	KOH	700	2	0.75:1	1255	-	(Khezami and Capart 2005)
Grape seeds	KOH	800	1	0.25:1	1222	30.5	(Okman et al. 2014)
Oil palm fruit bunch	KOH	814	1.9	2.8:1	1141	18	(Hameed et al. 2009)
Willow catkins	KOH	700	2	1:1	1067	-	(Wang et al. 2017)
Coconut shell	KOH	850	2	1:1	1026	-	(Mohd Din et al. 2009)
Rambutan peel	KOH	750	2	1:1	972	-	(Njoku et al. 2014a)
Pomegranate peel	KOH	700	0.5	1	941	-	(Ahmad et al. 2014)
Jatropha curcas	KOH	500	4	2:1	905	-	(Kurniawan and Ismadji 2011)
Oil Palm kernel shell	KOH	500	4	-	727	-	(Misnon et al. 2015)
Coconut pith (2-stage)	KOH	700	1	4:1	505	31.82	(Saman et al. 2015)
Popcorn	NaOH	650	2	4:1	3291		(Yu et al. 2019)
Coconut shell	NaOH	700	1.5	3:1	2825	18.8	(Cazetta et al. 2011)
Cotton linter fibers	NaOH	550	1	3:1	2143	-	(Sun et al. 2012)
Guava seed	NaOH	750	1.5	3:1	2574	14.9	(Pezoti et al. 2016)
Date press cake	NaOH	650	1.5	2	2026	26.2	(Norouzi et al. 2018)
Rice husk	NaOH	850	3	2:1	1958	-	(Youssef et al. 2012)
Plum kernel	NaOH	780	1	4:1	1887	37	(Tseng 2007)
Macadamia nutshell	NaOH	700	1.5	3:1	1524	19.8	(Martins et al. 2015)
Lacosperma secundiflorum	NaOH	600	1	3:1	1135	-	(Islam et al. 2017)
Tea	NaOH	800	1	3:1	368	-	(Islam et al. 2015)
Bamboo	K_2CO_3	750	0.5	3:1	2237	-	(Wang et al. 2015b)
Chickpea	K_2CO_3	850	-	1:1	1921	-	(Özsin et al. 2019)
Rice husk ash	K_2CO_3	1000	1.5	1.5:1	1713	-	(Liu et al. 2012b)
Albizia lebbeck	K_2CO_3	540W MW radiation	0.13	1.5:1	1676.6	26.19	(Ahmed and Theydan 2012)
Bean pods	K_2CO_3	700	1	4:1	1580	-	(Cabal et al. 2009)
Pine	K_2CO_3	800	2	3:1	1509	35	(Galhetas et al. 2014b)
Grape seeds	K_2CO_3	800	1	0.5:1	1238	26.4	(Okman et al. 2014)
Waste apricot	K_2CO_3	900	1	1:1	1214	-	(Erdoğan et al. 2005)
Pine gasification residues	K_2CO_3	900	1	3:1	1171	-	(Galhetas et al. 2014a)
Mangosteen shell	K_2CO_3	900	2	1:1	1123	20.7	(Chen et al. 2011)
Sisal waste	K_2CO_3	700	1	0.5:1	1038	-	(Mestre et al. 2011)
Pine	K_2CO_3	800	2	1:1	945	50	(Galhetas et al. 2014b)
Cork waste	K_2CO_3	700	1	1:1	891	-	(Mestre et al. 2007)
Peach stones	K_2CO_3	700	1	1:1	866	-	(Cabrita et al. 2010)

2.3.2.2 Neutral activating agents

Neutral activating agents include, but are not limited to, the chlorides of zinc, iron, calcium, magnesium, potassium, and ammonia, of which the most commonly used salt is zinc chloride ($ZnCl_2$). The main pore forming mechanism herein relies on condensation, gasification, and redox reactions (Gao et al. 2020). During the impregnation step, the activating agent reaches the interior of the precursor CM particles (Rodríguez-Reinoso and Molina-Sabio 1992). As a Lewis acid (Ahmadpour and Do 1997), $ZnCl_2$ acts as a dehydrating agent (Balci et al. 1994) by inducing acid catalyzed hydrolysis reactions as evident by the weight loss and evolution of gases and volatiles. Accordingly, molecular hydrogen is evolved leaving active sites that undergo further aromatic condensation, cyclization or polymerization reactions (Yahya et al. 2015). These reactions weaken the CM structure by breaking the lateral cellulosic bonds which, in turn, increases the intra and inter-voids and generates interspaces between the carbon layers to develop the material's microporosity (Heidarinejad et al. 2020, Rodríguez-Reinoso and Molina-Sabio 1992, Saka 2012) and enhance its elasticity and swelling properties (Rodríguez-Reinoso and Molina-Sabio 1992). In a study for developing AC from Canadian peat, $ZnCl_2$ has been shown to generate a microporous structure of AC with a high surface area (Donald et al. 2011).

Since $ZnCl_2$ does not react with carbon, activation by $ZnCl_2$ produces higher yield of AC relative to that obtained with KOH. At a low impregnation ratio, $ZnCl_2$ promotes cross-linking within a rigid structure of carbon matrix, reducing the generation of volatile components and thus leading to a high AC yield. Upon increasing the impregnation ratio, the AC yield increases as a result of the induced polymerization reactions and formation of larger polycyclic aromatic molecules (Anisuzzaman et al. 2016). However, further increase in the impregnation ratio induces further gasification reactions and develops more cracks in the AC structure, while deforming the micropores and enhancing the mesoporosity of the AC structure leading to a lower yield (Anisuzzaman et al. 2016, Donald et al. 2011). As with regard to the textural properties, AC generated from hazelnut bagasse exhibited higher surface area (1642 m^2/g) and pore volume (0.964 cm^3/g) when activated by KOH than when activated by $ZnCl_2$ (1489 m^2/g, 0.9329 cm^3/g) at the same activation temperature and impregnation ratio (Demiral et al. 2008).

Despite its hazardous ecological footprint, $ZnCl_2$ has been used to activate various AWs as shown in Table 3. AWs were activated within a temperature range of 400–900°C and for a time duration of 0.5–3 h with different impregnation ratios, to yield about 30–90% AC. $ZnCl_2$ produced AC of noticeable surface areas of 2869 (Kumar and Jena 2016), 2060 (Chiu and Ng 2012), 1718 (Ahmadpour and Do 1997), 1688 (Gao et al. 2014), 1689 (Erdem et al. 2016), and 1668 m^2/g (Akpen and Leton 2011) with AW precursors such as fox nutshell, cotton fiber, macadamia nut shell, *Enteromorpha prolifera*, vine shoots, and mango seed shell, respectively. Moreover, when used to produce AC from lignin, $ZnCl_2$ yielded AC at a lower activation temperature and with a higher surface area (i.e., 1000 m^2/g) than its counterpart generated by H_3PO_4 (i.e., 700 m^2/g) (Hayashi et al. 2000). Earlier work concerned with the activation of orange peels concluded that AC produced by $ZnCl_2$ activation possessed a higher pore volume and surface area relative to its counterpart generated by KOH activation at the same conditions (Wei et al. 2019). Similarly, other work demonstrated the favorability of $ZnCl_2$ activation over KOH activation of wood biomass with regard to yielding higher surface areas and pore volumes for the generated AC (Danish et al. 2018).

2.3.2.3 Acidic activating agents

Various acidic activating agents are utilized for the activation of CM, such as phosphoric acid (H_3PO_4), sulfuric acid (H_2SO_4), and nitric acid (HNO_3). Activation with phosphoric acid (H_3PO_4) has been favored to others due to its lower environmental footprint relative to other strong acids (Asadullah et al. 2007). In addition, it effectively produces mesoporous AC with larger pore volume and pore diameter than that generated using neutral activating agents (Donald et al. 2011). Thus, activation with H_3PO_4 has been widely used to chemically activate cellulosic and lignocellulosic wastes. H_3PO_4

Table 3. Previous studies reported on AWs activated by $ZnCl_2$. The table compiles the activation conditions, surface areas and yields of AC obtained.

AW	Activation conditions			Surface area	Yield	Reference
	Temperature (°C)	Time (h)	Impregnation ratio	(m^2/g)	(%)	
Fox nutshell	600	1	2:1	2869	32.8	(Kumar and Jena 2016)
Cotton fiber	500	1	-	2060	37.6	(Chiu and Ng 2012)
Sugar beet bagasse	700	1.5	3:1	1826	-	(Demiral and Gündüzoğlu 2010)
Sour cherry	700	2	3:1	1704	-	(Angin 2014)
Vine shoots	700	2	1.3:1	1689	18.2	(Erdem et al. 2016)
Enteromorpha prolifera	500	1	1:1	1688	-	(Gao et al. 2014)
Mango seed shell	500	1	2:1	1667.8	-	(Akpen and Leton 2011)
Hazelnut bagasse	700	2	3:1	1642	-	(Demiral et al. 2008)
Tea seed shells	500	1	1:1	1530	44.1	(Gao et al. 2013)
Licorice/Pistachio nut shells mixture	600	1	2:1	1492	-	(Asasian et al. 2012)
Glycyrrhiza glabra	450	1	2:1	1483	63	(Mohammadi et al. 2014a)
Posidonia oceanica	600	2	4.5:1	1483	40	(Asimakopoulos et al. 2021)
Hazelnut husk	700	4	1:1	1369	36	(Karaçetin et al. 2014)
Waste potato residue	600	1	1.5:1	1357	29.2	(Zhang et al. 2015)
Tamarind wood	439	0.67	1:1	1322	-	(Acharya et al. 2009)
Date pit	600	1	1:1	1322	-	(Bouchenafa-Saïb et al. 2005)
Terminalia arjuna nuts	500	1	3:1	1260	≈ 50	(Mohanty et al. 2005)
Groundnut shell	650	15	1.75:1	1200	67	(Malik et al. 2007)
Cashew nutshell	400	2	1.5:1	1100	59	(Spagnoli et al. 2017)
Tomato solid waste	600	1	6:1	1093	38.2	(Sayğılı and Güzel 2016a)
Hazelnut husk	700	4	1:2	1092	37.5	(Imamoglu and Tekir 2008)
Van apple pulp	500	1	1:1	1067	43	(Depci et al. 2012)
Waste apricot	500	1	1:1	1060	-	(Başar 2006)
Potato waste	700	2	2:1	1052	-	(Ma et al. 2015)
Date stones	500	1.2	2:1	1045	40.4	(Theydan and Ahmed, 2012)
Coffee beans	900	1	1:1	1019	-	(Rufford et al. 2008)
Rice husk	500	1	1:1	927	-	(Boonpoke et al. 2011)
Pineapple waste	500	1	1:1	915	76	(Mahamad et al. 2015)
Coir pith	700	-	2:1	910	-	(Namasivayam and Sangeetha 2006)

Table 3 contd. ...

... Table 3 contd.

AW	Activation conditions			Surface area	Yield	Reference
	Temperature (°C)	Time (h)	Impregnation ratio	(m^2/g)	(%)	
Canadian Peat	450	0.75	3:2	890	≈ 61	(Donald et al. 2011)
Coffee residue	600	1	1:1	889	-	(Boudrahem et al. 2011)
Sugarcane bagasse	700	0.5	1:0.25	864	-	(Kalderis et al. 2008)
Sea buckthorn stones	550	3	2:1	829	33	(Mohammadi et al. 2010)
Rice husk	700	0.5	1:0.25	811	-	(Kalderis et al. 2008)
Walnut shell	-	-	1:1	803	-	(Zabihi et al. 2010)
Safflower biochar	900	1	4:1	801	26	(Angın et al. 2013)
Olive stone	650	2	5:1	790	-	(Kula et al. 2008)
Walnut shell	-	-	0.5:1	780	-	(Zabihi et al. 2010)
Corncob	500	2	-	143	-	(Nethaji et al. 2013)

acts on the CM through the cleavage of aliphatic and ether linkages between the cyclic structural units in cellulose and lignin, followed by protonation, and consequent cross linking at sub-pyrolysis temperatures (Jagtoyen and Derbyshire 1993). At low temperatures, where hemicellulose, lignin and other polysaccharides are less resistant to acid hydrolysis, the acid hydrolyzes the glycosidic linkages in the polysaccharides and cleaves the aryl ether bonds in lignin, leading to the formation of ketones. Bond cleavage reactions account for the released carbon monoxide, CO_2, and methane gas. The release of the latter, in particular, indicates the cleavage of the aliphatic side chains. Further chemical reactions occur simultaneously including dehydration, condensation, and depolymerization. At temperatures ranging from 150–250°C, structure dilation and stabilization, and pore formation take place, along with crosslinking of the polyaromatic units via phosphate and polyphosphate bridges. Upon increasing the temperature above 250°C, cyclization and condensation reactions occur due to the scission of P-O-C bonds, which gives rise to higher aromaticity and produces larger polyaromatic units (Jagtoyen and Derbyshire 1998). Various studies report the activation of different AWs by H_3PO_4 as listed in Table 4. In these studies, activation took place at temperatures ranging from 80–800°C, within a time span of 0.25–24 h using different impregnation ratios, depending on the type of AW. AC was obtained with % yields ranging from 30–85%.

Since the surface area is an important parameter to consider in judging the performance of AC, the highest reported surface areas for AW-derived ACs are depicted in Fig. 2. These ACs were obtained by five different activation methods using various AWs. Clearly, the highest surface areas, of about 3000 m^2/g, belong to the ACs produced by KOH activation. Following are those generated by $ZnCl_2$ and H_3PO_4 which amount to 2000 m^2/g on average. On the other hand, ACs produced by physical activation showed the least surface areas. The yields of a few of these ACs were reported as shown in Fig. 2.

3. CONVERSION PARAMETERS

Efficient conversion of AWs to AC requires the management of various parameters pertinent to each method. Physical carbonization requires control over the carbonization temperature and time, in addition to the heating and gas flow rates. Chemical carbonization entails the appropriate selection of the activating agent, in addition to optimizing the mass ratio of the activating agent to the AW. Furthermore, the process's optimization is also dependent on the mixing methods and heating profiles. In what follows is a detailed discussion of the three main factors affecting the conversion process, namely, carbonization/activation temperature, time and impregnation factor.

Carbonization/activation temperature and time have a direct influence on the yield, porosity, and surface area of the CM. Increasing the carbonization temperature and prolonging its time duration allow for the gasification and elimination of volatile matter and, thus, lower the yield and increase the surface area (Li et al. 2008). However, these factors should be carefully optimized; otherwise, too high activation temperatures or too long durations may lead to a decline in surface area due to the decomposition of some volatile fractions that form intermediate melts in the char structure (Lua et al. 2004). Additionally, this may cause pore widening due to destruction or collapse of pore walls (Khadiran et al. 2015, Lua et al. 2004).

On one hand, the yield produced from various AW materials has been reported to decrease with raising the activation temperature (Guo and Lua 2003, Nahil and Williams 2012, Zakaria et al. 2021). This was attributed to the intensified dehydration and elimination reactions associated with higher temperatures which result in higher emissions of volatile matter that, ultimately, leads to lower yields (Guo and Lua 2003).With regard to the surface area and porosity, the effect of the activation temperature on the surface area, the mesopore (V_{meso}), micropore (V_{micro}), and total pore (V_{total}) volumes have been investigated for the activation of lignin using H_3PO_4. The produced AC comprised micropores along with mesopores and it was observed that surface area increased across a temperature range of 500–600°C to reach its maximum value of 1000 m^2/g at 600°C (Hayashi

Table 4. Previous studies on AWs activated by H_3PO_4. The table compiles the activation conditions, surface areas and yields of AC obtained.

AW	Activation conditions			Surface area	Yield (%)	Reference
	Temperature (°C)	**Time (h)**	**Impregnation ratio**	**(m²/g)**		
Coconut shell	800	2	1:1	2648	10.53	(Zhang et al. 2020)
Orange peel	700	1	2:1	2210	-	(Wei et al. 2019)
Cotton stalk	500	1	1.5:1	1720	53.7	(Nahil and Williams 2012)
Reedy grass leaves	500	2	0.88:1	1474	-	(Xu et al. 2014)
Grain Sorghum	600	0.25	-	1431	26	(Diao et al. 2002)
Peach stones	500	2	0.4:1	1393	41.8	(Attia et al. 2008)
Woody birch	700	2	1.5:1	1360	-	(Budinova et al. 2006)
Marigold straw	400	1	2:1	1344.2	-	(Qin et al. 2014)
Bamboo	600	1	1.5:1	1335	-	(Liu et al. 2010)
Vetiver roots	600	1	1:1	1271	48	(Altenor et al. 2009)
Sugarcane bagasse	400	2	2:1	1254.5	-	(Chen et al. 2012)
Pomelo peel	450	1	2.5:1	1252	-	(Sun et al. 2016)
Persian mesquite grain	600	2	1:1	1243	-	(Lemraski and Sharafinia 2016)
Peanut hull	500	3	1:1	1177	22	(Girgis et al. 2002)
Lemon peel	500	1	2:1	1158	52.1	(Mohammadiet al. 2014)
Kenaf	700	1	-	1154	-	(Meryemoglu et al. 2016)
Date tree frond	400	3	-	1139	-	(Al-Swaidan and Ashfaq 2011)
Almond shell	400	1.25	1:1	1128	68.4	(Izquierdo et al. 2011)
Pinecone	500	1	3:1	1094	55.7	(Momčilović et al. 2011)
Potato peels	600	2	0.75:1	1041	51	(Kyzas et al. 2016)
Durian shell	400	1	1:2	1024	-	(Chandra et al. 2007)
Olive waste cake	450	2	1.75:1	1020	42.9	(Baccar et al. 2009)
Rice husk	500	1	2:1	1016	-	(Li et al. 2015)
Coffee residue	600	1	1:1	1003	-	(Boudrahem et al. 2011)
Wild olive cores	805	2	-	969	-	(Kaouah et al. 2013)
Cattail leaves (*Typha orientalis*)	500	0.75	-	966	62.73	(Anisuzzaman et al. 2015)
Date palm pits	450	1.2	3:1	952	41	(Reddy et al. 2012)
Coffee grounds	450	1	1.8:1	925	32	(Reffas et al. 2010)

Olive tree wood	550	1	3.5:1	904	21.7	(Ould-Idriss et al. 2011)
Tunisian date stone	450	2	1.75:1	826	-	(Bouhamed et al. 2012)
Walnut shell	500	1.12	2:1	789	80	(Moreno-Barbosa et al. 2013)
Sugar beet bagasse	105	12	-	748	34.3	(Ghorbani et al. 2020)
Date pit	500	2	0.85:1	720	-	(Bouchenafa-Saïb et al. 2005)
Melon shell	500	1.12	2:1	710	85	(Moreno-Barbosa et al. 2013)
Cotton cake	450	1.5	2:1	584	29.8	(Ibrahim et al. 2013)
Lemna minor (duckweed)	500	2	3:1	531.9	-	(Huang et al. 2014)
Rice straw	450	2	1:1	522	51.9	(Fierro et al. 2010)
Date pits	500	1	3:1	502	-	(Abdulkarim and Al-Rub 2004)
Apricot nut shells	400	1.5	1:1	308	26	(Marzbali et al. 2016)
Stink bean (*Parkia speciosa*)	600		1:2	190	41.1	(Foo and Lee 2010)
Oil palm	500	1	1:1	-	-	(Rahman et al. 2014)
Coconut shells	500	1	1:1	-	-	(Rahman et al. 2014)
Plum stone (*Prunus nigra*)	500	1	3:1	-	-	(Parlayıcı and Pehlivan 2017)
Diplotaxis harra	400	1	1.5:1	-	-	(Tounsadi et al. 2016)

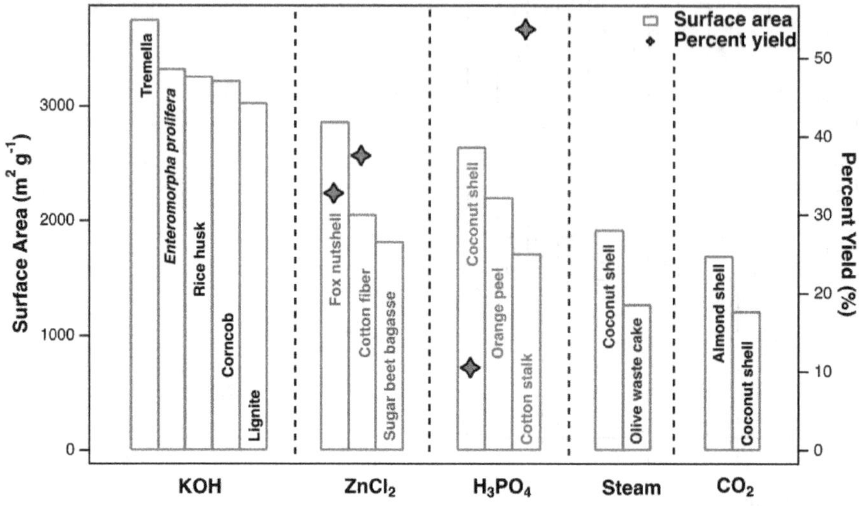

Figure 2. Surface areas and yields of various AW-derived ACs obtained by physical and chemical activation.

et al. 2000). In another study, AC from cotton stalks activated with H_3PO_4 showed a noticeable surface area of 1720 m^2/g at a temperature as low as 500°C and an impregnation ratio of 1.5, which declined concurrently with V_{micro} upon increasing the temperature to 700°C. Such decline has been attributed to the collapse or contraction of pores at the higher temperature(Nahil and Williams 2012).Similar behavior was observed for AC activated with KOH and $ZnCl_2$ (Demiral et al. 2008). However, increasing the activation temperature to 800°C resulted in an increase of V_{micro} and surface area, which was attributed to partial gasification by P_2O_5 and the elimination of $H_2P_2O_7^{2-}$ and $H_2PO_4^{-1}$ (Nahil and Williams 2012). Similarly, H_3PO_4 activation of sugarcane bagasse produced AC with a high porosity and a surface area of 1132 m^2/g at 500°C. It was found that increasing the activation temperature from 300 to 500°C increased V_{meso} by 17 times and V_{micro} by only three folds. Upon exceeding 500°C, the pore volumes decreased due to pore collapse and V_{micro} exhibited a sharper decline than V_{meso} (Castro et al. 2000).

The third parameter affecting the conversion process is the degree of impregnation which governs the porous structure of the resulting AC, and is, thus, imperative in investigating the optimum conditions for synthesizing AC. Increase in the impregnation ratio means an increase in the concentration of the activating agent with respect to the CM and, thus, more of this agent is incorporated into the CM. Such an increase results in more gasification reactions which consume the carbon precursor and thus lower the yield, enlarge the pore volume, and increase the surface area of the produced AC. As a result, V_{micro} and consequently the internal surface area increase. In a study for activating cotton stalks with H_3PO_4, increasing the impregnation ratio from 0.3 to 1.5 at an activation temperature of 500°C resulted in an increase in both V_{micro} and V_{meso} leading to an increase in V_{total} from 0.16 to 0.89 cm^3/g, in addition to an associated rise in surface area from 330 to 1720 m^2/g. Similar trends were exhibited at different activation temperatures (Nahil and Williams 2012). Such behavior has also been reported in other studies concerned with activation using H_3PO_4 (Castro et al. 2000), and other activating agents as KOH (Demiral et al. 2008, Okman et al. 2014, Manyà et al. 2018), K_2CO_3 (Okman et al. 2014), and $ZnCl_2$ (Demiral et al. 2008, Saka 2012).

However, further increase in the impregnation ratio results in more aggressive gasification reactions, which reversely yield a reduction in V_{micro} and the surface area. In a study for activating cotton stalks with H_3PO_4, increasing the impregnation ratio from 1.5 to 3 resulted in a slight decrease in V_{micro} by 1.7% combined with a significant increase in V_{meso} by 259 %, and a subsequent decrease in surface area by 2.9%. This behavior was ascribed to pore widening of the micropores by means of the activating agent (Nahil and Williams 2012). Similar behavior was also observed with other activating agents such as KOH (Chen et al. 2021, Demiral et al. 2008) and $ZnCl_2$ (Demiral et al. 2008).

4. AW-DERIVED AC ADSORBENTS IN WATER TREATMENT APPLICATIONS

Carbon based adsorbents have been extensively studied and applied in wastewater treatment. These include carbon nanotubes, carbon aerogels, fullerenes, graphene, and bio-based carbon. The following sections focus on carbon adsorbents derived from AWs and their performance in removing major classes of conventional contaminants (heavy metal ions and anions) and emerging contaminants (pharmaceuticals and pesticides).

4.1 Removal of conventional contaminants

Conventional water contaminants include heavy metals (such as Pb, Ni, As, Cr, Hg), nitrates, phosphates, sulfates, fluorides, calcium and magnesium ions, pathogens (e.g., viruses), and carcinogenic organic compounds (Pooja et al. 2020). Many methods have been applied for the removal of such contaminants, such as electrochemical, advanced oxidation, membrane filtration, precipitation, coagulation, dissolved air flotation, electrodialysis, photocatalysis, and ion exchange. Nevertheless, many of these methods suffer from serious drawbacks such as high cost, long operational times, and generation of hazardous by-products. Table 5 summarizes the drawbacks of such methods.

Adsorption is a surface phenomenon, by which the target species (i.e., adsorbates) are bound physically or chemically to the surface of an adsorbent. Adsorption has many advantages over other methods, since it is generally characterized by availability, ease of operation, high efficiency, low operating cost, and lack of generation of toxic products (Afroze and Sen 2018, Demirbas 2008). Commercial AC has been extensively reported for the removal of conventional contaminants from wastewater. However, it exhibits several disadvantages such as its difficulty of regeneration and its relatively high cost stemming from its expensive and non-renewable precursors (Acharya et al. 2018,

Table 5. Summary of the drawbacks of some removal methods for conventional contaminants (Acharya et al. 2018, Al-Qodah et al. 2017, Gupta et al. 2009).

Method	Drawbacks
Electrochemical	High capital and operational cost Formation of by products (possibly toxic)
Advanced oxidation	High cost Formation of by products (possibly toxic)
Ion exchange	High cost Not suitable for all heavy metal removal
Membrane filtration (RO, UF, NF)	High capital and operational cost Membrane fouling Limited flux
Chemical precipitation	Sludge formation, which requires additional processes to be disposed of
Chemical coagulation	High cost High levels of chemical doses
Electrodialysis	High operational cost Fouling of membrane
Photocatalysis	Long operational time Difficulty of scaling up
Dissolved air flotation	Multiple steps are required to enhance removal of heavy metals
Electrokinetic coagulation	Sludge formation

Al-Qodah et al. 2017, Amirza et al. 2017). Hence, utilizing abundant AWs for the production of AC is considered an economic solution that would result in cost reduction (Demirbas 2008, Nguyen et al. 2013). In addition, AW-derived AC–one type of bio-based carbons—has been proven in previous studies to possess a higher adsorption capacity for heavy metals than that of commercial AC (El-Naggar et al. 2016, Periasamy and Namasivayam 1995). It has also been reported that AW-derived AC can be regenerated and reused (Ghasemi et al. 2014). Table 6 summarizes previous studies that deployed AW-derived AC in conventional water treatment in the past decades.

The performance of adsorbents can be evaluated by two means, the maximum adsorption capacity (Q_{max}) and the removal efficiency (percentage removal). Q_{max} is measured under equilibrium conditions and is a function of pH and temperature. It can be predicted by theoretical mathematical models such as Langmuir, Freundlich and Sips, etc. The removal efficiency, on the other hand, is dependent on various factors including pH, initial concentration, contact time, temperature, type of water to be treated, and the characteristics of the adsorbent (AW-derived AC) (Nguyen et al. 2013). The latter, in particular, depends on the process of activation and the nature of the AW precursor.

Table 6. Summary of the studies reported on the performance of AW-derived AC as an adsorbent for conventional contaminants.

AW	Activation	Target Contaminant	pH	Q_{max} (mg/g)*	Reference
Walnut shell	KOH	Cd(II)	6	68.31	(Ansari et al. 2018)
Conocarpus pruning	$ZnCl_2$ / KOH	Cd(II)	5	21.5	(El-Naggar et al. 2016)
Phaseolus aureus hulls	Steam	Cd(II)	8	15.7	(Rao et al. 2009a)
Olive stone	KOH	Cd(II)	5	11.72	(Alslaibi et al. 2013)
Date pits	Steam	Co(II)	810	1317	(Awwad et al. 2013)
Coconut shell fibers	Physical activation	Cr(III)	5	16.10	(Mohan et al. 2006)
Chickpea husks (*Cicer arietinum*)	KOH/K_2CO_3	Cr(VI)	8	59.6	(Özsin et al. 2019)
Sugar beet bagasse	H_3PO_4	Cr(VI)	4	52.87	(Ghorbani et al. 2020)
Oil palm and coconut shells	H_3PO_4	Cr(VI)	-	46.30	(Rahman et al. 2014)
Barley straw	no activation	Cr(VI)	2	87.36	(Chand et al. 2009)
Wheat straw	no activation	Cr(VI)	2	86.84	(Chand et al. 2009)
Coconut coir	N_2	Cr(VI)	1-2	38.5	(Chaudhuri and Azizan 2012)
Coconut shell	Physical activation	Cr(VI)	2	32.61	(Mohan et al. 2005)
Coconut shell fibers	Physical activation	Cr(VI)	2	24.11	(Mohan et al. 2005)
Sugarcane bagasse	$FeCl_3.6H_2O$/ CTMAB	Cr(VI)	>8	13.72	(Zhu et al. 2012)
Terminalia arjuna nuts	$ZnCl_2$	Cr(VI)	1	28.43	(Mohanty et al. 2005)
Coconut buttons	Steam	Cu(II)	6	73.6	(Anirudhan and Sreekumari 2011)
Chickpea husks (*Cicer arietinum*)	KOH/K_2CO_3	Cu(II)	8	56.2	(Özsin et al. 2019)
Plum stone (*Prunus nigra*)	H_3PO_4	Cu(II)		48.31	(Parlayıcı and Pehlivan, 2017)
Conocarpus pruning	$ZnCl_2$/KOH	Cu(II)	5	25.36	(El-Naggar et al. 2016)
Phaseolus aureus hulls	Steam	Cu(II)	7	19.5	(Raoet al. 2009)

Table 6 contd. ...

... *Table 6 contd.*

AW	Activation	Target Contaminant	pH	Q_{max} (mg/g)*	Reference
Hazelnut husks	$ZnCl_2$	Cu(II)	5.7	6.645	(Imamoglu and Tekir 2008)
Date pits	Steam	Fe(III)	8- 10	1555	(Awwad et al. 2013)
Coirpith	$H_2SO_4/(NH_4)_2S_2O_8$	Hg(II)	5	154	(Namasivayam and Kadirvelu 1999)
Coconut buttons	Steam	Hg(II)	7	78.84	(Anirudhan and Sreekumari 2011)
Ceiba pentandra hulls	Steam	Hg(II)	6	25.88	(Rao et al. 2009b)
Phaseolus aureus hulls	Steam	Hg(II)	7	23.66	(Rao et al. 2009b)
Cicer arietinum	Steam	Hg(II)	7	22.88	(Rao et al. 2009b)
Areca nut waste	N_2	Hg(II)	4	11.23	(Lalhmunsiama et al. 2017)
Lapsi seed	H_2SO_4	Ni(II)	6	85.5	(Shrestha et al. 2013)
Coirpith	H_2SO_4/ $(NH_4)_2S_2O_8$	Ni(II)	5	62.5	(Kadirvelu et al. 2001b)
Oil palm and Coconut shells	H_3PO_4	Ni(II)	-	19.6	(Rahman et al. 2014)
Date pits	Steam	Pb	8–10	1261	(Awwad et al. 2013)
Lapsi seed	H_2SO_4	Pb	5	424	(Shrestha et al. 2013)
Chickpea husks (*Cicer arietinum*)	KOH / K_2CO_3	Pb	8	135.8	(Özsin et al. 2019)
Coconut buttons	Steam	Pb	6	92.72	(Anirudhan and Sreekumari 2011)
Plum stone (*Prunus nigra*)	H_3PO_4	Pb	5.5	80.65	(Parlayıcı and Pehlivan 2017)
Oil palm and Coconut shells	H_3PO_4	Pb	-	74.6	(Rahman et al. 2014)
Conocarpus pruning	$ZnCl_2$ / KOH	Pb	5	44.15	(El-Naggar et al. 2016)
Ceiba pentandra hulls	Steam	Pb	6	25.5	(Rao et al. 2008)
Phaseolus aureus hulls	Steam	Pb	6	21.8	(Rao et al. 2009a)
Hazelnut husks	$ZnCl_2$	Pb	5.7	13.05	(Imamoglu and Tekir 2008)
Date pits	Steam	Zn(II)	8–10	1594	(Awwad et al. 2013)
Ceiba pentandra hulls	Steam	Zn(II)	6	24.1	(Rao et al. 2008)
Phaseolus aureus hulls	Steam	Zn(II)	7	21.2	(Raoet al. 2009)
Sugar beet bagasse	$ZnCl_2$	NO_3^-	6.58	27.55	(Demiral and Gündüzoğlu 2010)
Citrus limetta peels	$FeCl_3$	F^-	6.6	9.709	(Siddique et al. 2020)
Pisum sativum peel	$FeCl_3$	F^-	7	4.71	(Sahu et al. 2021)
Aegle marmelos	-	F^-	6	2.087	(Singh et al. 2017)
Crocus sativus leaves	$CaCl_2$	F^-	4.5	2.01	(Dehghani et al. 2018)
Wheat straw	H_2SO_4	F^-	6	1.93**	(Yadav et al. 2013)
Sugarcane bagasse	H_2SO_4	F^-	6	1.15**	(Yadav et al. 2013)
Chinese chaste tree *(Vitex negundo)*	HNO_3	F^-	7	1.15	(Suneetha et al. 2015)
Rice straw	Steam	NO_3^-	2	18.9	(Daifullah et al. 2007)

*Q_{max} as predicted by Langmuir isotherm model
**Q_{max} as predicted by Freundlich isotherm model

Therefore, to compare the performance of different AW-derived ACs as adsorbents, Q_{max} is used. In Table 6, all the reported studies predicted Q_{max} fairly well using the Langmuir model, except for one study in which the adsorption of fluoride ions (F^-) onto each of the wheat straw and sugarcane bagasse followed the Freundlich isotherm.

Table 6 also shows that most targeted contaminants were heavy metals along with some anions like nitrates and fluorides. The sources of AWs vary among seeds, husks, shells, stones and peels of fruits, vegetables, and flowering plants. The suggested mechanisms for binding of heavy metals onto AC vary depending on the nature of AW and their surface characteristics. For instance, Rao et al. (2008) attributed the removal of Zn (II) and Pb (II) by AC derived from *Ceiba pentandra* hulls to the functional groups present on the surface of the carbon, examples of which are sulfoxide, carbonyl, and hydroxyl functional groups (Rao et al. 2008). Kadirvelu et al. (2001a) suggested that adsorption of Pb (II) and other heavy metals onto coirpith- derived AC was accomplished by ion exchange (Kadirvelu et al. 2001a), while others ascribed their adsorption to ion exchange or complexation processes (Anirudhan and Sreekumari 2011). For the removal of Pb (II) and Cu (II) onto AC derived by H_3PO_4 activation of plum stone, Parlayici et al. (2017) suggested electrostatic interaction, ion exchange, and complex formation as the underlying mechanisms (Zhu et al. 2012).

Regarding the adsorption capacity, Fig. 3a summarizes the highest reported values pertaining to the adsorption of conventional contaminants onto AW-derived ACs. ACs derived from date pits showed remarkable adsorption capacities that exceeded 1000 mg/g for different heavy metals, namely, Zn (II), Fe(III), Co (II), and Pb (II). Other heavy metals such as Hg (II), Ni (II), and Cu (II) were efficiently removed by AW-derived ACs with substantial capacities that are, however, lower by one order of magnitude. The least adsorption capacities were shown for the adsorption of F^- and nitrate ions (NO_3^-).

4.2 Removal of emerging contaminants

Alongside the conventional contaminants, there are other emerging contaminants that cause water pollution. Some of these contaminants have been recently introduced into the ecosystem, while others existed in the environment a long time ago and affected the water quality; yet, they have not been well monitored (Rodriguez-Narvaez et al. 2017). Hence, these contaminants are popularly known as 'contaminants of emerging concern (CECs)'. Such contaminants are characterized by their very low concentrations, that range from micro to nano grams (Rivera-Utrilla et al. 2013b). Current advancements in the analytical tools like liquid chromatography coupled with mass spectrometry and tandem mass spectrometry have enabled the detection and monitoring of these contaminants. Even though their health adverse effects have not been fully elucidated, the accumulation of CECs have been reported to pose a serious health and environmental threat (Patel et al. 2019, Pooja et al. 2020).

CECs include pharmaceuticals, personal-care products, artificial sweeteners, surfactants, pesticides, and industrial compounds (Dulio et al. 2018, Pooja et al. 2020). Due to the tremendous number of already registered chemicals (100 million), and the daily increasing number of chemicals (4000 new chemicals/day), the list is significantly increasing. Daily used chemicals are estimated to be in the range of 30,000 to 50,000 which, ultimately, find its way into the environment (Dulio et al. 2018). Owing to the growing number of new chemicals that are being consumed by humans and are, eventually, disposed of and introduced into the water streams, these pollutants are difficult to group into one category (Rodriguez-Narvaez et al. 2017). Herein, we focus on two major classes of these CECs, pharmaceuticals and pesticides.

Pharmaceutical contaminants make up a large portion of the CECs (Patel et al. 2019). The ingredients of these pharmaceuticals that are pharmacologically active, degradation resistant, and harmful to human health are known as active pharmaceutical ingredients (Rivera-Utrilla et al. 2013b), nearly 4000 of which are considered CECs. In the past decade or so, there has been a significant rise in the global consumption of pharmaceuticals, for example, the global antibiotics

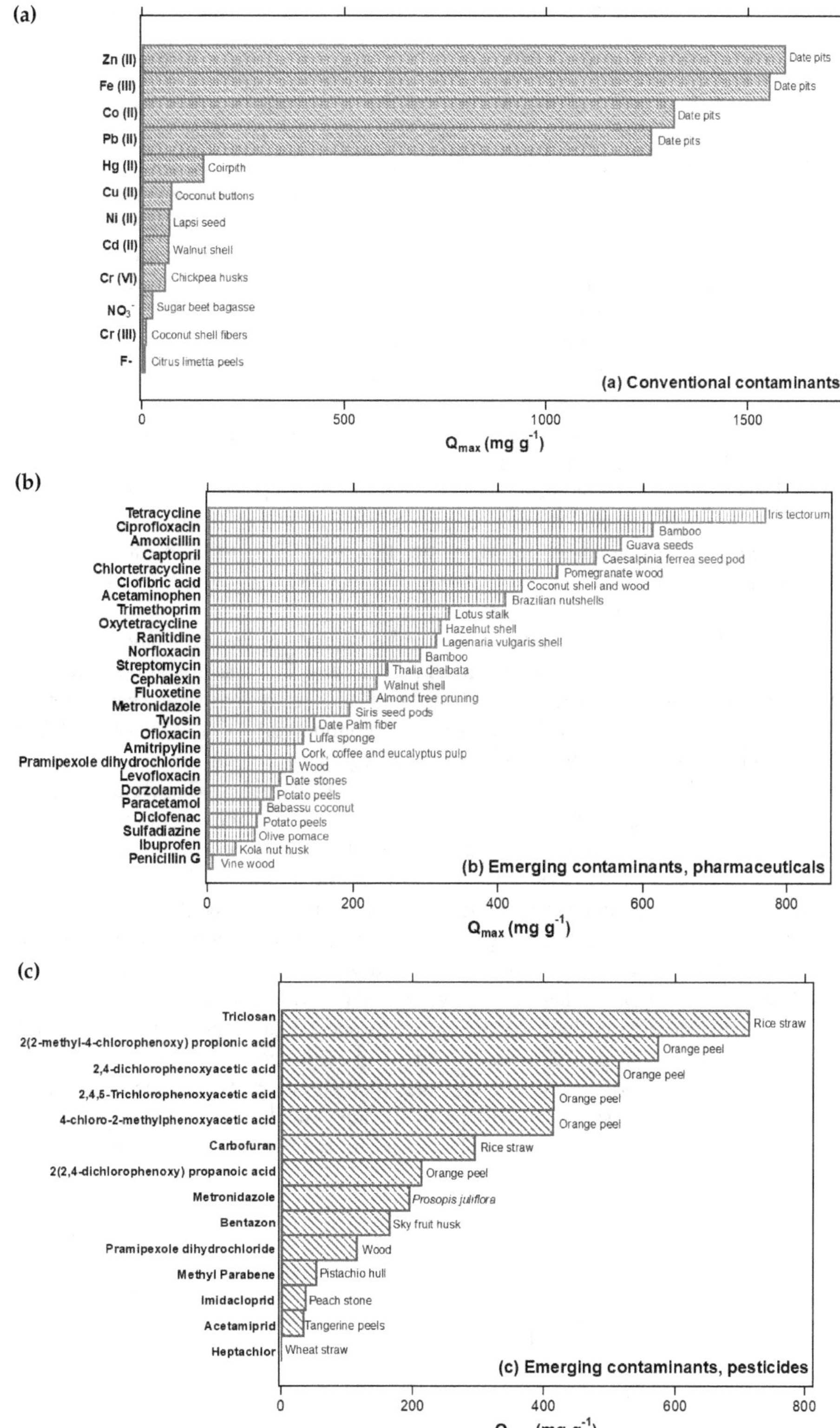

Figure 3. Adsorption capacities of (a) conventional contaminants, (b) pharmaceuticals, and (c) pesticides onto various AW-derived ACs.

consumption in 2015 was 34.8 billion doses, as opposed to only about 12 billion some 15 years earlier (i.e., 65 % increase) and was expected to reach 4.5 trillion in 2020 (Patel et al. 2019). In addition to the ever increasing consumption, 90–95% of the administered pharmaceutical drugs are excreted un-metabolized in feces and/or urine into the environment (Milić et al. 2013, Rasheed et al. 2019). In addition to their high stability, these excreted active pharmaceutical ingredients pose a threat to all living organisms as well as to the environment due to their potential toxicity and health hazards. Effluents from livestock farms also play a substantial role, because in many incidents these effluents are discharged without treatment (Patel et al. 2019). Other factors that garner our attention to pharmaceutical contaminants are their varying structure and functionality, their ionizability and their persistence in the environment for relatively long durations, either metabolized or un-metabolized (Rivera-Utrilla et al. 2013).

A significant number of studies have been reported for the removal of CECs. In addition to the methods mentioned in Table 6, other methods are also being used, yet they suffer drawbacks as summarized in Table 7. Hence, and for the same reasons mentioned in the previous section (Section 4.1), adsorption-based removal still holds superior hand when compared to these methods. Table 8 and Figs. 3b, c summarize the studies that tackled adsorption of emerging contaminants (mainly pharmaceuticals, along with some pesticides) in the last two decades. As can be observed, various AW-derived ACs were successful in removing antibiotics of the fluoroquinolones, tetracycline and penicillin families, achieving adsorption capacities that reached up to almost 800 mg/g. Similarly, for pesticides, the adsorption capacities reached up to about 750 mg/g with ACs derived from orange peel and rice straw recording the highest values.

As mentioned earlier with conventional contaminants, AW-derived ACs have shown superiority over commercial ACs in their adsorption capacities for heavy metals. In addition, Alahabadi et al. (2017) and Chayid and Ahmed (2015) both reported higher respective adsorption capacities (482.5 and 273.9 mg/g, respectively) for AW-derived AC than commercial AC (377.5 and 249.3 mg/g) in removing chlortetracycline and amoxicillin using pomegranate peel and *Arundo donax* (giant reed). According to Bernardo et al. (2016), however, commercial AC exhibited a higher adsorption capacity, 146 mg/g, compared to 69 mg/g for the AW-derived AC. The effect of using different AW precursors for deriving AC on the removal of antibiotics can be clearly deduced from the works of Balarak et al.(2017) and Lima et al. (2019a,b), who utilized the same $ZnCl_2$ activating agent under approximately the same conditions to produce AC. The former removed 265.2 mg/g of amoxicillin while the latter removed 454.7 mg/g using *Azolla filiculoides* and cashew of para, respectively. The chemical structure of the activating agent is another important influencing factor. For example, Liu and coworkers (2012) reported that activating lotus stalk with different oxyacids of phosphorus produced ACs with varying surface functional groups and texture. Specifically, ACs produced by orthophosphoric acid or pyrophosphoric acid activation exhibited increased Brunauer, Emmet and Teller (BET) surface area and total pore volume than their counterparts generated via activation with metaphosphoric acid and phosphorous acid, probably due to the higher degree of protonation of the former oxyacids. This was, in turn, reflected in their ability to remove the trimethoprim antibiotic (Liu et al. 2012).

Table 7. Summary of the drawbacks of some removal methods for emerging contaminants (Afroze and Sen 2018, Patel et al. 2019).

Method	Drawbacks
Chlorination	generation of toxic byproducts
Constructed wetland	land requirement, extended retention time, low removal efficiency for many emerging contaminants
Ozonation	high energy consumption and limitation to certain contaminants
Advanced oxidation process	complexity, production of toxic compounds and high operational cost
Fenton and photo-Fenton	sludge formation

Table 8. Summary of the studies reported on the performance of AW-derived AC as an adsorbent for emerging contaminants.

AW precursor	Activating Agent	Target Contaminant Pharmaceuticals group	pH	Q_{max} (mg/g)*	Reference
Brazilian nutshells	$ZnCl_2$	Acetaminophen	7	309.3	(Lima et al. 2019a)
Pomegranate wood	NH_4Cl	Acetaminophen	2–9	233	(Mashayekh-Salehi and Moussavi 2016)
Pine cone and deoiled canola meal	H_3PO_4	Acyclovir	8	-	(Jain et al. 2014)
Cork, coffee and	CO_2	Amitripyline	13	120	(Nabais et al. 2012)
Eucalyptus pulp	CO_2	Amitripyline	13	110	(Nabais et al. 2012)
Guava seeds	NaOH	Amoxicillin	4	570.48	(Pezoti et al. 2016)
Pomegranate wood	NH_4Cl	Amoxicillin	6	437	(Moussavi et al. 2013)
Date pits Cashew of Para	Physical activation $ZnCl_2$	Amoxicillin Amoxicillin	2–4 7	424 392.1	(Belhachemi and Djelaila 2017) (Lima et al. 2019b)
Arundo donax (giant reed)	KOH	Amoxicillin	7	273.9	(Chayid and Ahmed 2015)
Azolla filiculoides	$ZnCl_2$	Amoxicillin	7	265.2	(Balarak et al. 2017)
Olive stones	H_3PO_4	Amoxicillin	-	57	(Limousy et al. 2017)
Vine wood	NaOH	Amoxicillin	2	2.69	(Pouretedal and Sadegh 2014)
Grape slurry	HCl	Amoxicillin	-	2.28	(Chitongo et al. 2019)
Grape slurry	HCl	Ampicillin		1.87	(Chitongo et al. 2019)
Bamboo	H_3PO_4/ K_2CO_3	Ciprofloxacin	5	613	(Wang et al. 2015)
Rice husk	Steam	Ciprofloxacin	7.9	475.74	(Zhang et al. 2017)
Pomelo peel	H_3PO_4	Ciprofloxacin	-	400	(Sun et al. 2016)
Bamboo	H_3PO_4	Ciprofloxacin	6	245.6	(Peng et al. 2018)
Arundo donax linn	H_3PO_4	Ciprofloxacin	-	244	(Sun et al. 2016)
Date palm leaves	H_2SO_4	Ciprofloxacin	6	133.3	(El-Shafey et al. 2012)
Albizia lebbeck seed pods	KOH	Ciprofloxacin	9	131.14	(Ahmed and Theydan 2014a)
Corylus avellana	$ZnCl_2$	Ciprofloxacin	6	73.64	(Balarak et al. 2016)
Caesalpinia ferrea seed pod	$ZnCl_2$	Captopril	7	535.5	(Kasperiski et al. 2018)
Walnut shell	$ZnCl_4$	Cephalexin	6.5	233.1	(Nazari et al. 2016)
Albizia lebbeck seed pods	KOH	Cephalexin	7	137.02	(Ahmed and Theydan 2012)
Albizia lebbeck seed pods	K_2CO_3	Cephalexin	7	118.08	(Ahmed and Theydan 2012)
Alligator weed	H_3PO_4	Cephalexin	6	45	(Miao et al. 2016)
Vine wood	NaOH	Cephalexin	2	7.08	(Pouretedal and Sadegh 2014)
Pomegranate wood	NH_4Cl	Chlortetracycline	6.5	482.5	(Alahabadi et al. 2017)
Hazelnut shell	H_3PO_4	Chlortetracycline	5	333.3	(Fan et al. 2016)
Corn stalk	H_3PO_4	Chlortetracycline	5	115	(Song et al. 2020)
Grape slurry	KOH	Chloramphenicol		2.55	(Chitongo et al. 2019)
Coconut shell	-	Clofibric acid	3	433.2	(Mestre et al. 2016)
Wood	-	Clofibric acid	3	321.3	(Mestre et al. 2016)
Potato peels	K_2CO_3	Diclofenac	5	69	(Bernardo et al. 2016)
Cyclamen persicum	N_2	Diclofenac	2	22.22	(Jodeh et al. 2016)
Olive stones	H_2SO_4	Diclofenac	2	11	(Larous and Meniai 2016)
Cocoa pod husks	H_2SO_4	Diclofenac	7	0.47	(de Luna et al. 2017)
Potato peels	KOH	Dorzolamide	2	92	(Kyzas and Deliyanni 2015)

Table 8 contd. ...

... *Table 8 contd.*

AW precursor	Activating Agent	Target Contaminant Pharmaceuticals group	pH	Q_{max} (mg/g)*	Reference
Almond tree pruning	Steam/CO_2	Fluoxetine	6-7	224.4	(Román et al. 2012)
Kola nut husk	H_3PO_4	Ibuprofen		39.68	(Bello et al. 2020)
Date stone Siris seed pods	KOH KOH	Levofloxacin Metronidazole	9 7	100.4 196.31	(Darweesh and Ahmed 2017) (Ahmed and Theydan 2013)
Prosopis juliflora	KOH	Metronidazole	6-8	25.06	(Manjunath et al. 2020)
Lotus stalk	H_3PO_4	Norfloxacin	5.5	456.19	(Liu et al. 2011)
Bamboo	H_3PO_4	Norfloxacin	6	293.2	(Peng et al. 2018)
Albizia lebbeck seed pods	KOH	Norfloxacin	5	166.99	(Ahmed and Theydan 2014a)
Moringa oleifera pod husks	NH_4Cl	Norfloxacin	5	2.033	(Wuana et al. 2016)
Hazelnut shell	H_3PO_4	Oxytetracycline	5	322.6	(Fan et al. 2016)
Corn stalk	-	Oxytetracycline	5	90.9	(Song et al. 2020)
Luffa sponge	H_3PO_4	Ofloxacin	6	131.93	(Kong et al. 2017)
Moringa oleifera pod husks	NH_4Cl	Ofloxacin	5	5.051	(Wuana et al. 2015)
Dende coconut	no activation	Paracetamol	6.5	90.813	(De Lima et al. 2015)
Babassu coconut Potato Peel	no activation KOH	Paracetamol Pramipexole dihydrochloride	6.5 2	80.87 98	(De Lima et al. 2015) (Kyzas and Deliyanni 2015)
Vine wood	NaOH	Penicillin G	2	8.41	(Pouretedal and Sadegh 2014)
Lagenaria vulgaris shell	Steam	Ranitidine	11	317.4	(Bojić et al. 2017)
Mung bean husk	Steam	Ranitidine hydrochloride	2	28	(Mondal et al. 2015)
Thalia dealbata	H_3PO_4	Streptomycin	11	270.27	(Huang et al. 2015)
Olive pomace	KOH	Sulfadiazine	7	66.2	(Aslan and Şirazi 2020)
Prosopis juliflora	KOH	Sulfadiazine	6–8	18.48	(Manjunath et al. 2020)
Iris tectorum	H_3PO_4/ $Fe(NO_3)_3$	Tetracycline	2	769.2	(Li et al. 2013)
Tomato	$ZnCl_2$	Tetracycline	5.7	500	(Sayğılı and Güzel 2016b)
Macadamia nut shells Hazelnut shell	NaOH H_3PO_4	Tetracycline Tetracycline	3 5	455.8 312.5	(Martins et al. 2015) (Fan et al. 2016)
Apricot nut shells	H_3PO_4	Tetracycline	6.5	308.3	(Marzbali et al. 2016)
Beet pulp	Steam	Tetracycline	4	288	(Torres-Pérez et al. 2012)
Corn stover	KOH	Tetracycline	7.4	132.9	(Haghighat et al. 2020)
Corn stalk	H_3PO_4	Tetracycline	5	104	(Song et al. 2020)
Prosopis juliflora	KOH	Tetracycline	6–8	28.81	(Manjunath et al. 2020)
Peanut hulls	Steam	Tetracycline	-	28	(Torres-Pérez et al. 2012)
Vine wood	NaOH	Tetracycline	2	1.98	(Pouretedal and Sadegh 2014)
Lotus stalk	$H_4P_2O_7$	Trimethoprim	-	345.4	(Liu et al. 2012a)
Lotus stalk	o-H_3PO_4	Trimethoprim	-	333.2	(Liu et al. 2012a)
Lotus stalk	HPO_3	Trimethoprim	-	118.97	(Liu et al. 2012a)
Lotus stalk	H_3PO_3	Trimethoprim	-	118.57	(Liu et al. 2012a)
Date Palm Fiber	$ZnCl_2$	Tylosin	5.5	147	(Melliti et al. 2021)
Tangerine peels	H_3PO_4	Acetamiprid	5.6	35.7	(Mohammad et al. 2020)
Sky fruit husk	H_3PO_4	Bentazon	6	166.67	(Njokuet al. 2014)
Banana stalk	KOH	Bentazon	2	115.07	(Salman et al. 2011)

Table 8. contd. ...

... *Table 8 contd.*

AW precursor	Activating Agent	Target Contaminant Pharmaceuticals group	pH	Q_{max} (mg/g)*	Reference
Rice straw	KOH	Carbofuran	4.46	296.52	(Chang et al. 2011)
Orange peel	KOH	4-chloro-2-methylphenoxyacetic acid	3	414.94	(Pandiarajan et al. 2018)
Orange peel	KOH	2,4-dichlorophenoxyacetic acid	3	515.46	(Pandiarajan et al. 2018)
Langsat empty fruit bunch	H_3PO_4	2,4-dichlorophenoxyacetic acid	2	332	(Njoku et al. 2015)
Banana stalk	KOH	2,4-dichlorophenoxyacetic acid	2	196.33	(Salman et al. 2011)
Date-palm coir	KOH	2,4-dichlorophenoxyacetic acid	2	50.25	(Rambabu et al. 2021)
Orange peel	KOH	2(2,4-dichlorophenoxy) propanoic acid	3	215.52	(Pandiarajan et al. 2018)
Wheat straw	H_3PO_4	Heptachlor	4	2.22	(Seyhi et al. 2014)
Peach stone	H_3PO_4	Imidacloprid	5.2	39.37	(Mohammad and El-Sayed 2020)
Orange peel	KOH	2(2-methyl-4-chlorophenoxy) propionic acid	3	574.71	(Pandiarajan et al. 2018)
Pistachio hull	NaOH	Methyl Parabene	6	55.52	(Nodeh et al. 2020)
Pistachio hull	NaOH	Propyl Parabene	6	50.1	(Nodeh et al. 2020)
Orange peel	KOH	2,4,5-Trichlorophenoxyacetic acid	3	416.67	(Pandiarajan et al. 2018)
Rice straw	K_2CO_3	Triclosan	3	714	(Liu et al. 2014b)
Rice husk	N_2	Triclosan	4	72.7	(Triwiswara et al. 2020)
Coconut pulp (*Cocos nuciefera*)	$ZnCl_2$	Triclosan	5.6	38.91	(Khori et al. 2020)

*Q_{max} as predicted by Langmuir isotherm model

The adsorption mechanism of pharmaceuticals and pesticides depends upon the chemical and physical surface properties of the adsorbent and adsorbate, as well as the structure of the adsorbent. Hence, various adsorption mechanisms are involved in their removal from water using AW-derived ACs. Peng et al. (2018) attributed the adsorption of norfloxacin and ciprofloxacin to hydrophobicity as a dominant mechanism. In addition, they suggested other mechanisms alongside the former, which include π–π interaction, hydrogen bonding and electrostatic interaction depending on the effect of solution pH. Similarly, the adsorption mechanism of acetaminophen is proposed to include π–π interaction, hydrogen bonding and dispersion interactions (Limaet al. 2019). The hydrogen bond is usually formed due to the presence of functional groups (such as –COO, –O, –OH, C=N) on the surface of the AC which interact with other functional groups on the surface of the pharmaceuticals (such as –OH, -NH, -F). The π–π interaction can occur between π-electrons of the aromatic rings on both the AC and the pharmaceutical, or even between AC and a strong electron withdrawing functional group on the pharmaceutical (such as –F in Fluoroquinolones) (Lima et al. 2019, Peng et al. 2018). The graphitic structure of AC provides hydrophobic sites leading to hydrophobic interaction with the hydrophobic surfaces of the pharmaceuticals (Liu et al. 2011, Peng et al. 2018). In many cases, pharmaceuticals are characterized by a zwitterionic behavior, which can be exploited to create an electrostatic attraction with the AC (Peng et al. 2018). Other mechanisms reported in literature include surface complexation (Manjunath et al. 2020), cation exchange (Liu et al. 2011), and Lewis acid-base interaction (Liu et al. 2012).

The adsorption mechanisms for pesticides are mainly governed by electrostatic interactions as indicated by (Njoku et al. 2014b, Nodeh et al. 2020) for herbicides, bactericides and fungicides. In addition, Nodeh et al. (2020) suggested other adsorption mechanisms for parabens, which include π–π interaction and hydrogen bonding. Triwiswara et al. (2020) reported that the adsorption mechanism of Triclosan on AW-derived AC included π–π interaction, hydrogen bonding, hydrophobic interaction and electrostatic interaction. According to the former study, the hydrophobic interaction takes place due to the highly hydrophobic nature of Triclosan (Triwiswara et al. 2020). In another study, the adsorption of imidacloprid onto AC derived from peach stone was attributed to the π–π interaction and hydrogen bonding (Mohammad and El-Sayed 2020). In general, the adsorption of pharmaceuticals and pesticides to AC occurs through more than one mechanism taking place simultaneously, yet certain mechanisms are more dominant than the others.

5. CONCLUSIONS

Waste valorization is gaining momentum as a reliable approach for producing value-added products from waste materials. Owing to their carbonaceous chemical structure, numerous AWs have been utilized for the preparation of AC adsorbents to be used in water treatment. Conversion of AWs to AC adsorbents has two sustainable aspects-waste recovery, on one hand, and water pollution prevention, on the other.

AWs have been activated through different activation means either physically or chemically. The dual stage physical activation relies on carbonization and activation of steam or CO_2, while the single stage chemical activation employs any of the various chemical activating agents, ranging from acidic, basic, or neutral compounds. Of the most commonly used ones are KOH, H_3PO_4, and $ZnCl_2$. Physical activation requires control over the temperature and time of activation, whereas chemical activation is dictated by the temperature, time, and impregnation ratio. In terms of its environmental impact, physical activation is favored to chemical activation due to its greener approach in which no secondary waste is produced. However, chemical activation is more economical due to the lower activation temperatures used, and the higher obtained yields of ACs that possess higher porosity and surface area than their counterparts prepared by physical activation. The selection of activating agents and their activation conditions also provides more control over the internal porous structure of the activated material, and hence its suitability for the designated application. The ratio between micro- and meso-pores during chemical activation is dictated by the type of the activating agent as well as the activation conditions. A higher porosity of AC entails a higher surface area and a lower yield of the resulting material.

With respect to their potential for water treatment, AW-derived ACs were efficiently used to remove major conventional contaminants, particularly heavy metals, in addition to CECs such as pharmaceuticals and pesticides. These ACs showed remarkable adsorption capacities of approximately 1500 mg/g and 800 mg/g for heavy metals and pharmaceuticals/pesticides, respectively. The mechanisms by which these contaminants were bound to AC included electrostatic interaction, ion exchange, π–π stacking, hydrophobic interaction,and complex formation.

To conclude, this chapter provides a comprehensive critical review on the state-of-the-art physical and chemical processes for converting AWs to ACs that were efficiently used for water treatment. The chapter also addresses the challenges of the existing water treatment methods and highlights the opportunities that AW-derived AC adsorption can offer as an alternative method with potential economic and environmental merits. Work presented herein has implications for solid waste management and water treatment applications in alignment with the waste management 4 Rs strategy which encompasses principles of waste reduction, reuse, recycling, and recovery.

6. CHALLENGES AND FUTURE DIRECTIONS

To compare the performance of different AW-derived ACs in removing a particular contaminant, ACs should be produced from the same AWs and activated using the same physical or chemical agent under the same conditions. In spite of the mounting body of literature that reported the synthesis of AW-derived ACs, there remains a potential for further exploration of AWs that can produce new forms of ACs with tailored characteristics. For example, carbon nanotubes (CNTs) have been prepared from AWs (Fathy et al. 2017, Lotfy et al. 2018); however, no studies utilized AW-derived CNTs for removing contaminants, although some studies used chemically synthesized CNTs for this purpose (Ahmadzadeh Tofighy 2012, Ihsanullah et al. 2016, Ouni et al. 2019).

The planet is facing profound global challenges due to water scarcity, waste accumulation, and climate change; hence, researchers are seeking long-term solutions to cope with such challenges. The state-of-the-art processes for producing ACs from AWs are yet to be evaluated from the sustainability point of view. Applying the three pillars of sustainability, comprising the environmental, economic, and social aspects, is key, though complex; hence, there is an urgent need for performing life-cycle analyses on processes and applying the cradle-to-cradle concept to products to minimize their overall environmental impacts. The concepts of green processing including selection of non-hazardous raw materials, reduction of energy and solvent consumption, and use of renewable energy sources, should be applied.

Despite its green approach, physical activation of AWs using steam or CO_2 consumes high energy input which renders it unsustainable. On the other hand, chemical activation generates secondary waste that increases its ecological footprint. Accordingly, future research should look into deploying more sustainable alternatives to develop efficient, cost-effective and eco-friendly activation processes. To move to a greener approach, renewable energy sources could be used, and energy consumption along with operational times could potentially be reduced by adopting new activation methods. For instance, literature lacks studies that deploy microwave heating (Ahmed and Theydan 2012) or ultrasonication to replace the conventional heating. Selecting less hazardous activating agents or searching for eco-friendly alternatives like ionic liquids should also be investigated. One very recent study was reported on generating an AC adsorbent for CO_2 from rubber seed shell using a pyridinium-based ionic liquid (Mokti et al. 2021). In addition, recovery of the secondary waste or converting it to products of lower hazard should be adopted.

From a holistic standpoint, the process of converting AWs to ACs should be investigated from various technical, economic, environmental, and social perspectives. Aspects of energy reduction, waste minimization, abundance and location of raw materials, and cost effectiveness should hence be considered.

LIST OF ABBREVIATIONS

AC	:	Activated carbon
AW	:	Agricultural waste
BET	:	Brunauer, Emmet and Teller
CECs	:	Contaminants of emerging concern
CM	:	Carbonaceous material
CNTs	:	Carbon nanotubes
V_{meso}	:	Mesopore volume
V_{micro}	:	Micropore volume
V_{total}	:	Total volume
Q_{max}	:	Maximum adsorption capacity

REFERENCES

Abdulkarim, M. and Al-Rub, F.A. 2004. Adsorption of Lead Ions from Aqueous Solution onto Activated Carbon and Chemically-Modified Activated Carbon Prepared from Date Pits. Adsorption Science and Technology 22(2): 119–134. https://doi.org/10.1260/026361704323150908.

Acharya, J., Kumar, U. and Rafi, P.M. 2018. Removal of Heavy Metal Ions from Wastewater by Chemically Modified Agricultural Waste Material as Potential Adsorbent—A Review. International Journal of Current Engineering and Technology 8: 526–530.

Acharya, J., Sahu, J.N., Mohanty, C.R. and Meikap, B.C. 2009. Removal of lead(II) from wastewater by activated carbon developed from Tamarind wood by zinc chloride activation. Chemical Engineering Journal 149(1): 249–262. https://doi.org/10.1016/j.cej.2008.10.029

Afroze, S. and Sen, T.K. 2018. A Review on Heavy Metal Ions and Dye Adsorption from Water by Agricultural Solid Waste Adsorbents. Water, Air, and Soil Pollution 229(7): 225. https://doi.org/10.1007/s11270-018-3869-z

Ahmad, M.A., Ahmad Puad, N.A. and Bello, O.S. 2014. Kinetic, equilibrium and thermodynamic studies of synthetic dye removal using pomegranate peel activated carbon prepared by microwave-induced KOH activation. Water Resources and Industry 6: 18–35. https://doi.org/10.1016/j.wri.2014.06.002

Ahmadpour, A. and Do, D.D. 1997. The preparation of activated carbon from macadamia nutshell by chemical activation. Carbon 35(12): 1723–1732. https://doi.org/10.1016/S0008-6223(97)00127-9

Ahmadzadeh Tofighy, M. and Mohammadi, T. 2012. Nitrate removal from water using functionalized carbon nanotube sheets. Chemical Engineering Research and Design, 90(11): 1815–1822. https://doi.org/10.1016/j.cherd.2012.04.001

Ahmed, M.J. and Theydan, S.K. 2012. Adsorption of cephalexin onto activated carbons from Albizia lebbeck seed pods by microwave-induced KOH and K2CO3 activations. Chemical Engineering Journal 211–212: 200–207. https://doi.org/10.1016/j.cej.2012.09.089

Ahmed, M.J. and Theydan, S.K. 2013. Microporous activated carbon from Siris seed pods by microwave-induced KOH activation for metronidazole adsorption. Journal of Analytical and Applied Pyrolysis 99: 101–109. https://doi.org/10.1016/j.jaap.2012.10.019

Ahmed, M. J., and Theydan, S. K. 2014a. Fluoroquinolones antibiotics adsorption onto microporous activated carbon from lignocellulosic biomass by microwave pyrolysis. Journal of the Taiwan Institute of Chemical Engineers 45(1): 219–226. https://doi.org/10.1016/j.jtice.2013.05.014

Ahmed, M.J. and Theydan, S.K. 2014b. Optimization of microwave preparation conditions for activated carbon from Albizia lebbeck seed pods for methylene blue dye adsorption. Journal of Analytical and Applied Pyrolysis, 105: 199–208. https://doi.org/10.1016/j.jaap.2013.11.005

Akpen, G.D., Leton, I.L.N. and T.G. 2011. Optimum conditions for the removal of colour from waste water by mango seed shell based activated carbon. Indian Journal of Science and Technology 4(8): 890–894. https://doi.org/10.17485/ijst/2011/v4i8.16

Alahabadi, A., Hosseini-Bandegharaei, A., Moussavi, G., Amin, B., Rastegar, A., Karimi-Sani, H., Fattahi, M. and Miri, M. 2017. Comparing adsorption properties of NH4Cl-modified activated carbon towards chlortetracycline antibiotic with those of commercial activated carbon. Journal of Molecular Liquids 232: 367–381. https://doi.org/10.1016/j.molliq.2017.02.077

Al-Qodah, Z., Yahya, M.A. and Al-Shannag, M. 2017. On the performance of bioadsorption processes for heavy metal ions removal by low-cost agricultural and natural by-products bioadsorbent: A review. Desalination and Water Treatment 85: 339–357. https://doi.org/10.5004/dwt.2017.21256

Alslaibi, T.M., Abustan, I., Ahmad, M.A. and Foul, A.A. 2013. Cadmium removal from aqueous solution using microwaved olive stone activated carbon. Journal of Environmental Chemical Engineering 1(3): 589–599. https://doi.org/10.1016/j.jece.2013.06.028

Al-Swaidan, H.M. and Ashfaq, A. 2011. Synthesis and Characterization of Activated Carbon from Saudi Arabian Dates Tree's Fronds Wastes. 3rd International Conference on Chemical, Biological and Environmental Engineering 20: 7.

Altenor, S., Carene, B., Emmanuel, E., Lambert, J., Ehrhardt, J.-J. and Gaspard, S. 2009. Adsorption studies of methylene blue and phenol onto vetiver roots activated carbon prepared by chemical activation. Journal of Hazardous Materials 165(1): 1029–1039. https://doi.org/10.1016/j.jhazmat.2008.10.133

Amirza, M.a.R., Adib, M.M.R. and Hamdan, R. 2017. Application of Agricultural Wastes Activated Carbon for Dye Removal—An Overview. MATEC Web of Conferences 103: 06013. https://doi.org/10.1051/matecconf/201710306013

Angin, D. 2014. Utilization of activated carbon produced from fruit juice industry solid waste for the adsorption of Yellow 18 from aqueous solutions. Bioresource Technology 168: 259–266. https://doi.org/10.1016/j.biortech.2014.02.100

Angın, D. 2013. Effect of pyrolysis temperature and heating rate on biochar obtained from pyrolysis of safflower seed press cake. Bioresource Technology 128: 593–597. https://doi.org/10.1016/j.biortech.2012.10.150

Angın, D., Altintig, E. and Köse, T.E. 2013. Influence of process parameters on the surface and chemical properties of activated carbon obtained from biochar by chemical activation. Bioresource Technology 148: 542–549. https://doi.org/10.1016/j.biortech.2013.08.164

Anirudhan, T.S. and Sreekumari, S.S. 2011. Adsorptive removal of heavy metal ions from industrial effluents using activated carbon derived from waste coconut buttons. Journal of Environmental Sciences 23(12): 1989–1998. https://doi.org/10.1016/S1001-0742(10)60515-3

Anisuzzaman, S.M., Joseph, C.G., Daud, W.M.A.B.W., Krishnaiah, D. and Yee, H.S. 2015. Preparation and characterization of activated carbon from Typha orientalis leaves. International Journal of Industrial Chemistry 6(1): 9–21. https://doi.org/10.1007/s40090-014-0027-3

Anisuzzaman, S.M., Joseph, C.G., Krishnaiah, D., Bono, A., Suali, E., Abang, S. and Fai, L. M. 2016. Removal of chlorinated phenol from aqueous media by guava seed (Psidium guajava) tailored activated carbon. Water Resources and Industry 16: 29–36. https://doi.org/10.1016/j.wri.2016.10.001

Ansari, M., Mahvi, A.H., Salmani, M.H., Ehrampoush, M.H., Ghaneiyan, M.T., Shiraniyan, M. and Tallebi, P. 2018. Removal of Cadmium from Aqueous Solutions by a Synthesized Activated Carbon. Journal of Environmental Health and Sustainable Development 3(3): 593–605.

Asadullah, M., Rahman, M.A., Motin, M.A. and Sultan, M.B. 2007. Adsorption Studies on Activated Carbon Derived from Steam Activation of Jute Stick Char. Journal of Surface Science and Technology 23(1–2): 73–80. https://doi.org/10.18311/jsst/2007/1956

Asasian, N., Kaghazchi, T. and Soleimani, M. 2012. Elimination of mercury by adsorption onto activated carbon prepared from the biomass material. Journal of Industrial and Engineering Chemistry 18(1): 283–289. https://doi.org/10.1016/j.jiec.2011.11.040

Asimakopoulos, G., Baikousi, M., Salmas, C., Bourlinos, A.B., Zboril, R. and Karakassides, M.A. 2021. Advanced Cr(VI) sorption properties of activated carbon produced via pyrolysis of the "Posidonia oceanica" seagrass. Journal of Hazardous Materials 405: 124274. https://doi.org/10.1016/j.jhazmat.2020.124274

Aslan, S. and Şirazi, M. 2020. Adsorption of Sulfonamide Antibiotic onto Activated Carbon Prepared from an Agro-industrial By-Product as Low-Cost Adsorbent: Equilibrium, Thermodynamic, and Kinetic Studies. Water, Air, and Soil Pollution 231(5): 222. https://doi.org/10.1007/s11270-020-04576-0

Ateia, M., Attia, M.F., Maroli, A., Tharayil, N., Alexis, F., Whitehead, D.C. and Karanfil, T. 2018. Rapid Removal of Poly- and Perfluorinated Alkyl Substances by Poly(ethylenimine)-Functionalized Cellulose Microcrystals at Environmentally Relevant Conditions. Environmental Science and Technology Letters, 5(12): 764–769. https://doi.org/10.1021/acs.estlett.8b00556

Attia, A.A., Girgis, B.S. and Fathy, N.A. 2008. Removal of methylene blue by carbons derived from peach stones by H3PO4 activation: Batch and column studies. Dyes and Pigments, 76(1): 282–289. https://doi.org/10.1016/j.dyepig.2006.08.039

Awwad, N.S., El-Zahhar, A.A., Fouda, A.M. and Ibrahium, H.A. 2013. Removal of heavy metal ions from ground and surface water samples using carbons derived from date pits. Journal of Environmental Chemical Engineering 1(3): 416–423. https://doi.org/10.1016/j.jece.2013.06.006

Baçaoui, A., Yaacoubi, A., Dahbi, A., Bennouna, C., Phan Tan Luu, R., Maldonado-Hodar, F. J., Rivera-Utrilla, J. and Moreno-Castilla, C. 2001. Optimization of conditions for the preparation of activated carbons from olive-waste cakes. Carbon 39(3): 425–432. https://doi.org/10.1016/S0008-6223(00)00135-4

Baccar, R., Bouzid, J., Feki, M. and Montiel, A. 2009. Preparation of activated carbon from Tunisian olive-waste cakes and its application for adsorption of heavy metal ions. Journal of Hazardous Materials 162(2): 1522–1529. https://doi.org/10.1016/j.jhazmat.2008.06.041

Bagheri, N. and Abedi, J. 2009. Preparation of high surface area activated carbon from corn by chemical activation using potassium hydroxide. Chemical Engineering Research and Design 87(8): 1059–1064. https://doi.org/10.1016/j.cherd.2009.02.001

Balarak, D., Mostafapour, F.K., Akbari, H. and Joghtaei, A. 2017. Adsorption of Amoxicillin Antibiotic from Pharmaceutical Wastewater by Activated Carbon Prepared from Azolla filiculoides. Journal of Pharmaceutical Research International 18(3): 1–13.

Balarak, D., Mostafapour, F. K., and Azarpira, H. 2016. Adsorption Kinetics and Equilibrium of Ciprofloxacin from Aqueous Solutions Using Corylus avellana (Hazelnut) Activated Carbon. British Journal of Pharmaceutical Research 13(3): 1–14.

Balci, S., Doğu, T. and Yücel, H. 1994. Characterization of activated carbon produced from almond shell and hazelnut shell. Journal of Chemical Technology and Biotechnology 60(4): 419–426. https://doi.org/10.1002/jctb.280600413

Başar, C.A. 2006. Applicability of the various adsorption models of three dyes adsorption onto activated carbon prepared waste apricot. Journal of Hazardous Materials 135(1): 232–241. https://doi.org/10.1016/j.jhazmat.2005.11.055

Basta, A. H., Fierro, V., El-Saied, H. and Celzard, A. 2009. 2-Steps KOH activation of rice straw: An efficient method for preparing high-performance activated carbons. Bioresource Technology 100(17): 3941–3947. https://doi.org/10.1016/j.biortech.2009.02.028

Beis, S.H., Onay, Ö. and Koçkar, Ö.M. 2002. Fixed-bed pyrolysis of safflower seed: Influence of pyrolysis parameters on product yields and compositions. Renewable Energy 26(1): 21–32. https://doi.org/10.1016/S0960-1481(01)00109-4

Belhachemi, M. and Djelaila, S. 2017. Removal of Amoxicillin Antibiotic from Aqueous Solutions by Date Pits Activated Carbons. Environmental Processes 4(3): 549–561. https://doi.org/10.1007/s40710-017-0245-8

Bello, O.S., Alao, O.C., Alagbada, T.C., Agboola, O.S., Omotoba, O.T. and Abikoye, O.R. 2020. A renewable, sustainable and low-cost adsorbent for ibuprofen removal. Water Science and Technology 83(1): 111–122. https://doi.org/10.2166/wst.2020.551

Bernardo, M., Rodrigues, S., Lapa, N., Matos, I., Lemos, F., Batista, M.K.S., Carvalho, A.P. and Fonseca, I. 2016. High efficacy on diclofenac removal by activated carbon produced from potato peel waste. International Journal of Environmental Science and Technology 13(8): 1989–2000. https://doi.org/10.1007/s13762-016-1030-3

Bojić, D., Momčilović, M., Milenković, D., Mitrović, J., Banković, P., Velinov, N. and Nikolić, G. 2017. Characterization of a low cost Lagenaria vulgaris based carbon for ranitidine removal from aqueous solutions. Arabian Journal of Chemistry 10(7): 956–964. https://doi.org/10.1016/j.arabjc.2014.12.018

Boonpoke, A., Chiarakorn, S., Laosiripojana, N., Towprayoon, S. and Chidthaisong, A. 2011. Synthesis of Activated Carbon and MCM-41 from Bagasse and Rice Husk and their Carbon Dioxide Adsorption Capacity. Journal of Sustainable Energy and Environment 2: 77–81.

Bouchelta, C., Medjram, M.S., Zoubida, M., Chekkat, F.A., Ramdane, N. and Bellat, J.-P. 2012. Effects of pyrolysis conditions on the porous structure development of date pits activated carbon. Journal of Analytical and Applied Pyrolysis 94: 215–222. https://doi.org/10.1016/j.jaap.2011.12.014

Bouchenafa-Saïb, N., Grange, P., Verhasselt, P., Addoun, F. and Dubois, V. 2005. Effect of oxidant treatment of date pit active carbons used as Pd supports in catalytic hydrogenation of nitrobenzene. Applied Catalysis A: General 286(2): 167–174. https://doi.org/10.1016/j.apcata.2005.02.022

Boudrahem, F., Soualah, A. and Aissani-Benissad, F. 2011. Pb(II) and Cd(II) Removal from Aqueous Solutions Using Activated Carbon Developed from Coffee Residue Activated with Phosphoric Acid and Zinc Chloride. Journal of Chemical and Engineering Data 56(5): 1946–1955. https://doi.org/10.1021/je1009569

Bouhamed, F., Elouear, Z. and Bouzid, J. 2012. Adsorptive removal of copper(II) from aqueous solutions on activated carbon prepared from Tunisian date stones: Equilibrium, kinetics and thermodynamics. Journal of the Taiwan Institute of Chemical Engineers 43(5): 741–749. https://doi.org/10.1016/j.jtice.2012.02.011

Brunner, P.H. and Roberts, P.V. 1980. The significance of heating rate on char yield and char properties in the pyrolysis of cellulose. Carbon 18(3): 217–224. https://doi.org/10.1016/0008-6223(80)90064-0

Budinova, T., Ekinci, E., Yardim, F., Grimm, A., Björnbom, E., Minkova, V. and Goranova, M. 2006. Characterization and application of activated carbon produced by H3PO4 and water vapor activation. Fuel Processing Technology 87(10): 899–905. https://doi.org/10.1016/j.fuproc.2006.06.005

Cabal, B., Budinova, T., Ania, C.O., Tsyntsarski, B., Parra, J.B. and Petrova, B. 2009. Adsorption of naphthalene from aqueous solution on activated carbons obtained from bean pods. Journal of Hazardous Materials 161(2): 1150–1156. https://doi.org/10.1016/j.jhazmat.2008.04.108

Cabrita, I., Ruiz, B., Mestre, A.S., Fonseca, I. M., Carvalho, A. P., and Ania, C. O. 2010. Removal of an analgesic using activated carbons prepared from urban and industrial residues. Chemical Engineering Journal 163(3): 249–255. https://doi.org/10.1016/j.cej.2010.07.058

Cao, Q., Xie, K.-C., Lv, Y.-K. and Bao, W.-R. 2006. Process effects on activated carbon with large specific surface area from corn cob. Bioresource Technology 97(1): 110–115. https://doi.org/10.1016/j.biortech.2005.02.026

Castro, J.B., Bonelli, P.R., Cerrella, E.G. and Cukierman, A.L. 2000. Phosphoric Acid Activation of Agricultural Residues and Bagasse from Sugar Cane: Influence of the Experimental Conditions on Adsorption Characteristics of Activated Carbons. Industrial and Engineering Chemistry Research 39(11): 4166–4172. https://doi.org/10.1021/ie0002677

Caturla, F., Molina-Sabio, M. and Rodríguez-Reinoso, F. 1991. Preparation of activated carbon by chemical activation with ZnCl2. Carbon 29(7): 999–1007. https://doi.org/10.1016/0008-6223(91)90179-M

Cazetta, A.L., Vargas, A.M.M., Nogami, E.M., Kunita, M.H., Guilherme, M.R., Martins, A.C., Silva, T.L., Moraes, J.C.G. and Almeida, V.C. 2011. NaOH-activated carbon of high surface area produced from coconut shell: Kinetics and equilibrium studies from the methylene blue adsorption. Chemical Engineering Journal 174(1): 117–125. https://doi.org/10.1016/j.cej.2011.08.058

Chand, R., Watari, T., Inoue, K., Torikai, T. and Yada, M. 2009. Evaluation of wheat straw and barley straw carbon for Cr(VI) adsorption. Separation and Purification Technology 65(3): 331–336. https://doi.org/10.1016/j.seppur.2008.11.002

Chandra, T.C., Mirna, M.M., Sudaryanto, Y. and Ismadji, S. 2007. Adsorption of basic dye onto activated carbon prepared from durian shell: Studies of adsorption equilibrium and kinetics. Chemical Engineering Journal 127(1): 121–129. https://doi.org/10.1016/j.cej.2006.09.011

Chang, J., Gao, Z., Wang, X., Wu, D., Xu, F., Wang, X., Guo, Y. and Jiang, K. 2015. Activated porous carbon prepared from paulownia flower for high performance supercapacitor electrodes. Electrochimica Acta 157: 290–298. https://doi.org/10.1016/j.electacta.2014.12.169

Chang, K.-L., Lin, J.-H. and Chen, S.-T. 2011. Adsorption Studies on the Removal of Pesticides(Carbofuran) using Activated Carbon from Rice Straw Agricultural Waste. International Journal of Agricultural and Biosystems Engineering 5(4): 210–213.

Chaudhuri, M. and Azizan, N.K.B. 2012. Adsorptive Removal of Chromium (VI) from Aqueous Solution by an Agricultural Waste-Based Activated Carbon. Water, Air, and Soil Pollution 223(4): 1765–1771. https://doi.org/10.1007/s11270-011-0981-8

Chayid, M.A. and Ahmed, M.J. 2015. Amoxicillin adsorption on microwave prepared activated carbon from Arundo donax Linn: Isotherms, kinetics, and thermodynamics studies. Journal of Environmental Chemical Engineering 3(3): 1592–1601. https://doi.org/10.1016/j.jece.2015.05.021

Chen, C., Huang, B., Li, T. and Wu, G. 2012. Preparation of phosphoric acid activated carbon from sugarcane bagasse by mechanochemical processing. BioRes 7(4): 5109–5116.

Chen, S., Liu, Z., Jiang, S. and Hou, H. 2020. Carbonization: A feasible route for reutilization of plastic wastes. Science of The Total Environment 710: 136250. https://doi.org/10.1016/j.scitotenv.2019.136250

Chen, Y., Huang, B., Huang, M. and Cai, B. 2011. On the preparation and characterization of activated carbon from mangosteen shell. Journal of the Taiwan Institute of Chemical Engineers 42(5): 837–842. https://doi.org/10.1016/j.jtice.2011.01.007

Chitongo, R., Opeolu, B.O. and Olatunji, O.S. 2019. Abatement of Amoxicillin, Ampicillin, and Chloramphenicol From Aqueous Solutions Using Activated Carbon Prepared From Grape Slurry. CLEAN – Soil, Air, Water 47(2): 1800077. https://doi.org/10.1002/clen.201800077

Chiu, K.-L. and Ng, D.H.L. 2012. Synthesis and characterization of cotton-made activated carbon fiber and its adsorption of methylene blue in water treatment. Biomass and Bioenergy 46: 102–110. https://doi.org/10.1016/j.biombioe.2012.09.023

Daifullah, A.A.M., Yakout, S.M. and Elreefy, S.A. 2007. Adsorption of fluoride in aqueous solutions using KMnO4-modified activated carbon derived from steam pyrolysis of rice straw. Journal of Hazardous Materials 147(1): 633–643. https://doi.org/10.1016/j.jhazmat.2007.01.062

Danish, M., Ahmad, T., Hashim, R., Said, N., Akhtar, M.N., Mohamad-Saleh, J. and Sulaiman, O. 2018. Comparison of surface properties of wood biomass activated carbons and their application against rhodamine B and methylene blue dye. Surfaces and Interfaces 11: 1–13. https://doi.org/10.1016/j.surfin.2018.02.001

Darweesh, T.M. and Ahmed, M.J. 2017. Batch and fixed bed adsorption of levofloxacin on granular activated carbon from date (Phoenix dactylifera L.) stones by KOH chemical activation. Environmental Toxicology and Pharmacology 50: 159–166. https://doi.org/10.1016/j.etap.2017.02.005

De Lima, H., Cristina Ferreira, R., Candido, A., Junior, O., Arroyo, P., Carvalho, K., Gauze, G. and Barros, M.A. 2015. Adsorption of Paracetamol Using Activated Carbon of Dende and Babassu Coconut Mesocarp.

de Luna, M.D.G., Murniati, Budianta, W., Rivera, K.K.P. and Arazo, R.O. 2017. Removal of sodium diclofenac from aqueous solution by adsorbents derived from cocoa pod husks. Journal of Environmental Chemical Engineering 5(2): 1465–1474. https://doi.org/10.1016/j.jece.2017.02.018

Dehghani, M.H., Farhang, M., Alimohammadi, M., Afsharnia, M. and Mckay, G. 2018. Adsorptive removal of fluoride from water by activated carbon derived from CaCl2-modified Crocus sativus leaves: Equilibrium adsorption isotherms, optimization, and influence of anions. Chemical Engineering Communications 205(7): 955–965. https://doi.org/10.1080/00986445.2018.1423969

Demiral, H., Demiral, İ., Tümsek, F. and Karabacakoğlu, B. 2008. Pore structure of activated carbon prepared from hazelnut bagasse by chemical activation. Surface and Interface Analysis 40(3–4): 616–619. https://doi.org/10.1002/sia.2631

Demiral, H. and Gündüzoğlu, G. 2010. Removal of nitrate from aqueous solutions by activated carbon prepared from sugar beet bagasse. Bioresource Technology 101(6): 1675–1680. https://doi.org/10.1016/j.biortech.2009.09.087

Demirbas, A. 2008. Heavy metal adsorption onto agro-based waste materials: A review. Journal of Hazardous Materials 157(2): 220–229. https://doi.org/10.1016/j.jhazmat.2008.01.024

Depci, T., Kul, A.R. and Önal, Y. 2012. Competitive adsorption of lead and zinc from aqueous solution on activated carbon prepared from Van apple pulp: Study in single- and multi-solute systems. Chemical Engineering Journal 200–202: 224–236. https://doi.org/10.1016/j.cej.2012.06.077

Devnarain, P.B., Arnold, D.R. and Davis, S.B. 2002. Production of Activated Carbon from South African Sugarcane Bagasse. Proceedings of the 76th Annual Congress of the South African Sugar Technologists' Association 477–489.

Diao, Y., Walawender, W.P. and Fan, L.T. 2002. Activated carbons prepared from phosphoric acid activation of grain sorghum. Bioresource Technology 81(1): 45–52. https://doi.org/10.1016/S0960-8524(01)00100-6

Donald, J., Ohtsuka, Y. and Xu, C. Charles. 2011. Effects of activation agents and intrinsic minerals on pore development in activated carbons derived from a Canadian peat. Materials Letters 65(4): 744–747. https://doi.org/10.1016/j.matlet.2010.11.049

Dulio, V., van Bavel, B., Brorström-Lundén, E., Harmsen, J., Hollender, J., Schlabach, M., Slobodnik, J., Thomas, K. and Koschorreck, J. 2018. Emerging pollutants in the EU: 10 years of NORMAN in support of environmental policies and regulations. *Environmental Sciences* Europe 30(1): 5. https://doi.org/10.1186/s12302-018-0135-3

El-Hendawy, A.-N.A. 2009. An insight into the KOH activation mechanism through the production of microporous activated carbon for the removal of Pb2+ cations. Applied Surface Science 255(6): 3723–3730. https://doi.org/10.1016/j.apsusc.2008.10.034

El-Hendawy, A.-N.A., Samra, S.E. and Girgis, B.S. 2001. Adsorption characteristics of activated carbons obtained from corncobs. Colloids and Surfaces A: Physicochemical and Engineering Aspects 180(3): 209–221. https://doi.org/10.1016/S0927-7757(00)00682-8

El-Naggar, A.H., Alzhrani, A.K.R., Ahmad, M., Usman, A.R.A., Mohan, D., Ok, Y.S. and Al-Wabel, M.I. 2016. Preparation of Activated and Non-Activated Carbon from Conocarpus Pruning Waste as Low-Cost Adsorbent for Removal of Heavy Metal Ions from Aqueous Solution. BioResources 11(1): 1092–1107.

El-Shafey, E.-S.I., Al-Lawati, H. and Al-Sumri, A.S. 2012. Ciprofloxacin adsorption from aqueous solution onto chemically prepared carbon from date palm leaflets. Journal of Environmental Sciences 24(9): 1579–1586. https://doi.org/10.1016/S1001-0742(11)60949-2

Erdem, M., Orhan, R., Şahin, M. and Aydın, E. 2016. Preparation and Characterization of a Novel Activated Carbon from Vine Shoots by ZnCl2 Activation and Investigation of Its Rifampicine Removal Capability. Water, Air, and Soil Pollution 227(7): 226. https://doi.org/10.1007/s11270-016-2929-5

Erdoğan, S., Önal, Y., Akmil-Başar, C., Bilmez-Erdemoğlu, S., Sarıcı-Özdemir, Ç., Köseoğlu, E. and İçduygu, G. 2005. Optimization of nickel adsorption from aqueous solution by using activated carbon prepared from waste apricot by chemical activation. Applied Surface Science 252(5): 1324–1331. https://doi.org/10.1016/j.apsusc.2005.02.089

Fan, H.-T., Shi, L.-Q., Shen, H., Chen, X. and Xie, K.-P. 2016. Equilibrium, isotherm, kinetic and thermodynamic studies for removal of tetracycline antibiotics by adsorption onto hazelnut shell derived activated carbons from aqueous media. *RSC* Advances 6(111): 109983–109991. https://doi.org/10.1039/C6RA23346E

Fan, M., Marshall, W., Daugaard, D. and Brown, R.C. 2004. Steam activation of chars produced from oat hulls and corn stover. Bioresource Technology 93(1): 103–107. https://doi.org/10.1016/j.biortech.2003.08.016

Fathy, N.A., Lotfy, V.F. and Basta, A.H. 2017. Comparative study on the performance of carbon nanotubes prepared from agro- and xerogels as carbon supports. Journal of Analytical and Applied Pyrolysis 128: 114–120. https://doi.org/10.1016/j.jaap.2017.10.019

Fierro, V., Muñiz, G., Basta, A.H., El-Saied, H. and Celzard, A. 2010. Rice straw as precursor of activated carbons: Activation with ortho-phosphoric acid. Journal of Hazardous Materials 181(1): 27–34. https://doi.org/10.1016/j.jhazmat.2010.04.062

Foo, K.Y. and Hameed, B.H. 2012. Factors affecting the carbon yield and adsorption capability of the mangosteen peel activated carbon prepared by microwave assisted K2CO3 activation. Chemical Engineering Journal 180: 66–74. https://doi.org/10.1016/j.cej.2011.11.002

Foo, P.Y.L. and Lee, L.Y. 2010. Preparation of Activated Carbon from Parkia Speciosa Pod by Chemical Activation. Proceedings of the World Congress on Engineering and Computer Science 2.

Galhetas, M., Mestre, A.S., Pinto, M.L., Gulyurtlu, I., Lopes, H. and Carvalho, A.P. 2014a. Carbon-based materials prepared from pine gasification residues for acetaminophen adsorption. Chemical Engineering Journal 240: 344–351. https://doi.org/10.1016/j.cej.2013.11.067

Galhetas, M., Mestre, A.S., Pinto, M.L., Gulyurtlu, I., Lopes, H. and Carvalho, A.P. 2014b. Chars from gasification of coal and pine activated with K2CO3: Acetaminophen and caffeine adsorption from aqueous solutions. Journal of Colloid and Interface Science 433: 94–103. https://doi.org/10.1016/j.jcis.2014.06.043

Gao, J., Qin, Y., Zhou, T., Cao, D., Xu, P., Hochstetter, D. and Wang, Y. 2013a. Adsorption of methylene blue onto activated carbon produced from tea (Camellia sinensis L.) seed shells: Kinetics, equilibrium, and thermodynamics studies. Journal of Zhejiang University Science B 14(7): 650–658. https://doi.org/10.1631/jzus.B12a0225

Gao, Y., Yue, Q., Gao, B. and Li, A. 2020. Insight into activated carbon from different kinds of chemical activating agents: A review. Science of The Total Environment 746: 141094. https://doi.org/10.1016/j.scitotenv.2020.141094

Gao, Y., Yue, Q., Gao, B., Sun, Y., Wang, W., Li, Q. and Wang, Y. 2013b. Preparation of high surface area-activated carbon from lignin of papermaking black liquor by KOH activation for Ni(II) adsorption. Chemical Engineering Journal 217: 345–353. https://doi.org/10.1016/j.cej.2012.09.038

Gao, Y., Zhang, W., Yue, Q., Gao, B., Sun, Y., Kong, J. and Zhao, P. 2014. Simple synthesis of hierarchical porous carbon from Enteromorpha prolifera by a self-template method for supercapacitor electrodes. Journal of Power Sources 270: 403–410. https://doi.org/10.1016/j.jpowsour.2014.07.115

Ghasemi, M., Naushad, Mu., Ghasemi, N. and Khosravi-fard, Y. 2014. A novel agricultural waste based adsorbent for the removal of Pb(II) from aqueous solution: Kinetics, equilibrium and thermodynamic studies. Journal of Industrial and Engineering Chemistry 20(2): 454–461. https://doi.org/10.1016/j.jiec.2013.05.002

Ghorbani, F., Kamari, S., Zamani, S., Akbari, S. and Salehi, M. 2020. Optimization and modeling of aqueous Cr(VI) adsorption onto activated carbon prepared from sugar beet bagasse agricultural waste by application of response surface methodology. Surfaces and Interfaces, 18: 100444. https://doi.org/10.1016/j.surfin.2020.100444

Girgis, B.S., Yunis, S.S. and Soliman, A.M. 2002. Characteristics of activated carbon from peanut hulls in relation to conditions of preparation. Materials Letters 57(1): 164–172. https://doi.org/10.1016/S0167-577X(02)00724-3

González, J.F., Román, S., Encinar, J.M. and Martínez, G. 2009. Pyrolysis of various biomass residues and char utilization for the production of activated carbons. Journal of Analytical and Applied Pyrolysis 85(1): 134–141. https://doi.org/10.1016/j.jaap.2008.11.035

González-García, P. 2018. Activated carbon from lignocellulosics precursors: A review of the synthesis methods, characterization techniques and applications. Renewable and Sustainable Energy Reviews 82: 1393–1414. https://doi.org/10.1016/j.rser.2017.04.117

Guo, J. and Lua, A.C. 2003. Textural and chemical properties of adsorbent prepared from palm shell by phosphoric acid activation. Materials Chemistry and Physics 80(1): 114–119. https://doi.org/10.1016/S0254-0584(02)00383-8

Guo, N., Li, M., Sun, X., Wang, F. and Yang, R. 2017. Tremella derived ultrahigh specific surface area activated carbon for high performance supercapacitor. Materials Chemistry and Physics 201: 399–407. https://doi.org/10.1016/j.matchemphys.2017.08.054

Guo, S., Peng, J., Li, W., Yang, K., Zhang, L., Zhang, S. and Xia, H. 2009. Effects of CO2 activation on porous structures of coconut shell-based activated carbons. Applied Surface Science, 255(20): 8443–8449. https://doi.org/10.1016/j.apsusc.2009.05.150

Gupta, V.K., Carrott, P.J.M., Carrott, M.M.L.R. and Suhas. 2009. Low-Cost Adsorbents: Growing Approach to Wastewater Treatment—a Review. Critical Reviews in Environmental Science and Technology 39(10): 783–842. https://doi.org/10.1080/10643380801977610

Haghighat, G.A., Saghi, M.H., Anastopoulos, I., Javid, A., Roudbari, A., Talebi, S.S., Ghadiri, S. K., Giannakoudakis, D.A. and Shams, M. 2020. Aminated graphitic carbon derived from corn stover biomass as adsorbent against antibiotic tetracycline: Optimizing the physicochemical parameters. Journal of Molecular Liquids 313: 113523. https://doi.org/10.1016/j.molliq.2020.113523

Halder, G., Sinha, K. and Dhawane, S. 2015. Defluoridation of wastewater using powdered activated carbon developed from Eichhornia crassipes stem: Optimization by response surface methodology. Desalination and Water Treatment 56(4): 953–966. https://doi.org/10.1080/19443994.2014.942375

Hameed, B.H., Din, A.T.M. and Ahmad, A.L. 2007. Adsorption of methylene blue onto bamboo-based activated carbon: Kinetics and equilibrium studies. Journal of Hazardous Materials 141(3): 819–825. https://doi.org/10.1016/j.jhazmat.2006.07.049

Hameed, B.H., Tan, I.A.W. and Ahmad, A.L. 2009. Preparation of oil palm empty fruit bunch-based activated carbon for removal of 2,4,6-trichlorophenol: Optimization using response surface methodology. Journal of Hazardous Materials 164(2): 1316–1324. https://doi.org/10.1016/j.jhazmat.2008.09.042

Hayashi, J., Kazehaya, A., Muroyama, K. and Watkinson, A.P. 2000. Preparation of activated carbon from lignin by chemical activation. Carbon 38(13): 1873–1878. https://doi.org/10.1016/S0008-6223(00)00027-0

Heidarinejad, Z., Dehghani, M.H., Heidari, M., Javedan, G., Ali, I. and Sillanpää, M. 2020. Methods for preparation and activation of activated carbon: A review. Environmental Chemistry Letters 18(2): 393–415. https://doi.org/10.1007/s10311-019-00955-0

Heidarinejad, Z., Rahmanian, O., Fazlzadeh, M. and Heidari, M. 2018. Enhancement of methylene blue adsorption onto activated carbon prepared from Date Press Cake by low frequency ultrasound. Journal of Molecular Liquids 264: 591–599. https://doi.org/10.1016/j.molliq.2018.05.100

Huang, L., Li, G., Wang, B., Huang, J. and Zhang, B. 2015. Improved adsorption of streptomycin onto Thalia dealbata activated carbon modified by sodium thiosulfate. Desalination and Water Treatment 53(6): 1699–1709. https://doi.org/10.1080/19443994.2013.856348

Huang, Y., Li, S., Lin, H. and Chen, J. 2014. Fabrication and characterization of mesoporous activated carbon from Lemna minor using one-step H3PO4 activation for Pb(II) removal. Applied Surface Science 317: 422–431. https://doi.org/10.1016/j.apsusc.2014.08.152

Ibrahim, T., Moctar, B.L., Tomkouani, K., Gbandi, D.-B., Victor, D.K. and PhinthÃ¨, N. 2013. Kinetics of the Adsorption of Anionic and Cationic Dyes in Aqueous Solution by Low-Cost Activated Carbons Prepared from Sea Cake and Cotton Cake. American Chemical Science Journal 4(1): 38–57.

Ihsanullah, Abbas, A., Al-Amer, A.M., Laoui, T., Al-Marri, M.J., Nasser, M. S., Khraisheh, M., and Atieh, M. A. 2016. Heavy metal removal from aqueous solution by advanced carbon nanotubes: Critical review of adsorption applications. Separation and Purification Technology 157: 141–161. https://doi.org/10.1016/j.seppur.2015.11.039

Imamoglu, M. and Tekir, O. 2008. Removal of copper (II) and lead (II) ions from aqueous solutions by adsorption on activated carbon from a new precursor hazelnut husks. Desalination 228(1): 108–113. https://doi.org/10.1016/j.desal.2007.08.011

Ioannidou, O. and Zabaniotou, A. 2007. Agricultural residues as precursors for activated carbon production—A review. Renewable and Sustainable Energy Reviews 11(9): 1966–2005. https://doi.org/10.1016/j.rser.2006.03.013

Islam, M.A., Benhouria, A., Asif, M., and Hameed, B.H. 2015. Methylene blue adsorption on factory-rejected tea activated carbon prepared by conjunction of hydrothermal carbonization and sodium hydroxide activation processes. Journal of the Taiwan Institute of *Chemical Engineers*, *52*, 57–64. https://doi.org/10.1016/j.jtice.2015.02.010

Islam, Md. A., Ahmed, M.J., Khanday, W.A., Asif, M. and Hameed, B.H. 2017. Mesoporous activated carbon prepared from NaOH activation of rattan (Lacosperma secundiflorum) hydrochar for methylene blue removal. Ecotoxicology and Environmental Safety 138: 279–285. https://doi.org/10.1016/j.ecoenv.2017.01.010

Izquierdo, M.T., Martínez de Yuso, A., Rubio, B. and Pino, M.R. 2011. Conversion of almond shell to activated carbons: Methodical study of the chemical activation based on an experimental design and relationship with their characteristics. Biomass and Bioenergy 35(3): 1235–1244. https://doi.org/10.1016/j.biombioe.2010.12.016

Jagtoyen, M. and Derbyshire, F. 1993. Some considerations of the origins of porosity in carbons from chemically activated wood. Carbon 31(7): 1185–1192. https://doi.org/10.1016/0008-6223(93)90071-H

Jagtoyen, M. and Derbyshire, F. 1998. Activated carbons from yellow poplar and white oak by H3PO4 activation. Carbon 36(7): 1085–1097. https://doi.org/10.1016/S0008-6223(98)00082-7

Jain, S., Vyas, R.K., Pandit, P. and Dalai, A.K. 2014. Adsorption of antiviral drug, acyclovir from aqueous solution on powdered activated charcoal: Kinetics, equilibrium, and thermodynamic studies. Desalination and Water Treatment 52(25–27): 4953–4968. https://doi.org/10.1080/19443994.2013.810324

Jodeh, S., Abdelwahab, F., Jaradat, N., Warad, I. and Jodeh, W. 2016. Adsorption of diclofenac from aqueous solution using Cyclamen persicum tubers based activated carbon (CTAC). Journal of the Association of Arab Universities for Basic and Applied Sciences, 20(1): 32–38. https://doi.org/10.1016/j.jaubas.2014.11.002

Kadirvelu, K., Thamaraiselvi, K. and Namasivayam, C. 2001a. Removal of heavy metals from industrial wastewaters by adsorption onto activated carbon prepared from an agricultural solid waste. Bioresource Technology 76(1): 63–65. https://doi.org/10.1016/S0960-8524(00)00072-9

Kadirvelu, K., Thamaraiselvi, K. and Namasivayam, C. 2001b. Adsorption of nickel(II) from aqueous solution onto activated carbon prepared from coirpith. Separation and Purification Technology 24(3): 497–505. https://doi.org/10.1016/S1383-5866(01)00149-6

Kalderis, D., Koutoulakis, D., Paraskeva, P., Diamadopoulos, E., Otal, E., Valle, J.O. del and Fernández-Pereira, C. 2008. Adsorption of polluting substances on activated carbons prepared from rice husk and sugarcane bagasse. Chemical Engineering Journal 144(1): 42–50. https://doi.org/10.1016/j.cej.2008.01.007

Kaouah, F., Boumaza, S., Berrama, T., Trari, M. and Bendjama, Z. 2013. Preparation and characterization of activated carbon from wild olive cores (oleaster) by H3PO4 for the removal of Basic Red 46. Journal of Cleaner Production 54: 296–306. https://doi.org/10.1016/j.jclepro.2013.04.038

Karaçetin, G., Sivrikaya, S. and Imamoğlu, M. 2014. Adsorption of methylene blue from aqueous solutions by activated carbon prepared from hazelnut husk using zinc chloride. *Journal* of Analytical and Applied Pyrolysis 110: 270–276. https://doi.org/10.1016/j.jaap.2014.09.006

Kasperiski, F.M., Lima, E.C., Umpierres, C.S., dos Reis, G.S., Thue, P.S., Lima, D.R., Dias, S.L.P., Saucier, C. and da Costa, J.B. 2018. Production of porous activated carbons from Caesalpinia ferrea seed pod wastes: Highly efficient removal of captopril from aqueous solutions. Journal of Cleaner Production 197: 919–929. https://doi.org/10.1016/j.jclepro.2018.06.146

Khadiran, T., Hussein, M. Z., Zainal, Z. and Rusli, R. 2015. Activated carbon derived from peat soil as a framework for the preparation of shape-stabilized phase change material. *Energy*, *82*, 468–478. https://doi.org/10.1016/j.energy.2015.01.057

Khezami, L. and Capart, R. 2005. Removal of chromium(VI) from aqueous solution by activated carbons: Kinetic and equilibrium studies. Journal of Hazardous Materials 123(1): 223–231. https://doi.org/10.1016/j.jhazmat.2005.04.012

Khezami, L., Chetouani, A., Taouk, B. and Capart, R. 2005. Production and characterisation of activated carbon from wood components in powder: Cellulose, lignin, xylan. Powder Technology 157(1): 48–56. https://doi.org/10.1016/j.powtec.2005.05.009

Khori, N.K.E.M., Salmiati, Hadibarata, T. and Yusop, Z. 2020. A Combination of Waste Biomass Activated Carbon and Nylon Nanofiber for Removal of Triclosan from Aqueous Solutions. Journal of Environmental Treatment Techniques 8: 1036–1045.

Kong, Q., He, X., Shu, L. and Miao, M. 2017. Ofloxacin adsorption by activated carbon derived from luffa sponge: Kinetic, isotherm, and thermodynamic analyses. *Process Safety and* Environmental Protection 112: 254–264. https://doi.org/10.1016/j.psep.2017.05.011

Kula, I., Uğurlu, M., Karaoğlu, H. and Çelik, A. 2008. Adsorption of Cd(II) ions from aqueous solutions using activated carbon prepared from olive stone by ZnCl2 activation. Bioresource Technology 99(3): 492–501. https://doi.org/10.1016/j.biortech.2007.01.015

Kumar, A., and Jena, H. M. 2016. Removal of methylene blue and phenol onto prepared activated carbon from Fox nutshell by chemical activation in batch and fixed-bed column. Journal of Cleaner Production 137: 1246–1259. https://doi.org/10.1016/j.jclepro.2016.07.177

Kurniawan, A. and Ismadji, S. 2011. Potential utilization of Jatropha curcas L. press-cake residue as new precursor for activated carbon preparation: Application in methylene blue removal from aqueous solution. Journal of the Taiwan Institute of Chemical Engineers 42(5): 826–836. https://doi.org/10.1016/j.jtice.2011.03.001

Kwiatkowski, M. and Broniek, E. 2017. An analysis of the porous structure of activated carbons obtained from hazelnut shells by various physical and chemical methods of activation. Colloids and Surfaces A: Physicochemical and Engineering Aspects 529: 443–453. https://doi.org/10.1016/j.colsurfa.2017.06.028

Kyzas, G.Z. and Deliyanni, E.A. 2015. Modified activated carbons from potato peels as green environmental-friendly adsorbents for the treatment of pharmaceutical effluents. *Chemical* Engineering Research and Design 97: 135–144. https://doi.org/10.1016/j.cherd.2014.08.020

Kyzas, G.Z., Deliyanni, E.A. and Matis, K.A. 2016. Activated carbons produced by pyrolysis of waste potato peels: Cobalt ions removal by adsorption. Colloids and Surfaces A: Physicochemical and Engineering Aspects 490: 74–83. https://doi.org/10.1016/j.colsurfa.2015.11.038

Laksaci, H., Khelifi, A., Trari, M. and Addoun, A. 2017. Synthesis and characterization of microporous activated carbon from coffee grounds using potassium hydroxides. Journal of Cleaner Production 147: 254–262. https://doi.org/10.1016/j.jclepro.2017.01.102

Lalhmunsiama, Lee, S.M., Choi, S.S. and Tiwari, D. 2017. Simultaneous Removal of Hg(II) and Phenol Using Functionalized Activated Carbon Derived from Areca Nut Waste. *Metals*, *7*(7): 248. https://doi.org/10.3390/met7070248

Larous, S. and Meniai, A.-H. 2016. Adsorption of Diclofenac from aqueous solution using activated carbon prepared from olive stones. *International Journal of Hydrogen Energy*, 41(24): 10380–10390. https://doi.org/10.1016/j.ijhydene.2016.01.096

Lemraski, E.G. and Sharafinia, S. 2016. Kinetics, equilibrium and thermodynamics studies of Pb2+ adsorption onto new activated carbon prepared from Persian mesquite grain. Journal of Molecular Liquids 219: 482–492. https://doi.org/10.1016/j.molliq.2016.03.031

Li, G., Zhang, D., Wang, M., Huang, J. and Huang, L. 2013. Preparation of activated carbons from Iris tectorum employing ferric nitrate as dopant for removal of tetracycline from aqueous solutions. Ecotoxicology and Environmental Safety, 98, 273–282. https://doi.org/10.1016/j.ecoenv.2013.08.015

Li, H., Sun, Z., Zhang, L., Tian, Y., Cui, G. and Yan, S. 2016. A cost-effective porous carbon derived from pomelo peel for the removal of methyl orange from aqueous solution. Colloids and Surfaces A: Physicochemical and Engineering Aspects 489: 191–199. https://doi.org/10.1016/j.colsurfa.2015.10.041

Li, J., Ng, D.H.L., Song, P., Kong, C., Song, Y. and Yang, P. 2015. Preparation and characterization of high-surface-area activated carbon fibers from silkworm cocoon waste for congo red adsorption. Biomass and Bioenergy 75: 189–200. https://doi.org/10.1016/j.biombioe.2015.02.002

Li, W., Yang, K., Peng, J., Zhang, L., Guo, S. and Xia, H. 2008. Effects of carbonization temperatures on characteristics of porosity in coconut shell chars and activated carbons derived from carbonized coconut shell chars. Industrial Crops and Products 28(2): 190–198. https://doi.org/10.1016/j.indcrop.2008.02.012

Li, Y., Zhang, X., Yang, R., Li, G. and Hu, C. 2015. The role of H3PO4 in the preparation of activated carbon from NaOH-treated rice husk residue. *RSC Advances*, *5*(41): 32626–32636. https://doi.org/10.1039/C5RA04634C

Lima, D.R., Hosseini-Bandegharaei, A., Thue, P.S., Lima, E.C., de Albuquerque, Y. R. T., dos Reis, G.S., Umpierres, C.S., Dias, S.L.P. and Tran, H.N. 2019a. Efficient acetaminophen removal from water and hospital effluents treatment by activated carbons derived from Brazil nutshells. Colloids and Surfaces A: Physicochemical and Engineering Aspects 583: 123966. https://doi.org/10.1016/j.colsurfa.2019.123966

Lima, D.R., Lima, E.C., Umpierres, C.S., Thue, P. S., El-Chaghaby, G. A., da Silva, R. S., Pavan, F. A., Dias, S. L. P. and Biron, C. 2019b. Removal of amoxicillin from simulated hospital effluents by adsorption using activated carbons prepared from capsules of cashew of Para. Environmental Science and Pollution Research, 26(16): 16396–16408. https://doi.org/10.1007/s11356-019-04994-6

Limousy, L., Ghouma, I., Ouederni, A., and Jeguirim, M. 2017. Amoxicillin removal from aqueous solution using activated carbon prepared by chemical activation of olive stone. Environmental Science and Pollution Research, 24(11): 9993–10004. https://doi.org/10.1007/s11356-016-7404-8

Liu, B., Wang, W., Wang, N., and Au, (Peter) Chak Tong. 2014a. Preparation of activated carbon with high surface area for high-capacity methane storage. Journal of Energy Chemistry, 23(5): 662–668. https://doi.org/10.1016/S2095-4956(14)60198-4

Liu, D., Zhang, W., Lin, H., Li, Y., Lu, H., and Wang, Y. 2016. A green technology for the preparation of high capacitance rice husk-based activated carbon. Journal of Cleaner Production, 112, 1190–1198. https://doi.org/10.1016/j.jclepro.2015.07.005

Liu, H., Zhang, J., Bao, N., Cheng, C., Ren, L., and Zhang, C. 2012a. Textural properties and surface chemistry of lotus stalk-derived activated carbons prepared using different phosphorus oxyacids: Adsorption of trimethoprim. Journal of Hazardous Materials, 235–236, 367–375. https://doi.org/10.1016/j.jhazmat.2012.08.015

Liu, J., Li, Y., and Li, K. 2013. Optimization of preparation of microporous activated carbon with high surface area from Spartina alterniflora and its p-nitroaniline adsorption characteristics. Journal of Environmental Chemical Engineering, 1(3): 389–397. https://doi.org/10.1016/j.jece.2013.06.003

Liu, Q.-S., Zheng, T., Li, N., Wang, P., and Abulikemu, G. 2010. Modification of bamboo-based activated carbon using microwave radiation and its effects on the adsorption of methylene blue. Applied Surface Science, 256(10): 3309–3315. https://doi.org/10.1016/j.apsusc.2009.12.025

Liu, W., Zhang, J., Zhang, C., and Ren, L. 2011. Sorption of norfloxacin by lotus stalk-based activated carbon and iron-doped activated alumina: Mechanisms, isotherms and kinetics. Chemical Engineering Journal, 171(2): 431–438. https://doi.org/10.1016/j.cej.2011.03.099

Liu, Y., Guo, Y., Gao, W., Wang, Z., Ma, Y., and Wang, Z. 2012b. Simultaneous preparation of silica and activated carbon from rice husk ash. Journal of Cleaner Production, 32, 204–209. https://doi.org/10.1016/j.jclepro.2012.03.021

Liu, Y., Zhu, X., Qian, F., Zhang, S., and Chen, J. 2014b. Magnetic activated carbon prepared from rice straw-derived hydrochar for triclosan removal. RSC Advances, 4(109): 63620–63626. https://doi.org/10.1039/C4RA11815D

Lotfy, V. F., Fathy, N. A., and Basta, A. H. 2018. Novel approach for synthesizing different shapes of carbon nanotubes from rice straw residue. Journal of Environmental Chemical Engineering, 6(5): 6263–6274. https://doi.org/10.1016/j.jece.2018.09.055

Lua, A. C., and Yang, T. 2004. Effects of vacuum pyrolysis conditions on the characteristics of activated carbons derived from pistachio-nut shells. Journal of Colloid and Interface Science, 276(2): 364–372. https://doi.org/10.1016/j.jcis.2004.03.071

Lua, A. C., Yang, T., and Guo, J. 2004. Effects of pyrolysis conditions on the properties of activated carbons prepared from pistachio-nut shells. Journal of Analytical and Applied Pyrolysis, 72(2): 279–287. https://doi.org/10.1016/j.jaap.2004.08.001

Ma, G., Yang, Q., Sun, K., Peng, H., Ran, F., Zhao, X., and Lei, Z. 2015. Nitrogen-doped porous carbon derived from biomass waste for high-performance supercapacitor. Bioresource Technology, 197, 137–142. https://doi.org/10.1016/j.biortech.2015.07.100

Mahamad, M. N., Zaini, M. A. A., and Zakaria, Z. A. 2015. Preparation and characterization of activated carbon from pineapple waste biomass for dye removal. International Biodeterioration and Biodegradation, 102, 274–280. https://doi.org/10.1016/j.ibiod.2015.03.009

Malik, P. K. 2003. Use of activated carbons prepared from sawdust and rice-husk for adsorption of acid dyes: A case study of Acid Yellow 36. Dyes and Pigments, 56(3): 239–249. https://doi.org/10.1016/S0143-7208(02)00159-6

Malik, R., Ramteke, D. S., and Wate, S. R. 2007. Adsorption of malachite green on groundnut shell waste based powdered activated carbon. Waste Management, 27(9): 1129–1138. https://doi.org/10.1016/j.wasman.2006.06.009

Manjunath, S. V., Singh Baghel, R., and Kumar, M. 2020. Antagonistic and synergistic analysis of antibiotic adsorption on Prosopis juliflora activated carbon in multicomponent systems. Chemical Engineering Journal, 381, 122713. https://doi.org/10.1016/j.cej.2019.122713

Manyà, J. J., González, B., Azuara, M., and Arner, G. 2018. Ultra-microporous adsorbents prepared from vine shoots-derived biochar with high CO2 uptake and CO2/N2 selectivity. Chemical Engineering Journal, 345, 631–639. https://doi.org/10.1016/j.cej.2018.01.092

Marcilla, A., García-García, S., Asensio, M., and Conesa, J. A. 2000. Influence of thermal treatment regime on the density and reactivity of activated carbons from almond shells. Carbon, 38(3): 429–440. https://doi.org/10.1016/S0008-6223(99)00123-2

Márquez-Montesino, F., Torres-Figueredo, N., Lemus-Santana, A., and Trejo, F. 2020. Activated Carbon by Potassium Carbonate Activation from Pine Sawdust (Pinus montezumae Lamb.). Chemical Engineering and Technology, 43(9): 1716–1725. https://doi.org/10.1002/ceat.202000051

Martins, A. C., Pezoti, O., Cazetta, A. L., Bedin, K. C., Yamazaki, D. A. S., Bandoch, G. F. G., Asefa, T., Visentainer, J. V., and Almeida, V. C. 2015. Removal of tetracycline by NaOH-activated carbon produced from macadamia nut shells: Kinetic and equilibrium studies. Chemical Engineering Journal, 260, 291–299. https://doi.org/10.1016/j.cej.2014.09.017

Marzbali, M. H., Esmaieli, M., Abolghasemi, H., and Marzbali, M. H. 2016. Tetracycline adsorption by H3PO4-activated carbon produced from apricot nut shells: A batch study. Process Safety and Environmental Protection, 102, 700–709. https://doi.org/10.1016/j.psep.2016.05.025

Mashayekh-Salehi, A., and Moussavi, G. 2016. Removal of acetaminophen from the contaminated water using adsorption onto carbon activated with NH4Cl. Desalination and Water Treatment, 57(27): 12861–12873. https://doi.org/10.1080/19443994.2015.1051588

Melliti, A., Srivastava, V., Kheriji, J., Sillanpää, M., and Hamrouni, B. 2021. Date Palm Fiber as a novel precursor for porous activated carbon: Optimization, characterization and its application as Tylosin antibiotic scavenger from aqueous solution. Surfaces and Interfaces, 24, 101047. https://doi.org/10.1016/j.surfin.2021.101047

Meryemoglu, B., Irmak, S., and Hasanoglu, A. 2016. Production of activated carbon materials from kenaf biomass to be used as catalyst support in aqueous-phase reforming process. Fuel Processing Technology, 151, 59–63. https://doi.org/10.1016/j.fuproc.2016.05.040

Mestre, A. S., Bexiga, A. S., Proença, M., Andrade, M., Pinto, M. L., Matos, I., Fonseca, I. M., and Carvalho, A. P. 2011. Activated carbons from sisal waste by chemical activation with K2CO3: Kinetics of paracetamol and ibuprofen removal from aqueous solution. Bioresource Technology, 102(17): 8253–8260. https://doi.org/10.1016/j.biortech.2011.06.024

Mestre, A. S., Nabiço, A., Figueiredo, P. L., Pinto, M. L., Santos, M. S. C. S., and Fonseca, I. M. 2016. Enhanced clofibric acid removal by activated carbons: Water hardness as a key parameter. Chemical Engineering Journal, 286, 538–548. https://doi.org/10.1016/j.cej.2015.10.066

Mestre, A. S., Pires, J., Nogueira, J. M. F., and Carvalho, A. P. 2007. Activated carbons for the adsorption of ibuprofen. Carbon, 45(10): 1979–1988. https://doi.org/10.1016/j.carbon.2007.06.005

Miao, M.-S., Liu, Q., Shu, L., Wang, Z., Liu, Y.-Z., and Kong, Q. 2016. Removal of cephalexin from effluent by activated carbon prepared from alligator weed: Kinetics, isotherms, and thermodynamic analyses. Process Safety and Environmental Protection, 104, 481–489. https://doi.org/10.1016/j.psep.2016.03.017

Milić, N., Milanović, M., Letić, N. G., Sekulić, M. T., Radonić, J., Mihajlović, I., and Miloradov, M. V. 2013. Occurrence of antibiotics as emerging contaminant substances in aquatic environment. International Journal of Environmental Health Research, 23(4): 296–310. https://doi.org/10.1080/09603123.2012.733934

Misnon, I. I., Zain, N. K. M., Aziz, R. A., Vidyadharan, B., and Jose, R. 2015. Electrochemical properties of carbon from oil palm kernel shell for high performance supercapacitors. Electrochimica Acta, 174, 78–86. https://doi.org/10.1016/j.electacta.2015.05.163

Mohammad, S. G., Ahmed, S. M., Amr, A. E.-G. E., and Kamel, A. H. 2020. Porous Activated Carbon from Lignocellulosic Agricultural Waste for the Removal of Acetampirid Pesticide from Aqueous Solutions. Molecules, 25(10): 2339. https://doi.org/10.3390/molecules25102339

Mohammad, S. G., and El-Sayed, M. M. H. 2020. Removal of imidacloprid pesticide using nanoporous activated carbons produced via pyrolysis of peach stone agricultural wastes. Chemical Engineering Communications, 0(0): 1–12. https://doi.org/10.1080/00986445.2020.1743695

Mohammadi, S. Z., Hamidian, H., and Moeinadini, Z. 2014a. High surface area-activated carbon from Glycyrrhiza glabra residue by ZnCl2 activation for removal of Pb(II) and Ni(II) from water samples. Journal of Industrial and Engineering Chemistry, 20(6): 4112–4118. https://doi.org/10.1016/j.jiec.2014.01.009

Mohammadi, S. Z., Karimi, M. A., Afzali, D., and Mansouri, F. 2010. Removal of Pb(II) from aqueous solutions using activated carbon from Sea-buckthorn stones by chemical activation. Desalination, 262(1): 86–93. https://doi.org/10.1016/j.desal.2010.05.048

Mohammadi, S. Z., Karimi, M. A., Yazdy, S. N., Shamspur, T., and Hamidian, H. 2014b. Removal of Pb(II) ions and malachite green dye from wastewater by activated carbon produced from lemon peel. Química Nova, 37, 804–809. https://doi.org/10.5935/0100-4042.20140129

Mohan, D., Singh, K. P., and Singh, V. K. 2005. Removal of Hexavalent Chromium from Aqueous Solution Using Low-Cost Activated Carbons Derived from Agricultural Waste Materials and Activated Carbon Fabric Cloth. Industrial and Engineering Chemistry Research, 44(4): 1027–1042. https://doi.org/10.1021/ie0400898

Mohan, D., Singh, K. P., and Singh, V. K. 2006. Trivalent chromium removal from wastewater using low cost activated carbon derived from agricultural waste material and activated carbon fabric cloth. Journal of Hazardous Materials, 135(1): 280–295. https://doi.org/10.1016/j.jhazmat.2005.11.075

Mohanty, K., Jha, M., Meikap, B. C., and Biswas, M. N. 2005. Removal of chromium (VI) from dilute aqueous solutions by activated carbon developed from Terminalia arjuna nuts activated with zinc chloride. Chemical Engineering Science, 60(11): 3049–3059. https://doi.org/10.1016/j.ces.2004.12.049

Mohd Din, A. T., Hameed, B. H., and Ahmad, A. L. 2009. Batch adsorption of phenol onto physiochemical-activated coconut shell. Journal of Hazardous Materials, 161(2): 1522–1529. https://doi.org/10.1016/j.jhazmat.2008.05.009

Mokti, N., Borhan, A., Zaine, S. N. A., and Mohd Zaid, H. F. 2021. Development of Rubber Seed Shell–Activated Carbon Using Impregnated Pyridinium-Based Ionic Liquid for Enhanced CO2 Adsorption. Processes, 9(7): 1161. https://doi.org/10.3390/pr9071161

Momčilović, M., Purenović, M., Bojić, A., Zarubica, A., and Ranđelović, M. 2011. Removal of lead(II) ions from aqueous solutions by adsorption onto pine cone activated carbon. Desalination, 276(1): 53–59. https://doi.org/10.1016/j.desal.2011.03.013

Mondal, S., Sinha, K., Aikat, K., and Halder, G. 2015. Adsorption thermodynamics and kinetics of ranitidine hydrochloride onto superheated steam activated carbon derived from mung bean husk. Journal of Environmental Chemical Engineering, 3(1): 187–195. https://doi.org/10.1016/j.jece.2014.11.021

Moreno-Barbosa, J. J., López-Velandia, C., Maldonado, A. del P., Giraldo, L., and Moreno-Piraján, J. C. 2013. Removal of lead(II) and zinc(II) ions from aqueous solutions by adsorption onto activated carbon synthesized from watermelon shell and walnut shell. Adsorption, 19(2): 675–685. https://doi.org/10.1007/s10450-013-9491-x

Moussavi, G., Alahabadi, A., Yaghmaeian, K., and Eskandari, M. 2013. Preparation, characterization and adsorption potential of the NH4Cl-induced activated carbon for the removal of amoxicillin antibiotic from water. Chemical Engineering Journal, 217, 119–128. https://doi.org/10.1016/j.cej.2012.11.069

Nabais, J. M. V., Ledesma, B., and Laginhas, C. 2012. Removal of Amitriptyline from Aqueous Media Using Activated Carbons. Adsorption Science and Technology, 30(3): 255–263. https://doi.org/10.1260/0263-6174.30.3.255

Nahil, M. A., and Williams, P. T. 2012. Pore characteristics of activated carbons from the phosphoric acid chemical activation of cotton stalks. Biomass and Bioenergy, 37, 142–149. https://doi.org/10.1016/j.biombioe.2011.12.019

Namasivayam, C., and Kadirvelu, K. 1999. Uptake of mercury (II) from wastewater by activated carbon from an unwanted agricultural solid by-product: Coirpith. Carbon, 37(1): 79–84. https://doi.org/10.1016/S0008-6223(98)00189-4

Namasivayam, C., and Sangeetha, D. 2006. Recycling of agricultural solid waste, coir pith: Removal of anions, heavy metals, organics and dyes from water by adsorption onto ZnCl2 activated coir pith carbon. Journal of Hazardous Materials, 135(1): 449–452. https://doi.org/10.1016/j.jhazmat.2005.11.066

Nazari, G., Abolghasemi, H., and Esmaieli, M. 2016. Batch adsorption of cephalexin antibiotic from aqueous solution by walnut shell-based activated carbon. Journal of the Taiwan Institute of Chemical Engineers, 58, 357–365. https://doi.org/10.1016/j.jtice.2015.06.006

Nethaji, S., Sivasamy, A., and Mandal, A. B. 2013. Preparation and characterization of corn cob activated carbon coated with nano-sized magnetite particles for the removal of Cr(VI). Bioresource Technology, 134, 94–100. https://doi.org/10.1016/j.biortech.2013.02.012

Nguyen, T. A. H., Ngo, H. H., Guo, W. S., Zhang, J., Liang, S., Yue, Q. Y., Li, Q., and Nguyen, T. V. 2013. Applicability of agricultural waste and by-products for adsorptive removal of heavy metals from wastewater. Bioresource Technology, 148, 574–585. https://doi.org/10.1016/j.biortech.2013.08.124

Njoku, V. O., Foo, K. Y., Asif, M., and Hameed, B. H. 2014a. Preparation of activated carbons from rambutan (Nephelium lappaceum) peel by microwave-induced KOH activation for acid yellow 17 dye adsorption. Chemical Engineering Journal, 250, 198–204. https://doi.org/10.1016/j.cej.2014.03.115

Njoku, V. O., Islam, Md. A., Asif, M., and Hameed, B. H. 2014b. Utilization of sky fruit husk agricultural waste to produce high quality activated carbon for the herbicide bentazon adsorption. Chemical Engineering Journal, 251, 183–191. https://doi.org/10.1016/j.cej.2014.04.015

Njoku, V. O., Islam, Md. A., Asif, M., and Hameed, B. H. 2015. Adsorption of 2,4-dichlorophenoxyacetic acid by mesoporous activated carbon prepared from H3PO4-activated langsat empty fruit bunch. Journal of Environmental Management, 154, 138–144. https://doi.org/10.1016/j.jenvman.2015.02.002

Nodeh, H. R., Sereshti, H., Ataolahi, S., Toloutehrani, A., and Ramezani, A. T. 2020. Activated carbon derived from pistachio hull biomass for the effective removal of parabens from aqueous solutions: Isotherms, kinetics, and free energy studies. Desalination and Water Treatment, 155–164.

Nogueira, J., António, M., Mikhalev, S. M., Fateixa, S., Trindade, T., and Daniel-da-Silva, A. L. 2018. Porous Carrageenan-Derived Carbons for Efficient Ciprofloxacin Removal from Water. Nanomaterials, 8(12): 1004. https://doi.org/10.3390/nano8121004

Norouzi, S., Heidari, M., Alipour, V., Rahmanian, O., Fazlzadeh, M., Mohammadi-moghadam, F., Nourmoradi, H., Goudarzi, B., and Dindarloo, K. 2018. Preparation, characterization and Cr(VI) adsorption evaluation of NaOH-activated carbon produced from Date Press Cake; an agro-industrial waste. Bioresource Technology, 258, 48–56. https://doi.org/10.1016/j.biortech.2018.02.106

Okman, I., Karagöz, S., Tay, T., and Erdem, M. 2014. Activated Carbons From Grape Seeds By Chemical Activation With Potassium Carbonate And Potassium Hydroxide. Applied Surface Science, 293, 138–142. https://doi.org/10.1016/j.apsusc.2013.12.117

Olivares-Marín, M., Fernández, J. A., Lázaro, M. J., Fernández-González, C., Macías-García, A., Gómez-Serrano, V., Stoeckli, F., and Centeno, T. A. 2009. Cherry stones as precursor of activated carbons for supercapacitors. Materials Chemistry and Physics, 114(1): 323–327. https://doi.org/10.1016/j.matchemphys.2008.09.010

Ould-Idriss, A., Stitou, M., Cuerda-Correa, E. M., Fernández-González, C., Macías-García, A., Alexandre-Franco, M. F., and Gómez-Serrano, V. 2011. Preparation of activated carbons from olive-tree wood revisited. I. Chemical activation with H3PO4. Fuel Processing Technology, 92(2): 261–265. https://doi.org/10.1016/j.fuproc.2010.05.011

Ouni, L., Ramazani, A., and Taghavi Fardood, S. 2019. An overview of carbon nanotubes role in heavy metals removal from wastewater. Frontiers of Chemical Science and Engineering, 13(2): 274–295. https://doi.org/10.1007/s11705-018-1765-0

Ozdemir, I., Şahin, M., Orhan, R., and Erdem, M. 2014. Preparation and characterization of activated carbon from grape stalk by zinc chloride activation. Fuel Processing Technology, 125, 200–206. https://doi.org/10.1016/j.fuproc.2014.04.002

Özsin, G., Kılıç, M., Apaydın-Varol, E., and Pütün, A. E. 2019. Chemically activated carbon production from agricultural waste of chickpea and its application for heavy metal adsorption: Equilibrium, kinetic, and thermodynamic studies. Applied Water Science, 9(3): 56. https://doi.org/10.1007/s13201-019-0942-8

Pallarés, J., González-Cencerrado, A., and Arauzo, I. 2018. Production and characterization of activated carbon from barley straw by physical activation with carbon dioxide and steam. Biomass and Bioenergy, 115, 64–73. https://doi.org/10.1016/j.biombioe.2018.04.015

Pandiarajan, A., Kamaraj, R., Vasudevan, S., and Vasudevan, S. 2018. OPAC (orange peel activated carbon) derived from waste orange peel for the adsorption of chlorophenoxyacetic acid herbicides from water: Adsorption isotherm, kinetic modelling and thermodynamic studies. Bioresource Technology, 261, 329–341. https://doi.org/10.1016/j.biortech.2018.04.005

Parlayıcı, Ş., and Pehlivan, E. 2017. Removal of metals by Fe3O4 loaded activated carbon prepared from plum stone (Prunus nigra): Kinetics and modelling study. Powder Technology, 317, 23–30. https://doi.org/10.1016/j.powtec.2017.04.021

Patel, M., Kumar, R., Kishor, K., Mlsna, T., Pittman, C. U., and Mohan, D. 2019. Pharmaceuticals of Emerging Concern in Aquatic Systems: Chemistry, Occurrence, Effects, and Removal Methods. Chemical Reviews, 119(6): 3510–3673. https://doi.org/10.1021/acs.chemrev.8b00299

Peng, X., Hu, F., Zhang, T., Qiu, F., and Dai, H. 2018. Amine-functionalized magnetic bamboo-based activated carbon adsorptive removal of ciprofloxacin and norfloxacin: A batch and fixed-bed column study. Bioresource Technology, 249, 924–934. https://doi.org/10.1016/j.biortech.2017.10.095

Periasamy, K., and Namasivayam, C. 1995. Removal of nickel(II) from aqueous solution and nickel plating industry wastewater using an agricultural waste: Peanut hulls. Waste Management, 15(1): 63–68. https://doi.org/10.1016/0956-053X(94)00071-S

Pezoti, O., Cazetta, A. L., Bedin, K. C., Souza, L. S., Martins, A. C., Silva, T. L., Santos Júnior, O. O., Visentainer, J. V., and Almeida, V. C. 2016. NaOH-activated carbon of high surface area produced from guava seeds as a high-efficiency adsorbent for amoxicillin removal: Kinetic, isotherm and thermodynamic studies. Chemical Engineering Journal, 288, 778–788. https://doi.org/10.1016/j.cej.2015.12.042

Pooja, D., Kumar, P., Singh, P., and Patil, S. 2020. Sensors in Water Pollutants Monitoring: Role of Material. Springer.

Pouretedal, H. R., and Sadegh, N. 2014. Effective removal of Amoxicillin, Cephalexin, Tetracycline and Penicillin G from aqueous solutions using activated carbon nanoparticles prepared from vine wood. Journal of Water Process Engineering, 1, 64–73. https://doi.org/10.1016/j.jwpe.2014.03.006

Qin, C., Chen, Y., and Gao, J. 2014. Manufacture and characterization of activated carbon from marigold straw (Tagetes erecta L) by H3PO4 chemical activation. Materials Letters, 135, 123–126. https://doi.org/10.1016/j.matlet.2014.07.151

Rahman, M. M., Adil, M., Yusof, A. M., Kamaruzzaman, Y. B., and Ansary, R. H. 2014. Removal of Heavy Metal Ions with Acid Activated Carbons Derived from Oil Palm and Coconut Shells. Materials, 7(5): 3634–3650. https://doi.org/10.3390/ma7053634

Rambabu, K., AlYammahi, J., Bharath, G., Thanigaivelan, A., Sivarajasekar, N., and Banat, F. 2021. Nano-activated carbon derived from date palm coir waste for efficient sequestration of noxious 2,4-dichlorophenoxyacetic acid herbicide. Chemosphere, 282, 131103. https://doi.org/10.1016/j.chemosphere.2021.131103

Rao, M. M., Ramana, D. K., Seshaiah, K., Wang, M. C., and Chien, S. W. C. 2009a. Removal of some metal ions by activated carbon prepared from Phaseolus aureus hulls. Journal of Hazardous Materials, 166(2): 1006–1013. https://doi.org/10.1016/j.jhazmat.2008.12.002

Rao, M. M., Rao, G. P. C., Seshaiah, K., Choudary, N. V., and Wang, M. C. 2008. Activated carbon from Ceiba pentandra hulls, an agricultural waste, as an adsorbent in the removal of lead and zinc from aqueous solutions. Waste Management, 28(5): 849–858. https://doi.org/10.1016/j.wasman.2007.01.017

Rao, M. M., Reddy, D. H. K. K., Venkateswarlu, P., and Seshaiah, K. 2009b. Removal of mercury from aqueous solutions using activated carbon prepared from agricultural by-product/waste. Journal of Environmental Management, 90(1): 634–643. https://doi.org/10.1016/j.jenvman.2007.12.019

Rasheed, T., Bilal, M., Nabeel, F., Adeel, M., and Iqbal, H. M. N. 2019. Environmentally-related contaminants of high concern: Potential sources and analytical modalities for detection, quantification, and treatment. Environment International, 122, 52–66. https://doi.org/10.1016/j.envint.2018.11.038

Reddy, K. S. K., Al Shoaibi, A., and Srinivasakannan, C. 2012. A comparison of microstructure and adsorption characteristics of activated carbons by CO2 and H3PO4 activation from date palm pits. New Carbon Materials, 27(5): 344–351. https://doi.org/10.1016/S1872-5805(12)60020-1

Reffas, A., Bernardet, V., David, B., Reinert, L., Lehocine, M. B., Dubois, M., Batisse, N., and Duclaux, L. 2010. Carbons prepared from coffee grounds by H3PO4 activation: Characterization and adsorption of methylene blue and Nylosan Red N-2RBL. Journal of Hazardous Materials, 175(1): 779–788. https://doi.org/10.1016/j.jhazmat.2009.10.076

Rivera-Utrilla, J., Gómez-Pacheco, C. V., Sánchez-Polo, M., López-Peñalver, J. J., and Ocampo-Pérez, R. 2013. Tetracycline removal from water by adsorption/bioadsorption on activated carbons and sludge-derived adsorbents. Journal of Environmental Management, 131, 16–24. https://doi.org/10.1016/j.jenvman.2013.09.024

Rivera-Utrilla, J., Sánchez-Polo, M., Ferro-García, M. Á., Prados-Joya, G., and Ocampo-Pérez, R. 2013. Pharmaceuticals as emerging contaminants and their removal from water. A review. Chemosphere, 93(7): 1268–1287. https://doi.org/10.1016/j.chemosphere.2013.07.059

Rodriguez-Narvaez, O. M., Peralta-Hernandez, J. M., Goonetilleke, A., and Bandala, E. R. 2017. Treatment technologies for emerging contaminants in water: A review. Chemical Engineering Journal, 323, 361–380. https://doi.org/10.1016/j.cej.2017.04.106

Rodríguez-Reinoso, F., and Molina-Sabio, M. 1992. Activated carbons from lignocellulosic materials by chemical and/or physical activation: An overview. Carbon, 30(7): 1111–1118. https://doi.org/10.1016/0008-6223(92)90143-K

Román, S., Nabais, J. M. V., González, J. F., González-García, C. M., and Ortiz, A. L. 2012. Study of the Contributions of Non-Specific and Specific Interactions during Fluoxetine Adsorption onto Activated Carbons. CLEAN – Soil, Air, Water, 40(7): 698–705. https://doi.org/10.1002/clen.201100009

Rufford, T. E., Hulicova-Jurcakova, D., Zhu, Z., and Lu, G. Q. 2008. Nanoporous carbon electrode from waste coffee beans for high performance supercapacitors. Electrochemistry Communications, 10(10): 1594–1597. https://doi.org/10.1016/j.elecom.2008.08.022

Safdari, M.-S., Amini, E., Weise, D. R., and Fletcher, T. H. 2019. Heating rate and temperature effects on pyrolysis products from live wildland fuels. Fuel, 242, 295–304. https://doi.org/10.1016/j.fuel.2019.01.040

Sahu, N., Bhan, C., and Singh, J. 2021. Removal of fluoride from an aqueous solution by batch and column process using activated carbon derived from iron infused Pisum sativum peel: Characterization, Isotherm, kinetics study. Environmental Engineering Research, 26(4): https://doi.org/10.4491/eer.2020.241

Saka, C. 2012. BET, TG–DTG, FT-IR, SEM, iodine number analysis and preparation of activated carbon from acorn shell by chemical activation with ZnCl2. Journal of Analytical and Applied Pyrolysis, 95, 21–24. https://doi.org/10.1016/j.jaap.2011.12.020

Salman, J. M., Njoku, V. O., and Hameed, B. H. 2011. Adsorption of pesticides from aqueous solution onto banana stalk activated carbon. Chemical Engineering Journal, 174(1): 41–48. https://doi.org/10.1016/j.cej.2011.08.026

Saman, N., Abdul Aziz, A., Johari, K., Song, S.-T., and Mat, H. 2015. Adsorptive efficacy analysis of lignocellulosic waste carbonaceous adsorbents toward different mercury species. Process Safety and Environmental Protection, 96, 33–42. https://doi.org/10.1016/j.psep.2015.04.004

Savova, D., Apak, E., Ekinci, E., Yardim, F., Petrov, N., Budinova, T., Razvigorova, M., and Minkova, V. 2001. Biomass conversion to carbon adsorbents and gas. Biomass and Bioenergy, 21(2): 133–142. https://doi.org/10.1016/S0961-9534(01)00027-7

Sayğılı, H., and Güzel, F. 2016a. High surface area mesoporous activated carbon from tomato processing solid waste by zinc chloride activation: Process optimization, characterization and dyes adsorption. Journal of Cleaner Production, 113, 995–1004. https://doi.org/10.1016/j.jclepro.2015.12.055

Sayğılı, H. and Güzel, F. 2016b. Effective removal of tetracycline from aqueous solution using activated carbon prepared from tomato (Lycopersicon esculentum Mill.) industrial processing waste. Ecotoxicology and Environmental Safety 131: 22–29. https://doi.org/10.1016/j.ecoenv.2016.05.001

Şentorun-Shalaby, Ç., Uçak-Astarlıog˘lu, M.G., Artok, L. and Sarıcı, Ç. 2006. Preparation and characterization of activated carbons by one-step steam pyrolysis/activation from apricot stones. Microporous and Mesoporous Materials 88(1): 126–134. https://doi.org/10.1016/j.micromeso.2005.09.003

Seyhi, B., Drogui, P., Gortares-Moroyoqui, P., Estrada-Alvarado, M.I. and Alvarez, L.H. 2014. Adsorption of an organochlorine pesticide using activated carbon produced from an agro-waste material. Journal of Chemical Technology and Biotechnology 89(12): 1811–1816. https://doi.org/10.1002/jctb.4256

Shrestha, R.M., Varga, I., Bajtai, J. and Varga, M. 2013. Design of surface functionalization of waste material originated charcoals by an optimized chemical carbonization for the purpose of heavy metal removal from industrial waste waters. Microchemical Journal 108: 224–232. https://doi.org/10.1016/j.microc.2012.11.002

Siddique, A., Nayak, A.K. and Singh, J. 2020. Synthesis of FeCl3-activated carbon derived from waste Citrus limetta peels for removal of fluoride: An eco-friendly approach for the treatment of groundwater and bio-waste collectively. Groundwater for Sustainable Development 10: 100339. https://doi.org/10.1016/j.gsd.2020.100339

Singh, K., Lataye, D.H. and Wasewar, K.L. 2017. Removal of fluoride from aqueous solution by using bael (Aegle marmelos) shell activated carbon: Kinetic, equilibrium and thermodynamic study. Journal of Fluorine Chemistry 194: 23–32. https://doi.org/10.1016/j.jfluchem.2016.12.009

Song, Y.-X., Chen, S., You, N., Fan, H.-T. and Sun, L.-N. 2020. Nanocomposites of zero-valent Iron@Activated carbon derived from corn stalk for adsorptive removal of tetracycline antibiotics. Chemosphere 255: 126917. https://doi.org/10.1016/j.chemosphere.2020.126917

Spagnoli, A.A., Giannakoudakis, D.A. and Bashkova, S. 2017. Adsorption of methylene blue on cashew nut shell based carbons activated with zinc chloride: The role of surface and structural parameters. Journal of Molecular Liquids 229: 465–471. https://doi.org/10.1016/j.molliq.2016.12.106

Sun, Y., Li, H., Li, G., Gao, B., Yue, Q. and Li, X. 2016. Characterization and ciprofloxacin adsorption properties of activated carbons prepared from biomass wastes by H3PO4 activation. Bioresource Technology 217: 239–244. https://doi.org/10.1016/j.biortech.2016.03.047

Sun, Y., Yue, Q., Gao, B., Li, Q., Huang, L., Yao, F. and Xu, X. 2012. Preparation of activated carbon derived from cotton linter fibers by fused NaOH activation and its application for oxytetracycline (OTC) adsorption. Journal of Colloid and Interface Science 368(1): 521–527. https://doi.org/10.1016/j.jcis.2011.10.067

Suneetha, M., Sundar, B.S. and Ravindhranath, K. 2015. Removal of fluoride from polluted waters using active carbon derived from barks of Vitex negundo plant. Journal of Analytical Science and Technology 6(1): 15. https://doi.org/10.1186/s40543-014-0042-1

Talat, M., Mohan, S., Dixit, V., Singh, D.K., Hasan, S.H. and Srivastava, O.N. 2018. Effective removal of fluoride from water by coconut husk activated carbon in fixed bed column: Experimental and breakthrough curves analysis. Groundwater for Sustainable Development 7: 48–55. https://doi.org/10.1016/j.gsd.2018.03.001

Tay, J.H., Chen, X.G., Jeyaseelan, S. and Graham, N. 2001. Optimising the preparation of activated carbon from digested sewage sludge and coconut husk. Chemosphere 44(1): 45–51. https://doi.org/10.1016/S0045-6535(00)00383-0

Theydan, S. K., and Ahmed, M. J. 2012. Optimization of preparation conditions for activated carbons from date stones using response surface methodology. Powder Technology 224: 101–108. https://doi.org/10.1016/j.powtec.2012.02.037

Torres-Pérez, J., Gérente, C. and Andrès, Y. 2012. Sustainable Activated Carbons from Agricultural Residues Dedicated to Antibiotic Removal by Adsorption. Chinese Journal of Chemical Engineering 20(3): 524–529. https://doi.org/10.1016/S1004-9541(11)60214-0

Tounsadi, H., Khalidi, A., Abdennouri, M., and Barka, N. 2016. Activated carbon from Diplotaxis Harra biomass: Optimization of preparation conditions and heavy metal removal. Journal of the Taiwan Institute of Chemical Engineers, 59: 348–358. https://doi.org/10.1016/j.jtice.2015.08.014

Triwiswara, M., Kang, J., Moon, J. K., Lee, C.-G., and Park, S.-J. 2020. Removal of triclosan from aqueous solution using thermally treated rice husks. Desalination and Water Treatment, 202, 317–326. https://doi.org/10.5004/dwt.2020.26161

Tseng, R.-L. 2007. Physical and chemical properties and adsorption type of activated carbon prepared from plum kernels by NaOH activation. Journal of Hazardous Materials, 147(3): 1020–1027. https://doi.org/10.1016/j.jhazmat.2007.01.140

Tseng, R.-L., and Tseng, S.-K. 2006. Characterization and use of high surface area activated carbons prepared from cane pith for liquid-phase adsorption. Journal of Hazardous Materials, 136(3): 671–680. https://doi.org/10.1016/j.jhazmat.2005.12.048

Vicinisvarri, I., Kumar, S., and Aimi, N. 2014. Preparation and characterization of phosphoric acid activated carbon from Canarium Odontophyllum (Dabai) nutshell for methylene blue adsorption. Res J Chem Environ, 18, 57–62.

Wang, D., Geng, Z., Li, B., and Zhang, C. 2015a. High performance electrode materials for electric double-layer capacitors based on biomass-derived activated carbons. Electrochimica Acta, 173, 377–384. https://doi.org/10.1016/j.electacta.2015.05.080

Wang, K., Song, Y., Yan, R., Zhao, N., Tian, X., Li, X., Guo, Q., and Liu, Z. 2017. High capacitive performance of hollow activated carbon fibers derived from willow catkins. Applied Surface Science, 394, 569–577. https://doi.org/10.1016/j.apsusc.2016.10.161

Wang, Y. X., Ngo, H. H., and Guo, W. S. 2015b. Preparation of a specific bamboo based activated carbon and its application for ciprofloxacin removal. Science of The Total Environment, 533, 32–39. https://doi.org/10.1016/j.scitotenv.2015.06.087

Wei, Q., Chen, Z., Cheng, Y., Wang, X., Yang, X., and Wang, Z. 2019. Preparation and electrochemical performance of orange peel based-activated carbons activated by different activators. Colloids and Surfaces A: Physicochemical and Engineering Aspects, 574, 221–227. https://doi.org/10.1016/j.colsurfa.2019.04.065

Wu, F.-C., Tseng, R.-L., and Juang, R.-S. 2005. Preparation of highly microporous carbons from fir wood by KOH activation for adsorption of dyes and phenols from water. Separation and Purification Technology, 47(1): 10–19. https://doi.org/10.1016/j.seppur.2005.03.013

Wuana, R. A., Sha'Ato, R., and Iorhen, S. 2015. Aqueous phase removal of ofloxacin using adsorbents from Moringa oleifera pod husks. Advances in Environmental Research, 4(1): 49–68. https://doi.org/10.12989/aer.2015.4.1.049

Wuana, R. A., Sha'Ato, R., and Iorhen, S. 2016. Preparation, characterization, and evaluation of Moringa oleifera pod husk adsorbents for aqueous phase removal of norfloxacin. Desalination and Water Treatment, 57(25): 11904–11916. https://doi.org/10.1080/19443994.2015.1046150

Xing, B.-L., Guo, H., Chen, L.-J., Chen, Z.-F., Zhang, C.-X., Huang, G.-X., Xie, W., and Yu, J.-L. 2015. Lignite-derived high surface area mesoporous activated carbons for electrochemical capacitors. Fuel Processing Technology, 138, 734–742. https://doi.org/10.1016/j.fuproc.2015.07.017

Xu, J., Chen, L., Qu, H., Jiao, Y., Xie, J., and Xing, G. 2014. Preparation and characterization of activated carbon from reedy grass leaves by chemical activation with H3PO4. Applied Surface Science, 320, 674–680. https://doi.org/10.1016/j.apsusc.2014.08.178

Yadav, A. K., Abbassi, R., Gupta, A., and Dadashzadeh, M. 2013. Removal of fluoride from aqueous solution and groundwater by wheat straw, sawdust and activated bagasse carbon of sugarcane. Ecological Engineering, 52, 211–218. https://doi.org/10.1016/j.ecoleng.2012.12.069

Yahya, M. A., Al-Qodah, Z., and Ngah, C. W. Z. 2015. Agricultural bio-waste materials as potential sustainable precursors used for activated carbon production: A review. Renewable and Sustainable Energy Reviews, 46, 218–235. https://doi.org/10.1016/j.rser.2015.02.051

Yang, T. and Lua, A.C. 2003. Characteristics of activated carbons prepared from pistachio-nut shells by physical activation. Journal of Colloid and Interface Science 267(2): 408–417. https://doi.org/10.1016/S0021-9797(03)00689-1

Youssef, A.M., Ahmed, A.I., and El-Bana, U. A. 2012. Adsorption of cationic dye (MB) and anionic dye (AG 25) by physically and chemically activated carbons developed from rice husk. Carbon Letters, 13(2): 61–72. https://doi.org/10.5714/CL.2012.13.2.061

Yu, Y., Qiao, N., Wang, D., Zhu, Q., Fu, F., Cao, R., Wang, R., Liu, W., and Xu, B. 2019. Fluffy honeycomb-like activated carbon from popcorn with high surface area and well-developed porosity for ultra-high efficiency adsorption of organic dyes. Bioresource Technology, 285, 121340. https://doi.org/10.1016/j.biortech.2019.121340

Zabihi, M., Haghighi Asl, A. and Ahmadpour, A. 2010. Studies on adsorption of mercury from aqueous solution on activated carbons prepared from walnut shell. Journal of Hazardous Materials 174(1): 251–256. https://doi.org/10.1016/j.jhazmat.2009.09.044

Zakaria, R., Jamalluddin, N.A. and Abu Bakar, M.Z. 2021. Effect of impregnation ratio and activation temperature on the yield and adsorption performance of mangrove based activated carbon for methylene blue removal. Results in Materials 10: 100183. https://doi.org/10.1016/j.rinma.2021.100183

Zhang, B., Han, X., Gu, P., Fang, S. and Bai, J. 2017. Response surface methodology approach for optimization of ciprofloxacin adsorption using activated carbon derived from the residue of desilicated rice husk. Journal of Molecular Liquids 238: 316–325. https://doi.org/10.1016/j.molliq.2017.04.022

Zhang, Z., Lei, Y., Li, D., Zhao, J., Wang, Y., Zhou, G., Yan, C. and He, Q. 2020. Sudden heating of H3PO4-loaded coconut shell in CO2 flow to produce super activated carbon and its application for benzene adsorption. Renewable Energy 153: 1091–1099. https://doi.org/10.1016/j.renene.2020.02.059

Zhang, Z., Luo, X., Liu, Y., Zhou, P., Ma, G., Lei, Z., and Lei, L. 2015. A low cost and highly efficient adsorbent (activated carbon) prepared from waste potato residue. Journal of the Taiwan Institute of Chemical Engineers 49: 206–211. https://doi.org/10.1016/j.jtice.2014.11.024

Zhao, B., O'Connor, D., Zhang, J., Peng, T., Shen, Z., Tsang, D.C.W. and Hou, D. 2018. Effect of pyrolysis temperature, heating rate, and residence time on rapeseed stem derived biochar. Journal of Cleaner Production 174: 977–987. https://doi.org/10.1016/j.jclepro.2017.11.013

Zhu, Y., Zhang, H., Zeng, H., Liang, M. and Lu, R. 2012. Adsorption of chromium (VI) from aqueous solution by the iron (III)-impregnated sorbent prepared from sugarcane bagasse. International Journal of Environmental Science and Technology 9(3): 463–472. https://doi.org/10.1007/s13762-012-0043-9.

15

ENERGY RECOVERY FROM MUNICIPAL WASTEWATER SLUDGE

Muhammad N. Siddiquee,[1,2] *Saidur R. Chowdhury,*[3]
Muhammad M. Rahman,[4] *Syed Masiur Rahman*[5]
and *Shaikh A. Razzak*[1,6,*]

1. Introduction

Municipal wastewater treatment plants generate a significant amount of sludge throughout the initial, primary, secondary, and tertiary treatment processes (Rorat et al. 2019, Rulkens 2004). In 2020, Europe's annual production of sewage sludge was predicted to be between 10 and 13 million tons (dry matter), whereas China produced more than 25 million tons (Haghighat et al. 2020). Annually, the European Union generates approximately 50 million metric tons of liquid sludge, whereas the United States generates approximately 40 million metric tons (Bora et al. 2020). Sludge output is predicted to expand in tandem with population growth, urbanization, and industrialization. Not only is managing a large amount of sludge and disposing of it safely expensive, but it also presents issues due to the presence of high organic matter, wastewater-borne contaminants, hazardous compounds, and heavy metals. Inadequate management of municipal wastewater sludge (MWS) poses a substantial hazard to human health and contributes significantly to environmental contamination.

On the other hand, global energy demand is increasing in lockstep with population growth, urbanization, and industrialization. At the moment, the majority of energy is produced from non-renewable sources such as fossil fuels. However, one of the key issues in this area is the limited quantity of

[1] Department of Chemical Engineering, King Fahd University of Petroleum and Minerals, Dhahran 31261, Kingdom of Saudi Arabia.
[2] Interdisciplinary Research Center for Refining & Advanced Chemicals (IRC-RAC), King Fahd University of Petroleum and Minerals, Dhahran 31261, Kingdom of Saudi Arabia.
[3] Department of Civil Engineering, Prince Mohammad Bin Fahd University, Al Khobar 31952, Kingdom of Saudi Arabia.
[4] Department of Civil and Environmental Engineering, King Faisal University, Al-Ahsa 31982, Kingdom of Saudi Arabia.
[5] Interdisciplinary Research Center for Environment & Marine Studies (IRC-EMS), King Fahd University of Petroleum and Minerals, Dhahran 31261, Kingdom of Saudi Arabia.
[6] Interdisciplinary Research Center for Membranes & Water Security (IRC-MWS), King Fahd University of Petroleum and Minerals, Dhahran 31261, Kingdom of Saudi Arabia.
* Corresponding author: srazzak@kfupm.edu.sa

non-renewable resources required to meet expanding energy demand. The issue is that there is a limited supply of non-renewable energy sources available to meet the growing demand for energy. Research is being conducted to identify renewable energy sources that are environmentally benign, widely available, economically viable, and energy efficient (having a positive impact on society). Recently, the 'waste to energy' concept has gained popularity due to its ability to clean and manage garbage while extracting valuable chemicals and energy molecules. MWS, a type of waste, has the potential to play a significant role in energy recovery and other applications in assisting in the resolution of environmental challenges as well.

MWS is a by-product of wastewater treatment plants made up of various organic materials and nutrients. The raw MWS contains water, organic matter, minerals, nutrients such as nitrogen (N), phosphorus (P), and micronutrients, as well as toxic heavy metals, pesticides, endocrine disruptors, pathogens (such as bacteria, viruses, protozoa, and parasitic helminths, among others) and other microbiological pollutants (Rorat et al. 2019). MWS is a mixture of organic substances such as lipids, carbohydrates, proteins, and oils, as well as inorganic components with a high energy content (Siddiquee and Rohani 2021, Oladejo et al. 2019). Chemical analysis of this MWS is critical for evaluating alternative energy sources and valuable chemical compounds. Organic compound conversion to energy and chemicals drives the scientific community to research and exploit sludge for resource recovery, energy recovery, and other applications.

Energy recovery from MWS is being studied as a viable waste to energy alternative. Biodiesel, biochar, biooil, biogas, syngas, H2, bioethanol, biosolids, fertilizer, and electricity can all be produced from MWS using a variety of different conversion methods, including chemical (e.g., esterification); thermochemical (e.g., combustion, pyrolysis, and gasification); and biochemical conversion (e.g., anaerobic digestion, fermentation, or bio electrochemical conversion) (Fan et al. 2021, Siddiquee and Rohani 2021, 2011b, Oladejo et al. 2019, Supaporn et al. 2019). Each energy recovery strategy comes with its own set of problems and opportunities. However, it is vitally important to establish an integrated system for the efficient recovery of energy from MWS that minimizes these difficulties.

This chapter discusses sludge formation, sludge properties, existing management and handling options, several prospective energy recovery strategies, and an economic analysis of recovered energy from MWS. This chapter also discusses the obstacles and opportunities connected with energy recovery strategies. Additionally, the composition of sludge is compared to that of coal and biomass to determine the possibility of using sludge or residual sludge from cellulosic biomass for energy recovery.

2. Overview of Sludge Production

The sources of MWS in a wastewater treatment plant differ depending on the plant type and operation strategy (Al-Malack et al. 2008). In general, it is expected that the daily amount of solids entering the wastewater treatment plant as influent will fluctuate across a wide range. Therefore, it is difficult to predict solids generation in biological wastewater treatment plants, because all influent floating solids may not present in the final sludge. The biological wastewater treatment plant utilizes microorganisms to convert organic substances into the sludge that is isolated from the treated wastewater via the sedimentation processes. Indeed, the microbes consume the dissolved organic materials as food, resulting in far more sludge than chemical treatment of wastewater. In the biological sludge digestion process, some solids are converted to soluble or gaseous end-products. On the other hand, some soluble constituents in wastewater will be converted into biological solids, which can be digested and decreased (Al-Malack et al. 2008). The most common biological treatment process is Activated Sludge Process (ASP) (Fig. 1), which consists of four steps, i.e., preliminary, primary, secondary and tertiary treatments.

Typically, raw sewage is first strained to remove big debris using bar racks. Sand, grit, and stones settle to the bottom of the grit chamber, where they can be disposed of, but most of the suspended

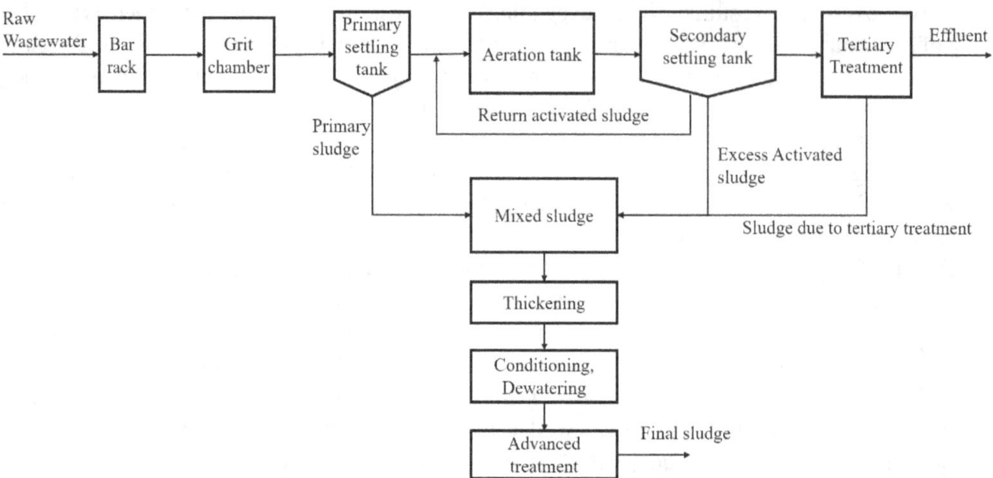

Figure 1. Production of sludge from Activated Sludge Process [Adopted from Verlicchi and Zambello 2015, Syed-Hassan et al. 2017].

organic solids remains in the wastewater. After that, the sewage is permitted to slowly travel through massive tanks known as main clarifiers/settling tanks. Grease and oil rise to the surface of the tanks due to gravitational differences and are skimmed off, whereas MWS settles at the bottom of the tank and is delivered to the sludge treatment process by pumps. The primary clarifier removes 50–70% of the total suspended particles, and the MWS produced at this step is referred to as primary sludge (Pathak et al. 2009). In the next step, microbes are used in an aeration tank to stabilize organic contents and remove non-colloidal particles. The mixed liquor (known as mixed liquor suspended solid, MLSS) is then pumped into a sedimentation tank, where the particles settle at the bottom and the clean effluent is collected as supernatant. Secondary MWS is the sludge that is formed at this point.

MWS from a treatment plant is typically processed in this order: thickening, stabilization/ conditioning, and dewatering (Cantinho et al. 2016). The main goals of the sludge treatment process are (i) to reduce weight and volume of the sludge, which makes the transportation easier, and (ii) to stabilize the material by killing pathogens, eliminating odors, and lowering volatile solid content for safer disposal (Yoshida et al. 2013, Zhang et al. 2014a). Some stabilization procedures, such as anaerobic digestion, allow for energy return as well. The primary sedimentation basin and secondary clarifiers are the main sources of sludge in an ASP at municipal wastewater treatment plants. The main sources of solids and sludge, as well as the varieties created, are listed in Table 1 (Tchobanoglous and Burton 1980).

According to Tchobanoglous and Burton(Tchobanoglous and Burton 1980), the amount of MWS forms during primary treatment may range from 0.25 to 0.35 percent of the total volume of wastewater treated. The quantities grow to 1.5 to 2.0% of the wastewater volume treated when treatment is improved to activated sludge. Chemical phosphorus reduction can add another 1.0 percent to the total. The yields of the anoxic reactors are 35 to 52 percent higher than those of the aerobic reactors. Under aerobic and anoxic circumstances, temperature has an effect on wastewater treatment. Increased phosphorus uptake results in an increase in inorganic MWS bulk, but no change in organic sludge mass. Table 2 displays the production of sludge by different operations. As can be seen from the Table 2, amount of MWS produced differs significantly based on the operation procedures. More specific data on wastewater properties, the treatment method, and operating parameters are needed to effectively estimate sludge generation.

Table 1. Sources of MWS from a Conventional Wastewater Treatment Plant (Davis and Masten 2004, Tchobanoglous and Burton1980).

Types of Sludge	Description
Grit	Material that is collected in the grit chamber is not true sludge in the sense that it is not fluid. Grit is non-biodegradable and generally trucked directly to a landfill without further treatment.
Primary sludge	At the bottom of the primary clarifiers, sludge has a solids content between 3 and 8 percent, and it is around 70% organic. The sludge turns anaerobic quickly, and it smells really foul.
Secondary sludge	Sludge made up of microorganisms and inert components that have been produced but discarded in the secondary treatment operations are the ingredients of this blend. As a result, the solids are approximately 90% organic. The solids concentration varies according to the source. Wasted activated sludge has around 0.5–2% particles, whereas trickling filter sludge contains approximately 2–5% solids.
Mixed sludge	A mixture of primary and secondary sludge
Digested sludge	The stabilized sludge generated in aerobic or anaerobic digesters are being mentioned here. Digestion is used to lower the level of organic material and microorganisms, as well as to remove odors.
Composted sludge	Organic substances decompose in an aerobic environment and the resulting sludge is known as a by product of this process.
Biosolids	After biological stabilization and dewatering, this is in the form of a solid.
Conditioned sludge	This sludge comes from systems which use chemicals and physical techniques to lower the water content in their systems.
Dried and semi-dried sludge	In semi-dried sludge, the water content is decreased by drying to 30-55 wt percent, dry matter, while in dried sludge, water content is reduced by thermal process by more than 80%.
Other types of sludges	Sludges obtained by disinfection, pasteurization, thermal hydrolysis and advanced oxidation

Table 2. Typical Data for the Physical Characteristics and Quantities of MWS Produced from Various Wastewater Treatment Operations and Processes (modified from Tchobanoglous and Burton (Tchobanoglous and Burton 1980)).

Treatment operation or Process	Dry Solids, g/10^3 L	
	Range	Typical
Primary sedimentation	107.84–167.76	149.78
Activated sludge (waste sludge)	71.90–95.86	83.88
Trickling Filtration (waste sludge)	59.91–95.86	71.90
Extended aeration (waste sludge)	83.88–119.83	95.86[a]
Aerated lagoons (waste sludge)	83.88–119.83	95.86[a]
Filtration	11.98–23.97	17.97
Algae removal	11.98–23.97	17.97
Chemical addition to primary sedimentation tanks for phosphorus removal		
Low lime (350–500 mg/l)	239.65–395.43	299.57[b]
High lime (800–1600 mg/l)	599.13–1318.09	790.85[b]
Suspended-growth nitrification	-	-[c]
Suspended-growth denitrification	11.98–29.96	17.97

[a] Assuming no primary treatment
[b] Sludge in addition to that normally removed by primary sedimentation
[c] Negligible

3. MWS Characteristic

The qualities of the MWS being measured are inextricably tied to its ultimate destination. For instance, for the gravity thickened sludge, the settling and compaction properties are crucial. On the other hand, volatile organic compounds and heavy metals concentrations are crucial if the sludge is to be used for energy production.

3.1 Physical characteristics

The first component, solids content, is almost certainly the most critical in determining the volume and the physical state of the MWS (i.e., liquid, semi-solid or solid). Inorganic solids have a specific gravity of between 2 and 2.5, whereas organic solids have a specific gravity of between 1.2 and 1.3. Rheological properties of MWS are significant because they are one of the few fundamental factors that accurately explain the physical properties of sludge (Markis et al. 2014). MWS can be classed as a Newtonian fluid (shear rate is proportional to the velocity gradient), or as a plastic fluid (sludge begins to move after applying a threshold shear). The majority of MWS are also being considered as pseudo plastic sludges (Vilardi et al. 2020). At low stress levels, below the yield stress, sludge behaves like a viscoelastic solid, with primary sludge yielding abruptly and secondary sludge flowing gently to steady state. Both sludges behave as shear thinning, yield stress fluids in the steady state, with primary sludge exhibiting extremely thixotropic behavior. Both primary and secondary sludge exhibit a rise in apparent viscosity, yield stress, and fluid consistency as the total solids content increases (Markis et al. 2014).

The water content of wet sludge must be determined precisely due to its fluctuating nature over time. According to Al-Malack et al.(2008), the initial water content of sludge may be between 95 and 99 percent, while the final water content may range between 13 and 48 percent. The constant decline in the water content of sludge can be attributed to two primary mechanisms: infiltration and evaporation. The same authors also reported a significant drop in the water content of the sludge after four days of drying. The relationship between sludge water content and drying time is further explained by the different forms of water associated with sludge flocs, namely bound and unbound waters (Syed-Hassan et al. 2017). Mechanical separation of unbound water, which has the highest water content, is possible. Dewatering systems remove largely free unbound water; some interstitial water can also be removed, but the majority of bound water is likely to be vicinal water that cannot be removed mechanically. Bound water, which contributes to the lowest water content, is physically and chemically most tightly bound to the particles and can be removed only thermally. While pyrolysis and gasification require significant drying of the feed prior to the reactions, hydrothermal methods can operate with moisture values of 70–95 weight percent (Bora et al. 2020). Table 3 shows the physical parameters of sludge following various treatment procedures.

Table 3. Physical Characteristics of Sludge (Oakley 2018, Tchobanoglous and Burton 1980).

Parameter	Primary sludge	Secondary sludge	Dewatered sludge
Dry solids	2–6%	0.5–2%	15–35%
Volatile solids	60–80%	50–70%	30–60%
Sludge specific gravity	1.02	1.05	1.1
Solids specific gravity	1.4	1.25	1.2–1.4
Shear strength (kN/m2)	< 5	< 2	< 20
Energy content (MJ/kg VS)	12-22	12–20	25–30
Particle size (90%)	< 200 mm	< 100 mm	< 100 mm

3.2 Chemical characteristics

Chemical characteristics of MWS heavily depend on source, wastewater treatment system, environmental condition, and manufacturing processes. These factors increase the unpredictability in the chemical composition of sludge in comparison to other fuel sources, such as biomass and coal. If sludge is to be used as fuel, three critical qualities must be understood:

- Energy content
- Proximate analysis; and
- Ultimate analysis.

3.2.1 Energy content

The fuel value of MWS is mostly determined by its composition and volatile solid content. Untreated primary sludge has a high fuel value if it contains a sufficient percentage of grease and skimming. Sludge that has been digested has a lower heating value than sludge that has not been digested. Table 4 contains typical heating parameters for various types of sludge.

Table 4. Typical heating values of different types of sludge (Chan and Wang 2016, Jiang et al. 2016, Tchobanoglous and Burton 1980).

Sludge Type	Heating value (MJ/kg)	
	Range	Typical
Primary (Raw)	23.26–29.08	25.59
Activated sludge	16.28–23.26	20.93
Anaerobically digested	9.30–13.96	11.63
Chemically precipitated	13.96–18.61	16.28
Biological filter	16.28–23.26	19.77

3.2.2 Proximate analysis

Proximate analysis of sludge's combustible components determines (i) moisture content (reduction of moisture after heated to 105°C for one hour) ii) volatile matter (VM) (reduction of weight after heated to 950°C) (iii) fixed carbon (FC) (combustible residue remaining after volatile materials has been removed) and (iv) ash (residue after open combustion) (Tchobanoglus et al. 2003). Figure 2(a) illustrates a typical distribution of proximate elements in sludge, including volatile matter, fixed carbon, and ash. According to (Bora et al. 2020), volatile matter is finally transformed to gaseous

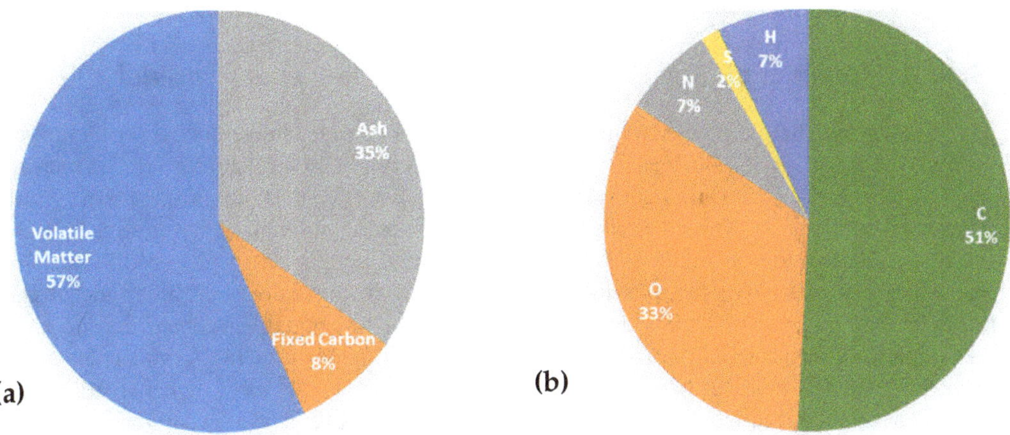

Figure 2. Results of typical (a) Proximate analysis and (b) Ultimate analysis; based on data reported by Bora et al. (2020) collected from thirteen studies.

and liquid fuels, with the carbon concentration and distribution determining the energy content of the products. Increased ash content is generally not favorable since it results in decreased energy conversion efficiency and may also cause equipment difficulties.

3.2.3 Ultimate analysis

The ultimate analysis of sludge comprises determining the percentages of carbon (C), hydrogen (H), nitrogen (N), sulfur (S), and ash. The results of the ultimate analysis are utilized to determine the chemical composition of the sludge's organic matter. This method is also used to determine the optimal mixture of sludges required to obtain the desired C/N ratios for biological conversion processes (Tchobanoglus et al. 2003). The distribution of ultimate elements in sludge is depicted in Fig. 2(b). As seen, carbon is the most abundant element in sludge, followed by oxygen, nitrogen, and sulfur. According to Chan and Wang(2016), inorganics contribute partially to the O and S contents of sludge samples, whereas organic matter contributes to the C, H, and N contents.

Some of the trace elements found in sludge have been referred to as "heavy metals." Table 5 depicts a typical heavy metal sludge composition.

Table 5. Concentrations of Heavy Metals in Sludge (Adopted from (Chan and Wang 2016, Mulchandani and Westerhoff 2016, Tchobanoglous and Burton 1980)) .

Metal	Dry sludge, mg/kg Range Median	
Arsenic	1.1–230	10
Cadmium	1–3410	10
Chromium	10–99000	500
Cobalt	11.3–2490	30
Copper	84–17000	800
Iron	1000–154000	17000
Lead	13–26000	500
Manganese	32–9870	260
Mercury	0.6–56	6
Molybdenum	0.1–214	4
Nickel	2–5300	80
Selenium	1.7–17.2	5
Tin	2.6–329	14
Zinc	101–49000	1700

3.3 Comparison of sludge to coal and biomass as a source of fuel

Figure 3 compares the proximate and final compositions of sewage sludge to coal (or lignite) and biomass, demonstrating the sludge's potential as a fuel. As can be observed from Fig. 3(a), sewage sludge contains far more ash and significantly less fixed carbon (FC) than biomass and coal. Sewage sludge has a substantially lower volatile matter (VM) content than biomass, but is slightly higher than lignite. According to (Syed-Hassan et al. 2017), sewage sludge has a higher VM/FC ratio (on average 6.4) than coal and biomass (on average 4.7 and 1.4). However, due to the high concentration of inorganics in sludge, some underestimating of FC and overestimation of ash content may occur as a result of inorganic element oxidation during ash analysis (Chan and Wang 2016). A similar observation was made by Oladejo et al. (2019). The authors discovered that biomass had a higher volatile matter content than coal, whereas coal had a lower volatile matter level than sludge. Additionally, coal and biomass had a larger fixed carbon content than sludge. Nonetheless, due to its high inorganic content, sludge has a higher ash content (prominently aluminum (Al), calcium (Ca), iron (Fe), magnesium (Mg), sodium (Na), phosphorus (P), and titanium(Ti)) than biomass and coal. In a recent study, Fan et

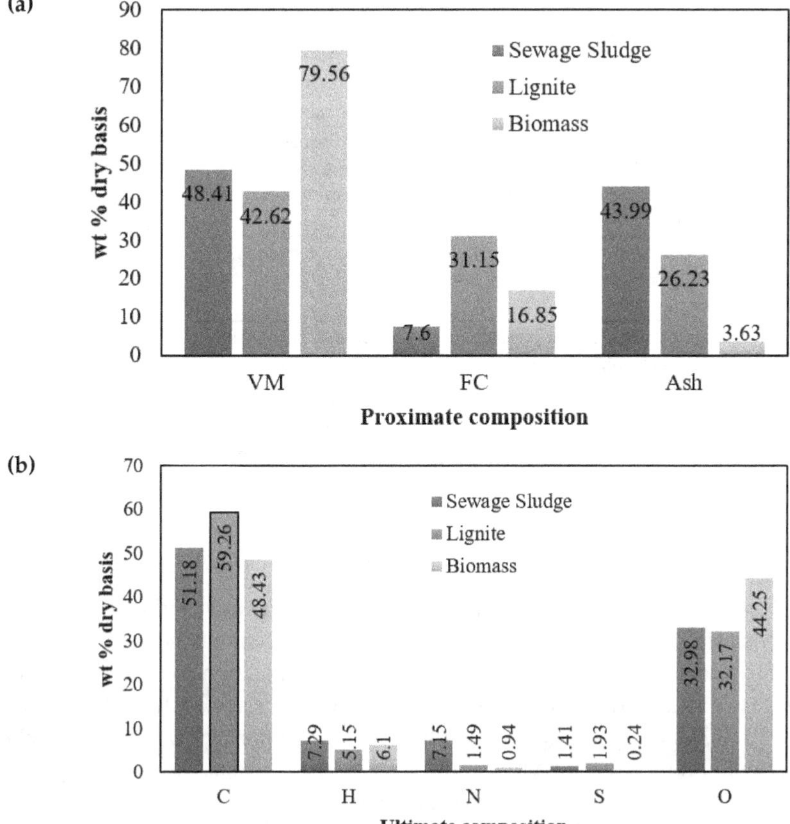

Figure 3. Comparison of sludge with coal and biomass, for (a) Proximate analysis (b) Ultimate analysis; based on data reported by (Syed-Hassan et al. 2017).

al. (2021) compared lipid-extracted sludge with regular sewage sludge. The authors observed protein as a major fraction (35–39 wt%) in both types of sludge, while lipid as a minor fraction constituting as 14 wt% of the sludge.

In terms of constituent compositions (Fig. 3(b)), Syed-Hassan et al. (2017) indicated much more N and somewhat more H than lignite and biomass. According to Debono and Villot (2015), the primary sources of nitrogen in sewage sludge are protein and peptides, but nitrogen can also be found in fatty acids and sugars. In comparison to coals, the diverse sources of nitrogen lead to a greater variety of nitrogen functions in sewage sludge. The S and O levels of sewage sludge are comparable to those of lignite but much greater than those of biomass, whereas the C content of sewage sludge is in the middle of the range when compared to lignite and biomass.

4. Sludge Management and Treatment

Proper sludge management and handling is a challenging task. Although a significant amount of sludge is produced in the wastewater treatment facilities, the treatment and management of sludge have been confronting significant technical, environmental and financial challenges. A major portion of the capital and operating cost has been allocated for sludge handling and treatment. The presence of toxic and nontoxic compounds in a single mixture is the main challenge of safe handling and disposal (Rulkens 2004). Figure 1 shows the process flow diagram of the activated sludge production in a municipal wastewater treatment facility. The conventional MSW treatment processes, procedures employed in each step, and their purposes are listed in Table 6.

Table 6. Conventional sludge treatment and processing.

Sludge treatment	Methods/types	Purposes/remarks	Sources
Thickening	Gravity thickening; centrifugation; dissolved air flotation	Water removal, stabilization	(Metcalf 2003, Rorat et al. 2019)
Conditioning	Adding chemicals (e.g., lime, sulfuric acid, alum, chlorinated copperas, ferrous sulfate, and ferric chloride); wet air oxidation; freeze-thawing	Improved dewatering characteristics of the sludge; stabilization;	(Metcalf 2003, Rorat et al. 2019)
Dewatering	Drying beds; filter press; filter belt press; vacuum disk filter; centrifuge; roto-plug concentrator	Volume reduction; stabilization	(Metcalf 2003, Novak 2006)
Advance treatment/ processing	Composting; thermal drying; incineration; pyrolysis; anaerobic digestion	Reuse, recycling and resource recovery	(Metcalf 2003, Rorat et al. 2019)
Final disposal and handling	Land application; combustion; sanitary landfill; coincineration	Sustainable management (e.g., waste to energy; recycling or waste to engineering materials, etc.)	(Campbell 2000, Rorat et al. 2019)

4.1 Current handling and environmental impact

The final disposal and handling of MSW is an integral aspect of the conventional sludge treatment and processing recovery process (Table 6). In the waste management industry, there are three broad categories for sludge management. These are: a) land application for soil enhancement (by incorporating sludges into the soil); b) landfill disposal; and c) energy recovery technologies (Campbell 2000, Rorat et al. 2019). One of the disposal alternate approaches for a resource usage field is land application for agricultural uses (Singh and Agrawal 2008). Land is considered a sludge repository, despite the fact that there are little restrictions on the activity (Campbell 2000). Sludge was applied as a fertilizer to the land at high rates without regard for agronomic nutrient requirements in certain cases, leading in increased contamination of the soil water continuum. Rather than being used for sludge utilization, the area was frequently used as a disposal site. Sludge disposal in a sanitary landfill was the easiest and cheapest method of sludge management, but it was questioned by scientists because other choices (such as biofuel or biogas production, composting, or resource recovery technologies, among others) were available. Sludge disposal at landfills can contribute to environmental pollution, greenhouse gas emissions, an obnoxious odor in the surrounding community, as well as serious health and environmental consequences. In addition to public perception, there were handling and transportation issues with sludge disposal in landfills. Because land application for disposal reasons is no longer allowed, treatment is now required before discharging onto the land to prevent further pollution (Rorat et al. 2019).

The environmental impact from improper disposal (such as landfills, composting or storage) of sludge can contaminate the soil, water and air as well as cause ecological degradation by releasing pathogens, heavy metals, toxic gases and organic pollutants into the nature (Oladejo et al. 2019). Unplanned usages of MWS produced bio-solids or bio-fertilizer can also cause environmental and public health issues such as potential trophic transfers (via cultivated plants) and possible contamination of groundwater by releasing metals and organic chemicals. Potential issues including groundwater and surface water contamination, pathogens and odors are some of the major problems associated with the use of sludges on cropland (United States Enviromental Protection Agency 1994). It was also reported that land application is a contributor to global warming, eutrophication, and acidification (Oladejo et al. 2019, Yoshida et al. 2018). Also, challenges associated with the public acceptability from sludge application especially to the land near residential area could be very common due to odor or nuisance issue or soil and groundwater pollution. The sludge contains biological agents that can be problematic for living organisms. Generally, four groups of pathogens (such as viruses, bacteria, parasites, and fungi) in the sludge may simply disturb natural ecosystems as well as local community (Fijalkowski et al. 2017).

4.2 Sludge treatment and solution

According to (Oladejo et al. 2019),the complex composition of the sludge generated from wastewater plant (as shown in Fig. 1) would be a very complex mixture due to high moisture, ash, toxic heavy metals and organic contaminants which influence the safety of final products.

Improvement of the quality of the sludge can be achieved by preventing the discharge of pollutants (such as heavy metals, pesticides, endocrine disrupters originate from households, industrial and agricultural activities) into the sewer. Prevention of discharge of these pollutants would decrease the toxic sludge production and minimize the handling cost. The quality of the sludge can also be improved by developing treatment methods in order to remove colloids and suspended particles from the influent in a preliminary treatment step.

MWS can be converted to chemical energy and alternative renewable energy sources (Puyol et al. 2017). Energy recovery technologies can be very promising as the energy extraction from the excess MWS can reduce the disposal of sludge in a sanitary landfill or other land applications that could cause contamination into the soil and water. Moreover, prevention from contamination, recycling and reuse of valuable products from the sludge, as well as energy recovery (such as bio-gas, flue gas, bio-oil and electricity, etc.) are the most desired pathways that can strictly limit the landfilling and disposal (Rorat and Kacprzak 2017). Thus, waste to energy methods address a worldwide call for renewable energy sources and is an important criterion in this development.

5. Energy Recovery from Municipal Wastewater Sludge

Energy recovery approaches include chemical conversion (catalytic transesterification and esterification), thermochemical conversion (combustion, pyrolysis, and gasification), biochemical conversion (anaerobic digestion, fermentation, and enzymatic catalytic esterification-transesterification), and bio-electrochemical as outlined in Fig. 4. This section highlights the approaches to recover energy from MWS.

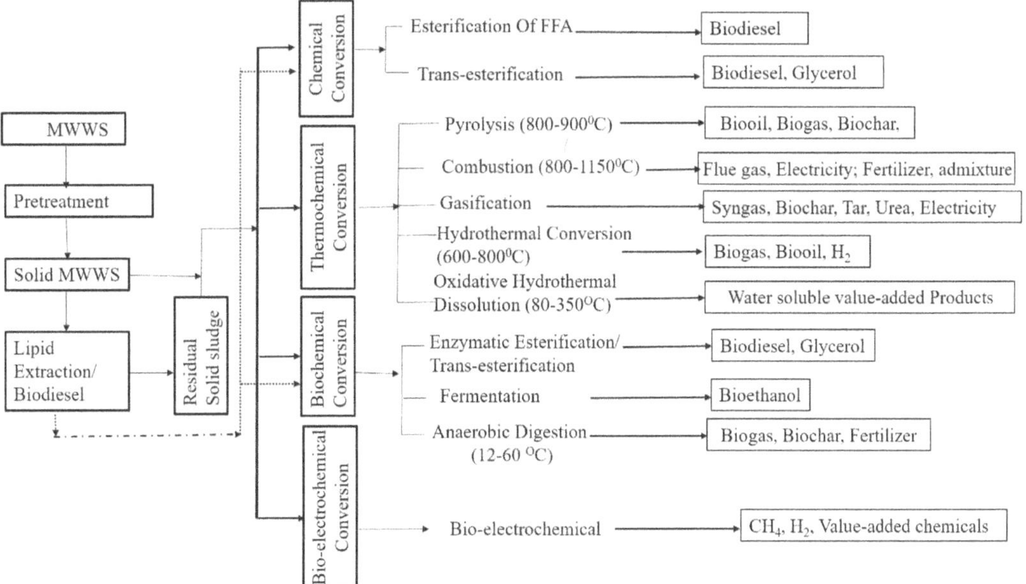

Figure 4. Energy recovery approaches from the municipal wastewater sludge (MWS).

5.1 Chemical conversion

The common chemical conversion is catalytic transesterification of lipid and catalytic esterification of free fatty acid (FFA) obtained from MWS to produce biodiesel. Biodiesel is the most potential renewable liquid fuel that can be used as either directly or as a blend with petro-diesel. MWS contains significant amount of lipids and is being considered as a potential option for biodiesel production to reduce raw material cost. It would also mitigate the challenges and operating cost associated with sludge management.

Biodiesel production from MWS is well described in literature (Siddiquee and Rohani 2021, 2011b, Capodaglio and Callegari 2018, Arazoet al. 2017, Kargbo 2010). Figure 5 illustrates the overall schematic of the biodiesel production from MWS. Firstly, the raw sludge is pretreated (Step I) by following few techniques such as settling, filtration, centrifugation, and drying to remove water present in the sludge to get solid sludge. This pre-treatment is one of the most difficult processes in the production of biodiesel or any other kind of energy recovery. The procedures, particularly drying, are energy-intensive and costly since they involve dealing with a large volume of biomass and lipid in the raw sludges. Biodiesel is then produced from the solid sludge either by a two-step process (lipid extraction (Step II) followed by biodiesel production (Step III) or by a single step *in-situ* process (Step IV). Both the lipid extraction and the *in-situ* process require solvents and are associated with the solvent recovery (Step V). Biodiesel, impure glycerol, and unconverted lipid are separated (Step VI). Solid residue from the lipid extraction and the *in-situ* process can be used to recover other form of energies such as biooil. The overall biodiesel production from MWS is involved with the challenges such as expensive pre-treatment and lipid extraction.

First step of biodiesel production process is the lipid extraction (Step II) which is influenced by afew factors including the type and quality of sludge, solvents, sludge-to-solvent ratio, extraction temperature, drying techniques used (such as oven drying, fluidized bed drying, freeze (lyophilisation) drying), vacuum drying, and use of the hydromatrix, and extraction time (Siddiquee and Rohani 2021, Dufreche et al. 2007). Organic solvents such as methanol, hexane, chloroform, toluene, and ethanol are usually used for the lipid extraction from the MWS (Siddiquee and Rohani 2021, Dufreche et al. 2007).Lipid extraction efficiency can be enhanced by using a combination of polar and nonpolar solvents. Also, the use of ultrasonication and ionic liquids (for instance, $[C_4mim][MeSO_4]$ and $[P(CH_2OH)_4]Cl$) enhance lipid extraction efficiency (Siddiquee and Rohani 2021, Olkiewicz et al. 2014, Zhang et al. 2014b).

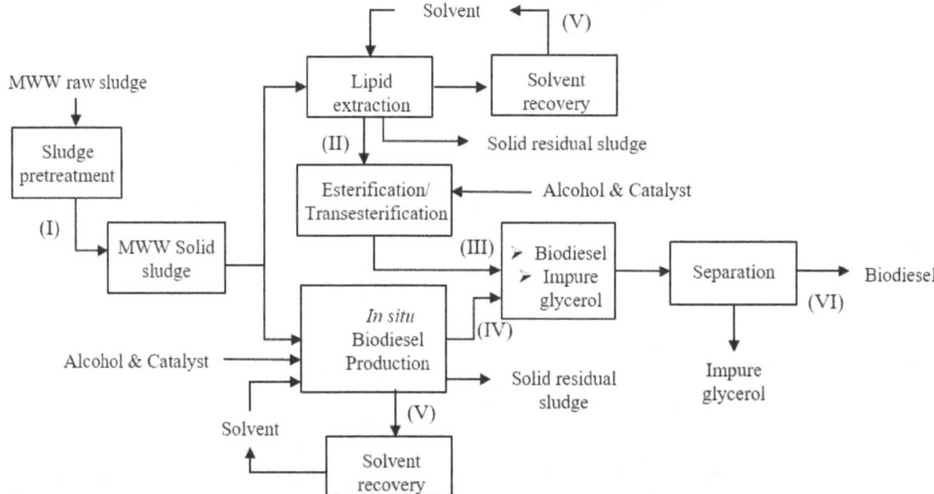

Figure 5. Overall biodiesel production scheme from municipal wastewater sludge (MWS) (Adapted from (Siddiquee and Rohani 2021, 2011a)).

$$\begin{array}{l} CH_2-OCOR_1 \\ | \\ CH-OCOR_2 \\ | \\ CH_2-OCOR_3 \end{array} + 3\,CH_3OH \xrightarrow[\Delta]{\text{Catalyst}} \begin{array}{l} CH_3COOR_1 \\ CH_3COOR_2 \\ CH_3COOR_3 \end{array} + \begin{array}{l} CH_2-OH \\ | \\ CH-OH \\ | \\ CH_2-OH \end{array}$$

Triglyceride FattyAcid Methyl Ester (FAME) Glycerol

(a) Transesterification

$$R-\overset{\overset{O}{\|}}{C}-OH + CH_3OH \xrightarrow[\Delta]{\text{Acid catalyst}} R-\overset{\overset{O}{\|}}{C}-OCH_3 + H_2O$$

Fatty Acids FattyAcid Methyl Ester (FAME)

(b) Esterification

Figure 6. The general form of biodiesel production by (a) transesterification and (b) esterification (Siddiquee and Rohani 2021).

As outlined in Fig. 5, biodiesel can be produced from MWS following either ex-situ process from the extracted lipid (Step III) or in-situ process directly from the solid sludge (Step IV). Biodiesel is chemically known as fatty acid alkyl (mainly methyl) ester and is produced by the catalytic transesterification of lipid (oil/fat) (Fig. 6(a)) and also by the catalytic esterification of free fatty acids (FFA) (Fig. 6(b)). R in Fig. 6 represents a long chain of aliphatic hydrocarbon (for example, C15–C18). In transesterification reaction (Fig. 6(a)), the triglyceride reacts first with primary alcohol to produce diglyceride which further reacts with another alcohol to produce monoglyceride and finally glycerol is produced.

Few factors such as catalyst type (homogeneous and heterogenous base and acid catalysts, zeolite-based catalysts, and enzymatic catalysts), alcohol/lipid molar ratio, temperature and reaction time, and quality of lipid (for example, moisture content and free fatty acid (FFA) content) influence the biodiesel production process. The use of excess methanol (typically 60–100% more) facilitate total conversion in transesterification reaction (Siddiquee and Rohani 2011b). Selection of catalyst is an important factor in biodiesel production. Homogeneous base catalyst, a widely used catalyst, is very fast compared to any other catalysts. But the homogenous base catalyst such as NaOH reacts with FFA present in the raw materials to produce soap which results in catalyst consumption and also inhibits the separation of glycerol from the biodiesel contributing in emulsion formation during the water wash.

Homogeneous acid catalyst (such as H_2SO_4) is very slow, requires excess alcohol, but is suitable for the simultaneous esterification and transesterification to produce more biodiesel. However, it is important to avoid water accumulation from the esterification process (Fig. 6(b)). A two-step process, acid catalytic followed by base catalytic, is usually practiced for the raw materials containing higher FFA. The use of heterogeneous catalysts (such as Mg/La mixed oxide, S-ZrO_2sulfated zirconia, KOH/Nax zeolite, Li/CaO, CaO, KI/Al_2O_3, (ZS/Si) zinc stearate immobilized on silica gel, KNO_3/Al_2O_3, SO_4^{2-}/TiO_2-SiO_2, SBA-15 impregnated with different percentage of heteropolyacid $H_3PO_4.12WO_3.xH_2O$, natural zeolite based catalyst, etc.), is promising in biodiesel production from the MWS to recover catalyst after reaction products, to avoid the undesired saponification reactions (Siddiquee et al. 2011, Melero et al. 2015). A non-catalytic supercritical methanol technique which requires extreme operating conditions (350°C and 43 MPa) is also promising for simultaneous transesterification and esterification reactions to produce biodiesel (Siddiquee and Rohani 2011b). As MWS contains lipids from various sources and contain higher FFA, two-steps catalytic processes (acid catalytic followed by base catalytic processes) or acid catalytic process could be used to produce biodiesel from MWS.

Approximately 57.1 wt% biodiesel was produced from the lipid of wastewater sludge at the methanol to lipid ratio of 125 ml/g, 14 h, 60°C, 3% H_2SO_4, and 50 mg natural zeolite as catalyst

Table 7. Summary of energy recovery from wastewater sludge following different techniques.

Type	Process	Source	Operating Parameters	Reported Energy	Reference
Chemical	Esterification/ transesterification	Lipid from wastewater sludge	Lipid in hexane, Methanol to lipid = 125 ml/g, 14 h, 60°C, 3% H_2SO_4, Natural zeolite 50 mg	57.1 wt% biodiesel	(Siddiquee and Rohani 2011a)
		Oven dried primary sludge	0.25 g sludge/ ml methanol, Zr-SBA-15 catalyst, 209°C, 3 h, 2000 rpm (*in-situ* process)	15.5 wt% biodiesel	(Melero et al. 2015)
		Oven dried secondary sludge		10.0 wt% biodiesel	
Thermochemical	Pyrolysis	MWS	500°C, 10°C min^{-1}	~ 52 wt% biooil, ~ 32 wt% biochar and ~ 16 wt% biogas	(Barry et al. 2019)
		Residual sewage sludge [a]	390°C, 5 minutes	33.3% biooil	(Supaporn et al. 2019)
	Gasification	Wastewater sludge	1000°C	43–46 vol% H_2	(Lee et al. 2018)
	Hydrothermal	Residual sludge after lipid extraction	350°C, pressure up to 40 MPa, water	21.26% bio-crude	(Fan et al. 2021)
	Oxidative hydrothermal dissolution	Fecal Sludge	470°C, 25 min, water, O_2	100% conversion of TOC [b]	(Hubner et al. 2016)
Biochemical	Enzymatic esterification/ transesterification	Lipid from secondary sludge	30°C, 12 h, 200 rpm 10 U immobilized lipase	~ 85% biodiesel	(Jeevan Kumar and Banerjee 2019)
	Fermentation	Dried Sewage sludge	30°C, pH 6.5, yeast 6% wt, 10 days incubation	Max ~ 40 mL/L bioethanol	(Manyuchi et al. 2018)
	Anaerobic digestion	Thickened sludge		~ 52% biogas	(Gherghel et al. 2019)

[a] Residual sewage sludge after lipid and carbohydrate extraction
[b] TOC: total organic carbon

(Table 7) (Siddiquee and Rohani 2011a). Approximately 15.5 wt% and 10.0 wt% biodiesel, respectively, were obtained from the oven dried primary and secondary sludge following *in-situ* esterification-transesterification process maintaining 0.25 g sludge/ml methanol, Zr-SBA-15 catalyst, 209°C, 3 h, and 2000 rpm (Table 7) (Melero et al. 2015). Additional biodiesel yield and FAME composition obtained from both the primary and secondary sludges following different operating conditions are tabulated in literature (Siddiquee and Rohani 2021). The composition of FAME obtained from the MWS is comparable to that of vegetable oils showing maximum palmitic acid (C16:0), followed by stearic acid (C18:0) and oleic acid (C18:1). However, biodiesel production from the MWS is promising. Potential benefits, key challenges and potential recommendations are summarized in Fig. 7.

5.2 Thermochemical conversion

Thermochemical conversion process is an important technique to convert biomass to energy and fuels. The common and widely used thermochemical conversion process includes pyrolysis, combustion, and gasification which are used to produce biooil, biochar, biosolid, and biogas. In addition to biodiesel production, energy can be recovered from the MWS and/or residual sludge after lipid extraction (Step II) or *in-situ* biodiesel production (IV) applying thermochemical conversion. This section highlights the different thermochemical conversions suitable to recover energy from the MWS.

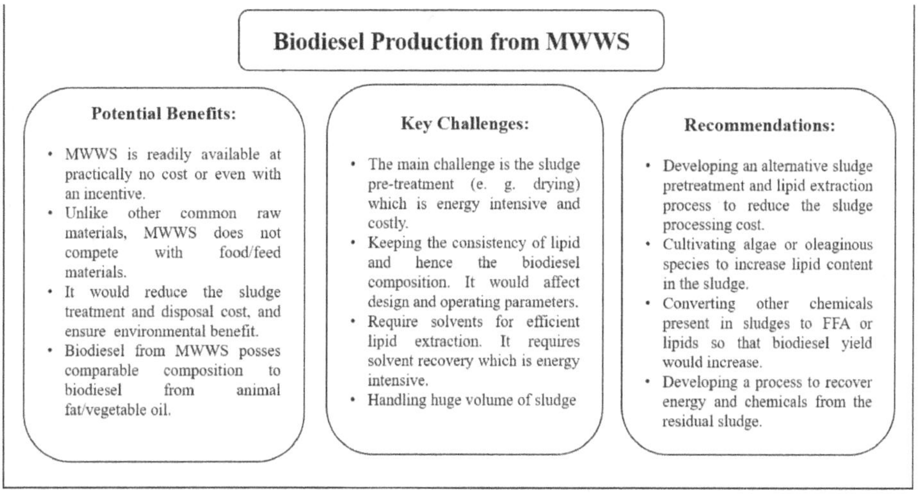

Biodiesel Production from MWWS
Potential Benefits:
• MWWS is readily available at practically no cost or even with an incentive.
• Unlike other common raw materials, MWWS does not compete with food/feed materials.
• It would reduce the sludge treatment and disposal cost, and ensure environmental benefit.
• Biodiesel from MWWS posses comparable composition to biodiesel from animal fat/vegetable oil.
Key Challenges:
• The main challenge is the sludge pre-treatment (e. g. drying) which is energy intensive and costly.
• Keeping the consistency of lipid and hence the biodiesel composition. It would affect design and operating parameters.
• Require solvents for efficient lipid extraction. It requires solvent recovery which is energy intensive.
• Handling huge volume of sludge
Recommendations:
• Developing an alternative sludge pretreatment and lipid extraction process to reduce the sludge processing cost.
• Cultivating algae or oleaginous species to increase lipid content in the sludge.
• Converting other chemicals present in sludges to FFA or lipids so that biodiesel yield would increase.
• Developing a process to recover energy and chemicals from the residual sludge.

Figure 7. Potential benefits, key challenges and recommendations for biodiesel production from the MWS.

5.2.1 Pyrolysis

Pyrolysis is an important thermochemical biomass conversion process in which inert atmosphere or very limited oxygen atmosphere is maintained at operating temperatures 300–900°C to recover fuels and chemicals from the wide range of biomasses. A typical pyrolysis process to recover energy from the MWS is illustrated in Fig. 8. Depending on the heating rates, residence time and vapor quenching rate, pyrolysis is recognized as fast pyrolysis and slow pyrolysis. Product composition varies depending on the type and composition of biomass used in pyrolysis. Typical heating rate in fast pyrolysis is 100 to 1000°C/min in the presence of nitrogen or limited O_2, typical vapor residence time is less than 2s and pyrolysis vapor quenched rapidly (Ganesapillai et al. 2020). Key products via fast pyrolysis process include biooil, biogas and biochar. In typical slow pyrolysis process, heating rate is less than 10°C, and longer vapor and solid residence time. Products are biooil, biogas and biochar, and products composition varies, as like fast pyrolysis, depending on the biomass sources and types. Details of the pyrolysis process is available in literature (Ganesapillai et al. 2020, Oladejo et al. 2019, Fonts et al. 2012).

Condensation
Bio-gas
CH_4, CO_2
Bio-oils
e.g. Heating Oil; Gasoline; Diesel
Pretreated sludge
Pyrolysis reactor < 900°C
Bio-char
Resource Recovery
Potential Energy input

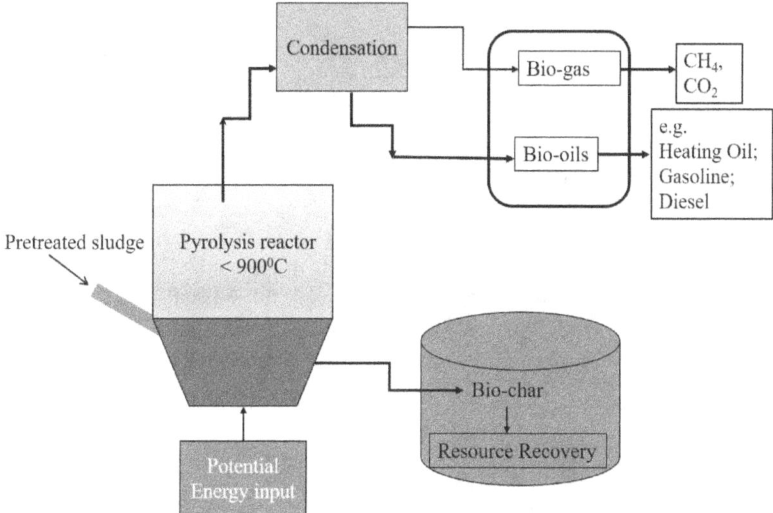

Figure 8. Typical pyrolysis process to recover energy from the MWS.

Fonts et al. (2012) reported biooil yield varied in the range of 27–54 wt% on sewage sludge fed basis. The biooil yields are lower compared to the lignocellulosic biomass as higher ash content in the sewage sludge (Trinh et al. 2013). Supaporn et al. (2019) investigated biooil production from the residual sewage sludge after lipid and carbohydrate extraction by using a micro-tube reactor at 390°C for 5 minutes. and obtained 33.3% biooil (Table 7). Barry et al. (2019) investigated slow and fast pyrolysis of MWS over temperature ranges 300–500°C (Table 7). They obtained ~ 52 wt% biooil, ~ 32 wt% biochar and ~ 16 wt% biogas at 500°C via slow pyrolysis at a heating rate of 10°C min^{-1}, and achieved ~ 53 wt% biooil, ~ 25 wt% biochar and ~ 22 wt% biogas at 500°C in case of fast pyrolysis. They also found that biochar yields decreased as the operating temperature increased regardless of fast or slow pyrolysis. However, biochar obtained via slow pyrolysis possessed higher heating value compared to the biochar obtained via fast pyrolysis.

5.2.2 Combustion

Combustion is another important thermo chemical biomass conversion process to recover fuels and chemicals in which biomass is thermally decomposed/burned in the presence of O_2. Figure 9 illustrates a typical combustion process to recover energy from the MWS. Combustion temperature ranges from 800°C to 1150°C. It can be used for the wide range of biomass including MWS and/or residual sludge after lipid extraction. Key advantage of such process is that the biomass/sludge can contain 20–50% moisture (Oladejo et al. 2019). Typical reactions of the combustion process include:

$$C + O_2 \rightarrow CO_2 \quad \text{(I)}$$

$$C + 0.5\ O_2 \rightarrow CO \quad \text{(II)}$$

$$H_2 + 0.5\ O_2 \rightarrow H_2O \quad \text{(III)}$$

$$CH_4 + 2O_2 \rightarrow CO_2 + 2H_2O \quad \text{(IV)}$$

Details of the combustion process and available technologies are reported in literature (Bora et al. 2020, Gherghel et al. 2019, Oladejo et al. 2019, Tyagi and Lo 2013). The main combustion

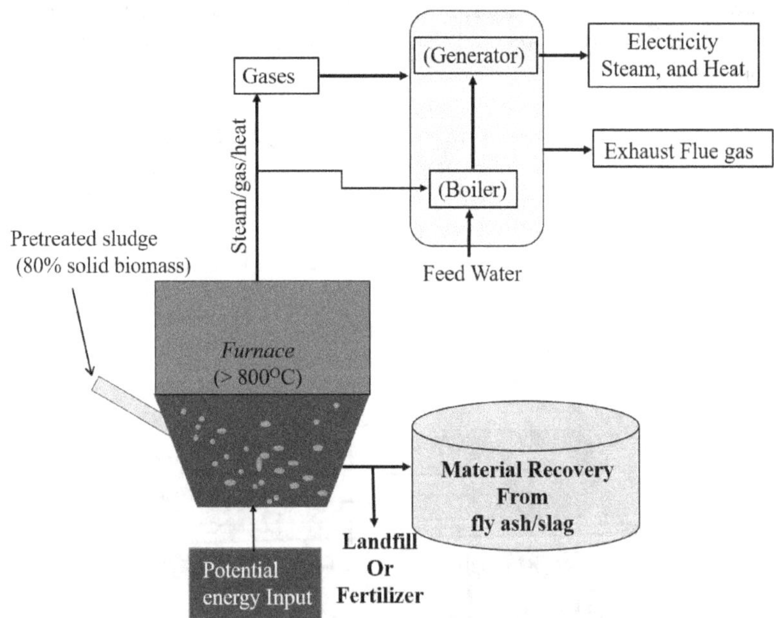

Figure 9. Typical combustion process to recover energy from the MWS.

products include flue gas that consists of CO, CO_2, NOx, SOx, PM, N_2, H_2O, which can be separated to produce value added fuels and chemicals. As this process also involves the oxidation of N, and S present in the biomass, it produces NOx, and SOx. It is required to investigate the formation and release of NOx and/or SOx depending on the environmental regulation.

5.2.3 Gasification

Gasification is an important thermochemical conversion process in which biomass is partially oxidized and gasified at 800°C–1000°C in the presence of air, O_2, CO_2, steam and/or mixture of these gases. A typical gasification process to recover energy from the pretreated MWS (biomass)is illustrated in Fig. 10. The main gasification products include syngas (H_2 and CO), CO_2, CH_4, H_2O and hydrocarbons. In addition to the products, energy can also be recovered as electricity and heat. Lee et al. (2018) investigated wastewater sludge gasification at 1000°C with steam to enhance reaction kinetics and H_2 yield and obtained 43–46 vol% H_2 (Table 7). Biomass should have < 15% moisture similar to combustion process and would also produce NOx and SOx based on the S and N content in biomass. Detail of the gasification process and advances are available in literature (Bora et al. 2020, Gherghel et al. 2019, Oladejo et al. 2019, Tyagi and Lo 2013).

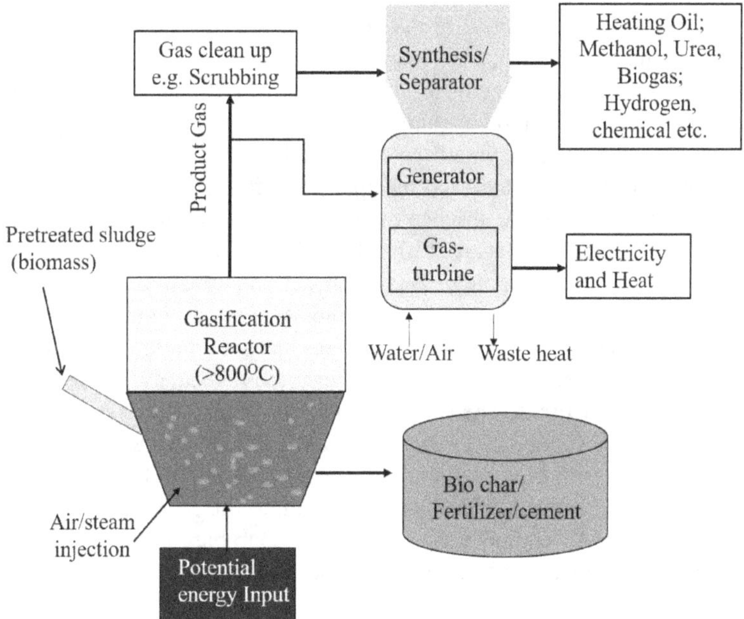

Figure 10. Typical gasification process to recover energy from the MWS (Oladejo et al. 2019).

5.2.4 Hydrothermal conversion

Hydrothermal conversion, also referred as hydrothermal liquefaction (HTL), is a thermochemical conversion process in which biomass is transformed chemically and physically in the absence of oxygen or other oxidants at 150–450°C and high pressure (5–40 MPa) to produce biooil. The main advantages of such process is that the sludge can contain water. However, maintaining high pressure and temperature could be an economical challenge. (Lühmann and Wirth 2020) hydrothermally carbonized sewage sludge at 160–200°C for 30–60 min maintaining initial pH in the range 1.93–8.08 to enhance dewaterability and to enhance the release of phosphorus in the liquid phase.

5.2.5 Oxidative hydrothermal dissolution

Oxidative hydrothermal dissolution, also known as hydrothermal oxidation, is a promising biomass conversion technology to recover value added chemicals from biomass, MWS, and/or residual sludge. In this process, biomass/sludge is treated in the presence of water and oxidant such as air, O_2, H_2O_2 maintaining subcritical or supercritical condition of water to produce value added water soluble compounds such as smaller organic acids, alcohol, and ketones. Process details areavailable in literature (Anderson et al. 2011, Shanableh and Shimizu 2000, Hayashi et al. 1997). Hubner et al.(2016) investigated supercritical hydrothermal oxidation of fecal sludge at 470°C and obtained complete conversion of total organic carbon present in the sludge in 25 minutes. This process is used to recover chemicals from the coal and other biomass, and could be potentially applied for chemical recovery from the MWS/residual sludge.

5.3 Biochemical conversion

5.3.1 Enzymatic catalytic esterification-transesterification

Enzymatic catalysis is an important biochemical technique to produce biodiesel from the extracted lipid from the MWS or from the MWS via *in-situ* esterification/transesterification. The key advantages are that enzymatic process is insensitive to the presence of FFA, requires less methanol, and easy glycerol separation, whereas the cost, slow reaction rate, the chance of biodiesel contamination by the residual enzyme are the main concerns for the enzymatic biodiesel production. Typical reactions of esterification and transesterification are similar to the reactions shown in Fig. 6, only difference being in catalyst type which uses enzymatic catalysts such as lipase.

Enzymatic transesterification catalyzed by lipase is a feasible process. It is distinguished by various characteristics, including the absence of side reactions, minimal energy consumption, and simplicity of product separation (Rawat et al. 2013, Vyas et al. 2010). In pseudo-homogeneous processes, lipase poses certain technological problems. Contamination of the product by leftover enzymes is an issue, as is the high cost of the enzyme. As a result, immobilization of the enzyme contributes to its cost reduction. This is because enzymes that are immobilized can be utilized several times (Vyas et al. 2010). Lipase is a fat hydrolyzing enzyme that catalyzes the breakdown of long-chain triglycerides into polar lipids. The enzyme's polarity (hydrophilic) and that of the substrates are diametrically opposing (lipophilic). As a result, a reaction occurs at the water-oil interface (Kumar and Banerjee 2019) reported maximum biodiesel yield of 85% from the lipid extracted from secondary sludge following enzymatic catalysts (10 U immobilized lipase) at 30°C, 12 h, 200 rpm (Table 7). The resulted FAME composition was also comparable to that of soybean oil produced FAME. It has been reported that the cultivation of the secondary (activated) sludge with oleaginosus species such as *Trichosporon* sp. enhanced the lipid content (Siddiquee and Rohani 2021).

5.3.2 Fermentation

Simultaneous saccharification and fermentation (SSF) is considered to be the most appropriate way to process MWS, as it contains negligible lignin. Fermentation is a biochemical process in which biomass or sludge is treated in the presence of enzyme to recover energy and chemicals such as bioethanol. In a typical fermentation process, enzymatic hydrolysis (saccharification) converts the cellulosic substance to sugar that further convertsinto ethanol in the presence of microbes. Fermentation takes place in the absence of oxygen and is being considered as anaerobic. Details of fermentation process is reported in literature (Alkasrawi et al. 2021, Gomes et al. 2021, Manyuchi et al. 2018).The fermentation of MWS to produce bioethanol is schematically represented in Fig. 11. The four unit operations that comprise biochemical conversion are pretreatment, hydrolysis, fermentation, and distillation. At first,

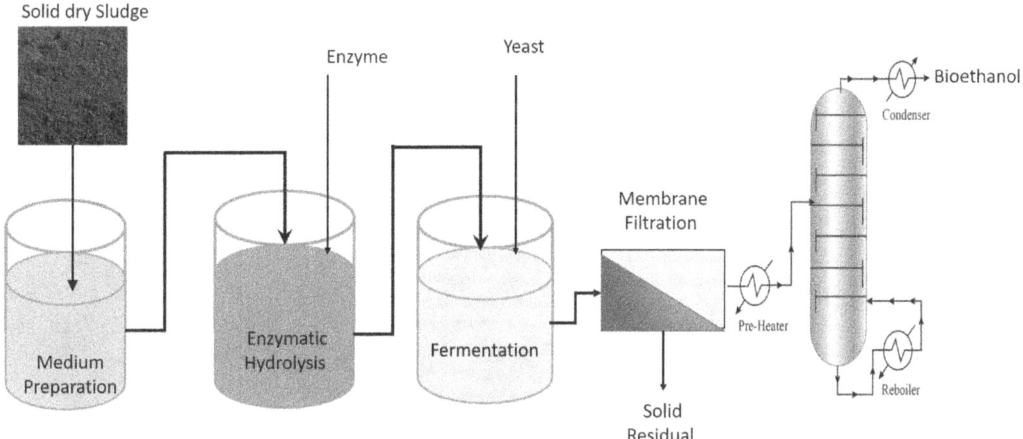

Figure 11. Schematic diagram of enzymatic hydrolysis (saccharification) followed by MWS fermentation to produce bioethanol.

dry MWS was mechanically processed to produce bioethanol. This is then saccharified in the media. This is a biological pretreatment approach that involves the destruction of cellulose and hemicellulose by hydrolytic enzymes, resulting in monomeric sugar. The upstream process includes the hydrolysis of cellulose and the breakdown of hemicellulose into soluble sugars. Following that, the sugars are fermented to produce bioethanol, and the ethanol is distilled to obtain pure ethanol.

Saccharification and fermentation can be performed simultaneously in a single biochemical reactor supplying appropriate enzyme and yeast and maintaining appropriate temperature (for instance, 37°C) and pH (such as 5.5) (Alkasrawi et al. 2021). Manyuchi et al. (2018) investigated bioethanol production from the sewage sludge via hydrolyzing sludge by Bacillus flexus, 10 g/L of peptone, 2 g/L of KH_2PO_4 and 1 g/L of $MgSO_4$ for saccharification following fermentation by using yeast (innoculum as bio catalyst) to produce bioethanol at different operating conditions: pH (4.0–7.0), temperature (15–45°C), incubation time (10–70 h) and yeast (Innoculum as bio catalyst) concentrations (2–10% (v/v). The maximum bioethanol yield of approximately 40 mL/L was obtained at 30°C, pH 6.5, yeast concentration 6 wt% at10 days of incubation period (Table 7). Bioethanol production from the MWS or residual MWS after lipid extraction is promising. However, it requires detailed techno-economic investigation.

5.3.3 Anaerobic digestion

Anaerobic digestion is an important biochemical process in which microorganism is used to maintain an inert environment to convert biomass to biogas and nutrient rich slurry in absence of air. Schematic diagram of integrated wastewater treatment and biogas production from MWS isshown in Fig. 12. The biogas (methane CH_4, carbon dioxide CO_2 and water H_2O) production in anaerobic digester involves the four consecutive stages including hydrolysis (formation of simple monomer, oligomer, sugar, fatty acids, and peptides), acidogenesis (fermentation to produce short chain organic acids), acetogenesis (formation of acetate) and methanogenesis (reduction of CO_2 to produce CH_4).Methanization is the transformation of organic matter into biogas (methane and carbon dioxide) by bacteria in anaerobic conditions (lack of oxygen):

$$Sludge\ (Organic) + microorganism \xrightarrow{anerobic} microorganism\ (Sludge) + CO_2 + CH_4$$

This degradation is only possible in the presence of fermentable materials: unstabilized sewage sludge, fresh fecal sludge and other organic waste such as manure, plants or food waste. Details of such processes are available in literature (Lee et al. 2011, Oladejo et al. 2019). This process is being

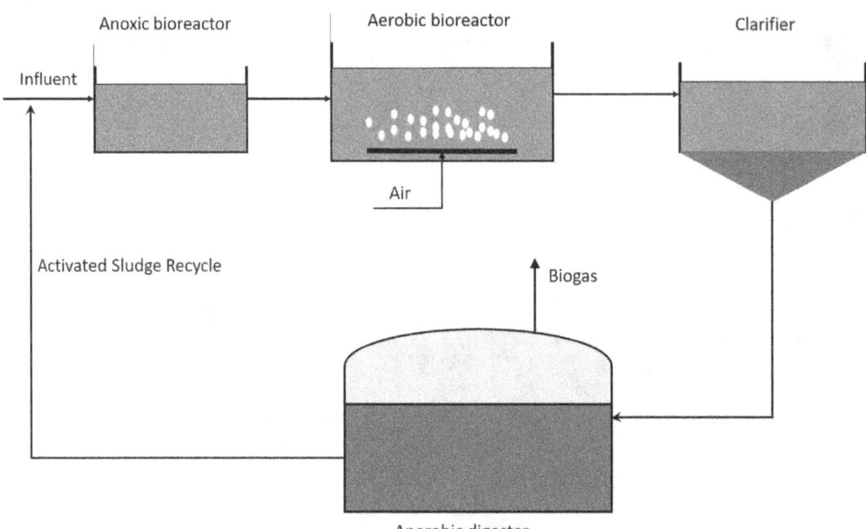

Figure 12. Schematic diagram of integrated wastewater treatment and biogas production from MWS.

considered as the cost effective process to recover energy from the MWS. The key advantages of this process are that the technique can be used for the sludge having higher moisture content, can generate biogas having higher energy content, and residual can be used as biofertilizer having higher nutrient contents. However, it requires long reaction time of 10–20 days, is associated with low overall conversion, and requires higher capital and maintenance costs.

5.4 Bioelectrochemical conversion

Bioelectrochemical process is being considered as an environment friendly and sustainable process that could potentially be used for the future energy recovery, electricity generation and biochemicals' production. This technique follows redox process in microbial electrolysis cells in the presence of biocatalyst to convert different organic compounds to energy such as hydrogen, methane, and different value-added chemicals including fatty acids, acetates, alcohols and ketones. Schematic diagram of biochemical conversion of MWS to bioenergy is shown in Fig. 13. In a bioelectrochemical system, water splitting or oxidation ($2\ H_2O \rightarrow 4\ H^+ + 4\ e^- + O_2$) at anode in the presence of external electric potential generates electrons (e^-) and protons (H^+) and are transported to the cathode. At cathode, electroactive microbes utilize the transported electrons (e^-) and protons (H^+) to facilitate the reduction reaction ($4H^+ + 4e^- \rightarrow 2H_2$). Details of this technique are available in literature (Kumar et al. 2017, Wang and Ren 2013, Desloover et al. 2012).

This potential technique to recover energy and chemicals from the MWS is comparatively new technology and requires proper understanding considering different factors such as electrode type, catalyst, microbial species, and product separations. It also requires detailed investigation of techno-economic prospect.

6. Economic Analysis

MWS management is a significant economic burden in municipal wastewater treatment facilities, accounting for 30% to 40% of total capital cost and 50% of plant operation cost (Cho et al. 2014). The economic analyses of sludge management focus on treatment methods and technologies ensuring reduction in energy requirement and enhancing value creation of the output through volume

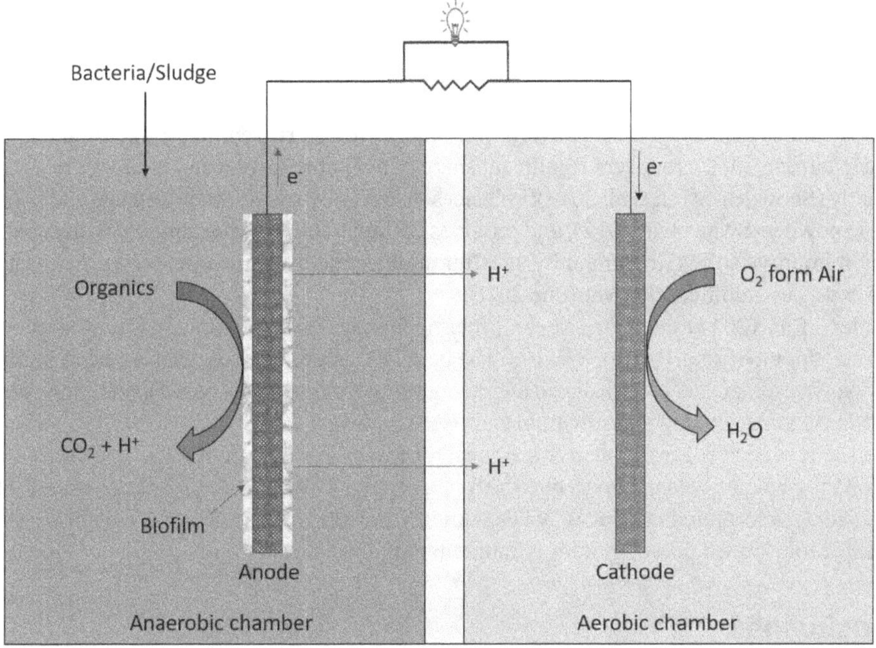

Figure 13. Schematic diagram of biochemical conversion of MWS to bioenergy.

reduction, energy recovery, nutrient recovery, and valuable raw materials to ensure reduction in capital and operation cost. These analyses also focus on reducing harmful environmental impacts. The environmental benefits can also be monetized to support the economic analysis. In the following paragraphs, different approaches to ensure economic benefits of sludge management practices are discussed.

Sludge volume reduction using anaerobic digestion (AD) is gaining popularity since it converts organic wastes to CH_4-rich biogas, inactivates pathogens, and improves sludge cake dewaterability (Appels et al. 2010). Lee et al. (2019) evaluated the economic cost and biodegradability of different AD-based sludge disintegration scenarios. According to the findings, alkali pre-treatment significantly improved sludge solubilization and organic acid accumulation, whereas heat treatment resulted in modest sludge solubilization along with superior energy recovery. The thermal pre-treatment at 90°C, followed by acidogenic fermentation and the AD process, resulted in a net economic advantage of 2.53 USD/ton of sludge, owing to the increased supply of external carbon source generated during acidogenic fermentation.

A study assessed the environmental and economic impacts of 13 sewage sludge-treatment scenarios in China adopting a cost-combined LCA. The results revealed that anaerobic digestion was a viable option for reducing both environmental and economic constraints. Energy production from the landfill and incinerator phases were critical for reducing environmental and economic costs. Because of the energy recovery, a sewage sludge-treatment scenario including anaerobic digestion, dewatering, and incineration technologies was found to be the most ecologically and economically viable way for treating sewage sludge (Xu et al. 2014). In Japan, Hong et al. (2009) conducted an integrated study of sewage sludge-treatment options and concluded that thickening, digestion, dewatering, and melting was the most economically affordable methods.

Municipal wastewater in the United States contains over 1.69×10^{13} kJ/ year (160 trillion Btu/ year) of chemical energy flow; however, only a small portion is recovered and used. Hydrothermal liquefaction (HTL), which is a thermochemical process that transforms wastewater solids and MSW into a biocrude product that may be upgraded into a range of liquid fuels including transportation biofuels. The modeling exercises were based on the deployment of HTL technology, and it was concluded that treatment facilities with a capacity of 6.35 × 10^6 m^3/ year (17.4 million liter/day) could supply 3.93 × 10^9 ton/year (9.77 Teragram/day) of dry solids feedstock and produce 3.67 × 10^6 m^3/ year (3.67 gigaliter/year) biocrude intermediate. It can be considered as a potential economically viable solution (Seiple et al. 2020). HTL converts wet sludge into sterilized, low-odor solids, and useful biocrude (Marrone 2016); recovers significant share of available carbon and reduces dry solids mass significantly (Snowden-Swan et al. 2017); enhances recovery by concentrating metals and phosphorus in solids (Snowden-Swan et al. 2017); and provides co-liquefaction opportunities (Biller et al. 2018). HTL might improve solids' treatment by substituting delicate biological processes, lowering solids disposal expenses significantly (Marrone 2016).

Lumley et al. (2014) evaluated several thermochemical conversion (TCC) systems of MWS from the standpoint of small urban WWTPs. The best TCC approach was determined to be air-blown gasification. To estimate net electrical and thermal outputs, a gasification-based generating system was built and simulated. Air-blown gasification was used to convert sludge to power with an efficiency of more than 17%, which is greater than that of anaerobic digester gas production. This form of power generation may counterbalance up to one-third of a regular WWTP's electrical demand. Finally, an economic study determined that for WWTPs with raw sewage flowing above 2.1 million gallons per day, a gasification-based power system is economically feasible.

7. Conclusions

MWS (municipal wastewater sludge) energy recovery is a promising approach to valorize waste materials. It would reduce sludge management and operating costs while also reducing the environmental impact. Furthermore, it would help meet some of the rising energy demand to some extent. In addition, as a by-product of the energy recovery processes, other value-added compounds would be produced. The development of a sustainable energy recovery technique from MWS is still being researched around the world. Chemical (e.g., esterification, transesterification), thermochemical (e.g., combustion, pyrolysis, and gasification), and biochemical conversion processes can all be used to recover various types of energy such as biodiesel, biochar, biooil, biogas, syngas, H_2, bioethanol, and bio-solids (e.g., anaerobic digestion, fermentation, and bio-electrochemical). Each process has its own set of obstacles and opportunities. The key challenge is the energy-intensive and costly sludge pre-treatment (e.g., drying). Other difficulties include the need for organic solvents and solvent recovery, sludge composition change, and the presence of various hazardous compounds. However, there are ways to reduce the difficulties. Developing an alternative sludge pretreatment and lipid extraction process to reduce sludge processing costs, for example, cultivating algae or oleaginous species to increase lipid content in the sludge. Also, developing a process to recover energy and chemicals from the residual sludge after lipid extraction would reduce the economic challenges. Furthermore, sludge and/or residual sludge can be coupled with other cellulosic biomass to produce biochar, biooil, biogas, syngas, H_2, bioethanol, and bio-solids, all of which can be used to recover energy. Overall, recovering energy from the MWS would benefit from an integrated and long-term procedure. It is necessary to establish an integrated method for efficient energy recovery from MWS, taking into account long-term goals as well as environmental benefits.

ACKNOWLEDGEMENTS

Authors would like to acknowledge the support received from King Fahd University of Petroleum and Minerals (KFUPM) along with internal direct funding grant and financial support for this work through project No. DF191050.

References

Al-Malack, Muhammad H., Nabil S. Abuzaid and Alaadin A. Bukhari. 2008. Physico-Chemical Characteristics of Municipal Sludge Produced at Three Major Cities of the Eastern Province of Saudi Arabia. Journal of King Saud University - Engineering Sciences 20(1): 15–26. doi: 10.1016/S1018-3639(18)30852-3.

Alkasrawi, Malek, Amani Al-Othman, Muhammad Tawalbeh, Shona Doncan, Raghu Gurram, Eric Singsaas, Fares Almomani and Sameer Al-Asheh. 2021. A Novel Technique of Paper Mill Sludge Conversion to Bioethanol toward Sustainable Energy Production: Effect of Fiber Recovery on the Saccharification Hydrolysis and Fermentation. Energy 223. doi: 10.1016/j.energy.2021.120018.

Anderson, K.B., Crelling, J.C., Huggett, W.W., Perry, D., Fullinghim, T., McGill, P. and Kaelin, P. 2011. Oxidative Hydrothermal Dissolution (OHD) of Coal and Biomass. Prepr. Pap.-Am. Chem. Soc., Div. Fuel Chem. 56(2): 310–11.

Appels, Lise, Jan Degrève, Bart Van der Bruggen, Jan Van Impe and Raf Dewil. 2010. Influence of Low Temperature Thermal Pre-Treatment on Sludge Solubilisation, Heavy Metal Release and Anaerobic Digestion. Bioresource Technology 101(15): 5743–48. doi: 10.1016/j.biortech.2010.02.068.

Arazo, Renato O., Mark Daniel G. de Luna and Sergio C. Capareda. 2017. Assessing Biodiesel Production from Sewage Sludge-Derived Bio-Oil. Biocatalysis and Agricultural Biotechnology 10: 189–96. doi: 10.1016/j.bcab.2017.03.011.

Barry, Devon, Chiara Barbiero, Cedric Briens and Franco Berruti. 2019. Pyrolysis as an Economical and Ecological Treatment Option for Municipal Sewage Sludge. Biomass and Bioenergy 122: 472–80. doi: 10.1016/j.biombioe.2019.01.041.

Biller, Patrick, Ib Johannsen, Juliano Souza dos Passos and Lars Ditlev Mørck Ottosen. 2018. Primary Sewage Sludge Filtration Using Biomass Filter Aids and Subsequent Hydrothermal Co-Liquefaction. Water Research 130: 58–68. doi: 10.1016/j.watres.2017.11.048.

Bora, Raaj R., Ruth E. Richardson and Fengqi You. 2020. Resource Recovery and Waste-to-Energy from Wastewater Sludge via Thermochemical Conversion Technologies in Support of Circular Economy: A Comprehensive Review. BMC Chemical Engineering 2(1). doi: 10.1186/s42480-020-00031-3.

Campbell, H.W. 2000. Sludge Management—Future Issues and Trends. Water Science and Technology 41(8): 1–8. doi: 10.2166/wst.2000.0135.

Cantinho, P., Matos, M., Trancoso, M.A. and Correi, M.M. dos Santos. 2016. Behaviour and Fate of Metals in Urban Wastewater Treatment Plants: A Review. International Journal of Environmental Science and Technology 13(1): 359–86. doi: 10.1007/s13762-015-0887-x.

Capodaglio, Andrea, G. and Arianna Callegari. 2018. Feedstock and Process Influence on Biodiesel Produced from Waste Sewage Sludge. Journal of Environmental Management 216: 176–82. doi: 10.1016/j.jenvman.2017.03.089.

Chan, Wei Ping and Jing Yuan Wang. 2016. Comprehensive Characterisation of Sewage Sludge for Thermochemical Conversion Processes—Based on Singapore Survey. Waste Management 54: 131–42. doi: 10.1016/j.wasman.2016.04.038.

Cho, Si Kyung, Hyun Jun Ju, Jeong Gyu Lee and Sang Hyoun Kim. 2014. Alkaline-Mechanical Pretreatment Process for Enhanced Anaerobic Digestion of Thickened Waste Activated Sludge with a Novel Crushing Device: Performance Evaluation and Economic Analysis. Bioresource Technology 165(C): 183–90. doi: 10.1016/j.biortech.2014.03.138.

Davis, Mackenzie Leo and Susan J. Masten. 2004. Principles of Environmental Engineering and Science. 3rd Editio. New York: McGraw-Hills Inc.

Debono, Olivier and Audrey Villot. 2015. Nitrogen Products and Reaction Pathway of Nitrogen Compounds during the Pyrolysis of Various Organic Wastes. Journal of Analytical and Applied Pyrolysis 114: 222–34. doi: 10.1016/j.jaap.2015.06.002.

Desloover, Joachim, Jan B. A. Arends, Tom Hennebel and Korneel Rabaey. 2012. Operational and Technical Considerations for Microbial Electrosynthesis. Biochemical Society Transactions 40(6): 1233–38. doi: 10.1042/BST20120111.

Dufreche, Stephen, Hernandez, R., French, T., Sparks, D., Zappi, M. and Alley, E. 2007. Extraction of Lipids from Municipal Wastewater Plant Microorganisms for Production of Biodiesel. JAOCS, Journal of the American Oil Chemists' Society 84(2): 181–87. doi: 10.1007/s11746-006-1022-4.

European Commission. 2008. Environmental, Economic and Social Impacts of the Use of Sewage Sludge on Land. Final Report. Part I: Overview Report.

Fan, Y., Fonseca, F.G., Gong, M., Hoffmann, A., Hornung, U. and Dahmen, N. 2021. Energy Valorization of Integrating Lipid Extraction and Hydrothermal Liquefaction of Lipid-Extracted Sewage Sludge. Journal of Cleaner Production 285. doi: 10.1016/j.jclepro.2020.124895.

Fijalkowski, Krzysztof, Agnieszka Rorat, Anna Grobelak and Malgorzata J. Kacprzak. 2017. The Presence of Contaminations in Sewage Sludge—The Current Situation. Journal of Environmental Management 203: 1126–36. doi: 10.1016/j.jenvman.2017.05.068.

Fonts, Isabel, Gloria Gea, Manuel Azuara, Javier Ábrego and Jesús Arauzo. 2012. Sewage Sludge Pyrolysis for Liquid Production: A Review. Renewable and Sustainable Energy Reviews 16(5): 2781–2805. doi: 10.1016/j.rser.2012.02.070.

Ganesapillai, Mahesh, Aruna Singh and Dhanaraj Sangeetha. 2020. "Biochar Technology for Environmental Sustainability. pp. 1–21. In: L.E. Gothandam K., Ranjan S. and Dasgupta, N. Cham: Springer.

Gherghel, Andreea, Carmen Teodosiu and Sabino De Gisi. 2019. A Review on Wastewater Sludge Valorisation and Its Challenges in the Context of Circular Economy. Journal of Cleaner Production 228: 244–63. doi: 10.1016/j.jclepro.2019.04.240.

Gomes, Daniel, Mariana Cruz, Miriam de Resende, Eloízio Ribeiro, José Teixeira and Lucília Domingues. 2021. Very High Gravity Bioethanol Revisited: Main Challenges and Advances. Fermentation 7(1): 1–18. doi: 10.3390/fermentation7010038.

Haghighat, Manouchehr, Nasrollah Majidian, Ahmad Hallajisani, and Mohammad samipourgiri. 2020. Production of Bio-Oil from Sewage Sludge: A Review on the Thermal and Catalytic Conversion by Pyrolysis. Sustainable Energy Technologies and Assessments 42. doi: 10.1016/j.seta.2020.100870.

Hayashi, Jun Ichiro, Yoshihiro Matsuo, Katsuki Kusakabe and Shigeharu Morooka. 1997. "Depolymerization of Lower Rank Coals by Low-Temperature O2 Oxidation. Energy and Fuels 11(1): 227–35. doi: 10.1021/ef960104o.

Hong, Jinglan, Jingmin Hong, Masahiro Otaki and Olivier Jolliet. 2009. Environmental and Economic Life Cycle Assessment for Sewage Sludge Treatment Processes in Japan. Waste Management 29(2): 696–703. doi: 10.1016/j.wasman.2008.03.026.

Hubner, Tobias, Markus Roth and Frédéric Vogel. 2016. Hydrothermal Oxidation of Fecal Sludge: Experimental Investigations and Kinetic Modeling. Industrial and Engineering Chemistry Research 55(46): 11910–22. doi: 10.1021/acs.iecr.6b03084.

Jeevan Kumar, S.P. and Rintu Banerjee. 2019. Enhanced Lipid Extraction from Oleaginous Yeast Biomass Using Ultrasound Assisted Extraction: A Greener and Scalable Process." Ultrasonics Sonochemistry 52: 25–32. doi: 10.1016/j.ultsonch.2018.08.003.

Jiang, Long Bo, Xing Zhong Yuan, Hui Li, Xiao Hong Chen, Zhi Hua Xiao, Jie Liang, Li Jian Leng, Zhi Guo and Guang Ming Zeng. 2016. Co-Pelletization of Sewage Sludge and Biomass: Thermogravimetric Analysis and Ash Deposits. Fuel Processing Technology 145: 109–15. doi: 10.1016/j.fuproc.2016.01.027.

Kargbo, David M. 2010. Biodiesel Production from Municipal Sewage Sludges. Energy and Fuels 24(5): 2791–94. doi: 10.1021/ef1001106.

Kumar, Gopalakrishnan, Rijuta Ganesh Saratale, Abudukeremu Kadier, Periyasamy Sivagurunathan, Guangyin Zhen, Sang Hyoun Kim, and Ganesh Dattatraya Saratale. 2017. "A Review on Bio-Electrochemical Systems (BESs) for the Syngas and Value Added Biochemicals Production. Chemosphere 177: 84–92. doi: 10.1016/j.chemosphere.2017.02.135.

Lee, Il Su, Prathap Parameswaran and Bruce E. Rittmann. 2011. Effects of Solids Retention Time on Methanogenesis in Anaerobic Digestion of Thickened Mixed Sludge. Bioresource Technology 102(22): 10266–72. doi: 10.1016/j.biortech.2011.08.079.

Lee, Mo-Kwon, Yeo-Myeong Yun and Dong-Hoon Kim. 2019. Enhanced Economic Feasibility of Excess Sludge Treatment: Acid Fermentation with Biogas Production. BMC Energy 1(1). doi: 10.1186/s42500-019-0001-x.

Lee, Uisung, Jun Dong and J.N. Chung. 2018. Experimental Investigation of Sewage Sludge Solid Waste Conversion to Syngas Using High Temperature Steam Gasification. Energy Conversion and Management 158: 430–36. doi: 10.1016/j.enconman.2017.12.081.

Lühmann, Taina and Benjamin Wirth. 2020. Sewage Sludge Valorization via Hydrothermal Carbonization: Optimizing Dewaterability and Phosphorus Release. Energies 13(17). doi: 10.3390/en13174417.

Manyuchi, M.M., Chiutsi, P., Mbohwa, C., Muzenda, E. and Mutusva, T. 2018. Bio Ethanol from Sewage Sludge: A Bio Fuel Alternative. South African Journal of Chemical Engineering 25: 123–27. doi: 10.1016/j.sajce.2018.04.003.

Markis, Flora, Jean Christophe Baudez, Rajarathinam Parthasarathy, Paul Slatter and Nicky Eshtiaghi. 2014. Rheological Characterisation of Primary and Secondary Sludge: Impact of Solids Concentration. Chemical Engineering Journal 253: 526–37. doi: 10.1016/j.cej.2014.05.085.

Marrone, P.A. 2016. Genifuel Hydrothermal Processing Bench-Scale Technology Evaluation Report. Vol. 15. IWA.

Martin, L., Alizadeh, V. and Meegoda, J. 2019. Electro-Osmosis Treatment Techniques and Their Effect on Dewatering of Soils, Sediments, and Sludge: A Review. Soils and Foundations.

Melero, J.A., Sánchez-Vázquez, R., Vasiliadou, I.A., Martínez Castillejo, F., Bautista, L.F., Iglesias, J., Morales, G. and Molina, R. 2015. Municipal Sewage Sludge to Biodiesel by Simultaneous Extraction and Conversion of Lipids. Energy Conversion and Management 103: 111–18. doi: 10.1016/j.enconman.2015.06.045.

Metcalf, Eddy. 2003. Waswater Engineering Treatment and Resource Recovery.

Mulchandani, Anjali and Paul Westerhoff. 2016. Recovery Opportunities for Metals and Energy from Sewage Sludges. Bioresource Technology 215: 215–26. doi: 10.1016/j.biortech.2016.03.075.

Novak, John T. 2006. Dewatering of Sewage Sludge. Drying Technology 24(10): 1257–62. doi: 10.1080/07373930600840419.

Oakley, Monica. 2018. Environmental Engineering. Vol. 110. London: McGraw-Hills Inc.

Oladejo, Jumoke, Kaiqi Shi, Xiang Luo, Gang Yang and Tao Wu. 2019. A Review of Sludge-to-Energy Recovery Methods. Energies 12(1). doi: 10.3390/en12010060.

Olkiewicz, Magdalena, Martin Pablo Caporgno, Agustí Fortuny, Frank Stüber, Azael Fabregat, Josep Font and Christophe Bengoa. 2014. Direct Liquid-Liquid Extraction of Lipid from Municipal Sewage Sludge for Biodiesel Production. Fuel Processing Technology 128: 331–38. doi: 10.1016/j.fuproc.2014.07.041.

Pathak, Ashish, Dastidar, M.G. and Sreekrishnan, T.R. 2009. Bioleaching of Heavy Metals from Sewage Sludge: A Review. Journal of Environmental Management 90(8): 2343–53. doi: 10.1016/j.jenvman.2008.11.005.

Puyol, Daniel, Damien J. Batstone, Tim Hülsen, Sergi Astals, Miriam Peces and Jens O. Krömer. 2017. Resource Recovery from Wastewater by Biological Technologies: Opportunities, Challenges, and Prospects. Frontiers in Microbiology 7(JAN). doi: 10.3389/fmicb.2016.02106.

Rawat, I., Ranjith Kumar, R., Mutanda, T. and Bux, F. 2013. Biodiesel from Microalgae: A Critical Evaluation from Laboratory to Large Scale Production. Applied Energy 103: 444–67. doi: 10.1016/j.apenergy.2012.10.004.

Rorat, Agnieszka, Pauline Courtois, Franck Vandenbulcke and Sébastien Lemiere. 2019. Sanitary and Environmental Aspects of Sewage Sludge Management. Industrial and Municipal Sludge: Emerging Concerns and Scope for Resource Recovery 155–80. doi: 10.1016/B978-0-12-815907-1.00008-8.

Rorat, Agnieszka and Małgorzata Kacprzak. 2017. Eco-Innovations in Sustainable Waste Management Strategies for Smart Cities. 221–37. doi: 10.1007/978-3-319-49899-7_13.

Rulkens, Wim H. 2004. Sustainable Sludge Management—What Are the Challenges for the Future? Water Science and Technology 49(10): 11–19. doi: 10.2166/wst.2004.0597.

Seiple, T.E., Skaggs, R.L., Fillmore, L. and Coleman, A.M. 2020. Municipal Wastewater Sludge as a Renewable, Cost-Effective Feedstock for Transportation Biofuels Using Hydrothermal Liquefaction. Journal of Environmental Management 270: 110852.

Shanableh, A. and Shimizu, Y. 2000. Treatment of Sewage Sludge Using Hydrothermal Oxidation - Technology Application Challenges. Water Science and Technology 41(8): 85–92. doi: 10.2166/wst.2000.0146.

Siddiquee, M.N., Kazemian, H. and Rohani, S. 2011. Biodiesel Production from the Lipid of Wastewater Sludge Using an Acidic Heterogeneous Catalyst. Chemical Engineering and Technology 34(12): 1983–88. doi: 10.1002/ceat.201100119.

Siddiquee, Muhammad N. and Sohrab Rohani. 2011a. Experimental Analysis of Lipid Extraction and Biodiesel Production from Wastewater Sludge. Fuel Processing Technology 92(12): 2241–51. doi: 10.1016/j.fuproc.2011.07.018.

Siddiquee, Muhammad, N. and Sohrab Rohani. 2011b. Lipid Extraction and Biodiesel Production from Municipal Sewage Sludges: A Review. Renewable and Sustainable Energy Reviews 15(2): 1067–72. doi: 10.1016/j.rser.2010.11.029.

Siddiquee, Muhammad Nurunnabi and Sohrab Rohani. 2021. Biodiesel Production from Municipal Wastewater Sludge. Biodiesel Fuels Based on Edible and Nonedible Feedstocks, Wastes, and Algae 623–41. doi: 10.1201/9780367456207-13.

Singh, R.P. and Agrawal, M. 2008. Potential Benefits and Risks of Land Application of Sewage Sludge. Waste Management 28(2): 347–58. doi: 10.1016/j.wasman.2006.12.010.

Snowden-Swan, Lesley J., Yunhua Zhu, Mark D. Bearden, Timothy E. Seiple, Susanne B. Jones, Andrew J. Schmidt, Justin M. Billing, Richard T. Hallen, Todd R. Hart, Jian Liu, Karl O. Albrecht, Samuel P. Fox, Gary D. Maupin and Douglas C. Elliott. 2017. Conceptual Biorefinery Design and Research Targeted for 2022: Hydrothermal Liquefaction Processing of Wet Waste to Fuels. Vol. 27186. Richland, WA (United States).

Supaporn, Pansuwan, Hoang Vu Ly, Seung Soo Kim and Sung Ho Yeom. 2019. Bio-Oil Production Using Residual Sewage Sludge after Lipid and Carbohydrate Extraction. Environmental Engineering Research 24(2): 202–10. doi: 10.4491/eer.2017.178.

Syed-Hassan, Syed Shatir A., Yi Wang, Song Hu, Sheng Su and Jun Xiang. 2017. Thermochemical Processing of Sewage Sludge to Energy and Fuel: Fundamentals, Challenges and Considerations. Renewable and Sustainable Energy Reviews 80: 888–913.

Tchobanoglous, George and Franklin L. Burton. 1980. Wastewater Engineering: Treatment, Disposal *and Reuse*. Vol. 3. 3rd editio. New York: McGraw-Hills Inc.

Tchobanoglus, G.H., Theisen, Y. and Vigil, S. 2003. Integrated Solid Waste Management: Engineering Principles and Management. New Jersey: McGraw-Hills Inc.

Trinh, Trung Ngoc, Peter Arendt Jensen, Dam Johansen Kim, Niels Ole Knudsen and Hanne Risbjerg Sørensen. 2013. Influence of the Pyrolysis Temperature on Sewage Sludge Product Distribution, Bio-Oil, and Char Properties. Energy and Fuels 27(3): 1419–27. doi: 10.1021/ef301944r.

Triplett, K.A., Ghiaasiaan, S.M., Abdel-Khalik, S.I. and Sadowski, D.L. 1999. Gas-Liquid Two-Phase Flow in Microchannels Part I: Two-Phase Flow Patterns. International Journal of Multiphase Flow 25(3): 377–94. doi: 10.1016/S0301-9322(98)00054-8.

Trowbridge, T.D. and Holcombe, T. 2006. The Carver-Greenfield Process: Dehydration/Solvent Extraction Technology for WasteTreatment. Environmental Progress.

Tyagi, Vinay Kumar and Shang Lien Lo. 2013. Sludge: A Waste or Renewable Source for Energy and Resources Recovery? Renewable and Sustainable Energy Reviews 25: 708–28. doi: 10.1016/j.rser.2013.05.029.

United States Enviromental Protection Agency. 1994. Guide to Septage Treatment and Disposal. Document EP/625/R-94/002.

Verlicchi, P. and Zambello, E. 2015. Pharmaceuticals and Personal Care Products in Untreated and Treated Sewage Sludge: Occurrence and Environmental Risk in the Case of Application on Soil—A Critical Review. Science of the Total Environment 538: 750–67. doi: 10.1016/j.scitotenv.2015.08.108.

Vilardi, Giorgio, Irene Bavasso, Marco Scarsella, Nicola Verdone and Luca Di Palma. 2020. Fenton Oxidation of Primary Municipal Wastewater Treatment Plant Sludge: Process Modelling and Reactor Scale-Up. Process Safety and Environmental Protection 140: 46–59. doi: 10.1016/j.psep.2020.05.002.

Vyas, Amish, P., Jaswant, L. Verma and Subrahmanyam, N. 2010. A Review on FAME Production Processes. Fuel 89(1): 1–9. doi: 10.1016/j.fuel.2009.08.014.

Wang, Heming and Zhiyong Jason Ren. 2013. A Comprehensive Review of Microbial Electrochemical Systems as a Platform Technology. Biotechnology Advances 31(8): 1796–1807. doi: 10.1016/J.BIOTECHADV.2013.10.001.

Xu, Changqing, Wei Chen and Jinglan Hong. 2014. Life-Cycle Environmental and Economic Assessment of Sewage Sludge Treatment in China. Journal of Cleaner Production 67: 79–87. doi: 10.1016/j.jclepro.2013.12.002.

Yoshida, Hiroko, Thomas H. Christensen and Charlotte Scheutz. 2013. Life Cycle Assessment of Sewage Sludge Management: A Review. Waste Management and Research 31(11): 1083–1101. doi: 10.1177/0734242X13504446.

Yoshida, Hiroko, Marieke ten Hoeve, Thomas H. Christensen, Sander Bruun, Lars S. Jensen and Charlotte Scheutz. 2018. "Life Cycle Assessment of Sewage Sludge Management Options Including Long-Term Impacts after Land Application. Journal of Cleaner Production 174: 538–47. doi: 10.1016/j.jclepro.2017.10.175.

Yuan, Ching, Weng and Chih-Huang. 2002. Enhancement of Sludge Dewatering by Electrokinetic Process. in TWA World Water Congress. Melbourne.

Zhang, Linghong, Chunbao Xu, Pascale Champagne and Warren Mabee. 2014a. Overview of Current Biological and Thermo-Chemical Treatment Technologies for Sustainable Sludge Management. Waste Management and Research 32(7): 586–600. doi: 10.1177/0734242X14538303.

Zhang, Xiaolei, Song Yan, Rajeshwar D. Tyagi, Patrick Drogui and Rao Y. Surampalli. 2014b. Ultrasonication Assisted Lipid Extraction from Oleaginous Microorganisms. Bioresource Technology 158: 253–61. doi: 10.1016/j.biortech.2014.01.132.

Index

O

P

R

S

T

V

W

For Product Safety Concerns and Information please contact our EU representative GPSR@taylorandfrancis.com
Taylor & Francis Verlag GmbH, Kaufingerstraße 24, 80331 München, Germany

www.ingramcontent.com/pod-product-compliance
Lightning Source LLC
Chambersburg PA
CBHW080904170526
45158CB00008B/1983

* 9 7 8 1 0 3 2 0 3 9 0 1 5 *